CAMBRIDGE LIBRARY COLLECTION

Books of enduring scholarly value

Mathematical Sciences

From its pre-historic roots in simple counting to the algorithms powering modern desktop computers, from the genius of Archimedes to the genius of Einstein, advances in mathematical understanding and numerical techniques have been directly responsible for creating the modern world as we know it. This series will provide a library of the most influential publications and writers on mathematics in its broadest sense. As such, it will show not only the deep roots from which modern science and technology have grown, but also the astonishing breadth of application of mathematical techniques in the humanities and social sciences, and in everyday life.

An Elementary Course of Infinitesimal Calculus

Sir Horace Lamb (1849–1934) the British mathematician, wrote a number of influential works in classical physics. A pupil of Stokes and Clerk Maxwell, he taught for ten years as the first professor of mathematics at the University of Adelaide before returning to Britain to take up the post of professor of physics at the Victoria University of Manchester (where he had first studied mathematics at Owens College). As a teacher and writer his stated aim was clarity: 'somehow to make these dry bones live'. The first edition of this work was published in 1897, the third revised edition in 1919, and a further corrected version just before his death. This edition, reissued here, remained in print until the 1950s. As with Lamb's other textbooks, each section is followed by examples.

Cambridge University Press has long been a pioneer in the reissuing of out-of-print titles from its own backlist, producing digital reprints of books that are still sought after by scholars and students but could not be reprinted economically using traditional technology. The Cambridge Library Collection extends this activity to a wider range of books which are still of importance to researchers and professionals, either for the source material they contain, or as landmarks in the history of their academic discipline.

Drawing from the world-renowned collections in the Cambridge University Library, and guided by the advice of experts in each subject area, Cambridge University Press is using state-of-the-art scanning machines in its own Printing House to capture the content of each book selected for inclusion. The files are processed to give a consistently clear, crisp image, and the books finished to the high quality standard for which the Press is recognised around the world. The latest print-on-demand technology ensures that the books will remain available indefinitely, and that orders for single or multiple copies can quickly be supplied.

The Cambridge Library Collection will bring back to life books of enduring scholarly value across a wide range of disciplines in the humanities and social sciences and in science and technology.

An Elementary Course of Infinitesimal Calculus

HORACE LAMB

CAMBRIDGE
UNIVERSITY PRESS

CAMBRIDGE UNIVERSITY PRESS

Cambridge New York Melbourne Madrid Cape Town Singapore São Paolo Delhi

Published in the United States of America by Cambridge University Press, New York

www.cambridge.org
Information on this title: www.cambridge.org/9781108005340

© in this compilation Cambridge University Press 2009

This edition first published 1956
This digitally printed version 2009

ISBN 978-1-108-00534-0

AN
ELEMENTARY COURSE
OF
INFINITESIMAL CALCULUS

AN ELEMENTARY COURSE

OF

INFINITESIMAL CALCULUS

BY

SIR HORACE LAMB

REVISED EDITION

CAMBRIDGE
AT THE UNIVERSITY PRESS
1956

PUBLISHED BY
THE SYNDICS OF THE CAMBRIDGE UNIVERSITY PRESS
London Office: Bentley House, N.W. I
American Branch: New York
Agents for Canada, India, and Pakistan: Macmillan

First Edition	1897
Second Edition	1902
Reprinted	1907
	1912
	1913
Third Edition (revised)	1919
Reprinted	1921
	1924
	1927
(with corrections)	1934
	1938
	1942
	1944
	1946
	1947
	1949
	1956

First printed in Great Britain at The University Press, Cambridge
Reprinted by Spottiswoode, Ballantyne & Co., Ltd., Colchester

PREFACE

THIS book was described in the original preface as an attempt to teach those portions of the Calculus which are of primary importance in the applications of the subject. The general arrangement of the work was, at the time, somewhat unusual, but appears to have been found convenient.

The present edition has been revised throughout, and a number of changes have been made. Apart from minor alterations and rearrangements, there are one or two points which call for remark.

A special chapter is devoted to the exponential and allied functions, the exponential function being now defined as the standard solution of the equation

$$\frac{dy}{dx} = y.$$

It is to this property, entirely, that the function owes its importance in Mathematics, and it seems therefore most natural to take this as the starting-point. No theory of the exponential series which has any pretensions to be rigorous can be said to be altogether elementary, but it is claimed that the method here followed is, from the standpoint of the Calculus, no more difficult than any other, whilst there can be no question as to its being the most appropriate.

Another considerable change is in the treatment of infinite series, their differentiation and integration. In previous editions these questions were discussed in a general manner, by the light of the theory of uniform convergence. There was perhaps some justification for including this theory, at a time when it was hardly accessible in any English manual, but it was out of perspective with the rest of the book, and is now omitted. It is replaced by a discussion restricted to *power*-series only, which are the only type which the student is likely to be concerned with until he reaches a more advanced stage.

Finally, some sections on mass-centres, quadratic moments, and the like, have been condensed or omitted. They have in the meantime been transferred, for the most part, to other books by the author.

HORACE LAMB.

June 1919.

CONTENTS

CHAPTER I

CONTINUITY

CHAPTER II

DERIVED FUNCTIONS

CHAPTER III

THE EXPONENTIAL AND LOGARITHMIC FUNCTIONS

CHAPTER IV

APPLICATIONS OF THE DERIVED FUNCTION

CHAPTER V

DERIVATIVES OF HIGHER ORDERS

CHAPTER VI

INTEGRATION

CHAPTER VII

DEFINITE INTEGRALS

CHAPTER VIII

GEOMETRICAL APPLICATIONS

CHAPTER IX

SPECIAL CURVES

CHAPTER X

CURVATURE

CHAPTER XI

DIFFERENTIAL EQUATIONS OF THE FIRST ORDER

CHAPTER XII

DIFFERENTIAL EQUATIONS OF THE SECOND ORDER

CHAPTER XIII

LINEAR EQUATIONS WITH CONSTANT COEFFICIENTS

CHAPTER XIV

DIFFERENTIATION AND INTEGRATION OF POWER-SERIES

CHAPTER XV

TAYLOR'S THEOREM

CHAPTER XVI

FUNCTIONS OF SEVERAL INDEPENDENT VARIABLES

APPENDIX

NUMERICAL TABLES

CHAPTER I

CONTINUITY

1. Continuous Variation.

In every problem of the Infinitesimal Calculus we have to deal with a number of magnitudes, or quantities, some of which may be constant, whilst others are regarded as variable, and (moreover) as admitting of *continuous* variation.

Thus in the applications to Geometry, the magnitudes in question may be lengths, angles, areas, volumes, &c.; in Dynamics they may be masses, times, velocities, forces, &c.

Algebraically, any such magnitude is represented by a letter, such as a or x, denoting the *ratio* which it bears to some standard or 'unit' magnitude of its own kind. This ratio may be integral, or fractional, or it may be 'incommensurable,' *i.e.* it may not admit of being exactly represented by any fraction whose numerator and denominator are finite integers. Its symbol will in any case be subject to the ordinary rules of Algebra.

A 'constant' magnitude, in any given process, is one which does not change its value. A magnitude to which, in the course of any given process, different values are assigned, is said to be 'variable.' The earlier letters $a, b, c, \ldots$ of the alphabet are generally used to denote constant, and the later letters $\ldots u, v, w, x, y, z$ to denote variable magnitudes.

Some kinds of magnitude, as for instance lengths, masses, densities, do not admit of variety of sign. Others, such as altitudes, rotations, velocities, may be either positive or negative. When we wish to designate the 'absolute' value of a magnitude of this latter class, without reference to sign, we enclose the representative symbol between two short vertical lines, thus

$$|x|, \ |\sin x|, \ \log|x|.$$

It is important to notice that, if a and b have the same sign,

$$|a+b| = |a| + |b|,$$

whilst, if they have opposite signs,

$$|a+b| < |a| + |b|.$$

The Infinitesimal Calculus had its origin in problems of Geometry, such as drawing tangents to curves, finding areas and lengths of curves, volumes of solids, and so on. It is therefore natural, and from the point of view of most applications even necessary, to adopt as a basis the geometrical notion of magnitude, with the various familiar assumptions, express or implied, which this involves.

A geometrical representation of any class of magnitudes is obtained by taking an unlimited straight line $X'X$, and in it a fixed origin O, and by measuring lengths OM proportional on any convenient scale to the various magnitudes considered. In the case of sign-less magnitudes (such as masses), these lengths are to be measured on one side only of O; in cases where there is a variety of sign, OM must be drawn to the right or left of O according as the magnitude to be represented is positive or negative. To each magnitude of the kind in question will then correspond a definite point M in the line $X'X$.

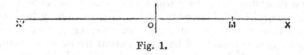

Fig. 1.

When we say that a magnitude admits of 'continuous' variation, we mean that the point M may occupy any position whatever in the line $X'X$ within (it may be) a certain range.

It will be observed that two things are postulated with respect to the magnitudes of the particular kind under consideration, viz. that every possible magnitude of the kind is represented by some point or other of the line $X'X$, and (conversely) that to every point on the line, within a certain range, there corresponds some magnitude of the kind. These conditions are fulfilled by all the kinds of magnitude with which we meet, either in Geometry, or in Mathematical Physics. It will be found on examination that these all involve in their specification a reference, direct or indirect, to linear magnitude. Thus an area may be represented by the altitude of an equivalent rectangle constructed on a given (unit) base; a velocity is represented by the length described in unit time, and so on.

2. Upper or Lower Limit of a Sequence.

The conception of a 'limit,' or 'limiting value,' occurs in various forms throughout the Calculus, and is of fundamental importance. In its primary form, now to be considered, it will be already more or less familiar to the student.

Suppose we have an endless ascending sequence of magnitudes of the same kind

$$x_1, \ x_2, \ x_3, \ ..., \ x_n, \ ..., \ \ \ \ ....(1)$$

i.e. each is greater than the preceding, so that the differences

$$x_2 - x_1, \ x_3 - x_2, \ ..., \ x_n - x_{n-1}, \ ...$$

are all positive. Suppose, further, that the magnitudes (1) are known to be all less than some fixed finite quantity a. The sequence will in this case have an 'upper limit,' *i.e.* there will exist a certain quantity μ, greater than any one of the magnitudes (1), but such that if we proceed far enough in the sequence its members will ultimately exceed any assigned magnitude which is less than μ. In other words, it is impossible to interpose a barrier *between* the members of the sequence and the quantity μ.

In the geometrical representation the magnitudes (1) are represented by a sequence of points

$$M_1, \ M_2, \ M_3, \ ..., \ \ \ \ ....(2)$$

each to the right of the preceding, but all lying to the left of some fixed point A. Hence every point on the line $X'X$, without exception, belongs to one or other of two mutually exclusive categories.

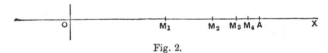

Fig. 2.

Either it has points of the sequence (2) to the right of it, or it has not. Moreover, every point in the former category lies to the left of every point of the latter. Hence there must be some point M, say, such that all points on the left of M belong to the former category and all points on the right of it to the latter. Hence if we put $\mu = OM$, μ fulfils the definition of an 'upper limit' above given.

In a similar manner we can shew that if we have an endless descending sequence of magnitudes

$$x_1, \ x_2, \ x_3, \ ..., \ \ \ \ ....(3)$$

i.e. each is less than the preceding, so that the differences

$$x_1 - x_2, \ x_2 - x_3, \ ..., \ x_{n-1} - x_n$$

are all positive, whilst the magnitudes all exceed some finite quantity b, there will be a lower limit ν, such that every magnitude in the sequence is greater than ν, whilst the members of the sequence ultimately become less than any assigned magnitude which is greater than ν.

The above argument would evidently apply if, occasionally, two or more successive members of the sequence were equal. In that case the sequences are still usually styled 'ascending' or 'descending,' respectively, although the terms 'non-decreasing' and 'non-increasing' would be more accurately descriptive*.

Ex. 1. The sequence

$$\frac{1}{2}, \frac{2}{3}, \frac{3}{4}, \dots, \frac{n}{n+1}, \dots \tag{4}$$

is ascending, with the upper limit 1. For

$$\frac{n}{n+1} = 1 - \frac{1}{n+1},$$

which can be made as nearly equal to 1 as we please by taking n great enough.

Ex. 2. If x be a positive quantity less than unity, the quantities

$$1, \ x, \ x^2, \ \dots, \ x^n, \ \dots \tag{5}$$

form a descending sequence, with the lower limit 0. For since $1/x$ is greater than unity we may write

$$1/x = 1 + y,$$

where y is positive.   Then

$$(1/x)^n = (1 + y)^n = 1 + ny + \dots + y^n,$$

by the Binomial Theorem.   Hence

$$1/x^n > 1 + ny,$$

and can therefore be made as great as we please by taking n great enough.   It follows that x^n can be made as small as we please.

Ex. 3. Consider the sequence defined by

$$x_1 = 1, \ x_{n+1} = \sqrt{(1 + x_n)}. \tag{6}$$

Since

$$x^2_{n+1} - x_n{}^2 = x_n - x_{n-1}, \tag{7}$$

x_{n+1} will be greater than x_n if x_n is greater than x_{n-1}. But x_2 is obviously greater than x_1. The sequence is therefore an ascending one. Again

$$x_{n+1} = \frac{1 + x_n}{x_{n+1}} < \frac{1 + x_{n+1}}{x_{n+1}} < 1 + \frac{1}{x_{n+1}}. \tag{8}$$

Since $x_{n+1} > 1$ it follows that $x_{n+1} < 2$, for all values of n. The sequence has therefore an upper limit. Denoting this by μ, it appears from (6) that μ is the positive root of the equation

$$x^2 = x + 1. \tag{9}$$

---

* In recent times the term 'monotonic' has been invented to include both types of sequence.

By actual calculation from (6) the first few members are found to be, to four figures,

$$1,\ 1\cdot414,\ 1\cdot554,\ 1\cdot598,\ 1\cdot612,\ 1\cdot618.$$

The number last written is the accurate value of μ, to the degree of approximation aimed at.

The matter may be illustrated graphically by tracing the loci

$$y = x + 1,\ y = x^2. \quad\ldots\ldots(10)$$

The figure shews how the successive values of x_n obtained from (6) converge towards the value of x at the intersection. A portion only of the graph is shewn.

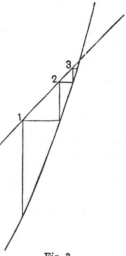

Fig. 3.

Evidently, the same result is arrived at if we start with any positive value of x_1 instead of 1. Only, if x_1 is greater than the positive root of (9) the sequence would be a descending one.

This graphical method has a wide application to the numerical solution of equations, both algebraic and transcendental.

3. Application to Infinite Series. Series with positive terms.

The above has been called the fundamental theorem of the Calculus. An important illustration is furnished by the theory of infinite series whose terms are all of the same sign. In strictness, there is no such thing as the 'sum' of an infinite series of terms, since the operations indicated could never be completed, but under a certain condition the series may be taken as defining a particular magnitude.

Consider a series

$$u_1 + u_2 + u_3 + \ldots + u_n + \ldots \ldots\ldots\ldots\ldots(1)$$

whose terms are all positive, and let

$$s_1 = u_1,\ \ s_2 = u_1 + u_2,\ \ \ldots\ldots s_n = u_1 + u_2 + \ldots + u_n. \ \ldots\ldots(2)$$

These quantities are called the 'partial sums.' If the sequence

$$s_1,\ s_2,\ \ldots\ldots s_n,\ \ldots \ \ldots\ldots\ldots\ldots\ldots(3)$$

has an upper limit S, the series (1) is said to be 'convergent,' and the quantity S is, by convention, called its 'sum.'

Again, if (1) be a series of positive terms which is known to be convergent, and if

$$u_1' + u_2' + u_3' + \ldots + u_n' + \ldots \quad\ldots\ldots\ldots\ldots(4)$$

be a series of positive terms which are respectively less than the corresponding terms in (1), *i.e.* $u_n' < u_n$ for all values of n, then (4) is also convergent. For if s_n' be the sum of the first n terms in (4), we have $s_n' < s_n$, and since the magnitudes s_n have by hypothesis an upper limit, the magnitudes s_n' will have one *à fortiori*.

Ex. 1.   The series

$$1 + \tfrac{1}{2} + \tfrac{1}{4} + \tfrac{1}{8} + \ldots$$

converges to the sum 2. For if in Fig. 2 (p 3) we make $OM_1 = 1$, $OA = 2$, and bisect M_1A in M_2, M_2A in M_3, and so on, the points M_1, M_2, M_3, ... will represent the magnitudes s_1, s_2, s_3, ... And since these points all lie to the left of A, whilst $M_nA = 1/2^{n-1}$ and can therefore be made as small as we please by taking n large enough, it appears that the sequence has the upper limit OA, $= 2$.

The case of any geometric progression whose common ratio is positive and less than unity may be illustrated in a similar manner.

Ex. 2.   Consider the series

$$\frac{1}{1 \cdot 2} + \frac{1}{2 \cdot 3} + \ldots + \frac{1}{n\,(n+1)} + \ldots.$$

If we write this in the form

$$\left(1 - \frac{1}{2}\right) + \left(\frac{1}{2} - \frac{1}{3}\right) + \ldots + \left(\frac{1}{n} - \frac{1}{n+1}\right) + \ldots,$$

we see that

$$s_n = 1 - \frac{1}{n+1},$$

which has the upper limit 1.

Ex. 3.   Further illustrations are supplied by every arithmetical process in which the digits of a non-terminating decimal are obtained in succession. For example, the ordinary process of extracting the square root of 2 gives the series

$$1\cdot414213\ldots$$

or

$$1 + \frac{4}{10} + \frac{1}{10^2} + \frac{4}{10^3} + \frac{2}{10^4} + \frac{1}{10^5} + \frac{3}{10^6} + \ldots.$$

Since s_n is always less than $1\cdot5$, there is an upper limit.

Ex. 4.   The terms of the series

$$1 + 1 + \frac{1}{2\,!} + \frac{1}{3\,!} + \ldots + \frac{1}{n\,!} + \ldots$$

are (after the first three) respectively less than those of the series

$$1 + 1 + \frac{1}{2} + \frac{1}{2^2} + \dots + \frac{1}{2^{n-1}} + \dots.$$

The latter series is convergent and has the sum 3. Hence the former is also convergent, and its sum is less than 3.

4. Limiting Value in a Sequence.

Suppose that we have an endless series of magnitudes

$$x_1, \ x_2, \ x_3, \ \dots, \ x_n, \ \dots \quad\dots\dots\dots(1)$$

arranged in a definite order. Suppose, further, that whatever quantity ϵ we choose to fix upon, however small, there will always be a point in the sequence beyond which every member of it differs from some *fixed* quantity μ by a quantity less in absolute value than ϵ. The sequence is then said to be 'convergent,' and to have the 'limiting value' μ. Statements of this kind occur so frequently in the present subject that it is convenient to have a condensed expression for them. We write

$$\lim_{n \to \infty} x_n = \mu. \quad\dots\dots\dots\dots(2)$$

We have had particular cases of the above relation in the upper and lower limits discussed in Art. 2, but in the present wider definition it is not implied that the members of the sequence are arranged in order of magnitude, or that they are all greater or all less than the limiting value μ.

The hypothesis is that a value of n can be found such that the members of the sequence which follow x_n, viz.

$$x_{n+1}, \ x_{n+2}, \ x_{n+3}, \ \dots,$$

all lie between the values $\mu - \epsilon$ and $\mu + \epsilon$. The value of n which is necessary to secure the fulfilment of this condition will be greater the smaller the value of ϵ, but it is implied that, however small ϵ be taken, such a value exists.

Ex. 1.   The sequence

$$\frac{1}{2}, \ \frac{3}{2}, \ \frac{2}{3}, \ \frac{4}{3}, \ \dots, \ 1 - \frac{1}{n}, \ 1 + \frac{1}{n}, \ \dots \quad\dots\dots\dots(3)$$

has obviously the limiting value 1.

Ex. 2.   In the sequence

$$\sin x, \ \frac{\sin 2x}{2}, \ \frac{\sin 3x}{3}, \ \dots, \ \frac{\sin nx}{n}, \ \dots \quad\dots\dots\dots(4)$$

the numerator lies always between ± 1, whilst the denominator increases indefinitely. The sequence has therefore the limiting value 0.

It is sometimes possible, as in the examples just given, to shew that a given sequence has a certain known quantity as its limit, and is therefore convergent. The question to be resolved is, however, in general less simple, and a criterion is required as to whether a proposed sequence has or has not a definite limiting value. There are in fact many important mathematical quantities which can only be defined as limits, and it is therefore necessary in such a case to satisfy ourselves that the limit exists.

It is obvious in the first place that if the sequence (1) has a limit, a value of n can always be found such that the members of the sequence which follow x_n, viz. $x_{n+1}, x_{n+2}, \ldots, x_{n+p}, \ldots$, will all differ from x_n by quantities not exceeding ϵ, where ϵ may be any assigned quantity, however small. Conversely, if this condition is fulfilled, the sequence has a definite limit.

To shew this let us construct in the first place a descending sequence of positive quantities $\epsilon_1, \epsilon_2, \epsilon_3, \ldots$ whose limit is 0. Such a sequence may be formed, for instance, by making each member one-half of the preceding one. By hypothesis, a number n_1 can be found such that all the members of the sequence which follow x_{n_1} lie between the values

$$x_{n_1} - \epsilon_1 \text{ and } x_{n_1} + \epsilon_1,$$

and will therefore have a lower limit (a_1) and an upper limit (β_1), such that

$$\beta_1 - a_1 \not> 2\epsilon_1.$$

Similarly, a number $n_2 (> n_1)$ can be found such that the members which follow x_{n_2} have a lower limit (a_2) and an upper limit (β_2), such that

$$\beta_2 - a_2 \not> 2\epsilon_2,$$

and so on. The quantities $a_1, a_2, a_3, \ldots$ form an ascending sequence, and, since they are all less than β_1, they have an upper limit μ, say. Similarly, the quantities $\beta_1, \beta_2, \beta_3$ form a descending sequence, with a lower limit ν. Moreover, since

$$\nu - \mu \leq \beta_p - a_p \leq 2\epsilon_p,$$

which may be as small as we please, these limits μ and ν cannot be different. Under the condition stated, the sequence (1) has the common value of μ and ν as its limit.

Ex. 3.   An illustration is furnished by any arithmetical process in which successive approximations to a result are obtained, provided these are adjusted in the usual manner, the last significant figure being increased by unity whenever the next following digit is 5 or any greater number. Thus the operation of finding the square root of 7 gives

$$2\cdot6457513\ldots$$

The successive approximations, adjusted as above, are

$$3, \ 2\cdot6, \ 2\cdot65, \ 2\cdot646, \ 2\cdot6458, \ 2\cdot64575, \ 2\cdot645751, \ \ldots,$$

forming a sequence of the kind now under discussion. The numbers which follow the first differ from it by less than ·5 ; those which follow the second differ from it by less than ·05 ; those which follow the third differ from it by less than ·005 ; and so on. The sequence has therefore a definite limit.

Ex. 4. Consider the sequence in which

$$x_1 = 1, \quad x_{n+1} = \frac{1}{1 + x_n} . \quad\ldots\ldots\ldots\ldots\ldots\ldots(5)$$

The members are all positive, and (after the first) less than unity. It follows that all members after the second are greater than $\frac{1}{2}$. Again, we have

$$x_{n+2} - x_{n+1} = \frac{1}{1 + x_{n+1}} - \frac{1}{1 + x_n} ,$$

or

$$\frac{x_{n+2} - x_{n+1}}{x_n - x_{n+1}} = \frac{1}{(1 + x_n)(1 + x_{n+1})} . \quad\ldots\ldots\ldots(6)$$

Each member of the sequence is therefore alternately greater and less than the one preceding it. Moreover, since the above ratio is, for $n > 1$, less than $\frac{4}{9}$, the intervals between successive members diminish indefinitely. It easily follows that the sequence must converge to a definite limit, which is obviously the positive root of

$$x^2 + x = 1. \quad\ldots\ldots\ldots\ldots\ldots\ldots\ldots(7)$$

By actual calculation from (5) we find in succession

1, ·5, ·6667, ·6, ·625, ·6154, ·6190, ·6176, ·6182,

the latter number being the correct value of the root in question, to four figures.

The character of the sequence may be illustrated graphically by means of the loci

$$y = x, \quad y = \frac{1}{1 + x} . \quad\ldots\ldots\ldots\ldots\ldots\ldots(8)$$

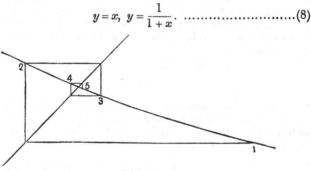

Fig. 4.

The figure shews the essential part of the graph.

In this Example, and in Art. 2, Ex. 3, we have simple illustrations of a method of approximating to the intersection of two curves which is often useful. The convergence is however slow if the curves have nearly the same inclination (in the same or in opposite senses) to the axis of x.

5. Application to Infinite Series.

If in the infinite series

$$u_1 + u_2 + \ldots + u_n + \ldots, \quad\quad\quad\ldots\ldots\ldots\ldots\ldots(1)$$

whose terms are no longer restricted to be all of the same sign, we write

$$s_1 = u_1, \quad s_2 = u_1 + u_2, \quad \ldots\ldots s_n = u_1 + u_2 + \ldots + u_n, \ldots \quad\ldots(2)$$

and if the sequence

$$s_1, \quad s_2, \quad \ldots\ldots s_n, \ldots \quad\quad\quad\ldots\ldots\ldots\ldots\ldots(3)$$

has a limiting value S, the series is said to be 'convergent,' and S is called its 'sum.'

It follows from Art. 4 that the necessary and sufficient condition for the convergence of (1) is that it should be possible to find a number n such that the partial sums $s_{n+1}, s_{n+2}, \ldots, s_{n+p}, \ldots$ all differ from s_n by less than ϵ, where ϵ may be any assigned quantity, however small.

An important theorem in the present connection is that if the series

$$|u_1| + |u_2| + \ldots + |u_n| + \ldots, \quad\quad\ldots\ldots\ldots\ldots(4)$$

formed by taking the absolute values of the several terms of (1), be convergent, the series (1) will be convergent.

For if (4) be convergent, the positive terms of (1) must à *fortiori* form a convergent series, and so also must the negative terms. Let the sum of the positive terms be p and that of the negative terms be $-q$. Also, let s_{m+n}, the sum of the first $m + n$ terms of (1), consist of m positive terms whose sum is p_m, and n negative terms whose sum is $-q_n$. We have, then,

$$(p - q) - s_{m+n} = (p - q) - (p_m - q_n)$$
$$= (p - p_m) - (q - q_n). \quad\ldots\ldots\ldots\ldots(5)$$

If $m + n$ be sufficiently great, $p - p_m$ and $q - q_n$ will both be less than ϵ, where ϵ is any assigned magnitude, however small; and the difference of these positive quantities will be à *fortiori* less than ϵ in absolute value. Hence s_{m+n} has the limiting value $p - q$.

When the series (4), composed of the absolute values of the several terms of (1), is convergent, the series (1) is said to be 'absolutely,' or 'essentially,' or 'unconditionally' convergent.

It is possible, however, for a series to be convergent, whilst the series formed by taking the absolute values of the terms has no upper limit. In this case, the convergence of the given series is said to be 'accidental,' or 'conditional.'

The following very useful theorem holds whether the series considered be essentially or only accidentally convergent:

If the terms of a series are alternately positive and negative, and continually diminish in absolute value, and moreover tend ultimately to the limit zero, the series is convergent, and its sum is intermediate between the sum of any finite odd number of terms and that of any finite even number, counting in each case from the beginning.

The proof will be familiar to the student, but as it is a good example of the kind of argument employed in the preceding Art., it is here repeated.

Let the series be

$$a_1 - a_2 + a_3 - a_4 + \dots, \quad \dots\dots\dots\dots(6)$$

where, by hypothesis,

$$a_1 > a_2 > a_3 > \dots.$$

In the figure, let

$$OM_1 = a_1, \quad M_1 M_2 = a_2, \quad M_2 M_3 = a_3, \quad M_3 M_4 = a_4, \dots.$$

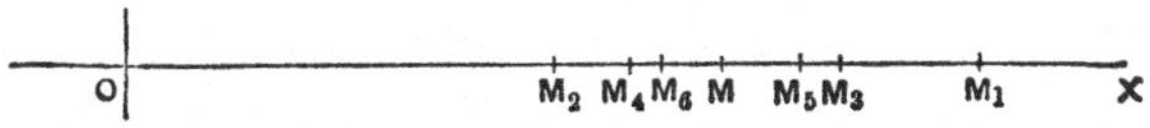

Fig. 5.

It is plain that the points M_1, M_3, M_5, ... form a descending sequence, and that the points M_2, M_4, M_6, ... form an ascending sequence. Also that every point of the former sequence lies to the right of every point of the latter. Hence each sequence has a limiting point, and since

$$M_{2n} M_{2n+1} = a_{2n+1},$$

and therefore is ultimately less than any assignable magnitude, these two limiting points must coincide, say in M. Then OM represents the sum of the given series (6).

Ex. The series

$$1 - \tfrac{1}{2} + \tfrac{1}{3} - \tfrac{1}{4} + \dots$$

converges to a limit between 1 and $1 - \tfrac{1}{2}$.

This series belongs to the 'accidentally' convergent class. It will be shewn later (Art. 175) that the sum of n terms of the series

$$1 + \tfrac{1}{2} + \tfrac{1}{3} + \tfrac{1}{4} + \dots$$

can be made as great as we please by taking n great enough.

It cannot be too carefully remembered that the word 'sum,' as applied to an infinite series, is used in a purely conventional sense, and that we are not at liberty to assume, without examination, that we may deal with such a series as if it were an expression consisting of a finite number of terms. For example, we may not assume that the sum is unaltered by any rearrangement of the terms. In the case of an essentially convergent series this assumption can be justified, but an accidentally convergent series can be made to converge to any limit we please by a suitable adjustment of the order in which the terms succeed one another. For the proofs of these theorems we must refer to books on Algebra ; they are hardly required in the present treatise.

The following simple theorems will, however, occasionally be referred to.

1°. If $$u_1 + u_2 + \ldots + u_n + \ldots \quad \ldots\ldots\ldots\ldots\ldots(7)$$
be a convergent series whose sum is S, the series

$$au_1 + au_2 + \ldots + au_n + \ldots, \quad \ldots\ldots\ldots\ldots\ldots(8)$$

obtained by multiplying the terms of (7) by a factor a, will converge to the sum aS. This hardly needs proof.

2°. If $$u_1 + u_2 + \ldots + u_n + \ldots \quad \ldots\ldots\ldots\ldots\ldots(9)$$
and $$u_1' + u_2' + \ldots + u_n' + \ldots \ldots\ldots\ldots\ldots\ldots(10)$$
be two convergent series whose sums are S and S', respectively, the series

$$(u_1 \pm u_1') + (u_2 \pm u_2') + \ldots + (u_n \pm u_n') + \ldots, \quad \ldots(11)$$

composed of the sums, or the differences, of corresponding terms in (9) and (10), will converge to the sum $S \pm S'$. This is easily proved. If s_n, s_n' denote the sums of the first n terms of (9) and (10) respectively, the sum of the first n terms of (11) will be $s_n \pm s_n'$. Now

$$(S \pm S') - (s_n \pm s_n') = (S - s_n) \pm (S' - s_n'). \quad \ldots\ldots(12)$$

By hypothesis, if ϵ be any assigned magnitude, however small, we can find a value of n such that for this and for all higher values we shall have

$$|S - s_n| < \tfrac{1}{2}\epsilon, \quad \text{and} \quad |S' - s_n'| < \tfrac{1}{2}\epsilon, \quad \ldots\ldots\ldots(13)$$

and therefore $$|(S \pm S') - (s_n \pm s_n')| < \epsilon, \quad \ldots\ldots\ldots\ldots(14)$$

which is the condition that $s_n \pm s_n'$ should have the limiting value $S \pm S'$.

3°. On the same hypothesis the series

$$(au_1 + bu_1') + (au_2 + bu_2') + \ldots + (au_n + bu_n') + \ldots \quad (15)$$

will converge to the sum $aS + bS'$. This follows easily from the two preceding theorems.

6. General Definition of a Function.

One variable quantity is said to be a 'function' of another when, other things remaining the same, if the value of the latter be fixed that of the former becomes determinate.

The two quantities thus related are distinguished as the 'dependent' and the 'independent' variable respectively.

The notion of a function of a variable quantity is one which presents itself in various branches of Mathematics. Thus, in Arithmetic, the number of permutations of n objects is a function of n; the number of balls in a square or a triangular pile of shot is a function of the number contained in each side of the base; the sum (s_n) of the first n terms of any given series is a function of the number n; and so on. In some of these cases there are definite mathematical formulæ for the functions in question, but it is to be noticed that the idea of functionality does not necessarily require this; for example, the sum of the first n terms of the series

$$\frac{1}{1^2} + \frac{1}{2^2} + \frac{1}{3^2} + \frac{1}{4^2} + \ldots$$

is a definite function of n, although no exact mathematical expression exists for it. So, again, the number of primes not exceeding a given integer n is a definite function of n, although it cannot be represented by a formula.

In these examples, the independent variable, from its nature, can only change by finite steps. The Infinitesimal Calculus, on the other hand, deals specially with cases where the independent variable is continuous, in the sense of Art. 1. For instance, in Geometry the area of a circle, or the volume of a sphere, is a function of the radius; in theoretical Physics the altitude, or the velocity, of a falling particle is regarded as a function of the time; the period of oscillation of a given pendulum as a function of the amplitude; the pressure of a given gas at a given temperature as a function of the density; the pressure of saturated steam as a function of the temperature; and so on. Here, again, the existence or non-existence of a mathematical expression for the function is not material; all that is necessary to establish a functional relation between two variables is that, when other things are unaltered, the value of one shall determine that of the other.

In general investigations it is usual to denote the independent variable by x, and the dependent variable by y. The relation between them is often expressed in such a form as

$$y = \phi(x), \quad \text{or} \quad y = f(x), \quad \&c.,$$

the symbol $\phi(x)$, for instance, meaning 'some particular function of x.'

When a quantity varies from one value to another, the amount (positive or negative) by which the new value exceeds the former value is called the 'increment' of the quantity. This increment is often denoted by prefixing δ or Δ (regarded as a symbol of *operation*) to the symbol which represents the variable magnitude. Thus we speak of the independent variable changing from x to $x + \delta x$, and of the dependent variable consequently changing from y to $y + \delta y$.

Hence if $\qquad y = \phi(x),$ ....(1)

we must have $\qquad y + \delta y = \phi(x + \delta x),$ ....(2)

and therefore $\qquad \delta y = \phi(x + \delta x) - \phi(x).$ ....(3)

At present there is no implication that δx or δy is small; the increments may have any values subject to the relation (2).

Ex. 1.   If $y = x^3$, then if $x = 100$, $\delta x = 1$, we have
$$\delta y = (101)^3 - (100)^3 = 30301.$$

Ex. 2.   If $y = \sin x^\circ$, then if $x = 60$, $\delta x = 1$, we have
$$\delta y = \cdot 87462 - \cdot 86603 = \cdot 00859,$$
within a certain degree of accuracy.

7.   Geometrical Representation of Functions.

We construct a graphical representation of the relation between two variables x, y, one of which is a function of the other, by taking rectangular coordinate axes OX, OY. If we measure OM along OX, to represent any particular value of the independent variable x, and ON along OY to represent the corresponding value of the function y, and if we complete the rectangle $OMPN$, the

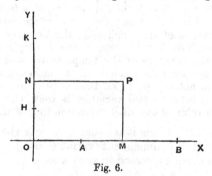

Fig. 6.

position of the point P will indicate the values of both the associated variables.

Since, by hypothesis, M may occupy any position on OX, between (it may be) certain fixed termini, we obtain in this way an infinite assemblage of points P.

A question arises as to the nature of this assemblage; whether, or in what sense, the points constituting it can be regarded as lying on a curve. In many cases, of course, there is no hesitation about the answer. For example, if, to represent the relation between the area of a circle and its radius, we make OM proportional to the radius, and PM proportional to the area, then $PM \propto OM^2$, and the points P lie on a parabola. The same curve will represent the relation between the space (s) described by a falling body and the time (t) from rest, since s varies as t^2.

The general question must, however, be answered in the negative. The definition of a function given at the beginning of Art. 6 stipulates that for each value of x there shall be a definite value of y; but there is no necessary relation between the values of y corresponding to different values of x, however close together these may be.

Without some further qualification the definition referred to is indeed far too wide for our present purposes, the functions ordinarily contemplated in the Calculus being subject to certain very important restrictions.

The first of these restrictions is that of 'continuity.' This implies that, as M ranges over any finite portion AB of the line OX, N ranges over a finite portion HK of the line OY, i.e. N occupies once at least every position between H and K. Further, that if the range AB be continually contracted, the range HK will also contract, and can be made as small as we please by taking AB small enough.

It will be seen presently (Art. 8) that the second of these properties includes the former. The formal definition which we proceed to give is slightly different, although, as will be seen, equivalent.

8. Definition of a Continuous Function.

Let x, y be any two corresponding values of the independent variable and the function. If, $x_1, x_2, \ldots, x_n, \ldots$ being any arbitrary sequence of admissible values of the independent variable, having x for its limit, the sequence $y_1, y_2, \ldots, y_n, \ldots$ of corresponding values of the function converges always to the limit y, the function is said to be 'continuous' for the particular value x of the independent variable.

It follows that if δx denote an increment of x, and δy the corresponding increment of y, we can always find a positive quantity ϵ, different from zero, such that, for all admissible* values of δx

---

* The restriction to 'admissible' values of δx means that $x + \delta x$ must be within the range of values of the independent variable for which the function is defined.

which are less in absolute value than ϵ, the value of δy will be less in absolute value than σ, where σ is any prescribed quantity, however small. This is often taken as the formal definition. If the condition which it involves were violated, that involved in the former definition could not be fulfilled. Hence the two definitions are really equivalent.

The second definition is sometimes summed up briefly, but imperfectly, in the statement that an infinitely small change in the independent variable produces an infinitely small change in the function. This means that if $\phi(x)$ be the function, it must be possible to find a quantity ϵ such that

$$| \phi(x+h) - \phi(x) | < \sigma,$$

for all admissible values of h such that $|h| < \epsilon$. The value of ϵ will in general depend upon that of σ, but it is implied that the condition can always be satisfied by *some* value of ϵ, however small σ may be.

9. Property of a Continuous Function.

If $\phi(x)$ be a function which is continuous from $x = a$ to $x = b$, inclusively, and if $\phi(a)$, $\phi(b)$ have opposite signs, there must be at least one value of x between a and b for which $\phi(x) = 0$.

In the annexed figure it is assumed for definiteness that $\phi(a)$ is positive and $\phi(b)$ negative. The points of the line $X'X$ for which $x = a$, $x = b$ are denoted by A, B, respectively, and the corresponding values of the function are represented by AH, BK. The proof consists in shewing that a series of diminishing intervals of lengths

$$AB, \quad \frac{1}{2}AB, \quad \frac{1}{2^2}AB, \quad ..., \quad \frac{1}{2^n}AB, \quad ...$$

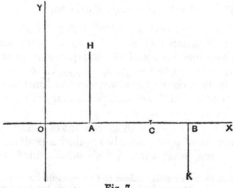

Fig. 7.

can be found, each forming part of the preceding, and each containing points at which $\phi(x)$ is positive and points at which it is negative.

Let AB be supposed bisected at M_1. If $\phi(x) = 0$ at M_1, the theorem is established for this particular case. We may therefore exclude this and similar contingencies in the sequel. If $\phi(x)$ does not vanish at M_1, then in one at least of the intervals AM_1, M_1B the function will have both positive and negative values. If the statement applies to only one of the intervals, we select that one; if to both, the selection may be made arbitrarily. The selected interval is next supposed bisected in M_2. Excluding the case where $\phi(x) = 0$ at M_2, one at least of the halves into which the selected interval has been divided will contain points at which $\phi(x)$ is positive and points at which it is negative. The process may be continued indefinitely, and since

$$M_n M_{n+1} = \frac{1}{2^{n+1}} AB,$$

it follows by Art. 4 that the sequence of dividing points

$$M_1, \; M_2, \; M_3, \; \ldots, \; M_n, \; \ldots$$

thus obtained has a definite limiting position, denoted, say, by C.

Moreover, the value of $\phi(x)$ at C must be zero. For if it were positive there would, in virtue of the assumed continuity of $\phi(x)$, be a finite range on each side of C throughout which $\phi(x)$ would be positive. This would be inconsistent with the result just proved. Similarly if the value of $\phi(x)$ at C were negative.

We may express the above theorem shortly by saying that a continuous function cannot change sign except by passing through the value zero.

It follows that if $\phi(x)$ be a function which is continuous from $x = a$ to $x = b$ inclusive, and if $\phi(a)$, $\phi(b)$ be unequal, there must be some value of x *between* a and b, such that $\phi(x) = \beta$, where β may be any quantity intermediate in value to $\phi(a)$ and $\phi(b)$. For, let

$$f(x) = \phi(x) - \beta\,;$$

since β is a constant, $f(x)$ also will be continuous. By hypothesis,

$$\phi(a) - \beta \;\; \text{and} \;\; \phi(b) - \beta$$

have opposite signs, and therefore $f(a)$ and $f(b)$ have opposite signs. Hence, by the above theorem, there is some value of x between a and b for which $f(x) = 0$, or $\phi(x) = \beta$.

In other words, a continuous function cannot pass from one value to another without assuming once (at least) every intermediate value.

10. Graph of a Continuous Function.

It follows from what precedes that the assemblage of points which represents, in the manner explained in Art. 7, any *continuous* function is a 'connected' assemblage. By this it is meant that a line cannot be drawn across the assemblage without passing through some point of it. For, denoting the function by $\phi(x)$, and the ordinate of any line by $f(x)$, then if $\phi(x)$ and $f(x)$ are both continuous the difference

$$\phi(x) - f(x)$$

will be continuous (Art. 12), and therefore cannot change sign without passing through the value zero.

The question whether any connected assemblage of points is to be regarded as lying on a *curve* is to some extent a verbal one, the answer depending upon what properties are held to be connoted by the term 'curve.' It is, however, obvious that a good representation of the general course or 'march' of any given continuous function can be obtained by actually plotting on paper the positions of a sufficient number of points belonging to the assemblage, and drawing a line through them with a free hand. A figure constructed in this way is called a 'graph' of the function.

The construction of graphs of functions of the types

$$y = Ax + B, \quad y = Ax^2 + Bx + C,$$

and their use to elucidate the theory of simple and quadratic equations, will be familiar to the reader.

The graphical method will be freely used in this book (as in other elementary treatises) for purposes of illustration. It may be worth while, however, to point out that, as applied to *mathematical* functions, it has certain limitations. In the first place, it is obvious that no finite number of isolated values can determine the function completely; and, indeed, unless some judgment is exercised in the choice of the values of x for which the function is calculated, the result may be seriously misleading. Again, the streak of ink, or graphite, by which we represent the course of the function, has (unlike the ideal mathematical line) a certain breadth, and the same is true of the streak which represents the axis of x; the distance between these streaks is therefore affected by a certain amount of vagueness. For the same reason, we cannot reproduce details of more than a certain degree of minuteness; the method is therefore intrinsically inadequate in the case of functions (such as can be proved to exist) in which new details reveal themselves *ad infinitum* as the scale is magnified*. Functions of this latter class are, however, seldom encountered in the ordinary applications of the Calculus.

* An instance is furnished by the function $x \sin(1/x)$ in the neighbourhood of the origin.

The method of graphical representation is often used in practice when the mathematical form of the function is unknown; a certain number of corresponding values of the dependent and independent variables being found by observation. The vagueness due to the breadth of the lines is then usually less serious than that due to the imperfections of our senses, errors of observation, and the like.

The reader may be reminded of the meteorological charts which exhibit the height of the barometer or thermometer as a function of the time.

The substratum of fact underlying a graph constructed in the above manner is of course no more than is contained in a numerical table giving a series of pairs of corresponding values of the variables, but the graphical form appeals far more effectively to the mind, by helping us to supply in imagination the intermediate values of the function.

11. Discontinuity.

Although the functions ordinarily met with in Mathematics are on the whole definite and continuous over the range of the independent variable considered, exceptions to this statement may occur at isolated points.

Thus it may happen that the original definition of the function fails to give a meaning for particular values of the independent variable.

Take, for instance, the function

$$\frac{\sin x}{x}.$$

For any value of x, other than 0, the numerator and denominator have certain values, and the quotient exists. But when $x = 0$, the fraction assumes the indeterminate form $0/0$. It is true that the value 1 is then usually attributed to it, but this is a matter of convention, and is not implied in the original definition. Many such instances will present themselves in the sequel.

Again, a function $\phi(x)$ may become 'infinite' for some particular value x_1 of x. The meaning of this is that by taking x sufficiently nearly equal to x_1 the value of the function can be made to exceed (in absolute value) any magnitude which we please to assign, however great. This is usually expressed by the formula

$$\lim_{x \to x_1} \phi(x) = \infty.$$

The above is the only meaning which the word 'infinite' has in Mathematics, and the only legitimate use of the symbol ∞ is in condensed statements of the above kind.

Examples are furnished by the function $1/x$, which becomes infinite for $x \to 0$, and $\tan x$, which becomes infinite for $x \to \tfrac{1}{2}\pi$, &c. See Fig. 15, p. 28.

Again, in Mechanics, the period of oscillation of a pendulum, regarded as a function of the amplitude (a), becomes infinite for $a \to \pi$.

Again, a function, though finite, may be discontinuous for a particular value x_1 of x, i.e. its values for $x = x_1 - \epsilon$ and $x = x_1 + \epsilon$ may be unequal, however small ϵ may be. In that case the original definition may or may not assign a definite value for $x = x_1$.

An illustration from Mechanics is furnished by the velocity of a particle which at a given instant receives a sudden impulse in the direction of motion. In this case the 'velocity' at the precise instant of the impulse is undefined, although it has a meaning immediately before and immediately after.

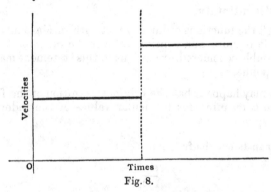

Fig. 8.

Other more general types of discontinuity are imaginable, but are not met with in the ordinary applications of the subject.

12. Theorems relating to Continuous Functions.

We may now proceed to investigate the continuity, or otherwise, of various functions which have an explicit mathematical definition, and to examine the character of their graphical representations.

For this purpose the following preliminary theorems are necessary:

1°. The *sum* of any finite number of continuous functions is itself a continuous function.

First suppose we have two functions u, v of the independent variable x. Then

$$\delta(u + v) = (u + \delta u + v + \delta v) - (u + v)$$
$$= \delta u + \delta v.$$

From the definition of continuity it follows that, whatever the value of σ, we can find a quantity ϵ such that for $|\delta x| < \epsilon$ we shall have $|\delta u| < \tfrac{1}{2}\sigma$ and $|\delta v| < \tfrac{1}{2}\sigma$, and therefore

$$|\delta u + \delta v| < \sigma.$$

Hence the function $u + v$ is continuous.

Next, if we have three continuous functions u, v, w, then $u + v$ is continuous, as we have just seen, and consequently $(u + v) + w$ is continuous. In this way the theorem may be extended, step by step, to the case of any *finite* number of functions.

2°. The *product* of any finite number of continuous functions is itself a continuous function.

First, take the case of two functions u, v. We have

$$\delta (uv) = (u + \delta u)(v + \delta v) - uv$$
$$= v\delta u + u\delta v + \delta u \delta v.$$

By hypothesis we can, by taking $|\delta x|$ small enough, make $|\delta u|$ and $|\delta v|$ less than any assigned quantity, however small. Hence, since u and v are finite, $|v\delta u|$ and $|u\delta v|$ can be made less than any assigned quantity, however small. The same is evidently true of $|\delta u \delta v|$. Hence, also, the value of

$$|v\delta u + u\delta v + \delta u \delta v|$$

can be made less than any assigned quantity, however small. That is, uv is a continuous function.

Next, suppose we have three continuous functions u, v, w. We have seen that uv is continuous; hence also $(uv)w$ is continuous. And so on for the product of any *finite* number of continuous functions.

3°. The quotient of two continuous functions is a continuous function, except for those values (if any) of the independent variable for which the divisor vanishes.

We have
$$\delta \left(\frac{u}{v}\right) = \frac{u + \delta u}{v + \delta v} - \frac{u}{v}$$
$$= \frac{v\delta u - u\delta v}{v(v + \delta v)}.$$

By hypothesis $v \neq 0$; there is therefore a lower limit M, different from zero, to the absolute magnitude of $v(v + \delta v)$. This makes

$$\left|\delta \left(\frac{u}{v}\right)\right| < \left|\frac{v}{M}\delta u - \frac{u}{M}\delta v\right|.$$

Since v/M and u/M are finite, we can, by taking δx small enough, make $|v/M . \delta u|$ and $|u/M . \delta v|$ less than any assigned magnitude, however small. The same will therefore be true of $|\delta(u/v)|$; *i.e.* the quotient u/v is continuous.

4°. If y be a continuous function of u, where u is a continuous function of x, then y is a continuous function of x.

For let δx be any increment of x, δu the consequent increment of u, and δy the consequent increment of y. Since y is a continuous function of u, we can find a quantity ϵ' such that if

$$|\,\delta u\,| < \epsilon', \text{ then } |\,\delta y\,| < \sigma,$$

where σ may be any assigned quantity, however small. And since u is a continuous function of x, we can find a quantity ϵ, such that if

$$|\,\delta x\,| < \epsilon, \text{ then } |\,\delta u\,| < \epsilon'.$$

Hence if $\qquad |\,\delta x\,| < \epsilon$, we have $|\,\delta y\,| < \sigma$,

which is the condition of continuity of y, considered as a function of x.

13. Algebraic Functions. Rational Integral Functions.

An 'algebraic' function is one which is obtained by performing with the variable and known constants any *finite* number of operations of addition, subtraction, multiplication, division, and extraction of integral roots.

All other functions are classed as 'transcendental'; they involve, in one form or another, the notion of a 'limiting value' (Arts. 4, 19).

A 'rational' algebraic function is one which is formed in like manner by operations of addition, subtraction, multiplication, and division, only. Any such function can be reduced to the form

$$\frac{F(x)}{f(x)}$$

where the numerator and denominator are rational 'integral' functions; *i.e.* each of them is of the type

$$A_n x^n + A_{n-1} x^{n-1} + A_{n-2} x^{n-2} + \ldots + A_1 x + A_0, \quad \ldots\ldots(1)$$

where n is a positive integer, and the coefficients are constants. Such an expression, when it consists of more than one term, is often more briefly referred to as a 'polynomial,' the algebraic character being understood.

A rational integral function is finite and continuous for all finite values of the variable. For x^m, being the product of a finite number (m) of continuous functions (each equal to x), is finite and continuous. Hence also $A x^m$ is finite and continuous; and the sum of any finite number of such terms is finite and continuous (Art. 12).

A rational integral function becomes infinite for $x \to \pm \infty$. Writing the function (1) in the form

$$x^n \left(A_n + \frac{A_{n-1}}{x} + \frac{A_{n-2}}{x^2} + \ldots + \frac{A_1}{x^{n-1}} + \frac{A_0}{x^n} \right),$$

we see that by taking x great enough (in absolute value) we can make the first factor (x^n) as great as we please, whilst the second factor can be made as nearly equal to A_n as we please. Hence the product can be made as great as we please. Moreover, if x be positive the sign of the product will be the same as that of A_n, whilst if x be negative the sign will be the same as that of A_n, or the reverse, according as n is even or odd.

It follows that in the graphical representation of a rational integral function $y = f(x)$ the curve is everywhere at a finite distance from the axis of x, but recedes from it without limit as x is continually increased, whether positively or negatively. In actually constructing the curve, it is convenient if possible to solve the equation $f(x) = 0$, as this gives the intersections of the curve with the axis of x.

Ex. To trace the curve

$$y = x (x^2 - 1).$$

This cuts the axis of x at the points $x = 0, \pm 1$. Since, when $x \to 0$, x^3 becomes infinitely small compared with x, the curve approximates near the origin to the straight line $y = -x$, which is in fact the tangent there.

Since y changes sign with x, we need only calculate the ordinates for positive values of x. We easily construct the following table, where only two significant figures are retained.

x	y	x	y
·1	− ·10	·8	− ·29
·2	− ·19	·9	− ·17
·3	− ·27	1·0	0
·4	− ·34	1·1	+ ·23
·5	− ·38	1·2	+ ·51
·6	− ·38	1·3	+ ·88
·7	− ·36	∞	∞

The graph is given on the next page.

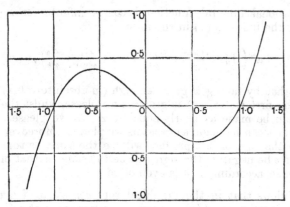

Fig. 9.

14. Rational Fractions.

A function
$$y = \frac{F(x)}{f(x)}, \quad \dots\dots\dots\dots\dots\dots\dots(1)$$

which is rational but not integral, is finite and continuous for all finite values of x except those which make $f(x) = 0^*$. For the rational integral functions $F(x)$ and $f(x)$ have been proved to be finite and continuous; and it follows, by Art. 12, that the quotient will be finite and continuous except when the denominator vanishes.

The curve represented by (1) will cut the axis of x in the points (if any) for which $F(x) = 0$. It will have asymptotes parallel to y wherever $f(x) = 0$, whilst, for all other finite values of x, y will be finite and continuous. The values of y for $x \to \pm\infty$ will depend on the relative order of magnitude of $F(x)$ and $f(x)$. If $F(x)$ be of higher degree than $f(x)$ the ordinates become infinite; if of lower degree the ordinates diminish indefinitely, the axis of x being an asymptote; if the degrees are the same, there is an asymptote parallel to x.

In cases where the degree of the numerator is not less than that of the denominator, it is convenient to perform the division indicated until the remainder is of lower degree than the divisor, and so express y as the sum of an integral function and a 'proper' fraction.

The following examples are chosen to illustrate some of the more important points which may arise.

* It is assumed that the fraction has been reduced to its lowest terms.

Ex. 1. $$y = \frac{1-x}{2x} = -\frac{1}{2} + \frac{1}{2x}.$$

This makes $y = 0$ for $x = 1$, and $y \to \pm \infty$ for $x \to 0$. Also y is positive for $1 > x > 0$, and negative outside this interval. From the second form of y it appears that for $x \to \pm \infty$ we have $y = -\frac{1}{2}$. We further find, as corresponding values of x and y:

$$\begin{cases} x = -\infty, & -3, & -2, & -1, & -\cdot5, & 0, & \cdot5, & 1, & 2, & 3, & +\infty, \\ y = -\cdot5, & -\cdot67, & -\cdot75, & -1, & -1\cdot5, & \mp\infty, & \cdot5, & 0, & -\cdot25, & -\cdot33, & -\cdot5. \end{cases}$$

The figure shews the curve.

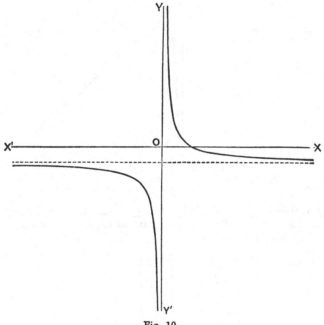

Fig. 10.

Ex. 2. $$y = \frac{x(1-x)}{1+x} = -x + 2 - \frac{2}{1+x}.$$

Here $y = 0$ for $x = 0$ and $x = 1$, and $y \to \pm \infty$ for $x \to -1$. Also y changes sign as x passes through each of these values. For numerically large values of x, whether positive or negative, the curve approximates to the straight line

$$y = -x + 2,$$

lying beneath this line for $x \to +\infty$, and above it for $x \to -\infty$. Fig. 11 on p. 26 shews the curve.

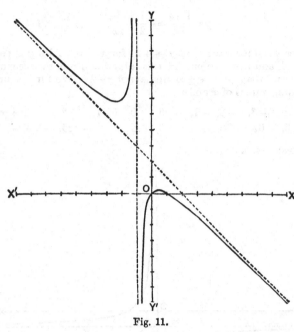

Fig. 11.

Ex. 3. $y = \dfrac{2x}{1-x^2}.$

Here y vanishes for $x = 0$, and for $x \to \pm\infty$, and becomes infinite for $x \to \pm 1$. Again, y is positive for $1 > x > 0$ and negative for $x > 1$. Also, y changes sign with x.

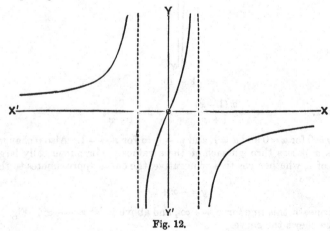

Fig. 12.

Ex. 4.
$$y = \frac{2x}{1 + x^2}.$$

As in the preceding Ex., y vanishes for $x = 0$ and $x \rightarrow \pm \infty$, and changes sign with x. But the denominator does not vanish for any real value of x, so that y is always finite.

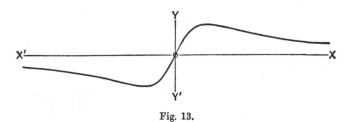

Fig. 13.

15. The Circular Functions.

The general definitions of the 'circular' functions

$$\sin x, \quad \cos x, \quad \tan x, \quad \&c.,$$

are given in books on Trigonometry.

The function $\sin x$ is continuous for all values of x. For

$$\delta (\sin x) = \sin (x + \delta x) - \sin x$$
$$= 2 \sin \tfrac{1}{2} \delta x \cdot \cos (x + \tfrac{1}{2} \delta x).$$

The last factor is always finite, and the product of the remaining factors can be made as small as we please by taking δx small enough.

In the same way we may shew that $\cos x$ is continuous. This result is, however, included in the former, since

$$\cos x = \sin (x + \tfrac{1}{2} \pi).$$

Again, since

$$\tan x = \frac{\sin x}{\cos x},$$

the continuity of $\sin x$ and $\cos x$ involves (Art. 12) that of $\tan x$, except for those values of x which make $\cos x = 0$. These are given by $x = (n + \tfrac{1}{2}) \pi$, where n is integral.

In the same way we might treat the cases of sec x, cosec x, cot x.

The annexed figures shew the graphs of sin x and tan x. The reader should observe how immediately such relations as

$$\sin(-x) = -\sin x, \quad \sin(\pi - x) = \sin x,$$
$$\sin(x + \pi) = -\sin x, \quad \tan(x + \pi) = \tan x$$

can be read off from the symmetries of the curves.

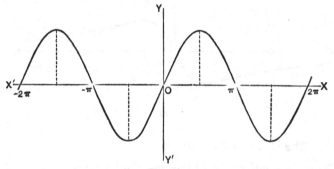

Fig. 14.

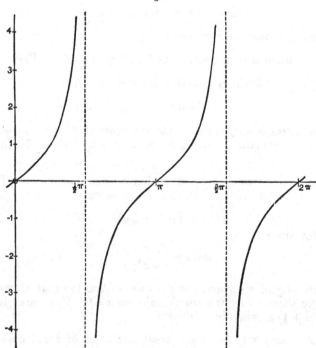

Fig. 15.

16. Inverse Functions.

If y be a continuous function of x, then under certain conditions x will be a continuous function of y. This will be the case whenever the range of x admits of being divided into portions (not infinitely small) such that within each the function y steadily increases, or steadily decreases, as x increases.

Let us suppose that as x increases from a to b the value of y steadily increases from α to β. Then corresponding to any given value of y between α and β there will be one and only one value of x between a and b. Hence if we restrict ourselves to values of x

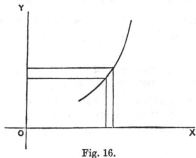

Fig. 16.

within this interval, x will be a single-valued function of y. It is also easily seen to be a continuous function of y.

For, if we give any positive increment ϵ to x, within the above interval, y will have a certain finite increment σ, and for all values of δy less than σ, we shall have $\delta x < \epsilon$. A similar argument holds if the increment of x be negative. Hence we can find a positive quantity σ such that, ϵ being any assigned positive quantity, however small, $|\delta x| < \epsilon$ for all values of δy such that $|\delta y| < \sigma$. But this is the condition for the continuity of x regarded as a function of y (Art. 8).

The same conclusion obviously holds if y steadily diminishes in the interval from $x = a$ to $x = b$.

If we do not limit ourselves to a range of x within which the function steadily increases, or steadily diminishes, then to any given value of y there may correspond more than one value of x; the inverse function is then said to be 'many-valued.' Again, it may (and in general will) happen that through some ranges of y there are no corresponding values of x, i.e. the inverse function does not exist.

If
$$y = f(x), \quad \dots\dots\dots\dots\dots\dots\dots(1)$$
the inverse functional relation is sometimes expressed by
$$x = f^{-1}(y). \quad \dots\dots\dots\dots\dots\dots(2)$$

We then have $f\{f^{-1}(y)\} = f(x) = y,$ ....(3)

i.e. the functional symbols f and f^{-1} cancel one another. This is the reason of the notation (2).

The graph of any inverse function is derived from that of the direct function by mere transposition of x and y.

Ex. 1. Let $y = x^2$. This is a continuous function of x, and, if x be positive, continuously increases with x. Hence $x, = \sqrt{y}$, is a continuous function of y. If x be unrestricted as to sign, we have two values of x for every positive value of y; these are usually denoted by $\pm \sqrt{y}$. If y be negative, the inverse function $\sqrt{y}$ does not exist.

Ex. 2. The 'goniometric' or 'inverse circular' functions

$$\sin^{-1}x, \cos^{-1}x, \tan^{-1}x, \&c.$$

are many-valued.

The functions $\sin^{-1}x$, $\cos^{-1}x$ exist for values of x ranging from -1 to $+1$, but not for values outside these limits.

The function $\tan^{-1}x$ exists for all values of x. It is many-valued, the values forming an arithmetical progression with the common difference π.

The curves for $\sin^{-1}x$ and $\tan^{-1}x$ are shewn in Figs. 21, 22, pp. 60, 62.

17. Upper or Lower Limit of an Assemblage.

Before proceeding further with the theory of continuous functions it is convenient to extend the definitions of the terms 'upper' and 'lower' limit, and 'limiting value,' given in Arts. 2 and 4.

Consider, in the first place, any assemblage of magnitudes, infinite in number, but all less than some finite magnitude β. The assemblage may be defined in any way; all that is necessary is that there should be some criterion by which it can be determined whether a given magnitude belongs to the assemblage or not. For instance, the assemblage may consist of the values which a given function (continuous or not) assumes as the independent variable ranges over any finite or infinite interval.

In such an assemblage there may or may not be contained a 'greatest' magnitude, *i.e.* one not exceeded by any of the rest; but there will in any case be an 'upper limit' to the magnitudes of the assemblage, *i.e.* there will exist a certain magnitude μ such that no magnitude in the assemblage exceeds μ, whilst one (at least) can be found exceeding any magnitude whatever which is less than μ. And if μ be not itself one of the magnitudes of the assemblage, then an infinite number of these magnitudes can be found exceeding any magnitude which is less than μ.

The proof of these statements follows, as in Art. 2, by means of the geometrical representation.

In the same way, if we have an infinite assemblage of magnitudes, all greater than some finite quantity a, there may or may not be a 'least' magnitude in the assemblage; but there will in any case be a 'lower limit' v such that no magnitude in the assemblage falls below v, whilst one (at least) can be found below any magnitude whatever which is greater than v. And if v be not itself one of the magnitudes of the assemblage, an infinite number of these magnitudes can be found less than any magnitude which is greater than v.

An important example occurs in the definition of the 'perimeter' of a circle.

If we have any number of points on the circumference of a given circle, then by joining them in order we obtain an inscribed polygon, and by drawing tangents at the points we obtain a circumscribed polygon. It is easily proved that the perimeter of any inscribed polygon formed in this way is less than that of any circumscribed polygon. Hence, considering the whole assemblage of possible inscribed polygons, their perimeters will have a definite upper limit. Similarly the perimeters of all possible circumscribed polygons will have a lower limit.

Moreover, these two limits must be identical. For, let PQ be a side of one of the inscribed polygons of the assemblage, PT and QT

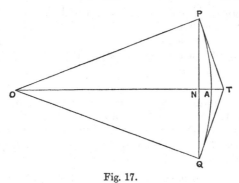

Fig. 17.

the tangents at P and Q; let O be the centre, and let PQ meet OT in N. Then PT and QT will be portions of sides of a circumscribed polygon, and if Σ be a sign of summation extending round the polygons the ratio of the perimeters of the two polygons will be

$$\frac{\Sigma(PQ)}{\Sigma(TP+TQ)} = \frac{\Sigma(PN)}{\Sigma(PT)}.$$

Hence, by a known theorem, the ratio in question will be intermediate in value between the greatest and least of the ratios

$$\frac{PN}{PT} \quad \text{or} \quad \frac{ON}{OP},$$

which would occur in the complete figure. But when the angles POQ are taken sufficiently small, the number of sides in the polygons being correspondingly increased, each of the ratios ON/OP can be made as nearly equal to unity as we please. Hence the upper and lower limits above mentioned must be the same.

This definite limit to which the perimeter of an inscribed (or circumscribed) polygon tends, as the angles which the sides subtend at the centre are indefinitely diminished, is adopted *by definition* as the 'perimeter' of the circle. The proof that the ratio (π) which this limit bears to the diameter of the circle is the same for all circles will be found in most books on Trigonometry.

The length of *any* arc of a circle less than the whole perimeter may be defined, and shewn to be unique, in a similar manner.

18. A Continuous Function has a Greatest and a Least Value.

An important property of a continuous function is that in any finite range of the independent variable the function has both a greatest and a least value.

More precisely, if y be a function which is continuous from $x = a$ to $x = b$, inclusively, and if μ be the upper limit of the values which y assumes in this range, there will be some value of x in the range for which $y = \mu$. Similarly for the lower limit.

The theorem is self-evident in the case of a function which steadily increases, or steadily decreases, throughout the range in question, greatest and least values obviously occurring at the extremities of the range. It is therefore true, further, when the function is such that the range can be divided into a *finite* number of intervals in each of which the function either steadily increases or steadily decreases.

The functions ordinarily met with in the applications of the subject are, as a matter of fact, found to be all of this character, but the general tests by which in any given case we ascertain this are established by reasoning which assumes the truth of the theorem of the present Art. See Art. 48. It is therefore desirable as a matter of logic to have a proof which shall assume nothing concerning the function considered except that it is continuous, according to the definition of Art. 8.

The following is an outline of such a demonstration. In the geometrical representation, let $OA = a$, $OB = b$. If at A the value of y is not equal to the upper limit μ, it will be less than μ; let us denote it by y_0. We can form, in an infinite number of ways, an ascending sequence of magnitudes

$$y_0, y_1, y_2, \ldots$$

whose upper limit is μ. For example, we may take y_1 equal to the arithmetic mean of y_0 and μ, y_2 equal to the arithmetic mean of y_1 and μ, and so on. Since, within the range AB, the value of y varies from y_0 to any quantity short of μ, there will (Art. 9) be at least one value of x for which y assumes the intermediate value y_1. Let x_1 denote this value, or (if there be more than one) the least of such values, of x.

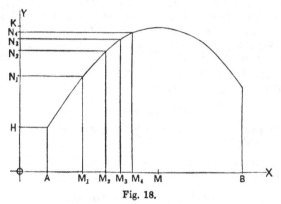

Fig. 18.

Similarly, let x_2 be the least value of x for which $y = y_2$, and so on. It is easily seen that the quantities

$$x_1, x_2, x_3, \ldots$$

(which are represented by the points M_1, M_2, M_3, ... in the figure*) must form an ascending sequence; let M represent the upper limit of this sequence. Since any range, however short, extending to the left of M contains points at which y differs from μ by less than any assignable magnitude, it follows from the continuity of the function that the value of y at the point M itself cannot be other than μ.

---

* The diagram is intended to be merely illustrative, and is not essential to the proof. It is of course evident that any function which can be *adequately* represented by a graph is necessarily of the special character above referred to, for which the present demonstration is superfluous.

In the figure, $OK = \mu$, $OH = y_0$, $ON_1 = y_1$, $ON_2 = y_2$, ... ; and, in the mode of forming the sequence

$$y_0, y_1, y_2, \ldots$$

which is suggested (as a particular case) in the text, N_1 bisects HK, N_2 bisects $N_1 K$, N_3 bisects $N_2 K$, and so on.

To see that the above theorem is not generally true of *discontinuous* functions, consider a function defined as follows. For values of x *other than* 0 let the value of the function be $(\sin x)/x$, and for $x = 0$ let the function have the value 0. This function has the upper limit 1, to which it can be made to approach as closely as we please by taking $|x|$ small enough; but it never actually attains this limit.

19. Limiting Value of a Function.

Consider the whole assemblage of values which a function y (continuous or not) assumes as the independent variable x ranges over some interval extending on one side of a fixed value x_1. Let us suppose that, as x approaches the value x_1, y approaches a certain fixed magnitude λ in such a way that by taking $|x - x_1|$ sufficiently small we can ensure that for this *and for all smaller values* of $|x - x_1|$ the value of $|y - \lambda|$ shall be less than σ, where σ may be any assigned magnitude however small. Under these conditions, λ is said to be the 'limiting value' of y as x approaches the value x_1 from the side in question.

The relation is often expressed thus:

$$\lim_{x \to x_1} y = \lambda,$$

but in strictness the side from which x approaches the value x_1 should be specified.

If we compare with the above the definition of Art. 8 we see that in the case of a *continuous* function we have

$$\lim_{x \to x_1} \phi(x) = \phi(x_1), \quad \dots \dots \dots \dots \dots \dots (1)$$

or the 'limiting value' of the function coincides with the value of the function itself, and that if x_1 lie *within* the range of the independent variable this holds from whichever side x approaches x_1. If, on the other hand, x_1 coincides with either terminus of the range, x must be supposed to approach x_1 from within the range.

Conversely, a function is not continuous unless the condition (1) be satisfied.

Let us next take the case of a function the range of whose independent variable is unlimited on the side of x positive. If as x is continually increased, y tends to a fixed value λ in such a way that by taking x sufficiently great we can ensure that for this and for all greater values of x we shall have $|y - \lambda|$ less than σ, where σ may be any assigned positive quantity, however small, then λ is called the limiting value of y for $x \to \infty$, and we write

$$\lim_{x \to \infty} y = \lambda.$$

There is a similar definition of

$$\lim_{x \to -\infty} y.$$

when it exists, in the case of an independent variable which is unlimited on the side of x negative.

20. General Theorems relating to Limiting Values.

1°. The limiting value of the sum of any *finite* number of functions is equal to the sum of the limiting values of the several functions, provided these limiting values be all finite.

2°. The limiting value of the product of any *finite* number of functions is equal to the product of the limiting values of the several functions, provided these limiting values be all finite.

3°. The limiting value of the quotient of two functions is equal to the quotient of the limiting values of the separate functions, provided these limiting values be finite, and that the limiting value of the divisor is not zero.

The proof is by the same method as in Art. 12, the theorems of which are in fact particular cases of the above.

Thus, let u, v be two functions of x, and let us suppose that as x approaches the value x_1, these tend to the limiting values u_1, v_1, respectively. If, then, we write

$$u = u_1 + a, \quad v = v_1 + \beta,$$

a and β will be functions of x whose limiting values are zero. Now

$$(u + v) - (u_1 + v_1) = a + \beta,$$

$$uv - u_1 v_1 = a v_1 + \beta u_1 + a\beta,$$

$$\frac{u}{v} - \frac{u_1}{v_1} = \frac{a v_1 - \beta u_1}{v_1 (v_1 + \beta)}.$$

And, as in Art. 12, it appears that by making x sufficiently nearly equal to x_1 we can, under the conditions stated, make the right-hand sides less in absolute value than any assigned magnitude however small.

21. Illustrations.

We have seen in Art. 19 that the limiting value of a continuous function for any value x_1 of the independent variable x, for which the function exists, is simply the value of the function itself for $x = x_1$. It may, however, happen that for certain isolated or extreme values of the variable the function does not exist, or is undefined, whilst it is defined for values of x differing infinitely little from these. It is in such cases that the conception of a 'limiting value' becomes of special importance.

Ex. 1. Take the function

$$\frac{1 - \sqrt{(1 - x^2)}}{x^2}.$$

The algebraical operations here prescribed can all be performed for any value of x between ± 1, except the value 0, which gives to the fraction the form $0/0$. Now the definition of a quotient a/b is that it is a quantity which, multiplied by b, gives the result a. Since *any* finite quantity, when multiplied by 0, gives the result 0, it is evident that the quotient $0/0$ may have any value whatever. It is therefore said to be 'indeterminate.'

We may, however, multiplying numerator and denominator of the given fraction by $1 + \sqrt{(1 - x^2)}$, put the function in the equivalent form

$$\frac{x^2}{x^2 \{1 + \sqrt{(1 - x^2)}\}},$$

and for all values of x between ± 1, *other than* 0, this is equal to

$$\frac{1}{1 + \sqrt{(1 - x^2)}}.$$

Since this function is continuous, and exists for $x = 0$, its limiting value for $x \to 0$ is $\frac{1}{2}$.

Ex. 2.   Consider the function

$$\sqrt{(1 + x)} - \sqrt{x}.$$

As x is continually increased this tends to assume the indeterminate form $\infty - \infty$. But, writing the expression in the equivalent form

$$\frac{1}{\sqrt{(1 + x)} + \sqrt{x}},$$

we see that its limiting value for $x \to \infty$ is 0.

Ex. 3.   The period of oscillation of a given pendulum, regarded as a function of the amplitude a, has a definite value for all values of a *between* 0 and π, but it does not exist, in any strict sense, for the extreme values 0 and π. There is, however, a definite limiting value to which the period tends as a approaches the value zero. This limiting value is known in Dynamics as the 'time of oscillation in an infinitely small arc.'

22.   Some Special Limiting Values.

The following examples are of special importance in the Differential Calculus.

1°.   To prove that

$$\lim_{x \to a} \frac{x^m - a^m}{x - a} = m a^{m-1}, \quad \ldots\ldots\ldots\ldots(1)$$

for all rational values of m.

If m be a positive integer, we have

$$\lim_{x \to a} \frac{x^m - a^m}{x - a} = \lim_{x \to a} (x^{m-1} + ax^{m-2} + \ldots + a^{m-2}x + a^{m-1})$$

$$= ma^{m-1},$$

since, the number (m) of terms being finite, the limiting value of the sum is equal to the sum of the limiting values of the several terms (Art. 20).

If m be a rational fraction, $= p/q$, say, we put

$$x = y^q, \quad a = b^q,$$

and therefore

$$\frac{x^m - a^m}{x - a} = \frac{y^{mq} - b^{mq}}{y^q - b^q} = \frac{y^p - b^p}{y^q - b^q}.$$

This fraction is equal to

$$\frac{\dfrac{y^p - b^p}{y - b}}{\dfrac{y^q - b^q}{y - b}}.$$

The limiting value of the numerator is pb^{p-1}, and that of the denominator is qb^{q-1}, by the preceding case. Hence the required limit is

$$\frac{p}{q} b^{p-q}, \quad = \frac{p}{q} a^{p/q-1}, \quad = ma^{m-1},$$

as before.

If m be negative, $= -n$, say, we have

$$\frac{x^m - a^m}{x - a} = \frac{x^{-n} - a^{-n}}{x - a} = -\frac{1}{x^n a^n} \cdot \frac{a^n - a^n}{x - a}.$$

If n be rational, the limiting value of this is

$$-\frac{1}{a^{2n}} \cdot na^{n-1}, \quad = -na^{-n-1}, \quad = ma^{m-1},$$

by the preceding cases.

2°. To prove that

$$\lim_{\theta \to 0} \frac{\sin \theta}{\theta} = 1, \quad \lim_{\theta \to 0} \frac{\tan \theta}{\theta} = 1. \quad \ldots\ldots\ldots\ldots(2)$$

If we recall the definition of the 'length' of a circular arc (Art. 17) these statements are seen to be little more than truisms. If, in Fig. 17, p. 31, the angle POQ be $1/n$th of four right angles, then $n \cdot PQ$ will be the perimeter of an inscribed regular polygon of n sides, and $n(TP + TQ)$ will be the perimeter of the corresponding circumscribed polygon. Now, if

$$\theta = \angle POA = \pi/n,$$

we shall have

$$\frac{\text{chord } PQ}{\text{arc } PQ} = \frac{PN}{\text{arc } PA} = \frac{\sin \theta}{\theta},$$

and

$$\frac{TP + TQ}{\text{arc } PQ} = \frac{PT}{\text{arc } PA} = \frac{\tan \theta}{\theta}.$$

Hence the fractions

$$\frac{\sin \theta}{\theta} \text{ and } \frac{\tan \theta}{\theta}$$

denote the ratios which the perimeters of the above-mentioned polygons respectively bear to the perimeter of the circle. Hence, as n is continually increased, each fraction tends to the limiting value unity (Art. 17).

In the above argument, it is assumed that θ is a submultiple of π. But, whatever the value of the angle POQ in the figure, we have

chord $PQ <$ arc PQ, and $TP + TQ >$ arc PQ;

i.e. $(\sin \theta)/\theta < 1$, and $(\tan \theta)/\theta > 1$. Hence these fractions must have respectively an upper and a lower limit; and it follows from the preceding that neither of these limits can be other than unity.

The following numerical table illustrates the way in which the above functions approach their common limiting value as θ is continually diminished.

n	θ/π	$(\sin \theta)/\theta$	$(\tan \theta)/\theta$
4	·25	·90032	1·27324
5	·20	·93549	1·15633
10	·10	·98363	1·03425
20	·05	·99589	1·00831
40	·025	·99897	1·00206
∞	·0	1·00000	1·00000

The third and fourth columns give the ratios which the perimeters of the inscribed and circumscribed regular polygons of n sides respectively bear to the perimeter of the circle.

23. Infinitesimals.

A variable quantity which in any process tends to the limiting value zero is said ultimately to vanish, or to be 'infinitely small.'

Two infinitely small quantities are said to be ultimately equal when the limiting value of the ratio of one to the other is unity. Thus, when θ tends to the limit 0, $\sin \theta$ and θ are ultimately equal, by Art. 22 (2).

It is sometimes convenient to distinguish between different orders of infinitely small quantities. Thus if u, v are two quantities which tend simultaneously to the limit zero, and if the limit of the ratio v/u be finite and not zero, then v is said to be an infinitely small quantity of the same order as u. But, if the limit of the ratio v/u be zero, then v is said to be an infinitely small quantity of a higher order than u. More particularly, if the limit of v/u^m be finite and not zero, v is said to be an infinitesimal of the mth order, the standard being u.

Ex. 1. When, in Fig. 17, p. 31, the angle POQ is indefinitely diminished, NA and AT are ultimately equal. For, by similar triangles,

$$\frac{OP}{ON} = \frac{OT}{OP},$$

and therefore

$$\frac{OP - ON}{ON} = \frac{OT - OP}{OP},$$

or

$$\frac{NA}{AT} = \frac{ON}{OP};$$

and the limiting value of the ratio ON/OP is unity.

Again, NT is an infinitesimal of the second order, if the standard be PN. For

$$PN^2 = ON \cdot NT,$$

whence

$$\frac{NT}{PN^2} = \frac{1}{ON}, \ = \frac{1}{OA} \text{ in the limit.}$$

Ex. 2. We have

$$1 - \cos \theta = 2 \sin^2 \tfrac{1}{2}\theta = \left(\frac{\sin \tfrac{1}{2}\theta}{\tfrac{1}{2}\theta}\right)^2 \cdot \tfrac{1}{2}\theta^2. \quad \ldots\ldots\ldots\ldots(1)$$

When $\theta \to 0$ the limit of the first factor is unity. Hence $1 - \cos \theta$ is an infinitesimal of the second order, the standard being θ.

Again

$$\tan \theta - \sin \theta = \left(\frac{\sin \tfrac{1}{2}\theta}{\tfrac{1}{2}\theta}\right)^2 \cdot \frac{\cos \tfrac{1}{2}\theta}{\cos \theta} \cdot \tfrac{1}{2}\theta^3. \quad \ldots\ldots\ldots\ldots(2)$$

When $\theta \to 0$, the first two fractions tend each to the limit 1. Hence $\tan \theta - \sin \theta$ is an infinitesimal of the third order. This is equivalent to the statement that in the figure referred to $PT - PN$ is ultimately of the third order, the standard being PN.

The following principle enables us to abbreviate many arguments, especially in the applications of the Calculus to Geometry and Mechanics:

If α and β be two infinitesimals of the same order, and if α' and β' be other infinitesimals which are ultimately equal to α and β respectively, then

$$\lim \left(\frac{\alpha'}{\beta'}\right) = \lim \left(\frac{\alpha}{\beta}\right). \quad \dots\dots\dots\dots\dots(3)$$

For
$$\frac{\alpha'}{\beta'} = \frac{\alpha'}{\alpha} \cdot \frac{\beta}{\beta'} \cdot \frac{\alpha}{\beta}, \quad \dots\dots\dots\dots\dots(4)$$

and the limits of the first two fractions on the right-hand are unity, by hypothesis. The result follows by Art. 20*.

A quantity which in any process finally exceeds any assignable magnitude is said to be 'infinitely great.' And if one such quantity u be taken as a standard, any other v is said to be infinitely great of the mth order when the limit of v/u^m is finite and not zero.

EXAMPLES. I.

(Algebraic Functions.)

1. Draw on the same diagram the graphs of x^n for the cases

$$n = 1, 2, 3, \tfrac{1}{2}, \tfrac{1}{3},$$

for values of x ranging from 0 to 1·2 †.

2. Draw graphs of:

(1) $(x-1)(x-2)$, $x^2 - x + 1$.

(2) $x(1-x)^2$, $x^2(1-x)^2$.

(3) $x + \dfrac{1}{x}$, $x - \dfrac{1}{x}$.

---

* A good example of the application of this principle will be found in Art. 63.
† The curves should be drawn carefully to scale ; for this purpose paper ruled into small squares is useful. The numerical tables of squares, square-roots, and reciprocals, given in the Appendix (Tables A, B, C), will occasionally help to shorten the arithmetical work.

(4) $\sqrt{(1-x^2)}, \quad \dfrac{1}{\sqrt{(1-x^2)}}.$

(5) $\dfrac{(x-1)(x-2)}{x}, \quad \dfrac{(x-1)(x-3)}{x-2}.$

(6) $\dfrac{x^2}{1-x^2}, \quad \dfrac{x^3}{1+x^2}.$

(7) $\dfrac{1-x+x^2}{1+x+x^2}.$

(8) $\dfrac{x}{1-x}, \quad \dfrac{x}{(1-x)^2}.$

3. Prove that the equation
$$2x^3 + 5x^2 - 5x - 3 = 0$$
has a root between $-\infty$ and -1, another between -1 and 0, and a third between 1 and 2.

4. Prove that the equation
$$2x^3 + 7x^2 + 3x - 5 = 0$$
has three real roots, and find roughly their situations.

5. Find roughly the situations of the roots of
$$2x^3 - 3x^2 - 36x + 10 = 0.$$

6. Prove that every algebraic equation of odd degree has at least one real root; and that every equation of even degree, whose first and last coefficients have opposite signs, has at least two real roots, one positive and one negative.

EXAMPLES. II.

(Circular Functions.)

1. Draw graphs of the following functions:

(1) $\cosec x, \quad \cot x, \quad \cot x + \tan x.$

(2) $\sin^2 x, \quad \cos^2 x, \quad \tan^2 x.$

(3) $\sin x + \sin 2x, \quad \sin x + \cos 2x.$

(4) $\sin x^2, \quad \sin \sqrt{x}, \quad \sin \dfrac{1}{x}.$

2. Prove that the equation
$$\sin x - x \cos x = 0$$
has a root between π and $\tfrac{3}{2}\pi.$

EXAMPLES. III.

(Sequences.)

1. Find the upper and lower limits of the magnitudes

$$\frac{2n}{n^2 + 1},$$

where $\qquad n = 1, \; 2, \; 3, \; \dots.$

2. If, in the sequence

$$a_1, \; a_2, \; a_3, \; \dots, \; a_n, \; \dots,$$
$$a_{n+1} = m a_n + c,$$

where m and c are positive, and $m < 1$, the sequence has the limit $c/(1 - m)$ whatever the value of a_1.

3. The quantities

$$\sqrt{2}, \quad \sqrt{(2 + \sqrt{2})}, \quad \sqrt{\{2 + \sqrt{(2 + \sqrt{2})}\}}, \; \dots,$$

where $\qquad a_{n+1} = \sqrt{(2 + a_n)},$

form an ascending sequence whose limit is 2.

4. If $\qquad a_{n+1} = \sqrt{(k + a_n)},$

where a_1 and k are positive, the sequence is ascending or descending, according as a_1 is less or greater than the positive root of $x^2 = x + k$, and has in either case this root as its limit.

5. Examine the character of the sequence where

$$a_{n+1} = \frac{k}{1 + a_n},$$

k being positive. Prove that if a_1 be positive it has as a limit the positive root of $x^2 + x = k$.

6. Find a sequence of quantities approximating to the positive root of the equation

$$x^3 = x + 1.$$

7. Prove that the sequence formed according to the law

$$x_{n+1} = \tan^{-1} x_n,$$

where x_1 lies between 1 and 2, has for its limit the least positive root of the equation $\tan x = x$.

8. If a_{n+1}, b_{n+1} are respectively the arithmetic and harmonic means between a_n and b_n, and a_1, b_1 be positive, the sequences whose nth terms are a_n and b_n respectively have the common limit $\sqrt{(a_1 b_1)}$.

9. If $\quad a_{n+1} = \tfrac{1}{2}(a_n + b_n), \quad b_{n+1} = \sqrt{(a_n b_n)},$

and a_1, b_1 are positive, the sequences whose nth terms are a_n, b_n converge to a common limit.

(This limit is called the 'arithmo-geometric mean' between a_1 and b_1.)

EXAMPLES. IV.

(Limiting Values of Functions.)

1. Find the limiting values, for $x \to 0$, of

$$\frac{\sin ax}{x}, \quad \frac{\tan ax}{x}.$$

2. Find the limiting values, for $x \to 0$, of

$$\frac{\sin^{-1} x}{x} \quad \text{and} \quad \frac{\tan^{-1} x}{x}.$$

3. Trace the curves

$$y = \frac{\sin x}{x}, \quad y = \frac{1 - \cos x}{x^2}.$$

4. Prove that

$$\lim_{x \to \frac{1}{2}\pi} (\sec x - \tan x) = 0.$$

5. Prove that

$$\lim_{x \to 0} \frac{\sqrt{(1 + x)} - \sqrt{(1 - x)}}{x} = 1.$$

6. Prove that

$$\lim_{x \to \infty} \{\sqrt{(x^2 + x + 1)} - x\} = \tfrac{1}{2}.$$

7. Regular polygons are inscribed and circumscribed to a given circle. Prove that when n is large the difference between the areas of the in- and circumscribed polygons of $2n$ sides is one-fourth the difference between the areas of the in- and circumscribed polygons of n sides.

8. A straight line AB moves so that the sum of its intercepts OA, OB on two fixed straight lines OX, OY is constant. If P be the ultimate intersection of two consecutive positions of AB, and Q the point where AB is met by the bisector of the angle XOY, then $AP = QB$.

9. Through a point A on a circle a chord AP is drawn, and on the tangent at A a point T is taken such that $AT = AP$. If TP produced meet the diameter through A in Q, the limiting value of AQ when P moves up to A is double the diameter of the circle.

10. A straight line AB moves so as to include with two fixed straight lines OX, OY a triangle AOB of constant area. Prove that the limiting position of the intersection of two consecutive positions of AB is the middle point of AB.

11. A straight line AB of constant length moves with its extremities on two fixed straight lines OX, OY which are at right angles to one another. Prove that if P be the ultimate intersection of two consecutive positions of AB, and N the foot of the perpendicular from O on AB, then $AP = NB$.

12. Tangents are drawn to a circular arc at its middle point and at its extremities; prove that the area of the triangle contained by the three tangents is ultimately one-half that of the triangle whose vertices are the three points of contact.

13. If PCP' be any fixed diameter of an ellipse, and QV any ordinate to this diameter ; and if the tangent at Q meet CP produced in T, the limiting value of the ratio $TP : PV$, when PV is infinitely small, is unity.

CHAPTER II

DERIVED FUNCTIONS

24. Introduction. Geometrical Illustrations.

The Differential Calculus originated in the problem of finding the direction of the tangent-line to a given curve at any given point.

Let P and Q be adjacent points on a continuous curve

$$y = \phi(x), \quad \dots\dots\dots\dots\dots\dots\dots(1)$$

and PM, QN their ordinates, and let PR be drawn parallel to OX. Let the chord PQ meet the axis of x in S. If the point P

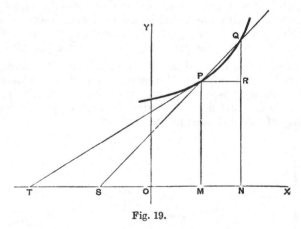

Fig. 19.

be fixed, whilst Q is made to approach P, this chord will, in the case of any ordinary geometrical curve, tend to assume, except possibly at one or more isolated points, a definite limiting position PT, which is adopted as the definition of the 'tangent-line' at P.

The direction of the tangent-line is determined by the angle which TP makes with OX, $i.e.$ by the limiting value of the angle PSX in the figure. If we put

$$OM = x, \quad PM = y, \quad ON = x + \delta x, \quad QN = y + \delta y,$$

we have

$$\tan PSX = \frac{PM}{SM} = \frac{QR}{PR} = \frac{\delta y}{\delta x}. \quad \dots\dots\dots\dots(2)$$

The problem is therefore to find the limiting value of the ratio $\delta y/\delta x$ as δx tends to the limit zero. This limiting value is denoted by

$$\frac{dy}{dx}. \quad \dots\dots\dots\dots\dots\dots\dots(3)$$

This is to be regarded as a single symbol, the fractional appearance being preserved merely in order to remind us how the limit is approached.

Analytically, the same thing is denoted by $\phi'(x)$, which is called the 'derived function' of $\phi(x)$.

It is convenient to have a geometrical name to denote the property of a curve which is indicated by the symbol (3). We shall use the term 'gradient' in this sense. If from any point P on the curve we draw the tangent-line *to the right*, the gradient is the trigonometrical tangent of the angle which the direction of this line makes with the positive direction of the axis of x. If this angle is negative, the gradient is negative; if the tangent-line is parallel to OX, the gradient vanishes. See Fig. 20.

In most cases with which we have to deal the gradient is itself a continuous function of x, except that it may occasionally become infinite at isolated points, where the tangent is perpendicular to OX. The figure includes the case of an isolated discontinuity of finite amount in the gradient.

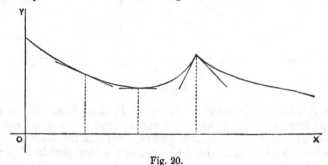

Fig. 20.

25. General Definition of the Derived Function.

As the notion of the derived function is important in almost all branches of Mathematics, we proceed to define it in a more formal manner, without special reference to Geometry.

Let y be a function which is continuous over a certain range of the independent variable x; let δx be any increment of x such that $x + \delta x$ lies within the above range, and let δy be the consequent increment of y. Then, x being regarded as fixed, the ratio

$$\frac{\delta y}{\delta x} \quad \dots\dots\dots\dots\dots\dots\dots\dots\dots\dots(1)$$

will be a function of δx. If as δx (and consequently also δy) assumes any series of values having zero as its limit, this ratio tends to a definite and unique limiting value, the value thus arrived at is called the 'derived function,' or the 'derivative,' or the 'differential coefficient,' of y with respect to x, and is denoted by the symbol

$$\frac{dy}{dx}. \quad \dots\dots\dots\dots\dots\dots\dots\dots\dots(2)$$

More concisely, the derived function (when it exists) is the limiting value of the ratio of the increment of the function to the increment of the independent variable, when both increments are indefinitely diminished.

It is to be carefully noticed that in the above definition we speak of the limiting value of a certain ratio, and not of the ratio of the limiting values of δy, δx. The latter ratio is indeterminate, being of the form $0/0$.

When we say that the ratio $\delta y/\delta x$ tends to a unique limiting value, it is implied that (when x lies *within* the range of the independent variable) this value is the same whether δx approach the value 0 from the positive or from the negative side. It may happen that there is one limiting value when δx approaches 0 from the positive, and another when δx approaches 0 from the negative side. In this case we may say that there is a 'right-derivative,' and a 'left-derivative,' but no proper 'derivative' in the sense of the above definition. See Fig. 20.

The question whether the ratio $\delta y/\delta x$ really has a determinate limiting value depends on the nature of the original function y. Functions for which the limit is determinate and unique (save for isolated values of x) are said to be 'differentiable.' All other functions are excluded *ab initio* from the scope of the Differential Calculus.

A differentiable function is necessarily continuous, but the converse statement is now known not to be correct. Functions which are continuous without being differentiable are, however, of very rare occurrence in Mathematics.

There are various other notations for the derived function, in place of dy/dx. The derived function is often indicated by attaching an accent to the symbol denoting the original function. Thus if

$$y = \phi(x), \quad \dots\dots\dots\dots\dots\dots\dots(3)$$

the derived function may be denoted by y' or, as already stated, by $\phi'(x)$.

Since
$$\frac{\delta y}{\delta x} = \frac{\phi(x + \delta x) - \phi(x)}{\delta x},$$

we have, writing h for δx,

$$\phi'(x) = \lim_{h \to 0} \frac{\phi(x + h) - \phi(x)}{h}. \quad \dots\dots\dots\dots(4)$$

This formula is often used in the sequel.

The operation of finding the differential coefficient of a given function is called 'differentiating.' If x be the independent variable, we may look upon d/dx as a symbol denoting this operation. It is often convenient to replace this by the single letter D; thus we may write, indifferently,

$$\frac{dy}{dx}, \quad \frac{d}{dx}y, \quad Dy,$$

for the differential coefficient of y with respect to x.

26. Physical Illustrations.

The importance of the derived function in the various applications of the subject rests on the fact that it gives us a measure of the *rate of increase* of the original function, per unit increase of the independent variable.

To illustrate this, we may consider, first, the rectilinear motion of a point. The distance s of the point from some fixed origin in the line of motion will be a function of the time t reckoned from some fixed epoch. The relation between these variables is often exhibited graphically by a 'curve of positions,' in which the abscissæ are proportional to t and the ordinates to s. If in the interval δt the space δs is described, the ratio $\delta s/\delta t$ is called the 'mean velocity' during the interval δt; *i.e.* a point moving with a constant velocity equal to this would accomplish the same space δs in the same time δt. In the limit, when δt (and consequently also δs) is indefinitely diminished, the limiting value to which this mean velocity tends is adopted, by definition, as the measure of the 'velocity at the instant t.' In the notation of the Calculus, therefore, this velocity v is given by the formula

$$v = \frac{ds}{dt}. \quad \dots\dots\dots\dots\dots\dots\dots\dots(1)$$

In the graphical representation aforesaid, v is the gradient of the curve of positions.

Again, the velocity v is itself a function of t. The curve representing this relation is called the 'curve of velocities.' If δv be the increase of velocity in the interval δt, then $\delta v/\delta t$ is called the 'mean rate of increase of velocity,' or the 'mean acceleration' in this interval. The limiting value to which the mean acceleration tends when δt is indefinitely diminished is called the 'acceleration at the instant t.' If this acceleration be denoted by a, we have

$$a = \frac{dv}{dt}. \quad\ldots\ldots\ldots\ldots\ldots\ldots\ldots\ldots(2)$$

In the graphical representation, a is the gradient of the curve of velocities.

In the case of a rigid body revolving about a fixed axis, if θ be the angle through which the body has revolved from some standard position, the 'mean angular velocity' in any interval δt is denoted by $\delta\theta/\delta t$, and the 'angular velocity at the instant t' by

$$\frac{d\theta}{dt}. \quad\ldots\ldots\ldots\ldots\ldots\ldots\ldots\ldots\ldots(3)$$

Again, if ω denote this angular velocity, the 'mean angular acceleration' in the interval δt is denoted by $\delta\omega/\delta t$, and the 'angular acceleration' at the instant t by

$$\frac{d\omega}{dt}. \quad\ldots\ldots\ldots\ldots\ldots\ldots\ldots\ldots(4)$$

Again, the length of a bar of given material is a function of the temperature (θ). If x be the length at temperature θ of a bar whose length at some standard temperature (say $0°$) is unity, then $\delta x/\delta\theta$ represents the mean coefficient of (linear) expansion from temperature θ to temperature $\theta + \delta\theta$, and $dx/d\theta$ represents the coefficient of expansion at temperature θ.

As another example, suppose we have a fluid which is free to assume a series of states such that the pressure (p) is a definite function of the volume (v) of unit mass. If the volume change from v to $v + \delta v$, the fraction $-\delta v/v$ measures the ratio of the diminution of volume to the original volume, and gives therefore what is called the 'compression.' The ratio of the increment of pressure δp required to produce this compression, to the compression, is $-v\delta p/\delta v$. The limiting value of this when δv is infinitely small, viz. $-v\,dp/dv$, is defined to be the 'elasticity of volume' of the fluid under the given conditions.

27.　Differentiations *ab initio*.

Before investigating general rules for calculating the derivatives of given analytical functions, we may discuss a few examples independently from first principles.

Ex. 1.   If $y = x$, we have $\delta y = \delta x$, and therefore

$$\frac{\delta y}{\delta x} = 1, \text{ whence } \frac{dy}{dx} = 1.$$

Ex. 2.   Let $\qquad\qquad y = x^2.$ ....(1)

We have, writing h for δx,

$$\frac{\delta y}{\delta x} = \frac{(x+h)^2 - x^2}{h} = 2x + h.$$

Proceeding to the limit $(h \to 0)$, we find

$$\frac{dy}{dx} = 2x.$$ ....(2)

Ex. 3.   Let $\qquad\qquad y = \frac{1}{x}.$ ....(3)

We have $\qquad\qquad \delta y = \frac{1}{x+h} - \frac{1}{x} = -\frac{h}{x(x+h)},$

and therefore $\qquad \frac{\delta y}{\delta x} = -\frac{1}{x(x+h)}.$

Hence $\qquad\qquad \frac{dy}{dx} = -\lim_{h \to 0} \frac{1}{x(x+h)} = -\frac{1}{x^2}.$ ....(4)

The negative sign is due to the fact that y diminishes as x increases.

Ex. 4.   If $\qquad\qquad y = \sqrt{x},$ ....(5)

we have $\qquad \delta y = \sqrt{(x+h)} - \sqrt{x} = \frac{h}{\sqrt{(x+h)} + \sqrt{x}},$

and $\qquad\qquad \frac{\delta y}{\delta x} = \frac{1}{\sqrt{(x+h)} + \sqrt{x}}.$

Proceeding to the limit $(h \to 0)$, we find

$$\frac{dy}{dx} = \frac{1}{2\sqrt{x}}.$$ ....(6)

28.   Differentiation of Standard Functions.

1°.   If $\qquad\qquad y = x^m,$ ....(1)

we have $\qquad\qquad \frac{\delta y}{\delta x} = \frac{(x+\delta x)^m - x^m}{(x+\delta x) - x}.$

It has been shewn in Art. 22, 1°, that, for all rational values of m, the limiting value of this fraction when $\delta x \to 0$ is mx^{m-1}. Hence

$$\frac{dy}{dx} = mx^{m-1}.$$ ....(2)

Ex. If $m = 2$, $dy/dx = 2x$; if $m = \frac{1}{2}$, $dy/dx = \frac{1}{2}x^{-\frac{1}{2}}$. Cf. Art. 27.

2°. If $\qquad\qquad\qquad y = \sin x,$ ....(3)

we have, writing h for δx,

$$\frac{\delta y}{\delta x} = \frac{\sin(x+h) - \sin x}{h} = \frac{\sin \frac{1}{2}h}{\frac{1}{2}h} \cdot \cos(x + \tfrac{1}{2}h).$$

If the angles be expressed in 'circular measure,' we have

$$\lim_{h \to 0} \left(\frac{\sin \frac{1}{2}h}{\frac{1}{2}h} \right) = 1,$$

by Art. 22, 2°; and the limiting value of the second factor is $\cos x$. Hence

$$\frac{dy}{dx} = \cos x. \qquad\qquad ....(4)$$

The student should refer to the graph of $\sin x$ on p. 28, and notice how the gradient of the curve varies in accordance with this formula.

3°. If $\qquad\qquad\qquad y = \cos x,$ ....(5)

we have $\qquad\qquad \dfrac{\delta y}{\delta x} = \dfrac{\cos(x+h) - \cos x}{h}$

$$= - \frac{\sin \frac{1}{2}h}{\frac{1}{2}h} \cdot \sin(x + \tfrac{1}{2}h);$$

the limiting value of which is, on the same understanding as before,

$$\frac{dy}{dx} = - \sin x. ....(6)$$

4°. If $\qquad\qquad\qquad y = \tan x,$ ....(7)

we have

$$\frac{\delta y}{\delta x} = \frac{\tan(x+h) - \tan x}{h} = \frac{\sin(x+h)\cos x - \cos(x+h)\sin x}{h \cos x \cos(x+h)}$$

$$= \frac{\sin h}{h} \cdot \frac{1}{\cos x \cos(x+h)}.$$

Hence, in the limit,

$$\frac{dy}{dx} = \frac{1}{\cos^2 x} = \sec^2 x. ....(8)$$

This shews that the gradient of the curve $y = \tan x$, between the points of discontinuity, is always positive; see Fig. 15, p. 28.

29. Rules for differentiating combinations of simple types. Differentiation of a Sum.

1°. Let $y = u + C,$ ....(1)

where u is a known function of x, and C is a constant.   We have

$$y + \delta y = u + \delta u + C,$$

and therefore $\delta y = \delta u,$

$$\frac{\delta y}{\delta x} = \frac{\delta u}{\delta x},$$

or, in the limit, $\dfrac{dy}{dx} = \dfrac{du}{dx}.$ ....(2)

This fact, that an *additive* constant disappears on differentiation, obvious as it is, is very important.   The geometrical meaning is that shifting a curve bodily parallel to the axis of y does not alter the gradient.

2°. Let $y = u + v,$ ....(3)

where u, v are given functions of x.   As in Art. 12, we find

$$\delta y = \delta u + \delta v,$$

and therefore $\dfrac{\delta y}{\delta x} = \dfrac{\delta u}{\delta x} + \dfrac{\delta v}{\delta x}.$

Hence, since the limiting value of a sum is the sum of the limiting values of the several terms,

$$\frac{dy}{dx} = \frac{du}{dx} + \frac{dv}{dx}.$$ ....(4)

Again, if $y = u + v + w,$ ....(5)

we have $\dfrac{dy}{dx} = \dfrac{d}{dx}(u+v) + \dfrac{dw}{dx}$

$$= \frac{du}{dx} + \frac{dv}{dx} + \frac{dw}{dx},$$ ....(6)

by a double application of the preceding result.   In this way we can prove, step by step, that the derived function of the sum of any *finite* number of given functions is the sum of the derivatives of the separate functions.

Ex.   The derived function of

$$A_0 x^m + A_1 x^{m-1} + \ldots + A_{m-1} x + A_m$$

is $m A_0 x^{m-1} + (m-1) A_1 x^{m-2} + \ldots + A_{m-1}.$

30. Differentiation of a Product.

1°. If
$$y = Cu, \dots\dots\dots\dots\dots\dots\dots(1)$$

where C is a constant, and u a function of x, we have

$$y + \delta y = C(u + \delta u),$$

and therefore
$$\delta y = C\,\delta u.$$

Hence
$$\frac{\delta y}{\delta x} = C\frac{\delta u}{\delta x}.$$

and, proceeding to the limit,

$$\frac{dy}{dx} = C\frac{du}{dx}. \quad\dots\dots\dots\dots\dots\dots(2)$$

Hence a constant factor remains attached after the differentiation.

The geometrical meaning of this result is that if all the ordinates of a curve be altered in a given ratio, the gradient is altered in the same ratio. Cf. Fig. 27, p. 84.

2°. Let
$$y = uv, \dots\dots\dots\dots\dots\dots\dots(3)$$

where u, v are both functions of x. As in Art. 12,

$$\delta y = (u + \delta u)(v + \delta v) - uv$$
$$= v\delta u + u\delta v + \delta u\delta v,$$

and therefore
$$\frac{\delta y}{\delta x} = v\frac{\delta u}{\delta x} + (u + \delta u)\frac{\delta v}{\delta x}.$$

Hence, proceeding to the limit, and making use of the principle that the limit of a product is the product of the limits, we have

$$\frac{dy}{dx} = v\frac{du}{dx} + u\frac{dv}{dx}. \dots\dots\dots\dots\dots(4)$$

If we divide both sides of this equation by y, $= uv$, we obtain the form

$$\frac{1}{y}\frac{dy}{dx} = \frac{1}{u}\frac{du}{dx} + \frac{1}{v}\frac{dv}{dx}.$$

This result is easily extended; thus if $y = uvw$, we have, writing $y = zw$, where $z = uv$,

$$\frac{1}{y}\frac{dy}{dx} = \frac{1}{z}\frac{dz}{dx} + \frac{1}{w}\frac{dw}{dx}$$

$$= \frac{1}{u}\frac{du}{dx} + \frac{1}{v}\frac{dv}{dx} + \frac{1}{w}\frac{dw}{dx}, \quad\dots\dots\dots\dots(5)$$

by a double application of the preceding result. And so on for any finite number of factors.

If we multiply both sides by y, $= uvw...$, the generalized form of the last result becomes

$$\frac{dy}{dx} = vw...\frac{du}{dx} + uw...\frac{dv}{dx} + uv...\frac{dw}{dx} + ..., \qquad ....(6)$$

or, in words:

The derived function of a product is found by differentiating with respect to x so far as it is involved in each factor separately, the other factors being treated as constants, and adding the results.

Ex. 1. If $\quad y = u.u.u ...... \text{ to } m \text{ factors} = u^m, \quad ....(7)$

we have $\qquad \dfrac{1}{y}\dfrac{dy}{dx} = \dfrac{1}{u}\dfrac{du}{dx} + \dfrac{1}{u}\dfrac{du}{dx} + ... \text{ to } m \text{ terms} = \dfrac{m}{u}\dfrac{du}{dx},$

whence $\qquad\qquad \dfrac{dy}{dx} = mu^{m-1}\dfrac{du}{dx}. \qquad ....(8)$

A general proof of this result, free from the restriction that m is a positive integer, is given in Art. 32.

Ex. 2. If $\qquad\qquad y = \sin x \cos x, \qquad ....(9)$

we have $\qquad \dfrac{dy}{dx} = \cos x \dfrac{d}{dx}(\sin x) + \sin x \dfrac{d}{dx}(\cos x)$

$$= \cos x . \cos x - \sin x . \sin x$$

$$= \cos^2 x - \sin^2 x = \cos 2x. \qquad ....(10)$$

Ex. 3. If $\qquad\qquad y = x^2 \sin x, \qquad ....(11)$

we have $\qquad \dfrac{dy}{dx} = x^2 \dfrac{d}{dx}(\sin x) + \sin x \dfrac{d}{dx}(x^2)$

$$= x^2 \cos x + 2x \sin x. \qquad ....(12)$$

31. Differentiation of a Quotient.

Let $\qquad\qquad\qquad y = \dfrac{u}{v}, \qquad ....(1)$

where u, v are given functions of x. As in Art. 12, we find

$$\delta y = \frac{u + \delta u}{v + \delta v} - \frac{u}{v} = \frac{v\delta u - u\delta v}{v(v + \delta v)},$$

whence $\qquad\qquad \dfrac{\delta y}{\delta x} = \dfrac{v\dfrac{\delta u}{\delta x} - u\dfrac{\delta v}{\delta x}}{v(v + \delta v)}.$

Hence, in the limit,

$$\frac{dy}{dx} = \frac{v\dfrac{du}{dx} - u\dfrac{dv}{dx}}{v^2} . \qquad \ldots\ldots\ldots\ldots\ldots\ldots(2)$$

In words: To find the derived function of a quotient, from the product of the denominator into the derived function of the numerator subtract the product of the numerator into the derived function of the denominator, and divide the result by the square of the denominator.

The particular case $\qquad y = \dfrac{1}{v}$ $\qquad\ldots\ldots\ldots\ldots\ldots\ldots\ldots\ldots(3)$

is worthy of separate notice. We then have

$$\delta y = \frac{1}{v + \delta v} - \frac{1}{v} = - \frac{\delta v}{v(v + \delta v)},$$

$$\frac{\delta y}{\delta x} = - \frac{\dfrac{\delta v}{\delta x}}{v(v + \delta v)},$$

$$\frac{dy}{dx} = - \frac{1}{v^2}\frac{dv}{dx}. \qquad \ldots\ldots\ldots\ldots\ldots\ldots\ldots\ldots\ldots(4)$$

This might of course have been deduced by putting $u = 1$, $du/dx = 0$ in the general formula (2).

Ex. 1. If $\qquad y = \dfrac{1 + x + x^2}{1 - x + x^2},$ $\qquad\ldots\ldots\ldots\ldots\ldots\ldots\ldots(5)$

we have $\qquad \dfrac{du}{dx} = 1 + 2x, \quad \dfrac{dv}{dx} = -1 + 2x,$

whence $\qquad \dfrac{dy}{dx} = \dfrac{(1 + 2x)(1 - x + x^2) + (1 - 2x)(1 + x + x^2)}{(1 - x + x^2)^2}$

$$= \frac{2(1 - x^2)}{(1 - x + x^2)^2}. \qquad\ldots\ldots\ldots\ldots\ldots\ldots\ldots\ldots(6)$$

Ex. 2. If $\qquad y = \dfrac{1}{u^m},$ $\qquad\ldots\ldots \ldots\ldots\ldots\ldots\ldots\ldots(7)$

where m is a positive integer, we have

$$\frac{dy}{dx} = - \frac{1}{u^{2m}} \cdot m u^{m-1}\frac{du}{dx} = - m u^{-m-1}\frac{du}{dx}; \qquad\ldots\ldots\ldots\ldots(8)$$

see Art. 32.

The formula of this Art. may also be deduced from Art. 30 (4). If $y = u/v$, we have $u = vy$, and therefore

$$\frac{1}{u}\frac{du}{dx} = \frac{1}{v}\frac{dv}{dx} + \frac{1}{y}\frac{dy}{dx} \dots\dots\dots\dots\dots(9)$$

whence

$$\frac{1}{y}\frac{dy}{dx} = \frac{1}{u}\frac{du}{dx} - \frac{1}{v}\frac{dv}{dx}; \dots\dots\dots\dots(10)$$

this is equivalent to (2) above.

The following examples are important:

1°. If

$$y = \tan x = \frac{\sin x}{\cos x}, \dots\dots\dots\dots\dots(11)$$

we find

$$\frac{dy}{dx} = \frac{\cos x \dfrac{d}{dx}(\sin x) - \sin x \dfrac{d}{dx}(\cos x)}{\cos^2 x}$$

$$= \frac{\cos^2 x + \sin^2 x}{\cos^2 x} = \sec^2 x. \dots\dots\dots\dots\dots(12)$$

This agrees with Art. 28, 4°.

Similarly, if

$$y = \cot x, \dots\dots\dots\dots\dots\dots(13)$$

we find

$$\frac{dy}{dx} = -\operatorname{cosec}^2 x \dots\dots\dots\dots\dots\dots(14)$$

2°. If

$$y = \sec x = \frac{1}{\cos x}, \dots\dots\dots\dots\dots(15)$$

we have

$$\frac{dy}{dx} = -\frac{1}{\cos^2 x}\frac{d}{dx}(\cos x) = \frac{\sin x}{\cos^2 x}. \dots\dots\dots(16)$$

Similarly, if

$$y = \operatorname{cosec} x, \dots\dots\dots\dots\dots(17)$$

we find

$$\frac{dy}{dx} = -\frac{\cos x}{\sin^3 x}. \dots\dots\dots\dots\dots(18)$$

If, as explained in Art. 25, we employ the symbol D to denote the operation of differentiating with respect to x, the results of Arts. 29—31 may be summed up as follows:

$$D(u + v) = Du + Dv, \dots\dots\dots\dots\dots(19)$$

$$D(uv) = vDu + uDv, \dots\dots\dots\dots\dots(20)$$

$$D\left(\frac{u}{v}\right) = \frac{vDu - uDv}{v^2}. \dots\dots\dots\dots(21)$$

32.   Differentiation of a Function of a Function.

If
$$y = F(u), \quad \dots\dots\dots\dots\dots\dots(1)$$

where
$$u = f(x), \quad \dots\dots\dots\dots\dots\dots(2)$$

the symbols F, f denoting given functions, then

$$\frac{dy}{dx} = \frac{dy}{du} \cdot \frac{du}{dx} = F'(u) \cdot f'(x). \quad \dots\dots\dots\dots(3)$$

For, if δx, δy, δu be simultaneous increments, we have

$$\frac{\delta y}{\delta x} = \frac{\delta y}{\delta u} \cdot \frac{\delta u}{\delta x},$$

identically; and therefore, since the limit of a product is the product of the limits,

$$\frac{dy}{dx} = \frac{dy}{du} \cdot \frac{du}{dx}.$$

A useful application of the formula (3) occurs in the theory of rectilinear motion. Thus if, as in Art. 26, we denote by v and a the velocity and the acceleration, respectively, of a moving point, we have

$$v = \frac{ds}{dt}, \qquad a = \frac{dv}{dt}. \quad \dots\dots\dots\dots\dots(4)$$

Hence if v be regarded as a function of the space described (s), we have

$$a = \frac{dv}{ds}\frac{ds}{dt} = v\frac{dv}{ds}. \quad \dots\dots\dots\dots\dots\dots(5)$$

Similarly, in the case of a rigid body rotating about an axis, the angular acceleration, when the angular velocity is regarded as a function of θ, will be given by

$$\frac{d\omega}{d\theta}\frac{d\theta}{dt}, \text{ or } \omega\frac{d\omega}{d\theta}. \quad \dots\dots\dots\dots \dots\dots(6)$$

The following deductions from (3) are important:

1°.   If
$$y = F(x + a), \quad \dots\dots\dots\dots\dots(7)$$

then, putting $u = x + a$, $du/dx = 1$, the formula (3) gives

$$\frac{dy}{dx} = F'(x + a). \quad \dots\dots\dots\dots\dots(8)$$

The geometrical meaning of this is that shifting a curve bodily parallel to the axis of x does not alter the gradient.

2°. If $\qquad y = F(kx),$ ....(9)

we have, putting $u = kx,$ $du/dx = k,$

$$\frac{dy}{dx} = kF'(kx).$$ ....(10)

3°. If $\qquad y = u^m,$ ....(11)

where m is any rational quantity, we have

$$F(u) = u^m, \quad F'(u) = mu^{m-1},$$

and therefore $\qquad \dfrac{dy}{dx} = mu^{m-1}\dfrac{du}{dx}.$ ....(12)

In particular, in the cases $m = \frac{1}{2}$, $m = -\frac{1}{2}$, we find

$$\left.\begin{array}{l} \dfrac{d}{dx}\sqrt{u} = \dfrac{1}{2\sqrt{u}}\dfrac{du}{dx}, \\[2ex] \dfrac{d}{dx}\dfrac{1}{\sqrt{u}} = -\dfrac{1}{2u^{\frac{3}{2}}}\dfrac{du}{dx}, \end{array}\right\}$$ ....(13)

respectively.

We add a few examples on the above rules.

Ex. 1. If $\qquad y = \sin^m x,$ ....(14)

that is, $\qquad y = u^m,$ where $u = \sin x,$

we have $\qquad Dy = mu^{m-1}\dfrac{du}{dx} = m\sin^{m-1}x\cos x.$ ....(15)

Ex. 2. If $\qquad y = \sqrt{(a^2 - x^2)},$ ....(16)

we have $\qquad Dy = D(a^2 - x^2)^{\frac{1}{2}} = \frac{1}{2}(a^2 - x^2)^{-\frac{1}{2}}.D(a^2 - x^2)$

$$= -\frac{x}{\sqrt{(a^2 - x^2)}}.$$ ....(17)

Ex. 3. Let $\qquad y = \dfrac{x}{\sqrt{(a^2 - x^2)}}.$ ....(18)

If we put $\qquad u = x,$ $v = \sqrt{(a^2 - x^2)},$

we have $\qquad Du = 1,$ and $Dv = \dfrac{-x}{\sqrt{(a^2 - x^2)}},$

by the preceding *Ex.* The rule for differentiating a fraction then gives

$$Dy = \frac{vDu - uDv}{v^2} = \frac{\sqrt{(a^2 - x^2)} + \dfrac{x^2}{\sqrt{(a^2 - x^2)}}}{a^2 - x^2}$$

$$= \frac{a^2}{(a^2 - x^2)^{\frac{3}{2}}}.$$ ....(19)

33. Differentiation of Inverse Functions.

If y be a continuous function of x, then under a certain condition (see Art. 16), which is fulfilled in the case of most ordinary mathematical functions, x will be a continuous function of y.

If δx, δy be corresponding increments of x and y, we have

$$\frac{\delta y}{\delta x} \cdot \frac{\delta x}{\delta y} = 1,$$

identically. Hence, since the limit of the product is equal to the product of the limits,

$$\frac{dy}{dx} \cdot \frac{dx}{dy} = 1. \quad\dots\dots\dots\dots\dots\dots(1)$$

Hence, it being presupposed that y is a differentiable function of x, it follows that x is in general a differentiable function of y, and that the two derived functions are reciprocals.

The geometrical meaning of this is that the tangent to a curve makes complementary angles with the axes of x and y.

The following cases are important:

1°.　If $\qquad\qquad y = \sin^{-1} x$, $\dots\dots\dots\dots\dots\dots(2)$

we have $\qquad\qquad x = \sin y, \ \dfrac{dx}{dy} = \cos y.$

Hence $\qquad\quad \dfrac{dy}{dx} = \dfrac{1}{\cos y} = \pm \dfrac{1}{\sqrt{(1-x^2)}} \cdot \ \dots\dots\dots\dots(3)$

2°.　If $\qquad\qquad y = \cos^{-1} x$, $\dots\dots\dots\dots\dots\dots(4)$

we have $\qquad\qquad y = \cos y, \ \dfrac{dx}{dy} = -\sin y,$

and therefore $\quad \dfrac{dy}{dx} = -\dfrac{1}{\sin y} = \mp \dfrac{1}{\sqrt{(1-x^2)}}. \dots\dots\dots\dots(5)$

The ambiguity of sign in these results is to be accounted for as follows. We have seen that if $y = \sin^{-1} x$, then y is a many-valued function of x; viz. for any assigned value of x (between the limits ± 1) there is a series of values of y. For some of these dy/dx is positive, for others negative; see Fig. 21 (p. 60). Similarly for $\cos^{-1} x$.

If, in accordance with a usual convention, we agree to understand by $\sin^{-1} x$ the angle between $-\frac{1}{2}\pi$ and $\frac{1}{2}\pi$ whose sine is x, we must write

$$\frac{d}{dx} \sin^{-1} x = +\frac{1}{\sqrt{(1-x^2)}} \cdot \ \dots\dots\dots\dots(6)$$

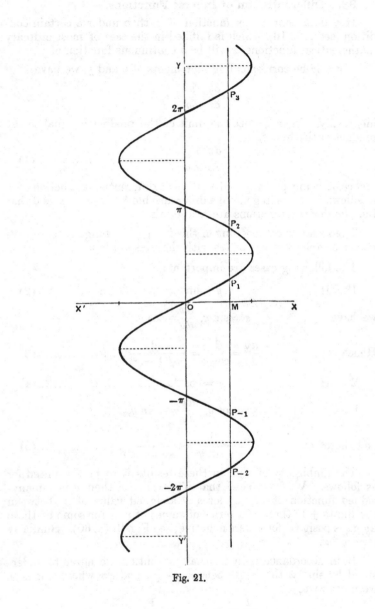

Fig. 21.

Similarly if $\cos^{-1} x$ be restricted to lie between 0 and π, we have

$$\frac{d}{dx} \cos^{-1} x = -\frac{1}{\sqrt{(1 - x^2)}}. \quad \dots\dots\dots\dots(7)$$

3°. If $\qquad\qquad y = \tan^{-1} x, \quad \dots\dots\dots\dots\dots\dots(8)$

we have $\qquad\qquad x = \tan y, \quad \dfrac{dx}{dy} = \sec^2 y,$

and therefore $\qquad \dfrac{dy}{dx} = \dfrac{1}{\sec^2 y} = \dfrac{1}{1 + x^2}. \quad \dots\dots\dots\dots\dots(9)$

There is here no ambiguity of sign. For each value of x there is an infinite series of values of y, but the value of dy/dx is the same for all, the tangent lines at the corresponding points of the curve $y = \tan^{-1} x$ being parallel. See Fig. 22, p. 62.

Ex. 1. Let $\qquad\qquad y = \sin^{-1} \dfrac{x}{\sqrt{(1 + x^2)}}, \quad \dots\dots\dots\dots(10)$

or $\qquad\qquad y = \sin^{-1} u, \quad \text{where } u = \dfrac{x}{\sqrt{(1 + x^2)}}.$

We have $\qquad \dfrac{dy}{dx} = \dfrac{dy}{du}\dfrac{du}{dx} = \dfrac{1}{\sqrt{(1 - u^2)}}\dfrac{du}{dx}.$

We easily find

$$\sqrt{(1 - u^2)} = \frac{1}{\sqrt{(1 + x^2)}}, \quad \frac{du}{dx} = \frac{1}{(1 + x^2)^{\frac{3}{2}}},$$

whence $\qquad\qquad \dfrac{dy}{dx} = \dfrac{1}{1 + x^2}. \quad \dots\dots\dots\dots\dots\dots(11)$

It is easily proved (putting $x = \tan \theta$) that

$$\sin^{-1} \frac{x}{\sqrt{(1 + x^2)}} = \tan^{-1} x,$$

so that the above result is in accordance with (9) above.

Ex. 2. Let $\qquad\qquad y = \tan^{-1} \dfrac{1 + x + x^2}{1 - x + x^2}. \quad \dots\dots\dots\dots(12)$

If we write $\qquad\qquad u = \dfrac{1 + x + x^2}{1 - x + x^2},$

we have $\qquad\qquad \dfrac{dy}{dx} = \dfrac{1}{1 + u^2}\dfrac{du}{dx}.$

We find $\quad 1 + u^2 = \dfrac{2\,(1 + 3x^2 + x^4)}{(1 - x + x^2)^2}, \quad \dfrac{du}{dx} = \dfrac{2\,(1 - x^2)}{(1 - x + x^2)^2},$

whence $\qquad\qquad \dfrac{dy}{dx} = \dfrac{1 - x^2}{1 + 3x^2 + x^4} .$ ....(13)

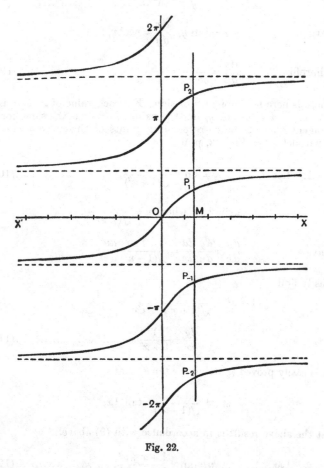

Fig. 22.

34. Functions of two or more independent variables. Partial Derivatives.

Although in this treatise we are primarily concerned with functions of a single independent variable, it will occasionally be useful, even at an early stage, to have at our command ideas and notations borrowed from the more general theory.

One quantity u is said to be a function of two or more independent variables x, y, ..., when its value is determined by those of the latter, which may be assigned arbitrarily, and independently, within (in each case) a certain range. Thus if P be any point of a given surface, and a perpendicular PN be drawn to any fixed horizontal plane, the altitude PN is a function of the coordinates (x, y) of the point N.

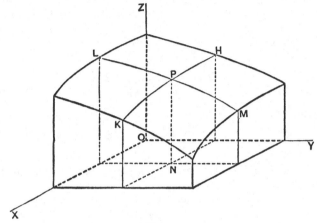

Fig. 23.

So again, in Physics, the pressure of a gas is a function of two independent variables, viz. the volume (per unit mass) and the temperature.

The functional relation is expressed by an equation of the form

$$u = \phi(x, y, \ldots). \qquad\ldots\ldots\ldots\ldots\ldots\ldots\ldots(1)$$

In particular, in the aforesaid case of a surface, if we denote the altitude PN by z, we have

$$z = \phi(x, y). \qquad\ldots\ldots\ldots\ldots\ldots\ldots\ldots(2)$$

The definition of continuity given in Art. 8 may be extended to the present case as follows. A function $\phi(x, y, \ldots)$ is said to be continuous for a particular set $x, y, \ldots$ of values of the independent variables if a quantity ϵ, different from zero, can always be found such that $\phi(x + \delta x, y + \delta y, \ldots)$ shall differ in absolute value from $\phi(x, y, \ldots)$ by a quantity less than any prescribed magnitude σ, however small, for *all* values of the increments $\delta x, \delta y, \ldots$ which are less in absolute value than ϵ.

Thus, in the case of two independent variables, illustrated by the figure, it is implied that a rectangle can be drawn in the plane xy, about N, such that the ordinates at all points within this rectangle shall differ from PN by less than σ, however small σ may be.

Let us now suppose, the function

$$u = \phi (x, y, \ldots) \quad \ldots\ldots\ldots\ldots\ldots\ldots\ldots(3)$$

being continuous, that all the independent variables save one (x) are kept constant. Then u being assumed to be a differentiable function of x, its derived function with respect to x is called the 'partial differential coefficient' or 'partial derivative' of u with respect to x, and is denoted by $\partial u/\partial x$. Thus

$$\frac{\partial u}{\partial x} = \lim_{\delta x \to 0} \frac{\phi (x + \delta x, y, \ldots) - \phi (x, y, \ldots)}{\delta x} . \quad \ldots\ldots(4)$$

In like manner

$$\frac{\partial u}{\partial y} = \lim_{\delta y \to 0} \frac{\phi (x, y + \delta y, \ldots) - \phi (x, y, \ldots)}{\delta y} . \quad \ldots\ldots(5)$$

In the case of the surface (2) it is plain that the partial derivatives

$$\frac{\partial z}{\partial x}, \quad \frac{\partial z}{\partial y}$$

are the gradients of the sections (HK, LM, in the figure) of the surface by planes parallel to the planes ZOX, ZOY respectively.

Ex. 1. If $\qquad z = x^m y^n, \qquad \ldots\ldots\ldots\ldots\ldots\ldots\ldots(6)$

we have $\qquad \dfrac{\partial z}{\partial x} = m x^{m-1} y^n, \quad \dfrac{\partial z}{\partial y} = n x^m y^{n-1}. \qquad \ldots\ldots\ldots\ldots(7)$

Ex. 2. Assuming that in a gas the pressure (p), volume (v), and temperature (θ) are connected by the relation

$$p = \frac{R\theta}{v} , \qquad \ldots\ldots\ldots\ldots\ldots\ldots\ldots\ldots(8)$$

we have $\qquad \dfrac{\partial p}{\partial v} = -\dfrac{R\theta}{v^2} , \quad \dfrac{\partial p}{\partial \theta} = \dfrac{R}{v} . \qquad \ldots\ldots\ldots\ldots\ldots\ldots(9)$

35. Implicit Functions.

An equation of the type

$$\phi (x, y) = 0 \qquad \ldots\ldots\ldots\ldots\ldots\ldots(1)$$

in general determines y as a function of x; for if we assign any arbitrary value to x, the resulting equation in y has in general one or more definite roots. These roots may be real or imaginary, but we shall only contemplate cases where, for all values of x within a certain range, one value (at least) of y is real. The term 'implicit' is applied to functions determined in this manner, by way of contrast with cases where y is given 'explicitly' in the form

$$y = f(x). \qquad \dots\dots\dots\dots\dots\dots\dots\dots(2)$$

If we regard $\qquad z = \phi(x, y) \qquad \dots\dots\dots\dots\dots\dots\dots\dots(3)$

as the equation of a surface, then (1) is the equation of the section of this surface by the plane $z = 0$. If the plane xy be regarded as horizontal, the sections $z = C$, where C may have different constant values, are the 'contour-lines.'

If we require to differentiate an implicit function, we may seek, first, to solve the equation (1) with respect to y, so as to bring it into the form (2). It is useful, however, to have a rule to meet cases where this process would be inconvenient or impracticable. It will be sufficient, for the present, to consider the case where $\phi(x, y)$ is a rational integral function of x and y, *i.e.* it is the sum of a finite series of terms of the type $A_{m,n} x^m y^n$, where m, n may have the values $0, 1, 2, 3, \dots$. Since, by hypothesis, $\phi(x, y)$ is constantly zero, its derived function with respect to x will be zero. Now by Arts. 28, 30, 32, we have

$$\frac{d}{dx}(x^m y^n) = m x^{m-1} y^n + n x^m y^{n-1} \frac{dy}{dx}.$$

Hence, if $\qquad \phi(x, y) = \Sigma A_{m,n} x^m y^n, \qquad \dots\dots\dots\dots\dots\dots(4)$

we have $\qquad \Sigma A_{m,n} m x^{m-1} y^n + \Sigma A_{m,n} n x^m y^{n-1} \dfrac{dy}{dx} = 0. \qquad \dots\dots\dots(5)$

In the notation of Art. 34, this may be written

$$\frac{\partial \phi}{\partial x} + \frac{\partial \phi}{\partial y} \frac{dy}{dx} = 0, \qquad \dots\dots\dots\dots\dots\dots(6)$$

or $\qquad \dfrac{dy}{dx} = -\dfrac{\dfrac{\partial \phi}{\partial x}}{\dfrac{\partial \phi}{\partial y}}. \qquad \dots\dots\dots\dots\dots\dots(7)$

It will be shewn in Chapter IV that the results (6) and (7) are not limited to the above special form of $\phi(x, y)$; but the present case is sufficient for most geometrical applications.

EXAMPLES. V.

(Differentiations *ab initio*.)

1.   Find, from first principles, the derived functions of

$$x^3, \quad x^4, \quad \frac{1}{x^2}, \quad \frac{1}{\sqrt{x}}.$$

2.   Also of $\qquad \dfrac{1}{x+a}, \quad \dfrac{x+a}{x-a}, \quad \dfrac{1}{x^2-a^2}.$

3.   Also of $\qquad \sqrt{(x^2+a^2)}, \quad \dfrac{1}{\sqrt{(x^2+a^2)}}.$

4.   Also of $\qquad \cot x, \quad \sec x, \quad \operatorname{cosec} x.$

5.   Also of $\qquad \sin^2 x, \quad \cos^2 x, \quad \sin 2x, \quad \cos 2x.$

6.   If, in the rectilinear motion of a point,

$$s = ut + \tfrac{1}{2}at^2,$$

where u, a are constants, prove that the velocity at time t is $u + at$, and that the acceleration is constant.

7.   If the pressure and the volume of a gas kept at constant temperature be connected by the relation

$$pv = \text{const.},$$

the cubical elasticity is equal to p.

8.   If the radius of a circle be increasing at the rate of one foot per second, find the rate of increase of the area, in square feet per second, at the instant when the radius is 10 feet.

9.   If the area of a circle increase at a uniform rate, the rate of increase of the perimeter varies inversely as the radius.

10. A is a fixed point on the circumference of a circle whose centre is O and radius one foot. A point P, starting from A, describes the circumference uniformly in one second. Find the rates of increase (1) of the arc AP, (2) of the chord AP, (3) of the sectorial area AOP, (4) of the triangular area AOP, at the instant when the angle AOP is 60°.

11.   If the volume of a gramme of water varies as

$$1 + \frac{(\theta - 4)^2}{144000},$$

where θ is the temperature centigrade, find the coefficients of cubical expansion for $\theta = 0°$ and $\theta = 20°$.

EXAMPLES. VI.

(Products and Quotients.)

Verify the following differentiations :

1. $y = x(1-x)$, $Dy = 1 - 2x$.

2. $y = x(1-x)^2$, $Dy = (1-x)(1-3x)$.

3. $y = x^m(1-x)^n$, $Dy = x^{m-1}(1-x)^{n-1}\{m - (m+n)x\}$.

4. $y = (x-1)(x-2)(x-3)$, $Dy = 3x^2 - 12x + 11$.

5. $y = x(1-x)^2(1+x)^3$, $Dy = (1-x)(1+x)^2(1+3x)(1-2x)$.

6. $y = (1+x^2)(1-2x^2)$, $Dy = -2x(1+4x^2)$.

7. $y = \left(x + \dfrac{1}{x}\right)^2$, $Dy = 2\left(x + \dfrac{1}{x}\right)\left(1 - \dfrac{1}{x^2}\right)$.

8. $y = \dfrac{x}{1-x}$, $Dy = \dfrac{1}{(1-x)^2}$.

9. $y = \dfrac{x}{1+x^2}$, $Dy = \dfrac{1-x^2}{(1+x^2)^2}$.

10. $y = \dfrac{1+x^2}{1-x^2}$, $Dy = \dfrac{4x}{(1-x^2)^2}$.

11. $y = \dfrac{x^m}{(1-x)^n}$, $Dy = \dfrac{x^{m-1}}{(1-x)^{n+1}}\{m - (m-n)x\}$.

12. $y = x\sin x$, $Dy = \sin x + x\cos x$.

13. $y = x^2\cos x$, $Dy = 2x\cos x - x^2\sin x$.

14. $y = \sin^2 x\cos x$ $Dy = 2\sin x - 3\sin^3 x$.

15. $y = \dfrac{\sin x}{x}$, $Dy = \dfrac{x\cos x - \sin x}{x^2}$.

16. $y = \dfrac{x}{\sin x}$, $Dy = \dfrac{\sin x - x\cos x}{\sin^2 x}$.

17. $y = \dfrac{\tan x}{x}$, $Dy = \dfrac{x - \sin x\cos x}{x^2\cos^2 x}$.

18. $y = \tan^2 x$, $Dy = 2\tan x\sec^2 x$.

19. $y = \sec^2 x$, $Dy = 2\tan x\sec^2 x$.

20. $y = \dfrac{\sin x}{1+\tan x}$, $Dy = \dfrac{\cos^3 x - \sin^3 x}{(\sin x + \cos x)^2}$.

21. $y = \dfrac{1+\sin x}{1-\sin x}$, $Dy = \dfrac{2\cos x}{(1-\sin x)^2}$.

22. $y = \dfrac{1-\cos x}{1+\cos x}$, $Dy = \dfrac{2\sin x}{(1+\cos x)^2}$.

EXAMPLES. VII.

(Functions of Functions.)

1. $y = (x + a)^m (x + b)^n,$ $Dy = (x + a)^{m-1} (x + b)^{n-1}$
$$\{(m + n) x + mb + na\}.$$

2. $y = \dfrac{x^n}{(1 + x)^n},$ $Dy = \dfrac{nx^{n-1}}{(1 + x)^{n+1}}.$

3. $y = \sqrt{(1 + x)},$ $Dy = \dfrac{1}{2\sqrt{(1 + x)}}.$

4. $y = \sqrt{\{(x + 1)(x + 2)\}},$ $Dy = \dfrac{2x + 3}{2\sqrt{\{(x + 1)(x + 2)\}}}.$

5. $y = (1 + x)\sqrt{(1 - x)},$ $Dy = \dfrac{1 - 3x}{2\sqrt{(1 - x)}}.$

6. $y = (1 - x)\sqrt{(1 + x^2)},$ $Dy = -\dfrac{(1 - x + 2x^2)}{\sqrt{(1 + x^2)}}.$

7. $y = \dfrac{1}{\sqrt{(1 - x^2)}},$ $Dy = \dfrac{x}{(1 - x^2)^{\frac{3}{2}}}.$

8. $y = \dfrac{x}{\sqrt{(1 + x^2)}},$ $Dy = \dfrac{1}{(1 + x^2)^{\frac{3}{2}}}.$

9. $y = \dfrac{\sqrt{(1 + x^2)}}{x},$ $Dy = -\dfrac{1}{x^2\sqrt{(1 + x^2)}}.$

10. $y = \sqrt{\left(\dfrac{1 + x}{1 - x}\right)}.$ $Dy = \dfrac{1}{(1 - x)\sqrt{(1 - x^2)}}.$

11. $y = \dfrac{1}{\sqrt{(1 + x^2)} + x},$ $Dy = \dfrac{x}{\sqrt{(x^2 + 1)}} - 1.$

12. $y = \dfrac{x}{\sqrt{(1 + x^2)} + x},$ $Dy = \dfrac{1 + 2x^2}{\sqrt{(1 + x^2)}} - 2x.$

13. $y = \sqrt{\left(\dfrac{1 + x + x^2}{1 - x + x^2}\right)},$ $Dy = \dfrac{1 - x^2}{(1 + x + x^2)^{\frac{3}{2}}(1 - x + x^2)^{\frac{3}{2}}}.$

14. $y = \sin 2(x - a),$ $Dy = 2\cos 2(x - a).$

15. $y = \sin^2 2x,$ $Dy = 2\sin 4x.$

16. $y = \sqrt{(1 + \sin x)},$ $Dy = \tfrac{1}{2}\sqrt{(1 - \sin x)}.$

17. $y = \tan^2 x,$ $Dy = \dfrac{2\sin x}{\cos^3 x}.$

18. $y = \sec^2 x,$ $Dy = \dfrac{2\sin x}{\cos^3 x}.$

19. $y = \tan^3 x$, $\qquad Dy = \dfrac{3 \sin^2 x}{\cos^4 x}$.

20. $y = \sin^3 x + \cos^3 x$, $\qquad Dy = 3 \sin x \cos x\,(\sin x - \cos x)$.

21. $y = \sin x - \frac{1}{3}\sin^3 x$, $\qquad Dy = \cos^3 x$.

22. $y = \tan x + \frac{1}{3}\tan^3 x$, $\qquad Dy = \sec^4 x$.

23. $y = \dfrac{\sin^2 x}{x^2}$, $\qquad Dy = \dfrac{2 \sin x}{x^3}\,(x \cos x - \sin x)$.

24. $y = \sin mx \sin nx$, $\qquad Dy = n \sin mx \cos nx$
$$+ \, m \cos mx \sin nx.$$

25. $y = \sqrt{(a \sin^2 x + \beta \cos^2 x)}$, $\quad Dy = \frac{1}{2}\,(a - \beta)\,\dfrac{\sin 2x}{\sqrt{(a \sin^2 x + \beta \cos^2 x)}}$.

EXAMPLES. VIII.

(Inverse Functions.)

1. $y = \sin^{-1}(1 - x)$, $\qquad Dy = -\dfrac{1}{\sqrt{(2x - x^2)}}$.

2. $y = x \sin^{-1} x$, $\qquad Dy = \sin^{-1} x + \dfrac{x}{\sqrt{(1 - x^2)}}$.

3. $y = \cot^{-1} x$, $\qquad Dy = -\dfrac{1}{1 + x^2}$.

4. $y = \sec^{-1} x$, $\qquad Dy = \dfrac{1}{x\sqrt{(x^2 - 1)}}$.

5. $y = \operatorname{cosec}^{-1} x$, $\qquad Dy = -\dfrac{1}{x\sqrt{(x^2 - 1)}}$.

6. $y = \sin^{-1} x$ $\qquad Dy = 0$.
$\qquad + \sin^{-1}\sqrt{(1 - x^2)}$,

7. $y = \tan^{-1} x + \tan^{-1}\dfrac{1}{x}$, $\qquad Dy = 0$.

8. $y = \sin^{-1}\{2x\sqrt{(1 - x^2)}\}$, $\quad Dy = \dfrac{2}{\sqrt{(1 - x^2)}}$.

9. $y = \tan^{-1}\dfrac{2x}{1 - x^2}$, $\qquad Dy = \dfrac{2}{1 + x^2}$.

10. $y = \cos^{-1}\dfrac{1 - x^2}{1 + x^2}$, $\qquad Dy = \dfrac{2}{1 + x^2}$.

11. $y = \tan^{-1}\dfrac{\tan a \sin x}{1 + \sec a \cos x}$, $\quad Dy = \dfrac{\tan a}{\sec a + \cos x}$.

12. $y = \tan^{-1}\{\sqrt{(x^2 + 1)} - x\}$, $Dy = -\dfrac{1}{2\,(x^2 + 1)}$.

13. $y = \sin^{-1}(\cos x)$, $Dy = -1$.

EXAMPLES. IX.

1.   What is the geometrical meaning of the theorem

$$\frac{d}{dx}\,\phi\,(kx) = k\phi'\,(kx)\,?$$

2.   If $y = \tan^{-1}\dfrac{u}{v}$,

prove that $Dy = \dfrac{v\,Du - u\,Dv}{u^2 + v^2}$.

3.   Assuming that $\dfrac{1 - x^{n+1}}{1 - x} = 1 + x + x^2 + \ldots + x^n$,

deduce, by differentiation, the sum of the series

$$1 + 2x + 3x^2 + \ldots + nx^{n-1},$$

and test the result by putting $x = 1$.

Hence shew that, if $|x| < 1$,

$$1 + 2x + 3x^2 + 4x^3 + \ldots \text{ to } \infty = (1 - x)^{-2}.$$

4.   If, in the rectilinear motion of a point, v^2 be a linear function of s, the acceleration is constant.

5.   If v^2 be a quadratic function of s, the acceleration varies as the distance from a fixed point in the line of motion.

6.   If the time be a quadratic function of the space described, the acceleration varies as the cube of the velocity.

7.   If $v^2 = A + \dfrac{B}{s}$,

the acceleration varies inversely as the square of the distance from a fixed point in the line of motion.

8.   If s^2 be a quadratic function of t, the acceleration varies as $1/s^3$.

9.   If the pressure and the volume of a gas be connected by the relation

$$pv^\gamma = \text{const.},$$

the cubical elasticity is γp.

EXAMPLES. X.

(Partial Differentiation.)

1. Sketch the contour-lines of the surface

$$az = x^2 + y^2,$$

and describe the general form of the surface.

2. Also of the surface $az = xy.$

3. If $z = f(x + y),$

prove that $\dfrac{\partial z}{\partial x} = \dfrac{\partial z}{\partial y},$

and give the geometrical interpretation of this result.

4. If $r = \sqrt{(x^2 + y^2)},$

prove that $r\dfrac{\partial r}{\partial x} = x, \quad r\dfrac{\partial r}{\partial y} = y.$

5. If $z = a \tan^{-1} \dfrac{y}{x},$

prove that $\dfrac{\partial z}{\partial x} = -\dfrac{ay}{x^2 + y^2}, \quad \dfrac{\partial z}{\partial y} = \dfrac{ax}{x^2 + y^2}.$

6. If $z = f(x^2 + y^2),$

prove that $\dfrac{\partial z}{\partial x} : \dfrac{\partial z}{\partial y} = x : y.$

7. If $z = f\left(\dfrac{y}{x}\right),$

prove that $x\dfrac{\partial z}{\partial x} + y\dfrac{\partial z}{\partial y} = 0.$

8. If $ax^2 + 2hxy + by^2 + 2gx + 2fy + c = 0,$

prove that $\dfrac{dy}{dx} = -\dfrac{ax + hy + g}{hx + by + f}.$

CHAPTER III

THE EXPONENTIAL AND LOGARITHMIC FUNCTIONS

36. The Exponential Function.

The functions now to be considered may be defined in various ways, but from the point of view of the Calculus, as well as of most applications, their fundamental property is that they satisfy equations of the type

$$\frac{dy}{dx} = ky, \qquad \dots\dots\dots\dots\dots\dots(1)$$

where k is a (positive or negative) constant. That is, the rate of increase bears always a constant ratio to the instantaneous value of the function.

The generalized 'exponential' function as thus defined*, may be contrasted with the 'linear' function

$$y = a + bx, \qquad \dots\dots\dots\dots\dots\dots(2)$$

so called because its graph is a straight line. It has in fact the same relation to the linear function which the law of compound bears to that of simple interest, provided we imagine the interest to accrue continually instead of at fixed intervals.

The general linear function involves two constants, viz. the gradient b, and the initial value a. If the matter were not already sufficiently simple, we might adopt as the standard linear function the one whose gradient and initial value are each $= 1$, so that

$$y = 1 + x. \qquad \dots\dots\dots\dots\dots\dots(3)$$

The general linear function may be derived from this by suitable alterations of the scales of x and y.

The general exponential function involves in like manner two constants, viz. the constant k in (1), and the initial value C, say. It will appear presently that these two constants completely determine the function. We choose as our *standard* function of this type the one for which $k = 1$, and whose initial value is unity.

---

* It will be seen later that it is necessarily of the form Ca^x if x is rational, whence the name, since the variable x appears as an index or 'exponent.'

In other words we define *the* exponential function *par excellence* as being that solution of the equation

$$\frac{dy}{dx} = y, \dots\dots\dots\dots\dots\dots\dots\dots(4)$$

which is equal to unity when $x = 0$.

37. The Exponential Series.

We have to shew in the first place that such a function exists, and also to find if possible a means of calculating it, to any desired degree of approximation, for any assigned value of x.

Let us assume, tentatively, that the equation

$$\frac{dy}{dx} = y \dots\dots\dots\dots\dots\dots\dots\dots(1)$$

can be satisfied by the sum of a power-series, say

$$y = 1 + A_1 x + A_2 x^2 + A_3 x^3 + \dots + A_n x^n + \dots, \dots\dots(2)$$

where the first term has been fixed by the condition that $y = 1$ for $x = 0$. On the hypothesis that this value of y can be differentiated by the same rule which applies (Art. 29) to a *finite* series of terms, we should have

$$\frac{dy}{dx} = A_1 + 2A_2 x + 3A_3 x^2 + \dots + nA_n x^{n-1} + \dots, \quad \dots(3)$$

and the equation (1) would therefore be satisfied, provided

$$A_1 = 1, \quad 2A_2 = A_1, \quad 3A_3 = A_2, \dots, \quad nA_n = A_{n-1}. \dots (4)$$

This requires

$$A_1 = 1, \quad A_2 = \tfrac{1}{2} A_1 = \frac{1}{2!}, \quad A_3 = \tfrac{1}{3} A_2 = \frac{1}{3!}, \dots,$$

and, generally,

$$A_n = \frac{1}{n!}. \dots\dots\dots\dots\dots\dots(5)$$

We are thus led to study the series

$$1 + x + \frac{x^2}{2!} + \frac{x^3}{3!} + \dots + \frac{x^n}{n!} + \dots\dots\dots\dots\dots(6)$$

This is convergent, and has therefore a definite 'sum,' for any given value of x. For the ratio of the $(n + 1)$th term to the nth, viz. x/n, can be made as small as we please (in absolute value) by taking n great enough. Hence a point in the series can always be found after which the successive terms will diminish more rapidly than those of any geometrical progression whatever. The series is therefore convergent, by Art. 5; it is moreover 'absolutely' convergent. We denote its sum by $E(x)$.

We proceed to shew that the function $E(x)$ as thus defined is continuous and differentiable, and that it does in fact satisfy (1). We write

$$C(x) = \tfrac{1}{2} \{E(x) + E(-x)\}$$

$$= 1 + \frac{x^2}{2!} + \frac{x^4}{4!} + \ldots + \frac{x^{2n}}{(2n)!} + \ldots, \quad \ldots\ldots\ldots\ldots(7)$$

$$S(x) = \tfrac{1}{2} \{E(x) - E(-x)\}$$

$$= x + \frac{x^3}{3!} + \frac{x^5}{5!} + \ldots + \frac{x^{2n+1}}{(2n+1)!} + \ldots, \quad \ldots\ldots\ldots\ldots(8)$$

so that

$$E(x) = C(x) + S(x), \quad E(-x) = C(x) - S(x). \quad \ldots\ldots(9)$$

We note that the terms of $C(x)$ are all positive, whilst those of $S(x)$ are of uniform sign, the same as that of x; this simplifies the subsequent discussion.

If x_1 and x be two values of the variable, which we will suppose to have the same sign, we have

$$C(x_1) - C(x) = \frac{x_1^2 - x^2}{2!} + \frac{x_1^4 - x^4}{4!} + \ldots + \frac{x_1^{2n} - x^{2n}}{(2n)!} + \ldots$$

$$= (x_1 - x) \left\{ \frac{x_1 + x}{2!} + \frac{x_1^3 + x_1^2 x + x_1 x^2 + x^3}{4!} + \ldots \right.$$

$$\left. + \frac{x_1^{2n-1} + x_1^{2n-2} x + \ldots + x^{2n-1}}{(2n)!} + \ldots \right\}, \quad \ldots(10)$$

the equalities resulting from the theorems 1° and 2° of Art. 5. Let ξ be a positive quantity equal to the absolute value of x_1 or x, whichever is the greater. We have, then,

$$|x_1^{2n-1} + x_1^{2n-2} x + \ldots + x^{2n-1}| < 2n\, \xi^{2n-1}, \quad \ldots\ldots(11)$$

and the series in { } is therefore less in absolute value than

$$\xi + \frac{\xi^3}{3!} + \ldots + \frac{\xi^{2n-1}}{(2n-1)!} + \ldots,$$

i.e. less than $S(\xi)$, which is finite. Hence

$$\lim_{x_1 \to x} \{C(x_1) - C(x)\} = 0. \quad \ldots\ldots\ldots\ldots(12)$$

The function $C(x)$ is therefore continuous for all values of x. In the same way we may shew that $S(x)$ is continuous. The continuity of $E(x)$ then follows from (9).

Again, from (10),

$$\frac{C(x_1) - C(x)}{x_1 - x} = \frac{x_1 + x}{2!} + \frac{x_1^3 + x_1^2 x + x_1 x^2 + x^3}{4!} + \cdots$$

$$+ \frac{x_1^{2n-1} + x_1^{2n-2}x + \cdots + x^{2n-1}}{(2n)!} + \cdots \cdots \cdots (13)$$

The terms on the right-hand side are of uniform sign, whether that of x and x_1 be positive or negative. The numerator of the term last written therefore lies between $2nx^{2n-1}$ and $2nx_1^{2n-1}$, and the sum of the series accordingly lies between $S(x)$ and $S(x_1)$. Since $S(x)$ is continuous, it follows that

$$\lim_{x_1 \to x} \frac{C(x_1) - C(x)}{x_1 - x} = S(x), \cdots \cdots \cdots (14)$$

or

$$C'(x) = S(x). \cdots \cdots \cdots (15)$$

It may be shewn, in the same way, that

$$S'(x) = C(x). \cdots \cdots \cdots (16)$$

Hence

$$E'(x) = \frac{d}{dx}\{C(x) + S(x)\} = C'(x) + S'(x)$$

$$= S(x) + C(x) = E(x), \cdots \cdots \cdots (17)$$

so that (1) is satisfied by $y = E(x)$, for all values of x.

Finally, we can shew that the solution of (1) thus obtained is unique, under the condition that $y = 1$ for $x = 0$. For if u, v denote two such solutions, we have

$$\frac{du}{dx} = u, \qquad \frac{dv}{dx} = v, \cdots \cdots \cdots (18)$$

and therefore

$$v\frac{du}{dx} - u\frac{dv}{dx} = 0, \cdots \cdots \cdots (19)$$

or

$$\frac{d}{dx}\left(\frac{u}{v}\right) = 0. \cdots \cdots \cdots (20)$$

The ratio v/u is therefore constant*, and if $u = 1$, $v = 1$ for $x = 0$, the constant must be unity, whence $u = v$.

---

* This assumes by anticipation an almost obvious theorem of which a formal proof is given in Art. 56.

38. Addition Theorem. Graph of $E(x)$.

The more general equation

$$\frac{dy}{dx} = ky \quad \dots\dots\dots\dots\dots\dots(1)$$

may be written

$$\frac{dy}{d(kx)} = y, \quad \dots\dots\dots\dots\dots\dots(2)$$

and its solution, under the condition that $y = 1$ for $x = 0$, is therefore

$$y = E(kx). \quad \dots\dots\dots\dots\dots\dots(3)$$

Now let

$$u = E(ax) . E(bx). \quad \dots\dots\dots\dots(4)$$

We have

$$\frac{du}{dx} = aE'(ax) . E(bx) + bE'(bx) . E(ax)$$

$$= (a + b) E(ax) . E(bx) = (a + b) u. \dots\dots\dots(5)$$

Also the initial value of u is unity. Hence

$$u = E\{(a + b) x\}, \dots\dots\dots\dots\dots(6)$$

or

$$E(a) . E(b) = E(a + b), \quad \dots\dots\dots\dots(7)$$

for all values of a and b. This constitutes the 'addition theorem' of the exponential function.

In particular we have

$$E(x) . E(-x) = E(0) = 1, \dots\dots\dots\dots(8)$$

or

$$E(-x) = \frac{1}{E(x)}. \quad \dots\dots\dots\dots(9)$$

We have seen that the function $E(x)$ is continuous. Moreover, when x is positive, every term of the series for $E(x)$ continually increases with x, and becomes infinite for $x \to +\infty$. The same holds à *fortiori* for the sum. Also, in virtue of (9), it appears that if x be positive $E(-x)$ is positive and continually diminishes in absolute value as x increases, and vanishes for $x \to \infty$. Hence as x increases from $-\infty$ to $+\infty$, the function $E(x)$ continually increases from 0 to $+\infty$, and assumes once, and only once, every intermediate value.

The accompanying figure shews the curve

$$y = E(x).$$

A column of numerical values of the function $E(x)$ is given in Table E at the end of the book.

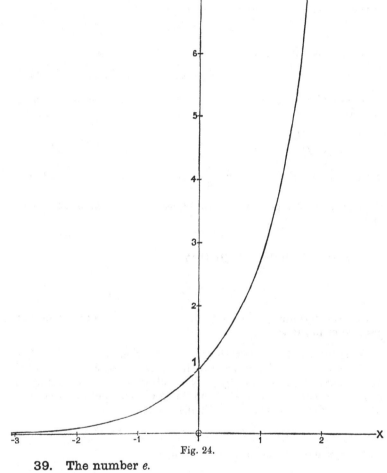

Fig. 24.

39. The number e.

The result (7) of Art. 38 may be extended. Thus

$$E(a) \cdot E(b) \cdot E(c) = E(a+b) \cdot E(c) = E(a+b+c), \quad (1)$$

and so on for any number of factors.

If we form the product of n factors, each equal to $E(1)$, we have

$$\{E(1)\}^n = E(1+1+\dots \text{ to } n \text{ terms}) = E(n). \quad \dots\dots(2)$$

It is usual to denote the quantity $E(1)$, or

$$1 + 1 + \frac{1}{2!} + \frac{1}{3!} + \ldots, \qquad \ldots\ldots\ldots\ldots\ldots(3)$$

by the symbol e. Its value to seven places of decimals is

$$e = 2\cdot7182818.$$

With this notation we have, if n be a positive integer,

$$E(n) = e^n. \qquad \ldots\ldots\ldots\ldots\ldots\ldots\ldots\ldots(4)$$

Again, if m/n be an arithmetical fraction (in its lowest terms), we have

$$\left\{ E\left(\frac{m}{n}\right) \right\}^n = E\left(\frac{m}{n} + \frac{m}{n} + \ldots \text{ to } n \text{ terms}\right) = E(m) = e^m,$$

and therefore

$$E\left(\frac{m}{n}\right) = e^{\frac{m}{n}}. \qquad \ldots\ldots\ldots\ldots\ldots\ldots(5)$$

Hence, if x be any positive rational quantity, integral or fractional, we have

$$E(x) = e^x. \qquad \ldots\ldots\ldots\ldots\ldots\ldots\ldots(6)$$

It follows, from Art. 38 (9), that

$$E(-x) = \frac{1}{e^x} = e^{-x},$$

so that the formula (6) holds for *all* rational values of x, whether positive or negative.

It is to be noticed that the symbol e^x, when x is *irrational*, is (so far) undefined. We may now define it as merely another symbol for the sum of the series $E(x)$. The advantage of this definition is that the notation serves to remind us of the algebraical laws to which the function is subject. Thus we have

$$e^x \cdot e^y = E(x) \times E(y) = E(x+y) = e^{x+y},$$

whether x and y be rational or irrational.

The actual calculation of e is very simple. The first 13 terms of the series (3) are as follows:

$$1 + 1 = 2 \qquad\qquad \frac{1}{3!} = \cdot166\ 666\ 667$$

$$\frac{1}{2!} = \cdot5 \qquad\qquad \frac{1}{4!} = \cdot041\ 666\ 667$$

$$\frac{1}{5!} = \cdot008\ 333\ 333 \qquad\qquad \frac{1}{9!} = \cdot000\ 002\ 756$$

$$\frac{1}{6!} = \cdot001\ 388\ 889 \qquad\qquad \frac{1}{10!} = \cdot000\ 000\ 276$$

$$\frac{1}{7!} = \cdot000\ 198\ 413 \qquad\qquad \frac{1}{11!} = \cdot000\ 000\ 025$$

$$\frac{1}{8!} = \cdot000\ 024\ 802 \qquad\qquad \frac{1}{12!} = \cdot000\ 000\ 002$$

The sum of these numbers is 2·718281830. The error involved in neglecting the remaining terms is

$$\frac{1}{13!} + \frac{1}{14!} + \frac{1}{15!} + \cdots,$$

which is less than

$$\frac{1}{13!}\left(1 + \frac{1}{13} + \frac{1}{13^2} + \frac{1}{13^3} + \cdots\right), \text{ or } \frac{1}{12 \cdot 12!},$$

and therefore does not affect the ninth place of decimals. Hence, allowing for the errors of the last figures in the above table, we may say with confidence that the result just found represents the value of e correctly to seven decimal places.

40. The Hyperbolic Functions.

There are certain combinations of exponential functions whose properties have a close formal analogy with those of the ordinary trigonometrical functions. They are called the hyberbolic sine, cosine, tangent, &c.*, and are defined and denoted as follows:

$$\left.\begin{array}{l} \sinh x = \tfrac{1}{2}\left(e^x - e^{-x}\right) = x + \dfrac{x^3}{3!} + \dfrac{x^5}{5!} + \cdots, \\[2mm] \cosh x = \tfrac{1}{2}\left(e^x + e^{-x}\right) = 1 + \dfrac{x^2}{2!} + \dfrac{x^4}{4!} + \cdots, \end{array}\right\} \quad\cdots\cdots\cdots(1)\dagger$$

$$\left.\begin{array}{ll} \tanh x = \dfrac{\sinh x}{\cosh x}, & \operatorname{sech} x = \dfrac{1}{\cosh x}, \\[2mm] \coth x = \dfrac{\cosh x}{\sinh x}, & \operatorname{cosech} x = \dfrac{1}{\sinh x}. \end{array}\right\} \quad\cdots\cdots\cdots\cdots(2)$$

We notice that $\cosh x$, like $\cos x$, is an 'even' function of x; i.e. it is unaltered by writing $-x$ for x, whilst $\sinh x$, like $\sin x$, is an

---

* They have in some respects the same relation to the rectangular hyperbola that the circular functions have to the circle. See Art. 100, Ex. 2.

† These functions have already appeared, under a slightly different notation, in Art. 37.

'odd' function, *i.e.* the function is unaltered in absolute value but reversed in sign by the same substitution of $-x$ for x.

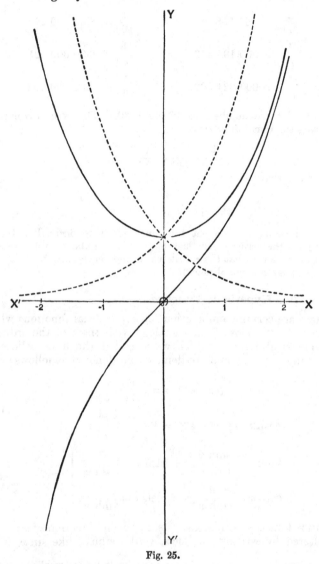

Fig. 25.

The figure shews the curves

$$y = e^x, \quad y = e^{-x},$$

together with the curves
$$y = \cosh x, \quad y = \sinh x,$$
which are derived from them by taking half the sum, and half the difference, of the ordinates, respectively *.

Since $\sinh x$ and $\cosh x$ are continuous, whilst $\cosh x$ never vanishes, it follows that $\tanh x$ is continuous for all values of *x*. Fig. 26 shews the curve
$$y = \tanh x.$$
This has the lines $y = \pm 1$ as asymptotes †.

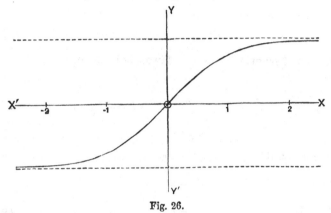

Fig. 26.

Since
$$\cosh x + \sinh x = e^{x}, \quad \cosh x - \sinh x = e^{-x}, \quad \ldots\ldots(3)$$
we have, by multiplication,
$$\cosh^2 x - \sinh^2 x = 1. \quad \ldots\ldots\ldots\ldots\ldots\ldots(4)$$
From this we derive, dividing by $\cosh^2 x$ and $\sinh^2 x$, respectively,
$$\left.\begin{array}{l} \operatorname{sech}^2 x = 1 - \tanh^2 x, \\ \operatorname{cosech}^2 x = \coth^2 x - 1. \end{array}\right\} \quad \ldots\ldots\ldots\ldots(5)$$
Again, we have
$$\begin{aligned} \cosh (x + y) &= \tfrac{1}{2} \left(e^{x} \cdot e^{y} + e^{-x} \cdot e^{-y} \right) \\ &= \tfrac{1}{2} \left\{ (\cosh x + \sinh x)(\cosh y + \sinh y) \right. \\ &\qquad \left. + (\cosh x - \sinh x)(\cosh y - \sinh y) \right\} \\ &= \cosh x \cosh y + \sinh x \sinh y, \quad \ldots\ldots\ldots(6) \end{aligned}$$

* The curve $y = \cosh x$ is known in Statics as the 'catenary,' from its being the form assumed by a chain of uniform density hanging freely under gravity.

† The numerical values (to three places) of the functions $\cosh x$, $\sinh x$, $\tanh x$, for values of x ranging from 0 to 2·5 at intervals of 0·1, are given in the Appendix, Table E.

and similarly
$$\sinh (x + y) = \sinh x \cosh y + \cosh x \sinh y. \quad \ldots\ldots(7)$$
As particular cases
$$\cosh 2x = \cosh^2 x + \sinh^2 x, \quad \sinh 2x = 2 \sinh x \cosh x. \ldots(8)$$
The formulæ (4) and (5) correspond to the trigonometrical formulæ
$$\cos^2 x + \sin^2 x = 1, \ldots\ldots\ldots\ldots\ldots\ldots\ldots(9)$$
$$\left.\begin{aligned} \sec^2 x &= 1 + \tan^2 x, \\ \cosec^2 x &= \cot^2 x + 1. \end{aligned}\right\} \quad \ldots\ldots\ldots\ldots\ldots(10)$$
Similarly the formulæ (8) are the analogues of
$$\cos 2x = \cos^2 x - \sin^2 x, \quad \sin 2x = 2 \sin x \cos x. \quad \ldots\ldots(11)$$

41.   Differentiation of the Hyperbolic Functions.

1°.   If $\qquad\qquad y = \sinh x, \ldots\ldots\ldots\ldots\ldots\ldots\ldots(1)$

we have $\qquad \dfrac{dy}{dx} = D\left(\dfrac{e^x - e^{-x}}{2}\right) = \tfrac{1}{2}\,(De^x - De^{-x})$

$$= \tfrac{1}{2}\,(e^x + e^{-x}) = \cosh x. \ldots\ldots\ldots\ldots\ldots(2)$$

Similarly, if $\qquad\qquad y = \cosh x, \ldots\ldots\ldots\ldots\ldots\ldots(3)$

we find $\qquad\qquad \dfrac{dy}{dx} = \sinh x. \ldots\ldots\ldots\ldots\ldots\ldots(4)$

2°.   If $\qquad\qquad y = \tanh x, \ldots\ldots\ldots\ldots\ldots\ldots(5)$

we have
$$\frac{dy}{dx} = D\,\frac{\sinh x}{\cosh x} = \frac{\cosh x\, D \sinh x - \sinh x\, D \cosh x}{\cosh^2 x}$$
$$= \frac{\cosh^2 x - \sinh^2 x}{\cosh^2 x} = \sech^2 x, \ldots\ldots\ldots\ldots(6)$$
by Art. 40 (4).

Similarly, if $\qquad\qquad y = \coth x, \ldots\ldots\ldots\ldots\ldots\ldots(7)$

we find $\qquad\qquad \dfrac{dy}{dx} = -\cosech^2 x. \ldots\ldots\ldots\ldots\ldots(8)$

3°.   If $\qquad\qquad y = \sech x, \ldots\ldots\ldots\ldots\ldots\ldots(9)$

we have, by Art. 31 (4),
$$\frac{dy}{dx} = D\left(\frac{1}{\cosh x}\right) = \frac{-D\cosh x}{\cosh^2 x}$$
$$= -\frac{\sinh x}{\cosh^2 x}. \quad \ldots\ldots\ldots\ldots\ldots\ldots\ldots(10)$$

Similarly, if $\qquad$ $y = \operatorname{cosech} x,$ ....(11)

we find $\qquad$ $\dfrac{dy}{dx} = -\dfrac{\cosh x}{\sinh^2 x}.$ ....(12)

42. The Logarithmic Function.

The 'logarithmic' function is defined as the inverse of the exponential function. Thus if

$$x = e^y,$$

we have $\qquad$ $y = \log x.$ ....(1)

It was seen in Art. 38 that as y ranges from $-\infty$ through 0 to $+\infty$, e^y steadily increases from 0 through 1 to $+\infty$. Hence for every *positive* value of x there is one and only one value of $\log x$; moreover this value will be positive or negative, according as $x \gtreqless 1$. Also for $x = 0$ we have $y = -\infty$, and for $x = +\infty$, $y = +\infty$. For negative values of x the logarithmic function does not exist.

The ordinary properties of the logarithmic function follow from the above definition in the usual manner.

The full line in Fig. 27 (p. 84) shews the graph of $\log x$. It is of course the same as that of e^x (Fig. 24, p. 77) with x and y interchanged*.

We can now define the symbol a^x, where a is positive, for the case of x irrational. Since

$$a = e^{\log a},$$

we have, if x be rational, $\qquad a^x = e^{x \log a},$ ....(2)

and the latter form may be adopted as the *definition* of a^x when x is irrational†.

Hence if $\qquad$ $y = a^x,$ ....(3)

we have $\qquad$ $\dfrac{dy}{dx} = a^x \log a.$ ....(4)

The logarithm of x, as above defined, is sometimes denoted by $\log_e x$ to distinguish it from the common or 'Briggian' logarithm $\log_{10} x$. The latter may be regarded as defined by the statement that

$$y = \log_{10} x, \text{ if } 10^y = x, \text{ or } e^{y \log_e 10} = x. \quad ....(5)$$

* The function $\log x$ is tabulated in the Appendix, Table F.

† The more usual algebraic definition is that if $x_1, x_2, x_3, \ldots$ be any sequence of *rational* quantities having the irrational x as its limit, a^x is the limit of the sequence

$$a^{x_1}, a^{x_2}, a^{x_3}, \ldots.$$

A proof is then required that the limit is definite, and is moreover a continuous function of x.

Hence $\qquad y \log_e 10 = \log_e x,$ or $\log_{10} x = \mu \log_e x,$ ....(6)

where $\qquad \mu = \dfrac{1}{\log_e 10} = \cdot 43429.$ ....(7)

The mode of calculating μ will be indicated in Chapter XIV.

Hence the graph of the function $\log_{10} x$ is obtained from that of $\log_e x$ by diminishing the ordinates in the constant ratio μ. See the dotted line in the figure.

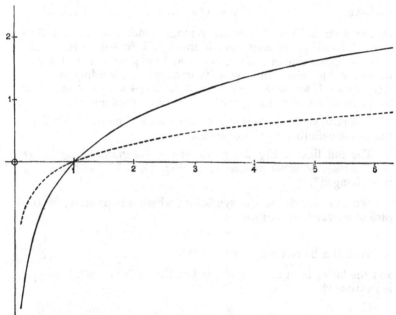

Fig. 27.

In this book we shall always use the symbol $\log x$ in the sense of $\log_e x$.

43. Some Limiting Values.

There are certain limiting values connected with the exponential and logarithmic functions which are of importance.

1°. To find $\qquad \lim_{x \to \infty} x e^{-x}.$ ....(1)

The function assumes the indeterminate form $\infty \times 0$. But since

$$\frac{x}{e^x} = 1 \bigg/ \left(\frac{1}{x} + 1 + \frac{x}{2!} + \frac{x^2}{3!} + \dots \right),$$

we see that the limit in question is 0.

In the same way we can prove that

$$\lim_{x \to \infty} x^m e^{-x} = 0, \quad \dots\dots\dots\dots\dots(2)$$

where m is any rational quantity. This shews that as x increases indefinitely, e^x becomes infinite in comparison with any power of x, however high.

Again, if a be positive and < 1, we have

$$\lim_{n \to \infty} na^n = 0. \quad \dots\dots\dots\dots\dots\dots(3)$$

For if $k = \log(1/a)$, and is therefore positive, we have

$$na^n = ne^{-kn}.$$

2°. If in (1) we put $z = e^x$, and therefore $x = \log z$, we infer that

$$\lim_{z \to \infty} \frac{\log z}{z} = 0. \quad \dots\dots\dots\dots\dots(4)$$

Hence as x increases indefinitely, $\log x$, though ultimately infinite, is infinitely small compared with x.

3°. Again, if in (1) we put $z = e^{-x}$, and therefore $x = -\log z$, we have

$$\lim_{z \to 0} z \log z = 0. \quad \dots\dots\dots\dots\dots(5)$$

4°. We have

$$\frac{e^x - 1}{x} = 1 + \frac{x}{2!} + \frac{x^2}{3!} + \dots$$

$$= 1 + \tfrac{1}{2}x\left(1 + \frac{x}{3} + \frac{x^2}{3 \cdot 4} + \dots\right).$$

The series within brackets is convergent and therefore has a finite sum. Hence

$$\lim_{x \to 0} \frac{e^x - 1}{x} = 1. \quad \dots\dots\dots\dots\dots(6)$$

If we write kx for x, we have

$$\lim_{x \to 0} \frac{e^{kx} - 1}{x} = k, \quad \dots\dots\dots\dots\dots(7)$$

or, putting $k = \log a$, where a is any positive quantity,

$$\lim_{x \to 0} \frac{a^x - 1}{x} = \log a. \quad \dots\dots\dots\dots\dots(8)$$

Another form of this result is obtained by putting $x = 1/n$. Thus

$$\lim_{n \to \infty} n\left(a^{\frac{1}{n}} - 1\right) = \log a. \quad \dots\dots\dots\dots(9)$$

5°. If in (6) we put $x = \log(1 + z)$, and therefore $e^x = 1 + z$, we deduce

$$\lim_{z \to 0} \frac{\log(1 + z)}{z} = 1. \qquad \ldots\ldots\ldots\ldots\ldots(10)$$

6°. If

$$u = \left(1 + \frac{x}{n}\right)^n, \qquad \ldots\ldots\ldots\ldots\ldots(11)$$

we have

$$\log u = n \log\left(1 + \frac{x}{n}\right).$$

Putting $x = nz$, we have

$$\lim_{n \to \infty} \log u = x \lim_{z \to 0} \frac{\log(1 + z)}{z} = x, \ldots\ldots\ldots(12)$$

by (10). Hence

$$\lim_{n \to \infty} \left(1 + \frac{x}{n}\right)^n = e^x. \qquad \ldots\ldots\ldots\ldots\ldots(13)$$

The limit on the left hand is sometimes adopted, in Algebra, as the definition of the exponential function.

44. Differentiation of a Logarithm.

1°. If

$$y = \log x, \qquad \ldots\ldots\ldots\ldots\ldots\ldots(1)$$

we have

$$x = e^y, \quad \frac{dx}{dy} = e^y = x,$$

and therefore

$$\frac{dy}{dx} = \frac{1}{x}. \qquad \ldots\ldots\ldots\ldots\ldots(2)$$

This diminishes as x increases, so that the representative curve becomes less and less inclined to the axis of x. See Fig. 27, p. 84.

2°. If

$$y = \log_a x, \qquad \ldots\ldots\ldots\ldots\ldots\ldots(3)$$

we have

$$x = a^y, \quad \frac{dx}{dy} = a^y . \log_e a = x . \log_e a,$$

whence

$$\frac{dy}{dx} = \frac{1}{\log_e a} \cdot \frac{1}{x}. \qquad \ldots\ldots\ldots\ldots\ldots(4)$$

For instance, if

$$y = \log_{10} x, \qquad \ldots\ldots\ldots\ldots\ldots\ldots(5)$$

we have

$$\frac{dy}{dx} = \frac{\mu}{x}, \qquad \ldots\ldots\ldots\ldots\ldots\ldots(6)$$

where $\mu = \cdot 43429\ldots$ as in Art. 42.

3°. If

$$y = \log u, \qquad \ldots\ldots\ldots\ldots\ldots\ldots(7)$$

where u is a given function of x, we have, by Art. 32,

$$\frac{dy}{dx} = \frac{dy}{du}\frac{du}{dx} = \frac{1}{u}\frac{du}{dx}. \qquad \ldots\ldots\ldots\ldots\ldots(8)$$

Ex. 1. If $\qquad y = \log \sin x,$ ....(9)

we have $\qquad \dfrac{dy}{dx} = \dfrac{1}{\sin x} . D \sin x = \cot x.$ ....(10)

Similarly, if $\qquad y = \log \sec x = -\log \cos x,$ ....(11)

we find $\qquad \dfrac{dy}{dx} = \tan x.$ ....(12)

Ex. 2. If $\qquad y = \log \tan \tfrac{1}{2} x,$ ....(13)

we have $\qquad \dfrac{dy}{dx} = \dfrac{1}{\tan \frac{1}{2} x} . D \tan \tfrac{1}{2} x = \dfrac{1}{\tan \frac{1}{2} x} . \sec^2 \tfrac{1}{2} x . \tfrac{1}{2}$

$$= \dfrac{1}{\sin x}.$$ ....(14)

Similarly, if $\qquad y = \log \tan (\tfrac{1}{4}\pi + \tfrac{1}{2} x),$ ....(15)

we should find $\qquad \dfrac{dy}{dx} = \dfrac{1}{\cos x}.$ ....(16)

Ex. 3. Let $\qquad y = \tfrac{1}{2} \log \dfrac{1+x}{1-x}$

$$= \tfrac{1}{2} \log (1 + x) - \tfrac{1}{2} \log (1 - x).$$ ....(17)

Hence $\qquad \dfrac{dy}{dx} = \dfrac{1}{2} \dfrac{1}{1+x} + \dfrac{1}{2} \dfrac{1}{1-x} = \dfrac{1}{1-x^2}.$ ....(18)

Ex. 4. Let $\qquad y = \log \{x + \sqrt{(x^2 \pm 1)}\}$ ....(19)

We have $\qquad \dfrac{dy}{dx} = \dfrac{1}{x + \sqrt{(x^2 \pm 1)}} . D \{x + \sqrt{(x^2 \pm 1)}\}$

$$= \dfrac{1}{x + \sqrt{(x^2 \pm 1)}} \left\{1 + \dfrac{x}{\sqrt{(x^2 \pm 1)}}\right\}$$

$$= \dfrac{1}{\sqrt{(x^2 \pm 1)}}.$$ ....(20)

45. Logarithmic Differentiation.

In the case of a function consisting of a number of factors it is sometimes convenient to take the logarithm before differentiating. Thus if

$$y = \dfrac{u_1 u_2 u_3 \dots}{v_1 v_2 v_3 \dots},$$ ....(1)

we have

$$\log y = \log u_1 + \log u_2 + \log u_3 + \dots - \log v_1 - \log v_2 - \log v_3 - \dots,$$ ...(2)

and therefore, by Art. 44, 3°,

$$\frac{1}{y}\frac{dy}{dx} = \frac{1}{u_1}\frac{du_1}{dx} + \frac{1}{u_2}\frac{du_2}{dx} + \frac{1}{u_3}\frac{du_3}{dx} + \ldots$$

$$- \frac{1}{v_1}\frac{dv_1}{dx} - \frac{1}{v_2}\frac{dv_2}{dx} - \frac{1}{v_3}\frac{dv_3}{dx} - \ldots \quad (3)$$

This is a generalization of the results of Arts. 30, 31.

The same method can be applied to the differentiation of

$$y = u^v. \quad \ldots\ldots\ldots\ldots\ldots\ldots\ldots\ldots\ldots\ldots(4)$$

We have

$$\log y = v \log u, \quad \ldots\ldots\ldots\ldots\ldots\ldots\ldots(5)$$

$$\frac{1}{y}\frac{dy}{dx} = \frac{dv}{dx}\log u + \frac{v}{u}\frac{du}{dx}. \quad \ldots\ldots\ldots\ldots(6)$$

Hence

$$\frac{dy}{dx} = u^v \frac{dv}{dx}\log u + vu^{v-1}\frac{du}{dx}. \quad \ldots\ldots\ldots(7)$$

That is, we differentiate as if each of the functions u, v, in turn, were constant, and add the results.

Ex. 1.　If　　　　$y = \sqrt{\left\{\frac{(a+x)(b+x)}{(a-x)(b-x)}\right\}}$,

we have

$$\log y = \tfrac{1}{2}\log(a+x) + \tfrac{1}{2}\log(b+x) - \tfrac{1}{2}\log(a-x) - \tfrac{1}{2}\log(b-x).$$

Hence

$$\frac{1}{y}\frac{dy}{dx} = \frac{1}{2}\left\{\frac{1}{a+x} + \frac{1}{b+x} + \frac{1}{a-x} + \frac{1}{b-x}\right\}$$

$$= \frac{a}{a^2-x^2} + \frac{b}{b^2-x^2} = \frac{(a+b)(ab-x^2)}{(a^2-x^2)(b^2-x^2)},$$

$$\frac{dy}{dx} = \frac{(a+b)(ab-x^2)}{(a-x)^{\frac{3}{2}}(b-x)^{\frac{3}{2}}(a+x)^{\frac{1}{2}}(b+x)^{\frac{1}{2}}}.$$

Ex. 2.　If　　　　$y = x^x$,

we find　　　　$\dfrac{dy}{dx} = x^x(1 + \log x)$.

46.　The Inverse Hyperbolic Functions.

The inverse hyperbolic functions

$$\sinh^{-1}x, \quad \cosh^{-1}x, \quad \tanh^{-1}x, \ \&c.$$

are defined on the principle explained in Art. 16; thus the meaning of $y = \sinh^{-1} x$ is that

$$x = \sinh y, \quad \dots\dots\dots\dots\dots\dots(1)$$

and so on.

These functions can all be expressed in terms of the logarithmic function. Thus if

$$x = \sinh y = \tfrac{1}{2}(e^y - e^{-y}), \quad \dots\dots\dots\dots(2)$$

we have

$$e^{2y} - 2xe^y - 1 = 0. \quad \dots\dots\dots\dots(3)$$

Solving this quadratic in e^y, we find

$$e^y = x \pm \sqrt{(x^2 + 1)}. \quad \dots\dots\dots\dots(4)$$

If y is to be real, e^y must be positive, and the upper sign must be taken. Hence

$$\sinh^{-1} x = \log \{x + \sqrt{(x^2 + 1)}\}. \quad \dots\dots\dots\dots(5)$$

In a similar manner we should find that, if $x > 1$,

$$\cosh^{-1} x = \log \{x \pm \sqrt{(x^2 - 1)}\}. \quad \dots\dots\dots\dots(6)$$

Either sign is here admissible; the quantities $x \pm \sqrt{(x^2 - 1)}$ are reciprocals, and their logarithms differ simply in sign. It appears on sketching the graph of $\cosh^{-1} x$ that for every value of x which is > 1 there are two values of y, equal in magnitude, but opposite in sign.

Again, if

$$x = \tanh y = \frac{e^y - e^{-y}}{e^y + e^{-y}}, \quad \dots\dots\dots\dots(7)$$

we have

$$e^{2y} = \frac{1 + x}{1 - x}. \quad \dots\dots\dots\dots\dots(8)$$

Hence

$$\tanh^{-1} x = \tfrac{1}{2} \log \frac{1 + x}{1 - x}. \quad \dots\dots\dots\dots(9)$$

This is real only if $|x| < 1$.

Similarly, we find

$$\coth^{-1} x = \tfrac{1}{2} \log \frac{x + 1}{x - 1}, \quad \dots\dots\dots\dots(10)$$

which is real only if $|x| > 1$.

47. Differentiation of the Inverse Hyperbolic Functions.

1°. If

$$y = \sinh^{-1} x, \quad \dots\dots\dots\dots\dots\dots(1)$$

we have

$$x = \sinh y, \quad \frac{dx}{dy} = \cosh y = \sqrt{(x^2 + 1)},$$

and therefore $\qquad \dfrac{dy}{dx} = \dfrac{1}{\sqrt{(1 + x^2)}}$ . ...(2)

There is no ambiguity of sign, for cosh y is essentially positive.

2°.   If $\qquad\qquad y = \cosh^{-1} x,$ ....(3)

we have $\qquad x = \cosh y, \quad \dfrac{dx}{dy} = \sinh y = \pm \sqrt{(x^2 - 1)},$

whence $\qquad\qquad \dfrac{dy}{dx} = \pm \dfrac{1}{\sqrt{(x^2 - 1)}}.$ ....(4)

For any given value of x, greater than unity, there are two values of y, and for these dy/dx has opposite signs. [Cf. Fig. 25, p. 80, interchanging x and y.]

3°.   If $\qquad\qquad y = \tanh^{-1} x,$ ....(5)

we have $\qquad x = \tanh y, \quad \dfrac{dx}{dy} = \operatorname{sech}^2 y = 1 - x^2,$

and therefore $\qquad \dfrac{dy}{dx} = \dfrac{1}{1 - x^2}.$ ....(6)

This agrees with Art. 44, Ex. 3. It is to be noticed that y is real only when $x^2 < 1$. See Fig. 26, p. 81.

Similarly, if $\qquad\qquad y = \coth^{-1} x,$ ....(7)

we find $\qquad\qquad \dfrac{dy}{dx} = -\dfrac{1}{x^2 - 1},$ ....(8)

x^2 being necessarily > 1, if y is real.

EXAMPLES. XI.

1.   Prove by calculation from the series for e^x that

$1/e = \cdot367879, \quad \cosh 1 = 1\cdot5430806, \quad \sinh 1 = 1\cdot1752012.$

2.   Prove that

$\sqrt{e} = 1\cdot6487213, \quad 1/\sqrt{e} = \cdot6065307,$

$\cosh \tfrac{1}{2} = 1\cdot1276260, \quad \sinh \tfrac{1}{2} = \cdot5210953.$

3.   Prove that if $|a| < |b|$ the equation

$a \cosh x + b \sinh x = 0,$

has one, and only one, real root.

4. Draw the graphs of

$$\operatorname{cosech} x, \quad \coth x, \quad \coth x - \tanh x.$$

5. Shew that the function $\tanh(1/x)$ is discontinuous for $x = 0$. Draw a graph of the function.

6. If a and b be positive, and $a > b$, the function

$$\frac{ae^x + be^{-x}}{e^x + e^{-x}}$$

has the upper limit a and the lower limit b.

Prove the following formulæ :

7. $\cosh 2x = 2\cosh^2 x - 1 = 1 + 2\sinh^2 x.$

8. $\sinh 2x = \dfrac{2\tanh x}{1 - \tanh^2 x}, \quad \cosh 2x = \dfrac{1 + \tanh^2 x}{1 - \tanh^2 x},$

$$\tanh 2x = \frac{2\tanh x}{1 + \tanh^2 x}.$$

9. $\cosh^2 x \cos^2 x + \sinh^2 x \sin^2 x = \tfrac{1}{2}(\cosh 2x + \cos 2x),$

$\cosh^2 x \sin^2 x + \sinh^2 x \cos^2 x = \tfrac{1}{2}(\cosh 2x - \cos 2x).$

10. $\cosh^2 x \cos^2 x - \sinh^2 x \sin^2 x = \tfrac{1}{2}(1 + \cosh 2x \cos 2x),$

$\cosh^2 x \sin^2 x - \sinh^2 x \cos^2 x = \tfrac{1}{2}(1 - \cosh 2x \cos 2x).$

11. $\cosh^2 u + \sinh^2 v = \sinh^2 u + \cosh^2 v = \cosh(u + v)\cosh(u - v),$

$\cosh^2 u - \cosh^2 v = \sinh^2 u - \sinh^2 v = \sinh(u + v)\sinh(u - v).$

12. $\sinh u + \sinh v = 2\sinh\tfrac{1}{2}(u + v)\cosh\tfrac{1}{2}(u - v),$

$\sinh u - \sinh v = 2\cosh\tfrac{1}{2}(u + v)\sinh\tfrac{1}{2}(u - v),$

$\cosh u + \cosh v = 2\cosh\tfrac{1}{2}(u + v)\cosh\tfrac{1}{2}(u - v).$

$\cosh u - \cosh v = 2\sinh\tfrac{1}{2}(u + v)\sinh\tfrac{1}{2}(u - v).$

13. $$\tanh\tfrac{1}{2}u = \frac{\sinh u}{\cosh u + 1} = \frac{\cosh u - 1}{\sinh u}$$

$$= \sqrt{\left(\frac{\cosh u - 1}{\cosh u + 1}\right)}.$$

14. $\sinh 3u = 4\sinh^3 u + 3\sinh u,$

$\cosh 3u = 4\cosh^3 u - 3\cosh u.$

15. $1 + 2\cosh u + 2\cosh 2u + 2\cosh 3u + \ldots + 2\cosh nu$

$$= \frac{\sinh(n + \tfrac{1}{2})u}{\sinh\tfrac{1}{2}u}.$$

EXAMPLES. XII.

Verify the following differentiations:

1. $y = e^{x^2}$, $\qquad\qquad\qquad Dy = 2x e^{x^2}$.

2. $y = \dfrac{e^{x^2}}{x}$, $\qquad\qquad\qquad Dy = e^{x^2}\left(2 - \dfrac{1}{x^2}\right)$.

3. $y = e^{\sin x}$, $\qquad\qquad\qquad Dy = \cos x\, e^{\sin x}$.

4. $y = e^{\sin^2 x}$, $\qquad\qquad\qquad Dy = \sin 2x\, e^{\sin^2 x}$.

5. $y = e^{ax} \sin \beta x$, $\qquad\qquad Dy = e^{ax}(a \sin \beta x + \beta \cos \beta x)$.

6. $y = e^{ax} \cos \beta x$, $\qquad\qquad Dy = e^{ax}(a \cos \beta x - \beta \sin \beta x)$.

7. $y = x e^x$, $\qquad\qquad\qquad Dy = (x + 1)\, e^x$.

8. $y = x e^{-x}$, $\qquad\qquad\qquad Dy = (1 - x)\, e^{-x}$.

9. $y = x^m e^x$, $\qquad\qquad\qquad Dy = (x + m)\, x^{m-1}\, e^x$.

10. $y = e^x \sin x$, $\qquad\qquad\quad Dy = e^x(\sin x + \cos x)$.

11. $y = e^x \cos x$, $\qquad\qquad\quad Dy = e^x(\cos x - \sin x)$.

12. $y = \dfrac{e^x - 1}{e^x + 1}$, $\qquad\qquad Dy = \dfrac{2e^x}{(e^x + 1)^2}$.

13. $y = \sinh^2 x$, $\qquad\qquad\quad Dy = \sinh 2x$.

14. $y = \cosh^2 x$, $\qquad\qquad\quad Dy = \sinh 2x$.

15. $y = \tanh^2 x$, $\qquad\qquad\quad Dy = \dfrac{2 \sinh x}{\cosh^3 x}$.

16. $y = \sinh x + \tfrac{1}{3} \sinh^3 x$, $\quad Dy = \cosh^3 x$.

17. $y = \tanh x - \tfrac{1}{3} \tanh^3 x$, $\quad Dy = \operatorname{sech}^4 x$.

18. $y = \cosh x \cos x$ $\qquad\quad Dy = 2 \sinh x \cos x$.
$\quad\ + \sinh x \sin x$,

19. $y = \cosh x \sin x$ $\qquad\quad Dy = 2 \cosh x \cos x$.
$\quad\ + \sinh x \cos x$,

20. $y = \dfrac{\cosh x - \cos x}{\sinh x + \sin x}$, $\quad Dy = \dfrac{2 \sin x \sinh x}{(\sinh x + \sin x)^2}$.

21. $y = \tan^{-1} \dfrac{\sinh a \sin x}{1 + \cosh a \cos x}$ $\quad Dy = \dfrac{\sinh a}{\cosh a + \cos x}$.

EXAMPLES. XIII.

Verify the following differentiations:

1. $y = x \log x$, $\quad\quad\quad Dy = 1 + \log x$.

2. $y = x^m \log x$, $\quad\quad Dy = x^{m-1}(1 + m \log x)$.

3. $y = \log \sin x$, $\quad\quad Dy = \cot x$.

4. $y = \log \cos x$, $\quad\quad Dy = -\tan x$.

5. $y = \log \tan x$, $\quad\quad Dy = 2 \operatorname{cosec} 2x$.

6. $y = \log \sinh x$, $\quad\quad Dy = \coth x$.

7. $y = \log \cosh x$, $\quad\quad Dy = \tanh x$.

8. $y = \log \tanh x$, $\quad\quad Dy = 2 \operatorname{cosech} 2x$.

9. $y = \log \dfrac{x}{1+x}$, $\quad\quad Dy = \dfrac{1}{x(1+x)}$.

10. $y = \log \dfrac{1-x}{1+x}$, $\quad\quad Dy = -\dfrac{2}{1-x^2}$.

11. $y = \log \dfrac{x}{\sqrt{(x^2+1)}}$, $\quad\quad Dy = \dfrac{1}{x(x^2+1)}$.

12. $y = \log \dfrac{1+\sqrt{x}}{1-\sqrt{x}}$, $\quad\quad Dy = \dfrac{1}{(1-x)\sqrt{x}}$.

13. $y = \log\{\sqrt{(x+1)} + \sqrt{(x-1)}\}$, $\quad Dy = \dfrac{1}{2\sqrt{(x^2-1)}}$.

14. $y = \log \dfrac{x}{\sqrt{(x^2+1)}-x}$, $\quad Dy = \dfrac{1}{x} + \dfrac{1}{\sqrt{(x^2+1)}}$.

15. $y = \log(x-1) - \dfrac{2x-1}{(x-1)^2}$, $\quad Dy = \dfrac{x^2+1}{(x-1)^3}$.

16. $y = \log \dfrac{1+x+x^2}{1-x+x^2}$, $\quad Dy = \dfrac{2(1-x^2)}{1+x^2+x^4}$.

17. $y = \sqrt{\left(\dfrac{1+x^2}{1-x^2}\right)}$, $\quad Dy = \dfrac{2x}{(1+x^2)^{\frac{1}{2}}(1-x^2)^{\frac{3}{2}}}$.

18. $y = \dfrac{\sqrt{(1+x^2)}-x}{\sqrt{(1+x^2)}+x}$, $\quad Dy = \dfrac{-2}{\sqrt{(1+x^2)}\{\sqrt{(1+x^2)}+x\}^2}$.

19. $y = \dfrac{\sqrt{(1+x)}+\sqrt{(1-x)}}{\sqrt{(1+x)}-\sqrt{(1-x)}}$, $\quad Dy = -\dfrac{1}{x^2\sqrt{(1-x^2)}} - \dfrac{1}{x^2}$.

20. $y = \dfrac{\sqrt{(1+x^2)}+\sqrt{(1-x^2)}}{\sqrt{(1+x^2)}-\sqrt{(1-x^2)}}$, $\quad Dy = -\dfrac{2}{x^3\sqrt{(1-x^4)}} - \dfrac{2}{x^3}$.

21. $y = \log \dfrac{\sqrt{(x^2+1)}-1}{\sqrt{(x^2+1)}+1}$, $\quad Dy = \dfrac{2}{x\sqrt{(x^2+1)}}$.

EXAMPLES. XIV.

1. Draw the graphs of $\log_{10} \sin x$ and $\log_{10} \tan x$ from $x = 0$ to $x = \pi$.

2. Prove that if $f(x)$ be any rational integral function of x,

$$\lim_{x \to \infty} \frac{f(x)}{e^x} = 0.$$

3. Prove that if a be positive

$$\lim_{n \to \infty} \sqrt[n]{a} = 1.$$

4. Prove that

$$\lim_{x \to 0} \frac{\log(1 + x) - x}{x^2} = -\tfrac{1}{2}.$$

Prove the formulæ :

5. $\sinh^{-1} x = \cosh^{-1} \sqrt{(x^2 + 1)}, \quad \cosh^{-1} x = \sinh^{-1} \sqrt{(x^2 - 1)}.$

$\operatorname{sech}^{-1} x = \log \dfrac{1 + \sqrt{(1 - x^2)}}{x}, \quad \operatorname{cosech}^{-1} x = \log \dfrac{1 + \sqrt{(1 + x^2)}}{x}.$

6. Prove that $\qquad \tanh^{-1} \dfrac{x^2 - 1}{x^2 + 1} = \log x.$

7. Prove that the equation

$$\cosh^{-1} x + \cosh^{-1} y = m$$

represents a hyperbola ; and find its asymptotes.

$$[y = xe^m, \ y = xe^{-m}.]$$

Verify the following differentiations :

8. $y = \operatorname{sech}^{-1} x,$ $\qquad Dy = -\dfrac{1}{x \sqrt{(1 - x^2)}}.$

9. $y = \operatorname{cosech}^{-1} x,$ $\qquad Dy = -\dfrac{1}{x \sqrt{(1 + x^2)}}.$

10. $y = \sin^{-1}(\tanh x),$ $\qquad Dy = \operatorname{sech} x.$

11. $y = \tan^{-1}(\sinh x),$ $\qquad Dy = \operatorname{sech} x.$

12. $y = \tan^{-1}(\tanh \tfrac{1}{2} x),$ $\qquad Dy = \tfrac{1}{2} \operatorname{sech} x.$

13. $y = \tanh^{-1}(\tan \tfrac{1}{2} x),$ $\qquad Dy = \tfrac{1}{2} \sec x.$

14. $y = \tanh^{-1} \dfrac{x + a}{1 + ax},$ $\qquad Dy = \dfrac{1}{1 - x^2}.$

CHAPTER IV

APPLICATIONS OF THE DERIVED FUNCTION

48. Inferences from the sign of the Derived Function.

If $y = \phi(x)$, and if δx, δy be simultaneous increments of x and y, the limiting value of the ratio $\delta y/\delta x$ when δx is indefinitely diminished is, by definition, $\phi'(x)$. Hence, before the limit, we may write

$$\frac{\delta y}{\delta x} = \phi'(x) + \sigma. \quad \dots\dots\dots\dots\dots\dots(1)$$

where σ is an ultimately vanishing quantity.

A numerical example of the manner in which the ratio $\delta y/\delta x$ approximates to its limiting value may be of interest. We take the case of $y = \log_{10} x$, for the neighbourhood of $x = 1$. The limiting value is here

$$\frac{dy}{dx} = \frac{\mu}{x} = \cdot 43429. \dots$$

The numbers in the second column are taken from the printed tables.

δx	δy	$\delta y/\delta x$
·1000	·041393	·41393
·0500	·021189	·42379
·0100	·0043214	·43214
·0050	·0021661	·43321
·0010	·00043408	·43408
·0005	·00021709	·43419
·0001	·000043427	·43427

Let us first suppose that

$$\phi'(x) > 0.$$

Since the limiting value of σ is zero, we can by taking δx small enough ensure that

$$\phi'(x) + \sigma > 0.$$

That is, by (1), δy will have the same sign as δx for all admissible values of δx which are less in absolute value than a certain magnitude ϵ.

In the same way, if
$$\phi'(x) < 0,$$
δy will have the opposite sign to δx for all admissible values of δx which are less in absolute value than a certain quantity ϵ.

If the independent variable be represented geometrically as in Fig. 1, Art. 1, and if $x = OM$, where M is a point within the range considered, we may say that if $\phi'(x)$ be positive there is a certain interval to the right of M for every point of which the value of the function $\phi(x)$ is greater than its value at M, and a certain interval to the left of M at every point of which the value of the function is less than its value at M. If $\phi'(x)$ be negative, the words 'greater' and 'less' must be interchanged in this statement. When M is at the beginning or end of the range of x, the intervals referred to lie of course to the right or left of M, respectively.

It follows that if $\phi'(x)$ be positive over any finite range, the value of $\phi(x)$ will steadily increase with x throughout the range; i.e. if x_1 and x_2 be any two values of x belonging to the range, such that $x_2 > x_1$, then
$$\phi(x_2) > \phi(x_1).$$
For $\phi(x)$, being by hypothesis differentiable, and therefore continuous, must have (Art. 18) a greatest and a least value in the interval from x_1 to x_2 (inclusive). And the preceding argument shews that the *greatest* value cannot occur at the beginning of the interval, or in the interior; it must therefore occur at the end. Similarly the *least* value of $\phi(x)$ must occur at the beginning of the interval.

In the same way it appears that if $\phi'(x)$ be negative over any finite range, then $\phi(x)$ will steadily decrease as x increases, throughout this range; i.e. if x_1 and x_2 be any two values of x belonging to the range, such that $x_2 > x_1$, then
$$\phi(x_2) < \phi(x_1).$$

The geometrical meaning of these results is obvious. When the gradient of a curve is positive the ordinates increase with x; when the gradient is negative the ordinates decrease as x increases. The graphs of various functions given in Chapter I will serve as illustrations.

The converse statements that if $\phi(x)$ steadily increases with x throughout any range, $\phi'(x)$ cannot be negative for any value of x belonging to this range, and that, if $\phi(x)$ steadily decreases as x

increases, $\phi'(x)$ cannot be positive, follow immediately from the definition of $\phi'(x)$.

Again, even if $\phi'(x)$ vanish at a finite number of isolated points, provided it be elsewhere uniformly positive, $\phi(x)$ will steadily increase. Suppose, for example, that $\phi'(x_1) = 0$, and that with this exception $\phi'(x)$ is positive in the interval from $x = x_1$ to $x = x_2$, where $x_2 > x_1$. The least value of $\phi(x)$ cannot then occur within this interval, or at the upper extremity $(x = x_2)$. It must therefore occur at the lower extremity $(x = x_1)$. Hence

$$\phi(x_2) > \phi(x_1).$$

The same conclusion is arrived at if $\phi'(x)$ is positive from $x = x_1$ to $x = x_2$, where it vanishes.

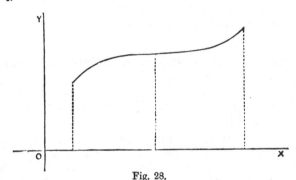

Fig. 28.

In the same way, if $\phi'(x)$ vanish at a finite number of isolated points, but is otherwise negative, $\phi(x)$ will steadily decrease.

Ex. 1. If $\qquad y = \cos x - (1 - \tfrac{1}{2}x^2),$

we have $\qquad \dfrac{dy}{dx} = x - \sin x,$

which is positive for positive values of x. Since y is an even function, and vanishes for $x = 0$, it follows that

$$1 > \cos x > 1 - \tfrac{1}{2}x^2.$$

Again if $\qquad y = \sin x - (x - \tfrac{1}{6}x^3),$

we have $\qquad \dfrac{dy}{dx} = \cos x - (1 - \tfrac{1}{2}x^2),$

which has been seen to be positive. Hence, since $y = 0$ for $x = 0$,

$$x > \sin x > x - \tfrac{1}{6}x^3,$$

for positive values of x. If x is negative the order of magnitude is reversed.

L. I. C. 7

Ex. 2. If $$y = \tan x - x,$$

we have $$\frac{dy}{dx} = \sec^2 x - 1 = \tan^2 x.$$

Hence dy/dx is positive, except for $x = 0$, π, 2π, ... Hence y steadily increases with x throughout any range which does not include one of the points of discontinuity $(x = \frac{1}{2}\pi, \frac{3}{2}\pi, ...)$.

It easily follows that the equation

$$\tan x - x = 0$$

has no root *between* 0 and $\frac{1}{2}\pi$; one, and only one, root between $\frac{1}{2}\pi$ and $\frac{3}{2}\pi$; and so on.

These results may be verified by a graphical construction. If we draw the lines

$$y = \tan x, \quad y = x,$$

their intersections will determine the values of x which make

$$\tan x = x.$$

49. The Derivative vanishes in the interval between two equal values of the Function.

If $\phi(x)$ vanish for $x = a$ and $x = b$, and if $\phi'(x)$ be finite for all values of x between a and b, then $\phi'(x)$ will vanish for some value of x *between* a and b.

For, either $\phi(x)$ is constantly zero throughout the interval from a to b, or it will have (Art. 18) a greatest or a least value for some value (x_1) of x within this interval. In the former case we shall have $\phi'(x) = 0$ throughout the interval; in the latter case $\phi'(x_1)$ cannot be either positive or negative (Art. 48) and must therefore vanish, since it is by hypothesis finite.

The geometrical statement of this theorem is that if a curve meets the axis of x at two points, and if the gradient is everywhere finite, there must be at least one intervening point at which the tangent is parallel to the axis of x. See, for example, the graph of $\sin x$ on p. 28; also Fig. 9, p. 24.

It is to be carefully noticed that, in the above argument, the conditions that $\phi(x)$ and $\phi'(x)$ should each have a definite (and therefore finite) value throughout the interval from $x = a$ to $x = b$ are essential. The annexed figures exhibit various cases where the

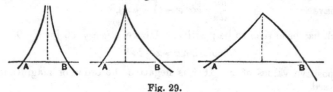

Fig. 29.

conclusion does not hold, owing to the violation of one or other of these conditions.

A slightly more general form of the theorem of this Art. is that if $\phi(x)$ has the same value (β) for $x = a$ and $x = b$, then under the same conditions as to the continuity and finiteness of $\phi(x)$ and $\phi'(x)$, the derived function $\phi'(x)$ will vanish for some intermediate value of x. This follows by the same argument, applied now to the function $\phi(x) - \beta$.

Ex. 1. If $\qquad \phi(x) = (x - a)(x - b),$

we have $\qquad \phi'(x) = 2x - (a + b).$

Hence $\phi'(x)$ vanishes for $x = \frac{1}{2}(a + b)$, which lies between a and b.

Ex. 2. If $\qquad \phi(x) = \dfrac{\sin x}{x},$

we have $\qquad \phi'(x) = \dfrac{x \cos x - \sin x}{x^2}.$

Here $\phi(x) = 0$ for $x = \pi$ and $x = 2\pi$; hence $\phi'(x)$ must vanish for some intermediate value of x. This is in agreement with Art. 48, Ex. 2, where it was shewn that the equation $x = \tan x$ has a root between π and $\frac{3}{2}\pi$.

50. Application to the Theory of Equations.

If $\phi(x)$ be a rational integral function of x, then $\phi(x)$ and its derivative $\phi'(x)$ are both of them continuous (and finite) for all finite values of x. Hence at least one real root of the equation

$$\phi'(x) = 0 \qquad \dots\dots\dots\dots\dots\dots(1)$$

will lie between any two real roots of

$$\phi(x) = 0. \qquad \dots\dots\dots\dots\dots\dots(2)$$

This result, which is known as 'Rolle's Theorem,' is important in the Theory of Equations. It is an immediate consequence that at most one real root of (2) lies between any two *consecutive* roots of (1). That is, the roots of (1) separate those of (2).

Ex. 1. If $\phi(x) = 4x^3 - 21x^2 + 18x + 20,$

we have $\qquad \phi'(x) = 12x^2 - 42x + 18 = 6(2x - 1)(x - 3).$

Hence the real roots of $\phi(x) = 0$, if any, will lie in the intervals between $-\infty$ and $\frac{1}{2}$, $\frac{1}{2}$ and 3, 3 and $+\infty$, respectively. Now, for

$$x = -\infty, \quad \tfrac{1}{2}, \quad 3, \quad +\infty,$$

the signs of $\phi(x)$ are $\qquad -, \quad +, \quad -, \quad +,$

respectively, so that $\phi(x)$ must in fact vanish once (by Art. 9) in each of the above intervals. Hence there are three real roots. The figure on the next page shews the graph of $\phi(x)$.

If by continuous modification of the form of $\phi(x)$, for example by the addition or subtraction of a constant, two roots are made to coalesce, the root of $\phi'(x) = 0$ which lies between must coalesce with them. Hence a double root of $\phi(x) = 0$ is also a root of $\phi'(x) = 0$.

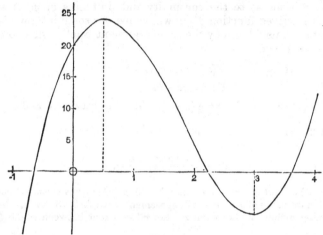

Fig. 30.

More generally, an r-fold root of $\phi(x) = 0$ being regarded as due to the coalescence of r distinct roots, the equation $\phi'(x) = 0$ will have $r - 1$ intervening roots which coalesce.

This suggests a method of ascertaining the multiple roots, if any, of a proposed algebraic equation. If α be an r-fold root of $\phi(x)$, we have

$$\phi(x) = (x - \alpha)^r \chi(x), \quad \dots\dots\dots\dots\dots\dots(3)$$

where $\chi(x)$ is a rational integral function. Hence

$$\phi'(x) = (x - \alpha)^{r-1} \{r\chi(x) + (x - \alpha)\chi'(x)\}; \quad \dots\dots\dots(4)$$

i.e. $(x - \alpha)^{r-1}$ will be a common factor of $\phi(x)$ and $\phi'(x)$. And it is easily seen that $(x - \alpha)^{r-1}$ will not be a common factor unless $\phi(x)$ is divisible by $(x - \alpha)^r$. Hence the multiple roots of $\phi(x)$, if any, are to be detected by finding the common factors of $\phi(x)$ and $\phi'(x)$ by the usual algebraical process.

Ex. 2. If $\qquad \phi(x) = x^4 - 9x^2 + 4x + 12,$

we have $\qquad \phi'(x) = 4x^3 - 18x + 4.$

The usual test leads to the conclusion that $x - 2$ is a common factor of $\phi(x)$ and $\phi'(x)$; whence we infer that $(x - 2)^2$ is a factor of $\phi(x)$. The remaining factors are then easily ascertained; thus we find

$$\phi(x) = (x - 2)^2 (x + 1) (x + 3).$$

Ex. 3.   To find the condition that the cubic

$$x^3 + qx + r = 0 \dots\dots\dots\dots\dots\dots\dots(5)$$

should have a double root.

The double root, if it exists, must satisfy

$$3x^2 + q = 0 \quad \text{or} \quad x = \pm \sqrt{(-\tfrac{1}{3}q)}. \quad \dots\dots\dots\dots(6)$$

Substituting in (5), we find

$$r = \pm \tfrac{2}{3} \sqrt{(-\tfrac{1}{3}q^3)}, \quad \text{or} \quad r^2 = -\tfrac{4}{27}q^3, \dots\dots\dots\dots(7)$$

which is the required condition.

51. Maxima and Minima.

A 'maximum' value of a continuous function is one which is greater, and a 'minimum' value is one which is less, than the values in the immediate neighbourhood, on either side.

More precisely, the function $\phi(x)$ is a maximum for $x = x_1$, if two positive quantities, ϵ and ϵ', can be found such that $\phi(x_1)$ is greater than the value which $\phi(x)$ assumes for any other value of x in the interval from $x = x_1 - \epsilon$ to $x = x_1 + \epsilon'$. Similarly for a minimum.

Since the comparison is made with values of the function in the immediate neighbourhood only of x_1, a maximum is not necessarily the *greatest*, nor a minimum the *least*, of all the values of the function. See Fig. 30.

We will limit ourselves for the present to the case, which includes all the more important applications, where $\phi(x)$ has a determinate and finite derivative at all points of the range considered. The argument of Art. 48 then shews that if $\phi(x_1)$ be a maximum or minimum, $\phi'(x_1)$ cannot differ from zero. For if it be either positive or negative, there will be points in the immediate neighbourhood of x_1 for which $\phi(x)$ will be greater, and others for which it will be less, than $\phi(x_1)$. Hence, in the case supposed, a first condition for a maximum or minimum value of $\phi(x)$ is that $\phi'(x)$ should vanish.

This condition is necessary, but it is not sufficient. To investigate the matter further, we will suppose that on each side of the point x_1 there is a certain interval throughout which $\phi'(x)$ is altogether positive or altogether negative*. Now if $\phi'(x)$ be positive for all values of x between $x_1 - \epsilon$ and x_1, $\phi(x)$ will (Art. 48) steadily increase throughout the interval thus defined; and if

---

* That is, we exclude cases where $\phi'(x)$ changes sign an infinite number of times within any interval including x_1, however short. The point $x = 0$ in the function $x^2 \sin 1/x$ is an instance.

$\phi'(x)$ be negative for all values of x between x_1 and $x_1 + \epsilon'$, $\phi(x)$ will steadily decrease throughout the corresponding interval. Hence if both these conditions hold, $\phi(x_1)$ is a maximum. And it is evident that if the signs be otherwise, $\phi(x_1)$ cannot be the greatest value which the function assumes within the interval extending from $x_1 - \epsilon$ to $x_1 + \epsilon'$.

We may express this shortly by saying that the necessary and sufficient condition in order that $\phi(x_1)$ may be a maximum value of $\phi(x)$ is that $\phi'(x)$ should change sign from $+$ to $-$ as x increases through the value x_1.

In the same way we find that the necessary and sufficient condition in order that $\phi(x_1)$ may be a minimum value of $\phi(x)$ is that $\phi'(x)$ should change sign from $-$ to $+$ as x increases through the value x_1.

In geometrical language, when the ordinate of a curve is a maximum the gradient must change from positive to negative; when the ordinate is a minimum the gradient must change from negative to positive. This is abundantly illustrated in our diagrams; see, for example, Figs. 9, 13, 14, 30.

Whenever the derived function $\phi'(x)$ vanishes, the rate of increase (Art. 26) of the original function $\phi(x)$ is momentarily zero, and the value of $\phi(x)$ is said to be 'stationary.' As already stated, a stationary value is not necessarily a maximum or minimum, for cases may of course occur in which $\phi'(x)$ vanishes without changing sign.

In most cases of interest, the derived function $\phi'(x)$ is continuous as well as determinate (and finite). It can then only change sign by passing through the value zero; and it is further evident from Art. 9 that the changes (if there are more than one) will take place from $+$ to $-$, and from $-$ to $+$, alternately. The maxima and minima will therefore occur alternately. See Fig. 14, p. 28.

Ex. 1. The distance (s), from an arbitrary origin, of a point moving in a straight line is a maximum when the velocity (ds/dt) changes from positive to negative, and is a minimum when the velocity changes from negative to positive.

Thus, in the case of a particle moving upwards under gravity, we have

$$ s = ut - \tfrac{1}{2}gt^2, \quad \frac{ds}{dt} = u - gt. $$

Hence ds/dt changes from positive to negative as t increases through the value u/g. The altitude (s) is therefore then a maximum.

Ex. 2. To find the rectangle of greatest area having a given perimeter.

Denoting the perimeter by $2a$, the lengths of two adjacent sides may be taken to be x and $a - x$; hence we have to find the maximum value of the function

$$x(a - x). \qquad \dots\dots\dots\dots\dots\dots\dots\dots(1)$$

The derivative of this is $a - 2x$, which changes sign from + to − as x increases through the value $\frac{1}{2}a$. The rectangle of greatest area is therefore a square.

Ex. 3. To find the maxima and minima of the function

$$\phi(x) \equiv 4x^3 - 21x^2 + 18x + 20. \qquad \dots\dots\dots\dots\dots(2)$$

We have $\qquad \phi'(x) = 12(x - \tfrac{1}{2})(x - 3). \qquad \dots\dots\dots\dots\dots(3)$

This can only change sign when x passes through the values $\frac{1}{2}$ and 3. Now when x is a little less than $\frac{1}{2}$, the signs of the second and third factors are −, −; whilst when x is a little greater than $\frac{1}{2}$ they are +, −. Hence as x increases through the value $\frac{1}{2}$, $\phi'(x)$ changes sign from + to −. In a similar manner we find that as x increases through the value 3, $\phi'(x)$ changes sign from − to +. Hence $\phi(x)$ is a maximum when $x = \frac{1}{2}$, and a minimum when $x = 3$. If we substitute in (2) we find that the maximum value is $24\frac{1}{4}$, and the minimum value -7. See Fig. 30, p. 100.

Ex. 4. If $\qquad \phi(x) = \dfrac{2x}{1 + x^2}, \qquad \dots\dots\dots\dots\dots\dots(4)$

we find $\qquad \phi'(x) = \dfrac{2(1 - x^2)}{(1 + x^2)^2}. \qquad \dots\dots\dots\dots\dots(5)$

This can only change sign for $x = \pm 1$. As x increases (algebraically) through the value -1, $1 - x^2$ changes sign from − to +. As x increases through $+1$, $1 - x^2$ changes sign from + to −. Hence for $x = -1$ we have a minimum value -1 of $\phi(x)$, and for $x = 1$ a maximum value 1. See Fig. 13, p. 27.

Ex. 5. If $\qquad \phi(x) = \dfrac{2x}{1 - x^2}, \qquad \dots\dots\dots\dots\dots\dots(6)$

we have $\qquad \phi'(x) = \dfrac{2(1 + x^2)}{(1 - x^2)^2}. \qquad \dots\dots\dots\dots\dots(7)$

Here $\phi'(x)$ is always positive, and the function $\phi(x)$ has no finite maxima or minima. See Fig. 11, p. 26.

Ex. 6. To find the right circular cylinder of least surface for a given volume.

If x denote the radius and y the altitude, the surface is

$$2\pi x^2 + 2\pi xy,$$

and if the given volume be $2\pi a^3$, we have

$$x^2 y = 2a^3.$$

Hence, eliminating y, the expression to be made a minimum is

$$x^2 + \frac{2a^3}{x},$$

the derived function of which is

$$2\left(x - \frac{a^3}{x^2}\right).$$

This changes sign as x increases through the value a, and the change is from $-$ to $+$. Hence $x = a$ makes the surface a minimum; and since y then $= 2a$, the height of the cylinder is equal to its diameter.

The reader may verify that with these proportions the surface is 1·1447... of that of a sphere of equal volume.

Ex. 7. To find the stationary values of the function

$$u = \frac{Ax^2 + 2Hx + B}{ax^2 + 2hx + b}. \qquad\qquad\dots\dots\dots\dots\dots(8)$$

We have $\qquad (ax^2 + 2hx + b)\,u = Ax^2 + 2Hx + B. \qquad\dots\dots\dots\dots(9)$

Differentiating, and putting $du/dx = 0$, we have

$$(ax + h)\,u = Ax + H. \qquad\qquad\dots\dots\dots\dots\dots(10)$$

Multiplying this by x, and subtracting from (9),

$$(hx + b)\,u = Hx + B. \qquad\qquad\dots\dots\dots\dots\dots(11)$$

Eliminating u between (10) and (11), the required values of x are given by the roots of the quadratic

$$(aH - Ah)\,x^2 + (aB - Ab)\,x + (hB - Hb) = 0. \qquad\dots\dots\dots(12)$$

If, on the other hand, we eliminate x, the stationary values of u are given by the quadratic

$$(ab - h^2)\,u^2 - (aB + bA - 2hH)\,u + AB - H^2 = 0. \qquad\dots\dots(13)$$

Ex. 8. The simplest instance of a stationary value which is not a maximum or minimum is furnished by the function

$$\phi(x) \equiv x^3. \qquad\qquad\dots\dots\dots\dots\dots\dots(14)$$

This makes $\phi'(x) = 3x^2$, which vanishes, but does not change sign, as x increases through the value 0. Hence $\phi(x)$, though 'stationary,' is not a maximum or minimum for $x = 0$. Fig. 31 shews the graph of x^3.

It may occasionally happen that $\phi'(x)$, though generally continuous, becomes discontinuous for some isolated value of x; and if the discontinuity be accompanied by a change of sign as x

increases through the value in question, we shall have a maximum or minimum, by the same argument as before.

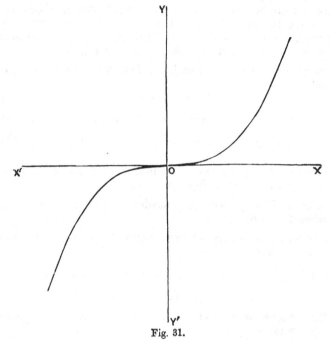

Fig. 31.

Ex. 9.   If $\phi(x) = a^{\frac{1}{3}} x^{\frac{2}{3}},$ ....(15)

we have $\phi'(x) = \dfrac{2}{3} \left(\dfrac{a}{x}\right)^{\frac{1}{3}}.$ ....(16)

As x increases through the value 0, this changes from $-\infty$ to $+\infty$. Hence $\phi(x)$ is a minimum for $x = 0$.   See Fig. 32.

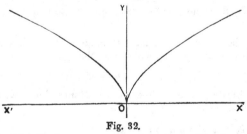

Fig. 32.

Again, in Fig. 29 there occurs a point where $\phi'(x)$ is discontinuous, passing abruptly from a finite positive to a finite negative value. The ordinate is then a maximum.

52. Algebraical Methods.

It is to be noticed that some important problems of maxima and minima can be solved by elementary algebraical methods, without recourse to the Calculus. This is especially the case with questions involving quadratic expressions. These are all easily treated by the method of 'completing the square.'

Again, the solution can often be made to depend upon identities such as

$$xy = \tfrac{1}{4}\{(x+y)^2 - (x-y)^2\}, \quad \dots\dots\dots(1)$$

$$(x+y)^2 = (x-y)^2 + 4xy, \quad \dots\dots\dots\dots(2)$$

$$x^2 + y^2 = \tfrac{1}{2}\{(x+y)^2 + (x-y)^2\}. \quad \dots\dots\dots(3)$$

Thus:

The product (xy) of two positive magnitudes, whose sum $(x+y)$ is given, is greatest when they are equal;

The sum of two positive magnitudes whose product is given is least when they are equal;

The sum of the squares of two magnitudes whose sum is given is least when they are equal.

Ex. 1.   Thus, in the problem of Ex. 2, Art. 51, we have

$$x(a-x) = \tfrac{1}{4}a^2 - (x - \tfrac{1}{2}a)^2.$$

Since the last term cannot fall below zero, this expression has its greatest value $(\tfrac{1}{4}a^2)$ when $x = \tfrac{1}{2}a$.

Ex. 2.   The expression $2x^2 - 3x + 2$,

may be put in the form

$$2(a^2 - \tfrac{3}{2}x + 1) = 2(x - \tfrac{3}{4})^2 + \tfrac{7}{8}.$$

Hence the expression has the minimum value $\tfrac{7}{8}$, corresponding to $x = \tfrac{3}{4}$.

Ex. 3.   To find the greatest rectangle which can be inscribed in a given circle.

If $2x$, $2y$ be the sides, we have to make xy a maximum subject to the condition that $x^2 + y^2 = a^2$, where a is the radius of the circle. Now

$$2xy = x^2 + y^2 - (x-y)^2 = a^2 - (x-y)^2,$$

which is obviously greatest when $x = y$. Hence the greatest inscribed rectangle is a square.

Ex. 4.   To find the minimum value of

$$a\cot\theta + b\tan\theta,$$

for values of θ between 0 and $\tfrac{1}{2}\pi$.

The product of $a \cot \theta$ and $b \tan \theta$ is constant, hence their sum is least when they are equal, *i.e.* when

$$\tan \theta = (a/b)^{\frac{1}{2}}.$$

The minimum value of the sum is therefore $2a^{\frac{1}{2}}b^{\frac{1}{2}}$.

Ex. 5.   To find the greatest cylinder which can be inscribed in a frustum of a paraboloid of revolution cut off by a plane perpendicular to the axis.

Supposing the paraboloid to be generated by the revolution of the curve

$$y^2 = 4ax, \quad \dots\dots\dots\dots\dots\dots\dots(4)$$

about the axis of x, then if h be the length of the axis, and x the abscissa of the end of the cylinder nearest the origin, the volume of the cylinder is

$$\pi y^2 (h - x) = \pi y^2 \left(h - \frac{y^2}{4a} \right). \quad \dots\dots\dots\dots\dots(5)$$

Now the sum of the quantities y^2 and $4ah - y^2$ is constant; their product is therefore greatest when they are equal, *i.e.* when

$$y^2 = 2ah, \text{ or } x = \tfrac{1}{2}h. \quad \dots\dots\dots\dots\dots\dots(6)$$

The height of the cylinder is therefore one-half that of the frustum.

53.   Maxima and Minima of Functions of several Variables.

We give a few indications concerning the extension of some of the preceding results to functions of two or more independent variables.

In the first place let us seek for the maxima and minima of a function

$$u = \phi(x, y). \quad \dots\dots\dots\dots\dots\dots\dots(1)$$

A first condition is that we must have simultaneously

$$\frac{\partial \phi}{\partial x} = 0, \quad \frac{\partial \phi}{\partial y} = 0, \quad \dots\dots\dots\dots\dots(2)$$

where the differential coefficients are 'partial,' as in Art. 34. For if u be greater (or less) than any other value of the function obtained by varying x, y within certain limits, u will *à fortiori* be a maximum (or minimum) when y is kept constant and x alone is varied. This requires in general (Art. 51) that $\partial \phi / \partial x = 0$. Similarly, u must be a maximum (or minimum) when x is kept constant and y alone varies; this requires that $\partial \phi / \partial y = 0$.

As before, these conditions, though necessary, are not sufficient. The further examination of the question, in its general form, is postponed till Chapter XVI; but it often happens that the existence of maxima and minima can be inferred, and the discrimination between them effected, by independent considerations. The conditions (2) then supply all that is analytically necessary.

Ex. To find the rectangular parallelepiped of least surface for a given volume.

Let x, y, z be the edges, and a^3 the given volume. Since

$$xyz = a^3, \quad \dots\dots\dots\dots\dots\dots\dots(3)$$

the function to be made a minimum is

$$u = xy + yz + zx = xy + \frac{a^3}{x} + \frac{a^3}{y}. \dots\dots\dots\dots(4)$$

The conditions $\partial u/\partial x = 0$, $\partial u/\partial y = 0$ give

$$x^2 y = a^3, \quad xy^2 = a^3,$$

the only real solution of which is $x = y = a$, whence, also, $z = a$.

It appears from (4) that, x and y being essentially positive in this problem, there is a lower limit to the surface of the parallelepiped. And the above investigation shews that this limit is not attained unless the figure be a cube.

As in Art. 52, the solutions of various problems can be deduced from known algebraical identities, such as

$$x^2 + y^2 + z^2 = \tfrac{1}{3}\{(x+y+z)^2 + (y-z)^2 + (z-x)^2 + (x-y)^2\}, \dots(5)$$

$$yz + zx + xy = x^2 + y^2 + z^2 - \tfrac{1}{2}\{(y-z)^2 + (z-x)^2 + (x-y)^2\}. \dots(6)$$

Thus:

If a straight line be divided into three segments, the sum of the squares on these is least when the segments are equal;

The surface of a parallelepiped inscribed in a given sphere $(x^2 + y^2 + z^2 = a^2)$ is greatest when the figure is a cube.

54. Notation of Differentials.

We return to the equation

$$\frac{\delta y}{\delta x} = \phi'(x) + \sigma \dots\dots\dots\dots\dots\dots(1)$$

of Art. 48. This is equivalent to

$$\delta y = \phi'(x)\,\delta x + \sigma\delta x. \dots\dots\dots\dots(2)$$

As δx approaches the value 0, the second term on the right hand becomes more and more insignificant compared with the first, since the limiting value of σ is zero. Hence it becomes more and more nearly true that

$$\delta y = \phi'(x)\,\delta x, \quad \dots\dots\dots\dots\dots(3)$$

not in the sense that both sides ultimately vanish, but in the sense that the *ratio* of the two sides approaches the value unity. In this artificial sense, the last equation is often written in the form

$$dy = \phi'(x)\,dx. \quad \dots\dots\dots\dots\dots(4)$$

The vanishing quantities dx, dy are called 'differentials.' *

The student need not take exception to the above mode of expression, which is purely conventional. Its use is simply to express the fact that in calculations involving the quantities δx and δy, which are afterwards made to approach the limit zero, we may at any stage replace δy by $\phi'(x)\,\delta x$, whenever it is plain that the omission of quantities of the second order will make no difference to the accuracy of the final result.

55. Calculation of Small Corrections.

The equation $\qquad \delta y = \phi'(x)\,\delta x \dots\dots\dots\dots\dots(1)$

may, moreover, be employed as an *approximate* formula to find the effect on the value of a function of a small change in the independent variable, since (as we have seen) the outstanding error will be merely a small fraction of $\phi'(x)\,\delta x$ provided δx be sufficiently small. An important practical application is to find the error, or the uncertainty, in a numerical result deduced from given data, owing to given errors or uncertainties in the data.

The above method is defective in one respect, in that there is no indication of the magnitude of the error involved in the approximation. This is supplied, however, by a theorem to be proved in Art. 56. It is there shewn that

$$\delta y = \phi'(x + \theta\,\delta x)\,\delta x, \quad \dots\dots\dots\dots(2)$$

where θ is some quantity between 0 and 1. Hence if A and B be the greatest and least values which the derived function assumes in the interval from x to $x + \delta x$, the error committed in (1) cannot be greater than $|(A - B)\,\delta x|$.

Ex. 1. To calculate the difference for one minute in a table of log sines.

If $y = \log_{10}\sin x$, we have $dy/dx = \mu \cot x$,

and $\qquad\qquad\qquad \delta y = \mu \cot x\,\delta x,$

---

* It is on account of the position which it occupies in the formula (4) that $\phi'(x)$ received the name 'differential coefficient.'

approximately, provided δx be expressed in circular measure. Putting

$$\delta x = \text{circular measure of } 1' = \frac{\pi}{10800} = \cdot 0002909,$$

we find $$\delta y = \cdot 0001263 \times \cot x.$$

The numerical factor agrees with the difference for $1'$, in the neighbourhood of $45°$, as given in the tables.

Ex. 2. Two sides a, b of a triangle and the included angle C are measured; to find the error in the computed length of the third side c due to a small error in the angle.

We have $$c^2 = a^2 + b^2 - 2ab \cos C, \dots\dots\dots\dots\dots\dots(3)$$

and therefore, supposing C and c alone to vary,

$$c\delta c = ab \sin C \delta C,$$

whence $$\delta c = \frac{ab}{c} \sin C \delta C = a \sin B \delta C. \quad\dots\dots\dots\dots(4)$$

This result may also be obtained geometrically; thus, if in the figure $\angle BCB' = \delta C$, and BN be drawn perpendicular to AB', we have, ultimately,

$$\delta c = B'N = BB' \cos BB'N = a\delta C . \sin CB'A = a\delta C . \sin B,$$

neglecting small quantities of the second order.

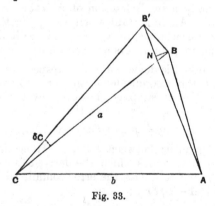

Fig. 33.

Again, to find the error in c due to a small error in the measured length of a, we have, on the hypothesis that a and c alone vary,

$$c\delta c = (a - b \cos C) \,\delta a = c \cos B \,\delta a,$$

or $$\delta c = \cos B \,\delta a, \quad\dots\dots\dots\dots\dots\dots\dots\dots\dots\dots(5)$$

a result which, like the former, admits of easy geometrical proof.

56.   Mean-Value Theorem.   Consequences.

The following very important theorem is an extension of that given in Art. 49.

If a function $\phi(x)$ be continuous, and have a determinate derivative, throughout the interval from $x = a$ to $x = b$, then

$$\frac{\phi(b) - \phi(a)}{b - a} = \phi'(x_1), \quad\dotsc\dotsc\dotsc\dotsc\dotsc(1)$$

where x_1 is some value of x *between* a and b.

Consider the function

$$\phi(x) - \phi(a) - \frac{\phi(b) - \phi(a)}{b - a}(x - a). \quad\dotsc\dotsc\dotsc\dotsc(2)$$

This is, under the conditions stated, continuous from $x = a$ to $x = b$, and it obviously vanishes for each of these values of x. Hence its derived function

$$\phi'(x) - \frac{\phi(b) - \phi(a)}{b - a} \quad\dotsc\dotsc\dotsc\dotsc\dotsc(3)$$

must vanish for some value (x_1, say) of x between a and b. This proves the statement (1).

The meaning of this result, and the nature of the proof, should be studied. The geometrical interpretation is as follows. In the annexed figure, we have

$$OA = a, \qquad OB = b,$$
$$PA = \phi(a), \qquad QB = \phi(b),$$

and

$$\tan QPN = \frac{QN}{PN} = \frac{\phi(b) - \phi(a)}{b - a}. \quad\dotsc\dotsc\dotsc\dotsc(4)$$

The theorem therefore asserts that (under the restrictions stated) there is some point between P and Q where the tangent to the curve $y = \phi(x)$ is parallel to the chord PQ.

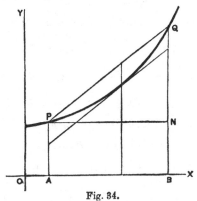

Fig. 34.

The equation of the chord PQ is

$$y = \phi(a) + \frac{\phi(b) - \phi(a)}{b - a}(x - a), \quad \dots\dots\dots\dots(5)$$

as is easily verified, and the expression (2) therefore measures the difference between the ordinate of the curve and that of the chord. This difference vanishes at P and Q, so that there must be one point at least between P and Q at which it is a maximum or a minimum.

Ex. If $\phi(x) = x^2$,

we have $\dfrac{\phi(b) - \phi(a)}{b - a} = b + a$,

which is equal to the value of $\phi'(x)$ for $x = \frac{1}{2}(a + b)$.

This is equivalent to the statement that any chord of a parabola is parallel to the tangent at the extremity of that diameter which bisects the chord.

The fraction $\dfrac{\phi(b) - \phi(a)}{b - a}, \quad\dots\dots\dots\dots\dots\dots(6)$

that is, the ratio of the increment of the function to that of the independent variable, measures what may be called the 'mean rate of increase' of the function in the interval $b - a$. Hence the theorem expresses that, under the conditions stated, the mean rate of increase in any interval is equal to the actual rate of increase at some point within the interval.

For instance, the mean velocity of a moving point in any interval of time is equal to the actual velocity at some instant within the interval.

Some other modes of stating the result (1) are to be noticed. The fact that x_1 lies between a and b may be expressed by putting

$$x_1 = a + \theta(b - a), \quad\dots\dots\dots\dots\dots\dots(7)$$

where θ stands for 'some quantity between 0 and 1.' The precise value of θ will in general depend on the values of a and b. If we further write $a + h$ for b, we get the very useful form

$$\frac{\phi(a + h) - \phi(a)}{h} = \phi'(a + \theta h), \quad\dots\dots\dots\dots(8)$$

or $$\phi(a + h) = \phi(a) + h\phi'(a + \theta h). \quad\dots\dots\dots\dots(9)$$

Again, if we write x for a, and δx for h, we have

$$\delta\phi(x) = \phi'(x + \theta \delta x) . \delta x. \quad\dots\dots\dots\dots(10)$$

A very important deduction from the preceding theorem is that if

$$\phi'(x) = 0 \quad\dots\dots\dots\dots\quad\dots\dots\dots\dots(11)$$

for *all* values of x within a certain range, then $\phi(x)$ must be constant throughout that range.

For if $\phi(x)$ vary, let a and b be two values of x for which it has unequal values. The fraction

$$\frac{\phi(b) - \phi(a)}{b - a} \qquad (12)$$

will then be different from zero, and there will therefore be some intermediate value of x for which $\phi'(x)$ will differ from zero, contrary to the hypothesis.

Moreover, if two functions $\phi(x)$ and $\psi(x)$ have equal derivatives for all values of x within a certain range, they can only differ by a constant. For, by hypothesis,

$$\phi'(x) - \psi'(x) = 0, \qquad (13)$$

or $$\frac{d}{dx} \{\phi(x) - \psi(x)\} = 0. \qquad (14)$$

Hence $$\phi(x) - \psi(x) = \text{const.}, \qquad (15)$$

by the preceding case.

If in place of (2) we consider the more general function

$$\phi(x) - \phi(a) - \frac{\phi(b) - \phi(a)}{\psi(b) - \psi(a)} \{\psi(x) - \psi(a)\}, \qquad (16)$$

we infer that under analogous conditions its derived function

$$\phi'(x) - \frac{\phi(b) - \phi(a)}{\psi(b) - \psi(a)} \psi'(x) \qquad (17)$$

will vanish for some value of x between a and b. The result may be written

$$\frac{\phi(a + h) - \phi(a)}{\psi(a + h) - \psi(a)} = \frac{\phi'(a + \theta h)}{\psi'(a + \theta h)}. \qquad (18)$$

Hence if $$\phi(a) = 0, \quad \psi(a) = 0, \qquad (19)$$

we have $$\lim_{h \to 0} \frac{\phi(a + h)}{\psi(a + h)} = \lim_{h \to 0} \frac{\phi'(a + h)}{\psi'(a + h)}. \qquad (20)$$

This is sometimes useful in evaluating the 'indeterminate form' $\frac{0}{0}$.

57. Total Variation of a Function of several Variables.

Let $$u = \phi(x, y) \qquad (1)$$

be a continuous function of x and y, and further let us suppose that the partial derivatives

$$\frac{\partial u}{\partial x}, \quad \frac{\partial u}{\partial y} \qquad (2)$$

are also continuous functions of x and y.

Let δu be the increment of u due to increments δx and δy of the independent variables; *i.e.*

$$\delta u = \phi\,(x + \delta x,\, y + \delta y) - \phi\,(x,\, y). \quad\ldots\ldots\ldots\ldots(3)$$

In the geometrical representation (Art. 34), δu is the difference of altitude of the two points of a surface which correspond to the two points $(x,\, y)$ and $(x + \delta x,\, y + \delta y)$ of the horizontal plane xy.

Now if x alone were varied, the corresponding increment of u would, by Art. 56 (10), be of the form

$$P\delta x, \quad\ldots\ldots\ldots\ldots\ldots\ldots\ldots\ldots(4)$$

where P is a certain function of x, y, and δx. And it appears from the same Art., and from the meaning of a partial derivative, that the limiting value of P when δx is indefinitely diminished is

$$P_0 = \frac{\partial u}{\partial x}. \quad\ldots\ldots\ldots\ldots\ldots\ldots\ldots\ldots(5)$$

Similarly, if y alone were varied, the increment of u would be

$$Q\delta y, \quad\ldots\ldots\ldots\ldots\ldots\ldots\ldots\ldots(6)$$

where the limiting value of Q, when δy is indefinitely diminished, is

$$Q_0 = \frac{\partial u}{\partial y}. \quad\ldots\ldots\ldots\ldots\ldots\ldots\ldots\ldots(7)$$

Let us now suppose that the actual variation from x, y to $x + \delta x$, $y + \delta y$ is made in two successive steps, in the first of which x alone, and in the second of which y alone is varied. The total increment of u will then be

$$\delta u = P\delta x + Q'\delta y, \quad\ldots\ldots\ldots\ldots\ldots\ldots(8)$$

where Q' differs from Q owing to the fact that the starting point of the second variation is now $(x + \delta x,\, y)$ instead of $(x,\, y)$.

To find the form which (8) assumes when δx and δy tend simultaneously to the value 0, preserving any assigned ratio to one another, we put

$$\delta x = \alpha\epsilon, \quad \delta y = \beta\epsilon, \quad\ldots\ldots\ldots\ldots\ldots(9)$$

where α, β are constants, and ϵ is infinitely small. We have, then,

$$\frac{\delta u}{P_0\delta x + Q_0\delta y} = \frac{P\alpha + Q'\beta}{P_0\alpha + Q_0\beta}\ldots\ldots\ldots\ldots\ldots(10)$$

In virtue of the assumed continuity of the derivatives (2), P and Q' tend, when $\epsilon \to 0$, to the limits P_0 and Q_0, respectively. Hence, the smaller δx and δy are taken, the more nearly does it become true that

$$\delta u = \frac{\partial u}{\partial x}\,\delta x + \frac{\partial u}{\partial y}\,\delta y, \quad\ldots\ldots\ldots\ldots\ldots(11)$$

in the sense that the *ratio* of the two sides is ultimately one of equality. This result is often expressed in the form

$$du = \frac{\partial u}{\partial x}\, dx + \frac{\partial u}{\partial y}\, dy. \quad\quad\quad\quad\quad (12)$$

The symbols dx, dy, du are then called 'differentials,' and du is called the 'total differential' of u.

The above reasoning may be amplified by writing down explicit values for the quantities which we have denoted by P and Q'. If we write

$$\frac{\partial u}{\partial x} = \phi_x(x, y), \quad\quad \frac{\partial u}{\partial y} = \phi_y(x, y), \quad\quad\quad (13)$$

we have

$$P = \frac{\phi(x + \delta x, y) - \phi(x, y)}{\delta x} = \phi_x(x + \theta_1 \delta x, y), \quad\quad (14)$$

$$Q' = \frac{\phi(x + \delta x, y + \delta y) - \phi(x + \delta x, y)}{\delta y} = \phi_y(x + \delta x, y + \theta_2 \delta y), \quad (15)$$

by Art. 56, where θ_1, θ_2 are some quantities lying between 0 and 1. Hence

$$\delta u = (\phi_x + \theta_1 \delta x, y)\, \delta x + \phi_y(x + \delta x, y + \theta_2 \delta y)\, \delta y. \quad\quad (16)$$

Since ϕ_x, ϕ_y are assumed to be continuous according to the definition of Art. 34, the limiting form of this equation is

$$\delta u = \phi_x \delta x + \phi_y \delta y, \quad\quad\quad\quad\quad\quad (17)$$

which is the same as (11).

The equation (11) shews that in the neighbourhood of a maximum or minimum, the variation of u is of the second (or higher) order of small quantities, since we then have

$$\frac{\partial u}{\partial x} = 0, \quad\quad \frac{\partial u}{\partial y} = 0, \quad\quad\quad\quad\quad (18)$$

by Art. 53. Thus, at a point of maximum or minimum altitude on a surface the tangent plane is in general horizontal. As already indicated, the converse is not necessarily true. See Art. 51.

The preceding theorem can be readily extended to the case of any number of independent variables, x, y, z .... We have

$$\delta u = \frac{\partial u}{\partial x}\, \delta x + \frac{\partial u}{\partial y}\, \delta y + \frac{\partial u}{\partial z}\, \delta z + \dots, \quad\quad\quad (19)$$

ultimately.

58. Application to Small Corrections.

The theorem of the preceding Art. can be applied after the manner of Art. 55 to the calculation of small corrections.

8—2

Ex. 1.   In the case of Art. 55, Ex. 2, the total error in c, due to errors δa, δb, δC in the observed values of the two sides and the included angle, is to be found from

$$\delta\left(c^2\right) = \frac{\partial\left(c^2\right)}{\partial a}\,\delta a + \frac{\partial\left(c^2\right)}{\partial b}\,\delta b + \frac{\partial\left(c^2\right)}{\partial C}\,\delta C, \quad\ldots\ldots\ldots\ldots(1)$$

which gives

$$c\,\delta c = (a - b\cos C)\,\delta a + (b - a\cos C)\,\delta b + ab\sin C\,\delta C,$$

or $\qquad \delta c = \cos B\,\delta a + \cos A\,\delta b + a\sin B\,\delta C. \quad\ldots\ldots\ldots\ldots\ldots(2)$

Ex. 2.   If Δ be the area of a triangle, as determined from a measurement of two sides a, b, and the included angle C, we have

$$\Delta = \tfrac{1}{2}ab\sin C, \quad\ldots\ldots\ldots\ldots\ldots\ldots\ldots(3)$$

whence $\qquad \log\Delta = \log\tfrac{1}{2} + \log a + \log b + \log\sin C. \quad\ldots\ldots\ldots(4)$

Hence, differentiating,

$$\frac{\delta\Delta}{\Delta} = \frac{\delta a}{a} + \frac{\delta b}{b} + \cot C\,\delta C. \quad\ldots\ldots\ldots\ldots\ldots(5)$$

This gives the 'proportional error,' *i.e.* the ratio of the error ($\delta\Delta$) to the whole quantity (Δ) whose value is sought.   In all measurements it is the proportional error, rather than the actual magnitude of the error, which is of importance.

An important point brought out by the investigation of Art. 57 is that the small variations of a quantity due to independent causes are *superposed*.   This follows from the linearity of the expression for δu in terms of δx, δy, δz, ...

Thus, in determining the weight of a body by the balance, the corrections for the buoyancy of the air, and for the inequality of the arms of the balance, may be calculated separately, and the (algebraic) sum of the results taken.   The error involved in this process will be of the second order.

59.   Differentiation of a Function of Functions, and of Implicit Functions.

Another important application of the formula (11) of Art. 57 is to the differentiation of a function of functions, and of implicit functions.

1°.   Thus if $\qquad u = \phi(x, y), \ldots\ldots\ldots\ldots\ldots\ldots\ldots\ldots(1)$

where x, y are given functions of a variable t, we have, ultimately,

$$\frac{\delta u}{\delta t} = \frac{\partial\phi}{\partial x}\frac{\delta x}{\delta t} + \frac{\partial\phi}{\partial y}\frac{\delta y}{\delta t}, \quad\ldots\ldots\ldots\ldots\ldots(2)$$

or $\qquad \dfrac{du}{dt} = \dfrac{\partial\phi}{\partial x}\dfrac{dx}{dt} + \dfrac{\partial\phi}{\partial y}\dfrac{dy}{dt}. \quad\ldots\ldots\ldots\ldots(3)$

This may be applied to reproduce various results obtained in Chapter II. To conform to previous notation we may write

$$y = \phi (u, v),$$

where u, v are given functions of x; the formula (3) then takes the shape

$$\frac{dy}{dx} = \frac{\partial\phi}{\partial u}\frac{du}{dx} + \frac{\partial\phi}{\partial v}\frac{dv}{dx}. \quad\dots\dots\dots\dots\dots\dots(4)$$

Thus, if

$$\phi (u, v) = uv, \dots\dots\dots\dots\dots\dots\dots(5)$$

we have

$$\partial\phi/\partial u = v, \quad \partial\phi/\partial v = u,$$

and therefore

$$\frac{d (uv)}{dx} = v\frac{du}{dx} + u\frac{dv}{dx}, \quad\dots\dots\dots\dots\dots\dots(6)$$

in agreement with Art. 30.

Again, if

$$\phi (u, v) = u^v, \dots\dots\dots\dots\dots\dots\dots(7)$$

we have

$$\partial\phi/\partial u = vu^{v-1}, \quad \partial\phi/\partial v = u^v \ \log u,$$

by Arts. 28, 42. Hence

$$\frac{d}{dx} (u^v) = vu^{v-1}\frac{du}{dx} + u^v \log u \frac{dv}{dx}, \quad\dots\dots\dots\dots(8)$$

in agreement with Art. 45 (7).

2°. Again, if y be an implicit function of x, defined by the equation

$$\phi (x, y) = 0, \dots\dots\dots\dots\dots\dots(9)$$

then differentiating this equation with respect to x, we have

$$\frac{\partial\phi}{\partial x}\frac{dx}{dx} + \frac{\partial\phi}{\partial y}\frac{dy}{dx} = 0,$$

or

$$\frac{\partial\phi}{\partial x} + \frac{\partial\phi}{\partial y}\frac{dy}{dx} = 0. \dots\dots\dots\dots\dots\dots(10)$$

This is an extension of a result given in Art. 35.

60. Geometrical Applications of the Derived Functions. Cartesian Coordinates.

We have seen (Art. 24) that if ψ denotes the angle which the tangent, drawn to the right, at any point of the curve

$$y = \phi (x), \quad\dots\dots\dots\dots\dots\dots\dots(1)$$

makes with the positive direction of the axis of x, then

$$\frac{dy}{dx} = \tan \psi. \dots\dots\dots\quad\dots\dots\dots\dots(2)$$

With the help of this formula, several magnitudes connected with a curve may be expressed in terms of x, y, and dy/dx.

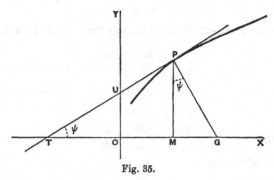

Fig. 35.

If the tangent and the normal at the point P meet the axis of x in T and G, respectively, and if M be the foot of the ordinate, then TM is called the 'subtangent' and MG the 'subnormal.' Hence we find

$$\text{subtangent} = TM = MP \cot \psi = y \div \frac{dy}{dx}, \dots\dots\dots\dots(3)$$

$$\text{subnormal} = MG = MP \tan \psi = y \frac{dy}{dx}, \dots\dots\dots\dots(4)$$

$$\text{tangent} = TP = MP \operatorname{cosec} \psi$$

$$= y \left\{1 + \left(\frac{dy}{dx}\right)^2\right\}^{\frac{1}{2}} \div \frac{dy}{dx}, \dots(5)$$

$$\text{normal} = PG = MP \sec \psi = y \left\{1 + \left(\frac{dy}{dx}\right)^2\right\}^{\frac{1}{2}}. \dots\dots(6)$$

Again, the intercepts of the tangent on the coordinate axes are

$$\left.\begin{aligned} OT &= OM - TM = \quad x - \frac{y}{\dfrac{dy}{dx}}, \\ OU &= TO \tan \psi = y - x \frac{dy}{dx}. \end{aligned}\right\} \dots\dots\dots(7)$$

Ex. 1. In the parabola $\quad y^2 = 4ax, \dots\dots\dots\dots\dots(8)$ we have, differentiating both sides with respect to x, and dividing by 2,

$$y \frac{dy}{dx} = 2a, \dots\dots\dots\dots\dots\dots(9)$$

which shews that the subnormal is constant and equal to $2a$.

Again, the subtangent is

$$y \div \frac{dy}{dx} = \frac{y^2}{2a} = 2x, \quad \dots\dots\dots\dots\dots(10)$$

and is therefore double the abscissa; in other words, the origin O bisects TM.

Ex. 2. In the hyperbola

$$xy = k^2, \quad \dots\dots\dots\dots\dots\dots(11)$$

we have

$$x\frac{dy}{dx} + y = 0. \quad \dots\dots\dots\dots\dots\dots(12)$$

Hence the formulæ (7) for the intercepts of the tangent on the co-ordinate axes give $2x$, $2y$ as the value of these intercepts, respectively. Hence M bisects OT, and therefore P bisects TU; *i.e.* the portion of the tangent included between the coordinate axes is bisected at the point of contact.

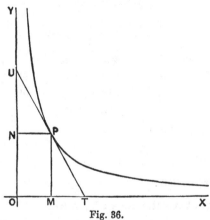

Fig. 36.

Again, the product of these intercepts is equal to $4xy$, or $4k^2$. Hence the area of the triangle OTU is constant and equal to $2k^2$.

Ex. 3. More generally, in the curve

$$x^m y^n = \text{const.}, \quad \dots\dots\dots\dots\dots\dots(13)$$

we find, taking logarithms of both sides and then differentiating,

$$\frac{m}{x} + \frac{n}{y}\frac{dy}{dx} = 0. \quad \dots\dots\dots\dots\dots\dots(14)$$

This makes $\quad OT = x - y \Big/ \frac{dy}{dx} = x + \frac{n}{m}x = \frac{m+n}{m}x.$

Hence $\quad UP : PT = OM : MT = x : \frac{n}{m}x = m : n, \quad \dots\dots\dots(15)$

that is, the tangent UT is divided in a constant ratio at the point of contact.

This includes the two preceding cases. In the parabola (**Ex. 1**) we had $m = -1$, $n = 2$; in the hyperbola (**Ex. 2**) $m = 1$, $n = 1$.

An important physical example is that of the 'adiabatic' relation between the pressure and the volume of a gas, viz.

$$pv^\gamma = \text{const.} \quad \dots\dots\dots\dots\dots\dots(16)$$

If a curve be constructed with v as abscissa and p as ordinate, the tangent is divided at the point of contact in the ratio $\gamma : 1$.

Ex. 4. In the ellipse

$$\frac{x^2}{a^2} + \frac{y^2}{b^2} = 1 \quad \dots\dots\dots\dots\dots\dots(17)$$

we find, on differentiating,

$$\frac{x}{a^2} + \frac{y}{b^2}\frac{dy}{dx} = 0,$$

whence

$$\frac{dy}{dx} = -\frac{b^2}{a^2}\frac{x}{y}. \quad \dots\dots\dots\dots\dots\dots(18)$$

The intercept made by the tangent on the axis of x is

$$OT = x - y\bigg/\frac{dy}{dx} = x + \frac{a^2}{x}\frac{y^2}{b^2} = x + \frac{a^2}{x}\left(1 - \frac{x^2}{a^2}\right) = \frac{a^2}{x},$$

whence

$$OM . OT = a^2. \quad \dots\dots\dots\dots\dots\dots(19)$$

The intercept made by the normal is

$$OG = x + y\frac{dy}{dx} = \left(1 - \frac{b^2}{a^2}\right)x = e^2 . OM, \quad \dots\dots\dots\dots(20)$$

where e is the eccentricity.

61. Coordinates determined by a Single Variable.

A curve is sometimes defined by means of two equations of the type

$$x = \phi(t), \quad y = \chi(t) \quad \dots\dots\dots\dots\dots(1)$$

giving the coordinates in terms of a subsidiary variable t.

For example, in Dynamics, the coordinates of a moving particle may be given as functions of the time.

If we take any convenient series of values of t, we can calculate the corresponding values of x and y, and so plot out as many points as we please on the curve.

If δx, δy, δt be simultaneous increments of x, y, t, we have

$$\frac{\delta y}{\delta x} = \frac{\delta y}{\delta t} \div \frac{\delta x}{\delta t},$$

and therefore in the limit, when δt is indefinitely diminished,

$$\tan \psi = \frac{dy}{dx} = \frac{dy}{dt} \div \frac{dx}{dt} \quad \text{.....................(2)}$$

Ex. 1. In the ellipse

$$x = a \cos \phi, \quad y = b \sin \phi, \quad \text{...................(3)}$$

we have
$$\tan \psi = \frac{dy}{d\phi} \Big/ \frac{dx}{d\phi} = -\frac{b}{a} \cot \phi. \quad \text{.................(4)}$$

Ex. 2. In the case of a projectile moving under gravity, we have

$$x = a + u_0 t, \quad y = b + v_0 t - \tfrac{1}{2} g t^2, \quad \text{.................(5)}$$

whence
$$\tan \psi = \frac{dy}{dt} \Big/ \frac{dx}{dt} = \frac{v - gt}{u_0}.$$

62. Equations of the Tangent and Normal at any point of a Curve.

1°. If (x, y) and $(x + \delta x, y + \delta y)$ be the coordinates of two points P and Q on a curve

$$y = \phi(x), \quad \text{..........................(1)}$$

the coordinates (ξ, η) of any other point on the line PQ satisfy the relation

$$\frac{\xi - x}{\delta x} = \frac{\eta - y}{\delta y}, \quad \text{.......................(2)}$$

or
$$\eta - y = \frac{\delta y}{\delta x}(\xi - x); \quad \text{....................(3)}$$

see Fig. 19, p. 45. In the limit, when Q approaches P indefinitely, this takes the form

$$\eta - y = \frac{dy}{dx}(\xi - x), \quad \text{.....................(4)}$$

which is the equation of the tangent-line at P.

Since the gradient of the normal is the negative reciprocal of the gradient of the tangent, the equation of the normal is

$$(\xi - x) + (\eta - y)\frac{dy}{dx} = 0. \quad \text{..................(5)}$$

2°. If the coordinates are expressed in the form

$$x = \phi(t), \quad y = \chi(t), \quad \text{....................(6)}$$

we have, from (2), at points on the secant PQ

$$\frac{\xi - x}{\delta x / \delta t} = \frac{\eta - y}{\delta y / \delta t}. \quad \text{.......................(7)}$$

The equation of the tangent at P is therefore

$$\frac{\xi - x}{\dfrac{dx}{dt}} = \frac{\eta - y}{\dfrac{dy}{dt}}. \qquad \ldots\ldots\ldots\ldots\ldots(8)$$

It easily follows that the equation of the normal is

$$(\xi - x)\frac{dx}{dt} + (\eta - y)\frac{dy}{dt} = 0 \ldots\ldots\ldots\ldots\ldots(9)$$

3°. If the equation of the curve is given in the form

$$\phi(x, y) = 0, \qquad \ldots\ldots\ldots\ldots\ldots(10)$$

we have, by Arts. 35, 59,

$$\frac{dy}{dx} = -\frac{\partial\phi}{\partial x}\Big/\frac{\partial\phi}{\partial y}. \qquad \ldots\ldots\ldots\ldots\ldots(11)$$

The equation of the tangent is therefore

$$(\xi - x)\frac{\partial\phi}{\partial x} + (\eta - y)\frac{\partial\phi}{\partial y} = 0. \qquad \ldots\ldots\ldots\ldots(12)$$

This follows also immediately from the fact that for an infinitesimal displacement of P along the curve we have $\delta\phi = 0$, or

$$\frac{\partial\phi}{\partial x}\delta x + \frac{\partial\phi}{\partial y}\delta y = 0. \qquad \ldots\ldots\ldots\ldots\ldots(13)$$

Since, from (2),

$$\delta x : \delta y = \xi - x : \eta - y, \qquad \ldots\ldots\ldots\ldots(14)$$

the form (12) results.

The equation of the normal is

$$\frac{\xi - x}{\dfrac{\partial\phi}{\partial x}} = \frac{\eta - y}{\dfrac{\partial\phi}{\partial y}} \qquad \ldots\ldots\ldots\ldots\ldots(15)$$

Ex. 1. In the parabola

$$y^2 = 4ax \qquad \ldots\ldots\ldots\ldots\ldots(16)$$

we have

$$\frac{dy}{dx} = \frac{2a}{y}. \qquad \ldots\ldots\ldots\ldots\ldots(17)$$

The equation of the tangent is therefore

$$\eta - y = \frac{2a}{y}(\xi - x), \qquad \ldots\ldots\ldots\ldots(18)$$

which reduces, by (16), to the usual form

$$\eta y = 2a(\xi + x). \qquad \ldots\ldots\ldots\ldots(19)$$

Ex. 2.   If the coordinates be given in the forms

$$x = at^2, \quad y = 2at, \quad \dots\dots\dots\dots\dots\dots(20)$$

the formula (8) for the tangent gives

$$\frac{\xi - x}{t} = \eta - y, \quad \dots\dots\dots\dots\dots\dots(21)$$

which reduces to

$$\eta = \frac{\xi}{t} + at. \quad \dots\dots\dots\dots\dots\dots(22)$$

The equation of the normal is

$$(\xi - x)\, t + \eta - y = 0, \quad \dots\dots\dots\dots\dots\dots(23)$$

or
$$t\xi + \eta = at^3 + 2at. \quad \dots\dots\dots\dots\dots\dots(24)$$

Since this is of the third degree in t, three real or imaginary normals can be drawn from any arbitrary point (ξ, η).

Ex. 3.   The equation of the tangent at any point of the central conic

$$Ax^2 + 2Hxy + By^2 = 1 \quad \dots\dots\dots\dots\dots(25)$$

is, by (12),

$$(\xi - x)\,(Ax + Hy) + (\eta - y)\,(Hx + By) = 0, \quad \dots\dots\dots(26)$$

or
$$(Ax + Hy)\,\xi + (Hx + By)\,\eta = 1. \quad \dots\dots\dots\dots(27)$$

63.   Polar Coordinates.

Let P, P' be two neighbouring points on a curve, and let r, θ be the polar coordinates of P, and $r + \delta r$, $\theta + \delta\theta$ those of P'.   If we join PP', and draw PN perpendicular to OP', we have

$$PN = OP \sin PON = r \sin \delta\theta,$$

$$P'N = OP' - ON = r + \delta r - r \cos \delta\theta = \delta r + r\,(1 - \cos \delta\theta).$$

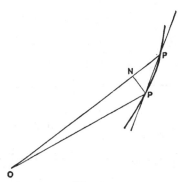

Fig. 37.

When $\delta\theta$ is indefinitely diminished, the ratio of $\sin \delta\theta$ to $\delta\theta$ tends to the limiting value unity, and $1 - \cos \delta\theta$, $= 2 \sin^2 \tfrac12\delta\theta$, is a small quantity of the second order.   Hence we may write

$$\tan PP'O = \frac{PN}{P'N} = \frac{r\delta\theta}{\delta r} + \sigma, \quad\dots\dots\dots\dots(1)$$

where σ is a quantity whose limiting value is zero.   Hence ultimately, when P' coincides with P, we have, if ϕ denotes the angle which the tangent to the curve at P, drawn on the side of θ increasing, makes with the positive direction of the radius vector,

$$\tan \phi = r\frac{d\theta}{dr}. \quad\dots\dots\dots\dots\dots(2)^*$$

Here θ is regarded as a function of r.   If r be regarded as a function of θ, the formula is

$$\cot \phi = \frac1r\frac{dr}{d\theta}. \quad\dots\dots\dots\dots\dots(3)$$

Ex. 1.   In the circle $r = 2a \sin \theta$ ....(4)

we have $\log r = \log 2a + \log \sin \theta,$

and therefore $\dfrac{1}{r}\dfrac{dr}{d\theta} = \cot \theta,$

whence $\cot \phi = \cot \theta, \quad\text{or}\quad \phi = \theta. \quad\dots\dots\dots(5)$

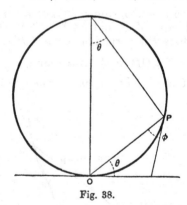

Fig. 38.

* The argument, which is an application of a principle stated in Art. 23, may be amplified as follows.   We have, exactly,

$$\tan PP'O = \frac{r \sin \delta\theta}{\delta r + 2r \sin^2 \tfrac12\delta\theta} = r\frac{\delta\theta}{\delta r} \cdot \frac{\sin \delta\theta}{\delta\theta} \cdot \frac{1}{1 + \dfrac{r\,\delta\theta}{\delta r} \cdot \dfrac{\sin \tfrac12\delta\theta}{\tfrac12\delta\theta} \cdot \sin \tfrac12\delta\theta},$$

and the limiting value of this is evidently $r\,d\theta/dr$.

Ex. 2. When the radius vector of a curve is a maximum or a minimum, it is in general normal to the curve.

For if $dr/d\theta = 0$, we have $\cot \phi = 0$, or $\phi = \frac{1}{2}\pi$.

Ex. 3. If the normals to a curve all pass through a fixed point, the curve must be a circle.

For, by hypothesis, if the fixed point be taken as pole, we have $\phi = \frac{1}{2}\pi$, and therefore $dr/d\theta = 0$, for all values of θ. Hence $r =$ const.

EXAMPLES. XV.

1. Verify the theorem of Art. 49 in the following cases:

  (1) $\phi(x) = (x-a)^m (x-b)^n$,

  (2) $\phi(x) = \log \dfrac{x^2 + ab}{(a+b)x}$,

  (3) $\phi(x) = \dfrac{(x-a)(x-b)}{x^2}$.

2. Prove that the curves

$$y = x^4 - 6x^3 + 9x^2 + 4x - 12,$$

and
$$y = x^4 - x^3 - 3x^2 + 5x - 2,$$

touch the axis of x, and find where they cut it. Trace the curves.

3. Prove that when x increases through a root of $\phi(x) = 0$, $\phi(x)$ and $\phi'(x)$ will have opposite signs just before, and the same sign just after, the passage. Does this hold in the case of a double root?

4. If, for $a > x > 0$, $\phi(x) = \dfrac{x^2}{a} - a$,

and, for $x > a$, $\phi(x) = a - \dfrac{a^2}{x^2}$,

whilst for $x = a$, $\phi(x) = 0$,

prove that $\phi(x)$ and $\phi'(x)$ are continuous from $x = 0$ to $x = \infty$. Trace the curve $y = \phi(x)$.

5. Examine whether the equation

$$x^3 - 12x + 16 = 0$$

has double roots. Draw the graph of the function on the left-hand side.

6. Shew that the curve

$$y = 8x^3 - 44x^2 + 78x - 45$$

touches the axis of x; and find where it cuts it.

Also find the points where the tangent is parallel to the straight line

$$2x + y = 0.$$

Sketch the curve.

7. Determine the coefficients A, B, C, D so that the curve

$$y = Ax^3 + Bx^2 + Cx + D$$

may pass through the point $(-2, 8)$, touch the axis of x at the point $(2, 0)$, and have its tangent parallel to the axis of x at the point for which $x = -1$.

8. Prove that the expression

$$(x - 1) e^x + 1$$

is positive for all positive values of x.

9. Prove that $\sin x$ lies between

$$x - \tfrac{1}{6} x^3 \quad \text{and} \quad x - \tfrac{1}{6} x^3 + \tfrac{1}{120} x^5.$$

Also that $\cos x$ lies between

$$1 - \tfrac{1}{2} x^2 \quad \text{and} \quad 1 - \tfrac{1}{2} x^2 + \tfrac{1}{24} x^4.$$

10. Prove that, if $x^2 < 1$, $\log(1 + x)$ lies between

$$x - \tfrac{1}{2} x^2 \quad \text{and} \quad x - \tfrac{1}{2} x^2 + \tfrac{1}{3} x^3.$$

11. Prove that, if $x^2 < 1$, $\tan^{-1} x$ lies between

$$x - \tfrac{1}{3} x^3 \quad \text{and} \quad x - \tfrac{1}{3} x^3 + \tfrac{1}{5} x^5.$$

EXAMPLES. XVI.
(Maxima and Minima.)

1. Prove that in the rectilinear motion of a point, the velocity is a maximum or a minimum when the acceleration changes sign.

Illustrate this from the simple-harmonic motion

$$s = a \cos nt.$$

2. Find the maxima or minima of the function

$$x^4 - 8x^3 + 22x^2 - 24x + 12.$$

3. Prove that the function

$$2x^3 - 3x^2 - 36x + 10$$

is a maximum when $x = -2$, and a minimum when $x = 3$.

4. The function $4x^3 - 18x^2 + 27x - 7$
has no maxima or minima.

5. Find the stationary points of the function

$$x^5 - 5x^4 + 5x^3 + 1,$$

and examine for which of them the function is a maximum or minimum.

6. Prove that the function

$$10x^6 - 12x^5 + 15x^4 - 20x^3 + 20$$

has a minimum value when $x = 1$, and no other maxima or minima.

7. Examine whether the equation

$$x^5 - x^3 - 4x^2 - 3x - 2 = 0$$

has a multiple root. Find the stationary points of the function on the left-hand side, and sketch its graph.

8. Prove that the curve

$$y = x^4 - 2x^3 - 3x^2 + 4x + 4$$

touches the axis of x at two points; and find its maximum ordinate.

Sketch the curve.

9. Prove that the function

$$x \sqrt{(ax - x^2)}$$

is a maximum when $x = \frac{3}{4}a$.

10. Prove that the function

$$\frac{(x-1)^2}{(x+1)^3}$$

has a maximum value $\frac{2}{27}$, and a minimum value 0.

11. Prove that the expression

$$\frac{1 + x + x^2}{1 - x + x^2}$$

has a maximum value 3, and a minimum value $\frac{1}{3}$.

12. The function $\dfrac{x(x^2+1)}{x^4 - x^2 + 1}$

has a maximum value 2, and a minimum value -2.

13. The function $\dfrac{x(x^2-1)}{x^4 - x^2 + 1}$

has two maxima, each $= \frac{1}{2}$, and two minima, each $= -\frac{1}{2}$.

14. Find the stationary points of the function

$$\frac{x^4 + 22x^2 + 9}{x(x^2+3)},$$

and draw its graph. [The stationary points are given by $x^2 = 1, 3, 9$]

15. Find the stationary points of

$$\frac{5x^2 - 18x + 45}{x^2 - 9},$$ $[x = 1, 9]$

and sketch the graph.

16. The function $\dfrac{x}{(a + x)(b + x)}$

is a maximum when $x = \sqrt{(ab)}$, and a minimum when $x = -\sqrt{(ab)}$.

17. Prove that the function

$$m_1(x - x_1)^2 + m_2(x - x_2)^2 + \ldots + m_n(x - x_n)^2$$

is a minimum when

$$x = \frac{m_1 x_1 + m_2 x_2 + \ldots + m_n x_n}{m_1 + m_2 + \ldots + m_n}.$$

18. The velocity of waves of length λ on deep water is proportional to

$$\sqrt{\left(\frac{\lambda}{a} + \frac{a}{\lambda}\right)},$$

where a is a certain linear magnitude; prove that the velocity is a minimum when $\lambda = a$.

19. If the power required to propel a steamer through the water vary as the cube of the speed, the most economical rate of steaming against a current will be at a speed equal to $1\frac{1}{2}$ times that of the current.

20. A copper wire is required to carry a given current from one electrical station to another. Prove that the most economical diameter of the wire is that which makes the interest on the cost equal to the value of the energy lost in heating the wire. (The rate of loss of energy varies inversely as the cross-section.)

21. The daily expenses of a steamer consist of wages, interest on capital, and coal. If the rate of coal-consumption vary as the cube of the speed, shew that if a voyage be performed at the most economical speed, the cost of coal will be half the amount of wages and interest.

22. Two ships are steaming, one directly towards, the other directly from, the same port, on courses making an angle of $60°$ with one another; and their speeds are as $2 : 1$. Prove that at the instant when they are nearest one another their distances from the port are as $4 : 5$.

23. The force exerted by a circular electric current of radius a on a small magnet whose axis coincides with the axis of the circle, varies as

$$\frac{x}{(a^2 + x^2)^{\frac{5}{2}}},$$

where x is the distance of the magnet from the plane of the circle. Hence prove that the force is a maximum when $x = \frac{1}{2}a$.

24. Prove that the expression

$$a \cos \theta + b \sin \theta$$

has the extreme values $\pm \sqrt{(a^2 + b^2)}$.

25. Prove that $\sin (\theta - a) \cos (\theta - \beta)$

is a maximum or a minimum when

$$\theta = \tfrac{1}{2} (a + \beta) + \tfrac{1}{4}\pi + \tfrac{1}{2}n\pi,$$

according as n is even or odd.

26. The inclination of a pendulum to the vertical, when the resistance of the air is taken into account, is given by the formula

$$\theta = ae^{-kt} \cos (nt + \epsilon) \,;$$

prove that the greatest elongations occur at equal intervals π/n of time, and that they form a series diminishing in geometrical progression.

27. Find the maximum ordinate of the curve

$$y = xe^{-x}.$$

Trace the curve.

28. The curve $y = x \log x$

has a minimum ordinate $- \cdot 3678....$

Trace the curve.

29. Prove that the ratio of the logarithm of a number (x) to the number itself is greatest when $x = e$.

30. Prove that if $a > b$ the expression

$$a \cosh x + b \sinh x$$

has the minimum value $\sqrt{(a^2 - b^2)}$, but that if $a < b$ it has neither a maximum nor a minimum.

31. Prove that the function

$$\cosh x + \cos x$$

has a minimum value when $x = 0$, but no other maxima or minima.

32. Prove that the function

$$\cosh x \cos x$$

has a maximum value when $x = 0$, a minimum value when $x = \tfrac{5}{4}\pi$ (nearly), and a series of alternate maxima and minima corresponding to $x = n\pi + \tfrac{1}{4}\pi$, approximately, where $n = 1, 2, 3....$

33. Find the maxima and minima of the function

$$\frac{\cosh x + \cos x}{1 + \cosh x \cos x}.$$

EXAMPLES. XVII.

(Geometrical Problems.)

1. The rectangle of least perimeter for a given area is a square.

2. The rectangle of given perimeter which has the shortest diagonal is a square.

3. The greatest rectangle which can be inscribed in a given triangle has one-half the area of the triangle.

4. A rectangle is inscribed in a right-angled triangle, so as to have one angle coincident with the right angle; prove that its area is a maximum when the opposite corner bisects the hypothenuse.

Shew also that under the same circumstances the perimeter of the rectangle has neither a maximum nor a minimum value.

5. Find the rectangle of greatest or least perimeter which can be inscribed in a given circle.

6. If through a given point A within a circle a chord PAQ be drawn, the sum of the squares of the segments PA, AQ is least when the chord is perpendicular to the diameter through A, and greatest when the chord coincides with the diameter.

7. Given a fixed straight line, and two fixed points A, B outside it, it is required to find a point P in the straight line such that $AP^2 + PB^2$ shall be a minimum.

8. Find the square of least area which can be inscribed in a given square; and the square of greatest area which can be circumscribed to a given square.

9. A quadrilateral $APQB$ is inscribed in a segment of a circle, AB being the base of the segment. Prove that when the area is a maximum

$$AP = PQ = QB.$$

10. A straight line drawn through a point (a, b) meets the (rectangular) coordinate axes in P and Q, respectively; prove that the minimum value of $OP + OQ$ is

$$a + 2\sqrt{(ab)} + b.$$

11. A straight line is drawn through a fixed point (a, b); prove that the minimum length intercepted between the coordinate axes (supposed rectangular) is

$$(a^{\frac{2}{3}} + b^{\frac{2}{3}})^{\frac{3}{2}}.$$

12. A rectangular sheet of metal has four equal square portions removed at the corners, and the sides are then turned up so as to form an open rectangular box. Shew that when the volume contained in the box is a maximum, the depth will be

$$\tfrac{1}{6}\{a + b - \sqrt{(a^2 - ab + b^2)}\},$$

where a, b are the sides of the original rectangle.

13. At what distance from the wall of a house must a man whose eye is 5 feet from the ground station himself in order that a window 5 feet high, whose sill is 20 feet from the ground, may subtend the greatest vertical angle?

14. It is required to cut from a cylindrical tree-trunk a beam of rectangular section of maximum flexural rigidity; prove that the breadth of the section must be $\tfrac{1}{2}$ the diameter, and its depth ·866 of the diameter. (Assume that the flexural rigidity varies as the breadth and as the cube of the depth.)

15. A straight road runs along the edge of a common, and a person on the common at a distance of one mile from the nearest point (A) of the road wishes to go to a distant place on the road in the least time possible. If his rates of walking on the common and on the road be 4 and 5 miles an hour, respectively, shew that he must strike the road at a point distant $1\tfrac{1}{3}$ miles from A.

16. Find at what height on the wall of a room a source of light must be placed in order to produce the greatest brightness at a point on the floor at a given distance a from the wall. (Assume that the brightness of a surface varies inversely as the square of the distance from the source, and directly as the cosine of the angle which the rays make with the normal to the surface.)

17. Two particles P, Q describe fixed straight lines intersecting in O, with constant velocities u, v. Prove that if A, B be simultaneous positions of the particles, and if $OA = a$, $OB = b$, $\angle AOB = \omega$, the distance PQ will be least after a time

$$\frac{au + bv - (av + bu)\cos\omega}{u^2 - 2uv\cos\omega + v^2},$$

and that the least distance will be

$$\frac{(av \sim bu)\sin\omega}{\sqrt{(u^2 - 2uv\cos\omega + v^2)}}.$$

18. Prove that the greatest rectangle which can be inscribed in a segment of a parabola bounded by a chord perpendicular to the axis has a length equal to $\tfrac{2}{3}$ that of the segment.

19. The greatest rectangle which can be inscribed in a given ellipse has its diagonals along the equi-conjugate diameters.

20. If the length of a tangent to an ellipse intercepted between the axes be a minimum, the tangent is divided at the point of contact into two portions equal to the semi-axes of the ellipse, respectively.

21. If a tangent to an ellipse includes with the principal axes (produced) a triangle of minimum area, it is parallel to one of the equi-conjugate diameters.

22. A circular sector has a given perimeter ; prove that the area is a maximum when the angle of the sector is 2 radians, and that the area is then equal to the square on the radius.

23. If a triangle have a given base, and if the sum of the other two sides be given, the area is greatest when these two sides are equal.

24. A quadrilateral has its four sides of given lengths, in a given order ; prove that its area is greatest when it can be inscribed in a circle.

EXAMPLES. XVIII.

[The following results may be assumed :

    (1)   The curved surface of a right circular cylinder of height h and radius a is $2\pi ah$;

    (2)   The volume of the same cylinder is $\pi a^2 h$;

    (3)   The curved surface of a right circular cone of height h, base-radius a, and slant side l is πal ;

    (4)   The volume of the same cone is $\frac{1}{3}\pi a^2 h$;

    (5)   The surface of a sphere of radius a is $4\pi a^2$;

    (6)   The volume of the same sphere is $\frac{4}{3}\pi a^3$.]

1. The cylinder of greatest volume which can be inscribed in a given sphere has a volume equal to ·5773 of that of the sphere.

2. The cylinder of greatest superficial area which can be inscribed in a given sphere has a surface equal to ·8090 of that of the sphere.

3. The cylinder of greatest volume for a given superficial area has its height equal to the diameter of the base, and its volume is ·8165 of that of a sphere having the given superficial area.

4. Find the cylinder of least surface for a given volume ; and prove that the ratio of its surface to that of a sphere of equal volume is 1·145.

5. Find the proportions of a thin open cylindrical vessel in order that the amount of material required may be the least possible for a given volume.

            [The height must equal the radius of the base.]

6. A cylinder is inscribed in a right circular cone; prove that its volume is a maximum when its altitude is $\frac{1}{3}$ that of the cone, and that its volume is then $\frac{4}{9}$ that of the cone.

7. If a cylinder be inscribed in a right circular cone the *curved* surface is a maximum when the altitude of the cylinder is $\frac{1}{2}$ that of the cone.

Shew also that the *total* surface of the cylinder cannot have a maximum value if the semi-angle of the cone exceeds $26° 34' \left[= \tan^{-1}\frac{1}{2} \right]$.

8. The cone of greatest volume which can be inscribed in a given sphere has an altitude equal to $\frac{2}{3}$ the diameter of the sphere.

Prove also, that the *curved* surface of the cone is a maximum for the same value of the altitude.

9. If a right circular cone be circumscribed to a given sphere, its volume will be a minimum when the altitude is double the diameter of the sphere. Shew also that the semi-vertical angle will be $19° 28'$ $\left[= \sin^{-1}\frac{1}{3} \right]$.

10. The right circular cone of greatest surface for a given volume has an altitude equal to $\sqrt{2}$ times the diameter of the base.

11. From a given circular sheet of metal it is required to cut out a sector so that the remainder can be formed into a conical vessel of maximum capacity; prove that the angle of the sector removed must be about $66°$.

12. An open rectangular tank is to contain a given volume of water, find what must be its proportions in order that the cost of lining it with lead may be a minimum.

[The length and breadth must each be double the depth.]

13. Given the sum of three concurrent edges of a rectangular parallelepiped, find its form in order that the surface may be a maximum.

14. Prove that the parallelepiped of greatest volume which can be inscribed in a given sphere is a cube.

15. Prove that the rectangular parallelepiped of greatest volume for a given surface is a cube.

16. If a triangle of maximum area be inscribed in any closed oval curve the tangents at the vertices are respectively parallel to the opposite sides.

17. If a triangle of minimum area be circumscribed to a closed oval curve, the sides are bisected at the points of contact.

18. The triangle of maximum area inscribed in a given circle is equilateral; and the triangle of minimum area circumscribed to the circle is also equilateral.

19. A polygon of maximum area, and of a given number (n) of sides, inscribed in a given circle is regular; and a polygon of minimum area, of n sides, circumscribed to the circle is also regular.

20. Assuming that the rectangle of greatest area for a given perimeter is a square, explain how it follows immediately that the rectangle of least perimeter for a given area is a square.

What inferences can be drawn in like manner from the results of Examples 13 and 15, above?

21. The polygon of n sides, which has maximum area for a given perimeter, or minimum perimeter for a given area, is regular. (Assume the result of Example 24, p. 132.)

Hence shew that the figure of maximum area for a given perimeter, or of minimum perimeter for a given area, is a circle.

22. By the regulations of the parcel post, a parcel must not exceed six feet in length and girth combined; prove that the most voluminous parcel which can be sent is a cylinder 2 feet long and 4 feet in girth, and that its volume is 2·546 cubic feet.

EXAMPLES. XIX.

(Small Variations.)

1. Prove that in a table of logarithmic tangents to base 10 the difference for one minute in the neighbourhood of 60° will be ·00029, approximately.

2. The height h of a tower is deduced from an observation of the angular elevation (a) at a distance a from the foot; prove that the error due to an error δa in the observed elevation is

$$\delta h = a \sec^2 a \, \delta a.$$

If $a = 100$ feet, $a = 30°$, and the error in the angle be $1'$, prove that $\delta h = ·47$ inch.

3. Find the cube root of 101, having given that the cube root of 100 is 4·6416. [4·6570.]

4. Having given $\log_e 10 = 2·3026$ find an approximate value of $\log_e 101$. [4·6151.]

5. In a table of anti-logarithms ($y = 10^x$) the entry opposite ·4 is 2·511886; find the anti-logarithm of ·40005 ($\mu = ·434294$). [2·512176.]

6. An angle is to be found from its log-tangent. Find in seconds of arc the error in the angle due to an error of ·0001 in the calculated log-tangent, the angle being in the neighbourhood of 30°. [20·6″.]

7. Prove that in a table of log-secants to base 10 the difference for one minute in the neighbourhood of 30° is ·00007, nearly.

8. Having given cosh $5 = 74\cdot2099$, calculate the value of cosh $5\cdot001$.
$$[74\cdot2841.]$$

9. Having given tanh $\cdot5 = \cdot46212$, find tanh $\cdot501$. $[\cdot46291.]$

10. Prove that if $\phi(x)$ be continuous and differentiable, except for $x = x_1$, when it becomes infinite, then $\phi'(x_1)$ is also infinite.

11. In a tangent galvanometer the tangent of the deflexion of the needle is proportional to the current; prove that the proportional error in the inferred value of the current, due to a given error of reading, is least when the deflexion is 45°.

12. The distances (x, x') of a point on the axis of a lens, and of its image, from the lens, are connected by the relation
$$\frac{1}{x} + \frac{1}{x'} = \frac{1}{f};$$
prove that the longitudinal magnification of a small object is $(x'/x)^2$.

13. A crank OP revolves about O with angular velocity ω, and a connecting rod PQ is hinged to it at P, whilst Q is constrained to move in a fixed groove OX. Prove that the velocity of Q is $\omega . OR$, where R is the point in which the line QP (produced if necessary) meets a perpendicular to OX drawn through O.

14. If the density (s) of a body be inferred from its weights (W, W') in air and in water respectively, the proportional error due to errors $\delta W, \delta W'$ in these weighings is
$$\frac{\delta s}{s} = -\frac{W'}{W - W'} \cdot \frac{\delta W}{W} + \frac{\delta W'}{W - W'}.$$

15. The radius r of a sphere is found by weighing it in air and in water. Prove that the proportional error, due to small errors in these weighings, is
$$\frac{\delta r}{r} = \frac{\delta W_1 - \delta W_2}{3(W_1 - W_2)},$$
where W_1, W_2 are the weights in air and water, respectively.

16. The error in the area (S) of an ellipse due to small errors in the lengths of the semi-axes a, b is given by
$$\frac{\delta S}{S} = \frac{\delta a}{a} + \frac{\delta b}{b}.$$

17. If the three sides a, b, c of a triangle are measured, the error in the angle A, due to given small errors in the sides, is
$$\delta A = \frac{\sin A}{\sin B \sin C} \frac{\delta a}{a} - \cot C \frac{\delta b}{b} - \cot B \frac{\delta c}{c}.$$

18. If the area (Δ) of a triangle be computed from measurements of one side (a) and the adjacent angles (B, C), shew that the proportional error in the area, due to small errors in the measurements, is given by

$$\frac{\delta\Delta}{\Delta} = 2\,\frac{\delta a}{a} + \frac{c}{a}\,\frac{\delta B}{\sin B} + \frac{b}{a}\,\frac{\delta C}{\sin C}.$$

Also, verify this result geometrically.

19. The area Δ of a triangle is calculated from the lengths of the sides a, b, c. If a be diminished, and b increased, by the same small amount a, prove that the consequent change in the area is given by

$$\frac{\delta\Delta}{\Delta} = \frac{2\,(a-b)\,a}{c^2 - (a-b)^2}.$$

20. The altitude of a triangle is computed from measurements of the base a and the base-angles B, C. If small errors δB, δC be made in the angles, the consequent proportional error in the altitude is

$$\frac{\sin C}{\sin A \sin B}\,\delta B + \frac{\sin B}{\sin A \sin C}\,\delta C.$$

21. If a triangle ABC be slightly varied, but so as to remain inscribed in the same circle, prove that

$$\frac{\delta a}{\cos A} + \frac{\delta b}{\cos B} + \frac{\delta c}{\cos C} = 0.$$

22. If $ABCD$ be a deformable plane quadrilateral of jointed rods, and if x, y be the lengths of the diagonals AC, BD, the infinitesimal variations of these lengths are connected by the relation

$$\sin A \sin C \cdot x\delta x + \sin B \sin D \cdot y\delta y = 0.$$

EXAMPLES. XX.

(Geometrical Applications.)

1. Prove that the condition that the tangent to a curve should pass through the origin is

$$\frac{y}{x} = \frac{dy}{dx}.$$

2. Prove that the straight line $y = 2x - 1$ is a tangent to the curve

$$y = x^3 - x + 1.$$

3. Prove that the straight line $y = 2x - 1$ touches the curve

$$y = x^4 + 2x^3 - 3x^2 - 2x + 3$$

at two distinct points.

4. Prove that the curve

$$y = x^4 - 6x^3 + 13x^2 - 11x + 4$$

touches the straight line $y = x$ twice; and find the abscissæ of the points of contact of the remaining tangents from the origin.

5. Find the points on the curve

$$y = x^4 - 6x^3 + 13x^2 - 10x + 5$$

where the tangent is parallel to $y = 2x$; and prove that two of these points have the same tangent.

6. Find the equations of the tangents which can be drawn from the origin to the curve

$$y = 4x^4 - 12x^3 + 9x^2 - 2x.$$

Also find where the tangent is parallel to the axis of x. Give a figure.

7. Prove that the perpendicular drawn from the foot of the ordinate to the tangent of a curve is

$$y \div \left\{ 1 + \left(\frac{dy}{dx}\right)^2 \right\}^{\frac{1}{2}}.$$

Hence shew that in the catenary $y = c \cosh x/c$ this perpendicular is constant.

8. Prove that the perpendicular from the origin on the tangent is

$$\left(y - x \frac{dy}{dx} \right) \div \left\{ 1 + \left(\frac{dy}{dx}\right)^2 \right\}^{\frac{1}{2}}.$$

Verify that in the circle

$$y = \pm \sqrt{(a^2 - x^2)}$$

this perpendicular is constant, and that in the rectangular hyperbola

$$xy = k^2$$

it is equal to

$$\frac{2k^2}{\sqrt{(x^2 + y^2)}}.$$

9. In the exponential curve (Fig. 24, p. 77)

$$y = be^{x/a},$$

the subtangent is constant, and the subnormal is y^2/a.

10. In the catenary $y = c \cosh x/c$,

the subtangent is $c \coth x/c$, the subnormal is $\frac{1}{2}c \sinh 2x/c$, and the normal is y^2/c.

11. The subtangent of the curve

$$y^n = a^{n-1} x$$

is nx.

12. Prove that the curve

$$\left(\frac{x}{a}\right)^n + \left(\frac{y}{b}\right)^n = 2$$

touches the straight line

$$\frac{x}{a} + \frac{y}{b} = 2$$

at the point (a, b), whatever the value of n.

13. In the curve of sines

$$y = b \sin \frac{x}{a},$$

the subtangent is $a \tan x/a$, the subnormal is $\frac{1}{2}b^2/a \cdot \sin 2x/a$, and the normal is

$$b \sin \frac{x}{a} \cdot \sqrt{\left(1 + \frac{b^2}{a^2} \cos^2 \frac{x}{a}\right)}.$$

14. Prove that the curves

$$y = e^{-ax} \sin \beta x, \quad y = e^{-ax}$$

touch at the points for which $\beta x = 2n\pi + \frac{1}{2}\pi$, where n is integral. Sketch the curves.

15. Prove that a pair of straight lines can be drawn through the origin, each of which touches all the curves obtained by giving c different values in the equation

$$y = c \cosh \frac{x}{c}.$$

16. If a curve be constructed with the velocity (v) of a moving point as ordinate, and the space described (s) as abscissa, the acceleration will be represented by the subnormal.

17. If a curve be constructed with the kinetic energy $(\frac{1}{2}mv^2)$ of a particle as ordinate, and the space s as abscissa, the force will be represented by the gradient of the curve.

18. Prove that the equations of the tangents to the curves

$$y^n = 2^n a^{n-1} x, \quad \frac{x^n}{a^n} + \frac{y^n}{b^n} = 1$$

may be put in the forms

$$yy_1 = 2a^{n-1}\{x + (n-1)x_1\}, \quad \frac{xx_1^{n-1}}{a^n} + \frac{yy_1^{n-1}}{b^n} = 1,$$

respectively.

19. Tangents are drawn to a conic $\phi(x, y) = 0$ parallel to $y = mx + c$; prove that the equation of the line joining their points of contact is

$$\frac{\partial \phi}{\partial x} + m \frac{\partial \phi}{\partial y} = 0.$$

20. Chords of a closed oval curve of finite curvature are drawn parallel to a fixed direction; prove that there is at least one chord of the system such that the tangents at its extremities are parallel.

21. Prove that in the ellipse
$$x = a \cos(\theta + a), \quad y = b \cos(\theta + \beta)$$
the tangent at the point θ_1 will be parallel to the radius drawn to the point θ_2 if $\theta_1 - \theta_2 = \pm \tfrac{1}{2}\pi$.

22. Find the values of θ corresponding to the principal axes of the ellipse
$$x = a \cos\theta + a' \sin\theta, \quad y = b \cos\theta + b' \sin\theta.$$
$$\left[\tan 2\theta = \frac{2(aa' + bb')}{a^2 + b^2 - a'^2 - b'^2}. \right]$$

23. Prove that the condition that the normal to the curve
$$\phi(x, y) = 0$$
should pass through the origin is
$$x \frac{\partial \phi}{\partial y} - y \frac{\partial \phi}{\partial x} = 0.$$
Deduce the equation of the principal axes of the conic
$$Ax^2 + 2Hxy + By^2 = 1.$$

24. Prove that if $b > 2a$ three real normals can be drawn from the origin to the parabola
$$x^2 = 4a(y + b).$$

25. Find the intercepts made on the coordinate axes by the normal at any point of the rectangular hyperbola
$$x^2 - y^2 = a^2;$$
and prove that the difference of their squares is constant.

26. Prove that the equation of the tangent to the hyperbola
$$x = kt, \quad y = k/t$$
is
$$x + t^2 y = 2kt.$$

27. If $x^4 + y^4 = a^4$, prove that $x^2 + y^2$ is a maximum when $x = \pm y$. Trace the curve; and prove that its greatest radial deviation from the circle $x^2 + y^2 = a^2$ is $\cdot 189a$.

28. Prove that in the parabola
$$r = \frac{a}{\sin^2 \tfrac{1}{2}\theta},$$
the focus being pole, $\qquad \phi = \pi - \tfrac{1}{2}\theta,$
and hence shew that the tangent makes equal angles with the focal distance and the axis.

29.   Two adjacent points P, P' on a curve being taken, straight lines PR, $P'R$ are drawn at right angles to the radii; prove that the limiting value of PR, when P' coincides with P, is $dr/d\theta$.

30.   If ϕ be the angle which the tangent to a curve makes with the radius vector drawn from the origin, prove that

$$\tan \phi = \frac{x\dfrac{dy}{dx} - y}{x + y\dfrac{dy}{dx}}.$$

31.   Prove that in the rectangular hyperbola

$$r^2 \cos 2\theta = a^2$$

the lines bisecting the angles between the radius and the tangent are constant in direction.

32.   Prove that the equation of the tangent to a curve

$$\frac{1}{r} = f(\theta)$$

at the point $\theta = a$ is

$$\frac{1}{r} = f(a) \cos(\theta - a) + f'(a) \sin(\theta - a).$$

Hence prove that the equation of the tangent to the conic

$$\frac{l}{r} = 1 + e \cos \theta$$

is

$$\frac{l}{r} = e \cos \theta + \cos(\theta - a).$$

CHAPTER V

DERIVATIVES OF HIGHER ORDERS

64. Definition, and Notations.

If y be a function of x, the derived function dy/dx will in general be itself a differentiable function of x. The result of differentiating dy/dx is called the 'second differential coefficient,' or 'second derivative.' If this, again, admits of differentiation, the result is called the 'third differential coefficient,' or 'third derivative'; and so on.

If we look upon d/dx as a symbol of operation, the first, second, third, ..., nth derivatives may be denoted by

$$\frac{d}{dx} \cdot y, \quad \left(\frac{d}{dx}\right)^2 \cdot y, \quad \left(\frac{d}{dx}\right)^3 y, \ldots, \left(\frac{d}{dx}\right)^n \cdot y,$$

respectively. The more usual forms are

$$\frac{dy}{dx}, \quad \frac{d^2y}{dx^2}, \quad \frac{d^3y}{dx^3}, \ldots, \frac{d^ny}{dx^n},$$

which may be regarded as contractions of the preceding, although (historically) they arose in a different manner.

Again, writing D for d/dx, as in Art. 25, we have the forms

$$Dy, \quad D^2y, \quad D^3y, \ldots, D^ny.$$

If
$$y = \phi(x),$$

the successive derivatives are also denoted by

$$\phi'(x), \quad \phi''(x), \quad \phi'''(x), \ldots, \phi^{(n)}(x).$$

Occasionally it is convenient to adopt the briefer notation

$$y', \quad y'', \quad y''', \ldots, y^{(n)}.$$

There are a few cases in which a general expression for the nth derivative of a function can be found. The more important of these are given in the following examples.

Ex. 1. If

$$y = A_0 + A_1 x + A_2 x^2 + A_3 x^3 + \ldots + A_m x^m, \quad\ldots\ldots\ldots\ldots(1)$$

we have

$$\left.\begin{aligned}
Dy &= A_1 + 2A_2 x + 3A_3 x^2 + \ldots + m A_m x^{m-1}, \\
D^2 y &= 2.1 A_2 + 3.2 A_3 + \ldots + m(m-1) A_m x^{m-2}, \\
&\ldots\ldots\ldots\ldots\ldots\ldots\ldots\ldots\ldots\ldots\ldots\ldots\ldots\ldots\ldots\ldots\ldots\ldots\ldots\ldots\ldots\ldots\ldots \\
D^m y &= m(m-1)(m-2)\ldots 2.1 A_m,
\end{aligned}\right\} \ldots(2)$$

and therefore $\qquad D^{m+1} y = 0, \quad D^{m+2} y = 0, \ \&c. \quad\ldots\ldots\ldots\ldots(3)$

Hence the mth derivative of a rational integral function of the mth degree is a constant, and all the higher derivatives vanish.

Ex. 2. If $\qquad\qquad\qquad y = e^{kx}, \quad\ldots\ldots\ldots\ldots\ldots\ldots\ldots\ldots\ldots\ldots\ldots(4)$

we have $\qquad\qquad Dy = k e^{kx}, \quad D^2 y = k^2 e^{kx}, \ldots,$

and, generally, $\qquad\qquad D^n y = k^n e^{kx}. \quad\ldots\ldots\ldots\ldots\ldots\ldots\ldots(5)$

Hence, putting $k = \log a$, we have

$$D^n a^x = (\log a)^n a^x. \quad\ldots\ldots\ldots\ldots\ldots\ldots\ldots(6)$$

Ex. 3. If $\qquad\qquad\qquad y = \sin \beta x, \quad\ldots\ldots\ldots\ldots\ldots\ldots\ldots\ldots(7)$

we have $\qquad\begin{aligned}
Dy &= \beta \cos \beta x, \quad D^2 y = -\beta^2 \sin \beta x, \\
D^3 y &= -\beta^3 \cos \beta x, \quad D^4 y = \beta^4 \sin \beta x,
\end{aligned}\biggr\} \ldots\ldots\ldots\ldots(8)$

and so on.

Otherwise, we have

$$Dy = \beta \sin (\beta x + \tfrac{1}{2}\pi),$$

and therefore $\qquad D^2 y = \beta^2 \sin (\beta x + \tfrac{1}{2}\pi + \tfrac{1}{2}\pi),$

and, generally, $\qquad D^n y = \beta^n \sin (\beta x + \tfrac{1}{2}n\pi). \quad\ldots\ldots\ldots\ldots\ldots(9)$

Ex. 4. If $\qquad\qquad\qquad y = \cos \beta x, \quad\ldots\ldots\ldots\ldots\ldots\ldots\ldots(10)$

we have $\qquad\begin{aligned}
Dy &= -\beta \sin \beta x, \quad D^2 y = -\beta^2 \cos \beta x, \\
D^3 y &= \beta^3 \sin \beta x, \quad D^4 y = \beta^4 \cos \beta x,
\end{aligned}\biggr\} \ldots\ldots\ldots\ldots(11)$

and so on.

Or, $\qquad\qquad\qquad Dy = \beta \cos (\beta x + \tfrac{1}{2}\pi),$

whence $\qquad\qquad D^2 y = \beta^2 \cos (\beta x + \tfrac{1}{2}\pi + \tfrac{1}{2}\pi),$

and, generally, $\qquad D^n y = \beta^n \cos (\beta x + \tfrac{1}{2}n\pi). \quad\ldots\ldots\ldots\ldots\ldots(12)$

Ex. 5. If $\qquad\qquad\qquad y = e^{ax} \cos \beta x, \ldots \quad\ldots\ldots\ldots\ldots\ldots(13)$

we find $\qquad\begin{aligned}
Dy &= e^{ax} (a \cos \beta x - \beta \sin \beta x), \\
D^2 y &= e^{ax} \{(a^2 - \beta^2) \cos \beta x - 2a\beta \sin \beta x\}.
\end{aligned}\biggr\} \ldots\ldots\ldots\ldots(14)$

Similarly, if $\qquad\qquad y = e^{ax} \sin \beta x,$ ....(15)

we have $\qquad Dy = e^{ax} (a \sin \beta x + \beta \cos \beta x),$

$\left. \begin{array}{l} D^2 y = e^{ax} \{(a^2 - \beta^2) \sin \beta x + 2a\beta \cos \beta x\}. \end{array} \right\}$ ....(16)

General formulæ may be obtained, in these cases, by putting

$$a = r \cos \theta, \quad \beta = r \sin \theta. \qquad ....(17)$$

This makes $\qquad D \cdot e^{ax} \cos \beta x = e^{ax} (a \cos \beta x - \beta \sin \beta x)$

$$= r e^{ax} \cos (\beta x + \theta),$$

and by repeated application of this result we find

$$D^n \cdot e^{ax} \cos \beta x = r^n e^{ax} \cos (\beta x + n\theta). \qquad ....(18)$$

Similarly $\qquad D^n \cdot e^{ax} \sin \beta x = r^n e^{ax} \sin (\beta x + n\theta). \qquad ....(19)$

Ex. 6. If $\qquad\qquad y = \log x,$ ....(20)

we have $\qquad Dy = x^{-1}, \quad D^2 y = - x^{-2}, \quad D^3 y = -1 \cdot - 2 \cdot x^{-3}, \quad ...,$

and, generally, $\qquad D^n y = -1 \cdot - 2 \cdot - 3 \ldots - (n-1) x^{-n}$

$$= (-)^{n-1} \frac{(n-1)!}{x^n} \cdot \qquad ....(21)$$

65. Successive Derivatives of a Product. Leibnitz' Theorem.

If u, v be functions of x, we have by Art. 31 (20),

$$D (uv) = Du \cdot v + u \cdot Dv. \qquad ....(1)$$

If we differentiate this again, we have

$$D^2 (uv) = D (Du \cdot v) + D (u \cdot Dv).$$

Now, by the rule referred to, we have

$$D (Du \cdot v) = D^2 u \cdot v + Du \cdot Dv,$$

$$D (u \cdot Dv) = Du \cdot Dv + u \cdot D^2 v,$$

whence $\qquad D^2 (uv) = D^2 u \cdot v + 2 Du \cdot Dv + u \cdot D^2 v. \qquad ....(2)$

The general formula for the nth derivative of a product is

$$D^n (uv) = D^n u \cdot v + n D^{n-1} u \cdot Dv + \frac{n (n-1)}{1 \cdot 2} D^{n-2} u \cdot D^2 v + \ldots$$

$$+ n Du \cdot D^{n-1} v + u \cdot D^n v, \qquad ......(3)$$

the coefficients being the same as in the Binomial Theorem. This formula is due to Leibnitz.

To see the truth of (3), consider the process of formation of the first few derivatives of uv. Using the accent notation, we have

$$D(uv) = u'v + uv'. \quad \dots\dots\dots\dots\dots\dots(4)$$

Differentiating this again,

$$D^2(uv) = u''v + u'v'$$
$$+ u'v' + uv''$$
$$= u''v + 2u'v' + uv''. \quad \dots\dots\dots\dots(5)$$

The next differentiation gives

$$D^3(uv) = u'''v + 2u''v' + u'v''$$
$$+ u''v' + 2u'v'' + uv''', \quad \dots\dots\dots\dots(6)$$

where in the first line we have differentiated the first variable factor in each term of (5), and in the second line the second variable factor. The result is

$$D^3(uv) = u'''v + 3u''v' + 3u'v'' + uv'''. \quad \dots\dots\dots\dots(7)$$

It appears that the numerical coefficient of the rth term in (7) is the sum of the coefficients of the rth and $(r-1)$th terms in (5); and it is evident from the nature of the successive steps that this law will obtain for all the subsequent derivatives. Now this is precisely the law of formation of the coefficients in the expansions of the successive powers of $a + b$; and since the coefficients of $D(uv)$ are the same as those of the first power of $a + b$, it follows that the coefficients in the expanded form of $D^n(uv)$ will be the same as those of $(a + b)^n$.

Ex. 1.   If $$y = xu, \quad \dots\dots\dots\dots\dots\dots\dots\dots(8)$$

we have $$D^n y = x D^n u + n Dx . D^{n-1} u$$
$$= x D^n u + n D^{n-1} u, \quad \dots\dots\dots\dots\dots(9)$$

since $D^2 x = 0$.

Thus if $$y = x \sin \beta x, \quad \dots\dots\dots\dots\dots(10)$$

we have $$D^2 y = x D^2 \sin \beta x + 2D \sin \beta x$$
$$= -\beta^2 x \sin \beta x + 2\beta \cos \beta x. \quad \dots\dots\dots(11)$$

Again, if $$y = x \log x, \quad \dots\dots\dots\dots\dots(12)$$

we have $$D^n y = x D^n \log x + n D^{n-1} \log x$$
$$= (-)^{n-1} \frac{(n-1)!}{x^{n-1}} + (-)^{n-2} \frac{n(n-2)!}{x^{n-1}},$$

by Art. 64 (21). Hence

$$D^n y = (-)^n \frac{(n-2)!}{x^{n-1}}. \quad \dots\dots\dots\dots\dots(13)$$

Ex. 2.   If $\qquad\qquad y = e^{ax}u,$ ....(14)

we have

$$D^n y = e^{ax} \cdot D^n u + n \cdot De^{ax} \cdot D^{n-1}u + \frac{n(n-1)}{1 \cdot 2} D^2 e^{ax} \cdot D^{n-2}u + \ldots$$

$$= e^{ax}\left(D^n u + na\, D^{n-1}u + \frac{n(n-1)}{1 \cdot 2} a^2 D^{n-2}u + \ldots\right). \qquad \ldots\ldots(15)$$

Thus, if $\qquad\qquad y = e^{ax} \sin \beta x,$ ....(16)

we have $\qquad D^2 y = e^{ax}\left(D^2 \sin \beta x + 2aD \sin \beta x + a^2 \sin \beta x\right)$

$$= e^{ax}\{(a^2 - \beta^2)\sin \beta x + 2a\beta \cos \beta x\}, \qquad \ldots\ldots\ldots(17)$$

in agreement with Art. 64 (16).

66.   Dynamical Illustrations.

The *second* derivative is especially predominant in the dynamical applications of the Calculus.

Thus, in the case of rectilinear motion, if s be the distance from a fixed origin, we have seen (Art. 26) that the velocity (v) and the acceleration (a) are given by the formulæ

$$v = \frac{ds}{dt}, \qquad a = \frac{dv}{dt}. \qquad \ldots\ldots\ldots\ldots\ldots\ldots(1)$$

Hence, in the present notation, we have

$$a = \frac{d}{dt}\left(\frac{ds}{dt}\right) = \frac{d^2 s}{dt^2}, \qquad \ldots\ldots\ldots\ldots\ldots\ldots(2)$$

i.e. the second derivative of s (with respect to the time) measures the *acceleration*.

So also the angular acceleration of a body about a fixed axis is given, in the notation of Art. 26, by

$$\frac{d\omega}{dt} = \frac{d^2\theta}{dt^2}. \qquad \ldots\ldots\ldots\ldots\ldots\ldots\ldots\ldots(3)$$

Ex. 1.   If s be a quadratic function of t, say

$$s = At^2 + Bt + C, \qquad \ldots\ldots\ldots\ldots\ldots\ldots\ldots(4)$$

we have $\qquad\qquad \dfrac{ds}{dt} = 2At + B,$

$$\frac{d^2 s}{dt^2} = 2A, \qquad \ldots\ldots\ldots\ldots\ldots\ldots\ldots\ldots(5)$$

i.e. the acceleration is constant.

Ex. 2.   In 'simple-harmonic' motion we have

$$s = a \cos(nt + \epsilon), \quad \dots\dots\dots\dots\dots\dots\dots(6)$$

whence

$$\frac{ds}{dt} = -na \sin(nt + \epsilon),$$

$$\frac{d^2s}{dt^2} = -n^2a \cos(nt + \epsilon) = -n^2s, \quad \dots\dots\dots(7)$$

i.e. the acceleration is directed always towards a fixed point (the origin of s) and varies as the distance from that point.

Ex. 3.   If $s = A \cosh nt + B \sinh nt, \quad \dots\dots\dots\dots\dots(8)$

we have $\dfrac{ds}{dt} = nA \sinh nt + nB \cosh nt,$

$$\frac{d^2s}{dt^2} = n^2A \cosh nt + n^2B \sinh nt = n^2s, \quad \dots\dots\dots(9)$$

i e. the acceleration is *from* a fixed point, and varies as the distance.

67.   Concavity and Convexity.   Points of Inflexion.

Just as $\phi'(x)$ measures (Art. 26) the rate of increase of $\phi(x)$, so $\phi''(x)$ measures the rate of increase of $\phi'(x)$.   Hence if $\phi''(x)$ be positive the gradient of the curve

$$y = \phi(x) \quad \dots\dots\dots\dots\dots\dots\dots(1)$$

increases with x; whilst if $\phi''(x)$ be negative the gradient decreases as x increases.

If $\phi''(x) = 0$, the rate of change of the gradient is momentarily zero, and we have a 'stationary tangent.'   The simplest case of this is at a 'point of inflexion,' *i.e.* a point at which the curve crosses its tangent; see Fig. 40.

A curve is said to be concave upwards at a point P when in the immediate neighbourhood of P it lies wholly above the tangent at P.   Similarly, it is said to be convex upwards when in the immediate neighbourhood of P it lies wholly below the tangent at P.

If the curve, to the right of P, lie above the tangent at P, as in Fig. 39, it is easily seen from Art. 56 that within any range (however short) extending to the right of P there will be points at which $\phi'(x)$ is greater than at P.   Hence, by Art. 48, the value of $\phi''(x)$ at P cannot be negative.   The same conclusion holds if the curve, to the left of P, lie above the tangent at P.

Similarly, if the curve, either to the right or left of P, lie below the tangent at P, the value of $\phi''(x)$ at P cannot be positive.

It follows that the curve is concave upwards when $\phi''(x)$ is positive, and convex upwards when $\phi''(x)$ is negative.

It appears, moreover, that at a point of inflexion, where the curve crosses its tangent, $\phi''(x)$ cannot be either positive or negative, and

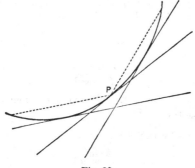

Fig. 39.

therefore (since it is assumed to be finite) must vanish. This condition, though essential, is not sufficient. It is further necessary that $\phi''(x)$ should change sign as x increases through the value in question. Suppose, for instance, that to the left of P the curve lies below the tangent at P, and that to the right of P it lies above it. It appears then from Art. 56 that there will be points of the curve both to the right and to the left, in the immediate neighbourhood of P, at which the gradient is greater than at P, *i.e.* the gradient is a minimum at P, and $\phi''(x)$ must therefore change (Art. 51) from negative to positive.

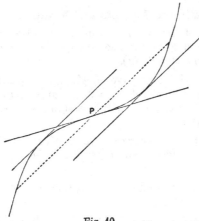

Fig. 40

If the crossing is in the opposite direction, the gradient is a maximum at P, and $\phi''(x)$ changes from positive to negative.

Ex. 1.   If $\qquad\qquad\qquad y = x^3,$ ....(2)

we have $\qquad\qquad\qquad\qquad y'' = 6x.$

This changes from − to + as x increases through 0.   Hence we have a point of inflexion; see Fig. 31, p. 105.

Ex. 2. $\qquad\qquad\qquad\qquad y = \dfrac{2x}{1 + x^2}.$ ....(3)

This makes $\qquad\qquad\qquad y'' = \dfrac{4x\,(x^2 - 3)}{(1 + x^2)^3}.$

Hence there are three points of inflexion, viz. when $x = 0$ and when $x = \pm \sqrt{3}$.   See Fig. 13, p. 27.

Ex. 3.   In the curve of sines

$$y = b \sin \frac{x}{a},$$ ....(4)

we have $\qquad\qquad y'' = -\dfrac{b}{a^2} \sin \dfrac{x}{a} = -\dfrac{y}{a^2}.$

Hence y'' changes sign, and there is a point of inflexion, whenever the curve crosses the axis of x.   See Fig. 14, p. 28.

Ex. 4.   In the curve $\qquad\quad y = x^4,$ ....(5)

we have $\qquad\qquad\qquad\qquad y'' = 12x^2.$

This vanishes, but does not change sign, when $x = 0$.   Hence we have a stationary tangent, but not a point of inflexion in the strict sense.   It is in fact obvious, since x^4 is essentially positive, that the curve lies wholly on one side of the tangent at the origin.

68.   Application to Maxima and Minima.

The criterion of Art. 51 for distinguishing maxima and minima values of a function $\phi(x)$ can also be expressed in general in terms of the second derivative $\phi''(x)$.

Since $\phi''(x)$ is the derivative of $\phi'(x)$, it appears that if, as x increases through a root of $\phi'(x) = 0$, $\phi''(x)$ is *positive*, $\phi'(x)$ must be increasing, and therefore changing sign from − to +.   Hence $\phi(x)$ is a *minimum*.

Similarly, if $\phi''(x)$ is *negative* when $\phi'(x) = 0$, $\phi'(x)$ must be decreasing, and therefore changing sign from + to −.   Hence $\phi(x)$ is a *maximum*.

The connection of these results with the criterion of concavity and convexity is obvious.

Ex. 1.   In rectilinear motion, the distance (s) from the origin, is a maximum or minimum when the velocity (ds/dt) vanishes, according as the acceleration (d^2s/dt^2) is then negative or positive.

Ex. 2.   Let $\phi(x) = \dfrac{2x}{1+x^2}$.

We have seen, Art. 51, Ex. 4, that $\phi'(x)$ vanishes for $x = 1$ and $x = -1$. Also from the value of $\phi''(x)$ given in Art. 67, Ex. 2, it appears that

$$\phi''(1) = -1, \quad \phi''(-1) = 1.$$

Hence the former value of x gives a maximum, and the latter a minimum, value of $\phi(x)$. See Fig. 13, p. 27.

It may happen, however, that a value of x which makes $\phi'(x) = 0$ also makes $\phi''(x) = 0$. It is easily shewn that in this case $\phi(x)$ is in general neither a maximum nor a minimum (cf. Fig. 31, p. 105), but it is hardly worth while to continue the discussion here. The complete rule will be given later (Chap. xv) as a deduction from Taylor's Theorem.

69.   Successive Derivatives in the Theory of Equations.

The successive derived functions play a great part in the Theory of Equations.

We have seen (Art. 50) that, if $\phi(x)$ be a rational integral function, at least one root of $\phi'(x) = 0$ will occur between any two roots of $\phi(x) = 0$. Similarly, at least one root of $\phi''(x) = 0$ will occur between any two roots of $\phi'(x) = 0$, and so on.

Moreover, since an r-fold root of $\phi(x) = 0$ is an $(r-1)$-fold root of $\phi'(x) = 0$, it will be an $(r-2)$-fold root of $\phi''(x) = 0$, ..., and finally a simple root of $\phi^{(r-1)}(x) = 0$. Hence the necessary and sufficient conditions for an r-fold root of $\phi(x) = 0$ are that the functions

$$\phi(x), \quad \phi'(x), \quad \phi''(x), \quad ..., \quad \phi^{(r-1)}(x) \quad ....(1)$$

should simultaneously vanish.

Ex.   If $\phi(x) = 2x^5 + 5x^4 + 4x^3 + 2x^2 + 2x + 1$,

we have $\phi'(x) = 10x^4 + 20x^3 + 12x^2 + 4x + 2$,

$$\phi''(x) = 4(10x^3 + 15x^2 + 6x + 1).$$

These all vanish for $x = -1$, which is therefore a triple root of $\phi(x) = 0$. We find, in fact, that

$$\phi(x) = (x+1)^3 (2x^2 - x + 1).$$

70.   Geometrical Interpretations of the Second Derivative.

In Art. 56 an important property of the first derived function was obtained by a process which consisted virtually in a comparison of the curve

$$y = \phi(x) \quad ....(1)$$

with a straight line $y = A + Bx, \quad ....(2)$

the constants A, B being determined so as to make (1) and (2) intersect for two given values of x.

We proceed, in a somewhat similar manner, to compare the curve (1) with a parabola

$$y = A + Bx + Cx^2, \quad \dots\dots\dots\dots\dots(3)$$

where the constants A, B, C are determined so as to make (1) and (2) intersect for three given values of x.

1°. We will first suppose these values of x to be equidistant; let them be $a - h$, a, $a + h$. The equations to determine the constants are then

$$\left. \begin{aligned} A + B(a-h) + C(a-h)^2 &= \phi(a-h), \\ A + Ba \quad\;\; + Ca^2 \quad\;\;\;\; &= \phi(a), \\ A + B(a+h) + C(a+h)^2 &= \phi(a+h). \end{aligned} \right\} \quad \dots\dots(4)$$

Let us now write

$$F(x) = \phi(x) - (A + Bx + Cx^2), \quad \dots\dots\dots(5)$$

i.e. $F(x)$ denotes the difference of the ordinates of the curves (1) and (3). By hypothesis, $F(x)$ vanishes for $x = a - h$, and for $x = a$; hence, by Art. 49, the derived function $F'(x)$ will vanish for some intermediate value of x, that is

$$F'(a - \theta_1 h) = 0, \quad \dots\dots\dots\dots\dots(6)$$

where $1 > \theta_1 > 0$. Again, since $F(x)$ vanishes for $x = a$, and for $x = a + h$, we shall have

$$F'(a + \theta_2 h) = 0, \quad \dots\dots\dots\dots\dots(7)$$

where $1 > \theta_2 > 0$.

By a further application of the same argument, since the function $F'(x)$ vanishes for $x = a - \theta_1 h$ and for $x = a + \theta_2 h$, its derived function $F''(x)$ will vanish for some intermediate value of x; we have therefore

$$F''(a + \theta h) = 0, \quad \dots\dots\dots\dots\dots(8)$$

where θ is some quantity lying between $-\theta_1$ and θ_2, and *à fortiori* between ± 1. Since, by (5),

$$F''(x) = \phi''(x) - 2C, \quad \dots\dots\dots\dots\dots(9)$$

it follows that, for some value of θ between ± 1,

$$\phi''(a + \theta h) = 2C. \quad \dots\dots\dots\dots\dots(10)$$

Now from (4) we find

$$\phi(a+h) - 2\phi(a) + \phi(a-h) = 2Ch^2, \quad \dots\dots(11)$$

and therefore

$$\frac{\phi(a+h) - 2\phi(a) + \phi(a-h)}{h^2} = \phi''(a+\theta h). \quad \ldots\ldots(12)$$

Hence

$$\lim_{h \to 0} \frac{\phi(a+h) - 2\phi(a) + \phi(a-h)}{h^2} = \phi''(a). \quad \ldots(13)$$

In the same way we could prove that

$$\lim_{h \to 0} \frac{\phi(a+2h) - 2\phi(a+h) + \phi(a)}{h^2} = \phi''(a). \quad \ldots(14)$$

If the difference $\phi(a+h) - \phi(a)$ be denoted by δy, the expression

$$\{\phi(a+2h) - \phi(a+h)\} - \{\phi(a+h) - \phi(a)\},$$

which is the difference of the differences, or the 'second difference,' for equal increments h of the independent variable, may be denoted by $\delta(\delta y)$ or $\delta^2 y$. Hence the formula (14) is equivalent to

$$\lim_{\delta x \to 0} \frac{\delta^2 y}{(\delta x)^2} = \frac{d^2 y}{dx^2}. \quad \ldots\ldots\ldots\ldots\ldots\ldots\ldots(15)$$

This is the origin of the notation $d^2 y/dx^2$, this being the limiting form of the second difference.

To interpret the theorem (13) geometrically, let, in Fig. 41

$$OA = a, \quad OH = a - h, \quad OH' = a + h,$$

and let AQ, HP, $H'P'$ be the corresponding ordinates of the curve (1). Join PP', and let AQ meet PP' in V. Then

$$VA = \tfrac{1}{2}(P'H + PH)$$
$$= \tfrac{1}{2}\{\phi(a+h) + \phi(a-h)\},$$

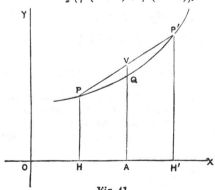

Fig. 41.

and therefore

$$VQ = VA - QA$$
$$= \tfrac{1}{2} \{\phi(a+h) - 2\phi(a) + \phi(a-h)\}.$$

Hence the theorem (13) asserts that

$$VQ = \tfrac{1}{2}HA^2 . \phi''(a), \dots\dots\dots\dots\dots(16)$$

ultimately.

It appears that the chord is above or below the arc according as $\phi''(a)$ is positive or negative.

2°. We will next suppose that two of the three points at which the curves (1) and (3) intersect are coincident. More precisely, we suppose that for $x = a$ the curves not only intersect but touch, and that they intersect again for $x = a + h$. The conditions that, for $x = a$, y and dy/dx should have the same values in the two curves, are

$$\left.\begin{array}{l} A + Ba + Ca^2 = \phi(a), \\ B + 2Ca = \phi'(a), \end{array}\right\} \dots\dots\dots\dots(17)$$

while the third condition gives

$$A + B(a+h) + C(a+h)^2 = \phi(a+h). \dots\dots(18)$$

With the same definition of $F(x)$ as before, we have

$$F(a) = 0, \quad F(a+h) = 0, \dots\dots\dots(19)$$

and therefore $\quad F'(a + \theta_1 h) = 0, \dots\dots\dots\dots(20)$

where $1 > \theta_1 > 0$. Again, since $F'(x)$ vanishes for $x = a$, and for $x = a + \theta_1 h$, we have

$$F''(a + \theta h) = 0, \dots\dots\dots\dots(21)$$

where $\theta_1 > \theta > 0$.

Now from (17) and (18) we find

$$\phi(a+h) - \phi(a) - h\phi'(a) = Ch^2. \dots\dots\dots(22)$$

Hence, by (9) and (21),

$$\phi(a+h) = \phi(a) + h\phi'(a) + \tfrac{1}{2}h^2\phi''(a+\theta h). \dots\dots(23)$$

This very important result will be recognized, later, as a particular case of Lagrange's form of Taylor's Theorem (see Chap. xv). It includes as much of this theorem as is ordinarily required in the dynamical and physical applications of the subject.

From (23) we deduce

$$\lim_{h \to 0} \frac{\phi(a+h) - \phi(a) - h\phi'(a)}{h^2} = \tfrac{1}{2}\phi''(a). \dots\dots(24)$$

In Fig. 42, let $OA = a$, $AH = h$, and let AP, HQ be the corresponding ordinates of the curve (1).   If QH meet the tangent at P in V, we have

$$QH = \phi(a+h), \quad VH = \phi(a) + h\phi'(a).$$

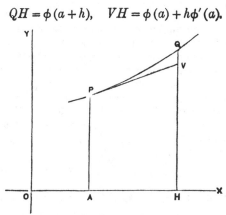

Fig. 42.

Hence (24) asserts that

$$QV = \tfrac{1}{2}AH^2 . \phi''(a), \quad \dots\dots\dots\dots(25)$$

ultimately.

Hence the deviation of a curve from a tangent, in the neighbourhood of the point of contact, is in general a small quantity of the second order.

If $\phi''(a) \neq 0$, QV does not change sign with h, and the curve in the immediate neighbourhood of P lies altogether above, or altogether below, the tangent-line, according as $\phi''(a)$ is positive or negative.   Cf. Art. 67.

The formulæ (16) and (25) have an interesting application in the theory of Curvature.   See Chap. x.

71.   Theory of Proportional Parts.

Let us make the curves

$$y = \phi(x), \quad \dots\dots\dots\dots\dots\dots(1)$$

and

$$y = A + Bx + Cx^2, \quad \dots\dots\dots\dots(2)$$

intersect for $x = a$, $x = a + zh$, $x = a + h$,

where $1 > z > 0$.

We find

$$(1-z)\phi(a) + z\phi(a+h) - \phi(a+zh) = z(1-z)h^2C; \dots(3)$$

and consequently, by the method of the preceding Art.,

$$(1-z)\phi(a)+z\phi(a+h)-\phi(a+zh)=\tfrac{1}{2}z(1-z)h^2\phi''(a+\theta h),\ (4)$$

where $1>\theta>0$.

This result, which includes the theorems of Art. 70 as particular cases, is here introduced for the sake of its bearing on the theory of 'proportional parts.' Suppose that $\phi(x)$ is a function which has been tabulated for a series of values of x at equal intervals h. Let a be one of these values, and suppose that $\phi(x)$ is required for some value of x between this and the next tabular value $a+h$; say for $a+zh$, where $1>z>0$. In the method of 'proportional parts,' the interpolation is made as if the function increased *uniformly* from $x=a$ to $x=a+h$, *i.e.* we assume

$$\frac{\phi(a+zh)-\phi(a)}{\phi(a+h)-\phi(a)}=\frac{z}{1},\qquad\dots\dots\dots\dots\dots(5)$$

or $\qquad\qquad \phi(a+zh)=(1-z)\phi(a)+z\phi(a+h).\ \dots\dots\dots(6)$

The formula (4) gives the error involved in this process, which is equivalent to assuming that the arc of the curve (1) between $x=a$ and $x=a+h$ may be replaced without sensible error by its chord.

The maximum value of $z(1-z)$ is $\tfrac{1}{4}$, by Art. 51, Ex. 2. Hence if R denote the greatest value which $\phi''(x)$ assumes in the interval from $x=a$ to $x=a+h$, the formula (4) shews that the error

$$\not> \tfrac{1}{8}h^2R.\ \dots\dots\dots\dots\dots\dots\dots\dots\dots(7)$$

Ex. 1.   In a seven-figure logarithmic table, the logarithms of all numbers from 10000 to 100000 are given at intervals of unity. Now if

$$\phi(x)=\log_{10}x,\ \dots\dots\dots\dots\dots\dots\dots(8)$$

we have $\qquad\qquad \phi''(x)=-\frac{\mu}{x^2}.\ \dots\dots\dots\dots\dots\dots\dots(9)$

Hence, putting $h=1$ in (7), we find that in the interpolation between $\log_{10}n$ and $\log_{10}(n+1)$ the error involved in the method of proportional parts is not greater than

$$\cdot05429\div n^2.\ \dots\dots\dots\dots\dots\dots\dots\dots(10)$$

Thus for $n=10000$, where it is greatest, the error does not exceed

$$\cdot000000000543,$$

and is therefore quite insensible from the standpoint of a seven-figure table.

It appears from (4) that the method may be expected to fail whenever $\phi''(x)$ is large. The differences are then said to be 'irregular.'

Ex. 2. If $\phi(x) = \log_{10} \sin x,$ ....(11)

we have $\phi''(x) = -\mu \operatorname{cosec}^2 x.$ ....(12)

Hence, putting $h = \dfrac{\pi}{10800} = \cdot000291,$

we find $\frac{1}{8}h^2 \phi''(x) = -\cdot00000000460 \operatorname{cosec}^2 x.$ ....(13)

Since $\operatorname{cosec}^2 18° = 10\cdot47$, it appears that in a table of log-sines at intervals of $1'$ the error of interpolation may amount to half a unit in the seventh place when the angle falls below $18°$.

EXAMPLES. XXI.

Verify the following differentiations :

1. $y = x^2(1-x)^2,$ $D^2 y = 2 - 12x + 12x^2.$

2. $y = \frac{1}{2}\mu x^2(l - \frac{1}{3}x),$ $D^2 y = \mu(l - x).$

3. $y = \frac{1}{24}\mu x(l-x)(l^2 + lx - x^2),$ $D^2 y = \frac{1}{2}\mu x(x - l).$

4. $y = \frac{1}{48}\mu x^2(3l^2 - 4lx + 2x^2),$ $D^2 y = \frac{1}{2}\mu(x - \frac{1}{2}l)^2.$

5. $y = \dfrac{1}{x^2 - x - 2},$ $D^2 y = \frac{2}{3}\left\{\dfrac{1}{(x-2)^3} - \dfrac{1}{(x+1)^3}\right\}.$

6. $y = \dfrac{1}{1+x^2},$ $D^4 y = \dfrac{24(1 - 10x^2 + 5x^4)}{(1+x^2)^5}.$

7. $y = \dfrac{2x}{1-x^2},$ $D^2 y = \dfrac{4x(x^2+3)}{(1-x^2)^3}.$

8. $y = (1-x)^{-m},$ $D^n y = m(m+1)\ldots(m+n-1)(1-x)^{-m-n}.$

9. $y = \dfrac{1+x}{1-x},$ $D^n y = \dfrac{2 \cdot n!}{(1-x)^{n+1}}.$

10. $y = \sin^2 x,$ $D^2 y = 2\cos 2x.$

11. $y = \cos^2 x,$ $D^n y = 2^{n-1}\cos(2x + \frac{1}{2}n\pi).$

12. $y = \sec x,$ $D^2 y = 2\sec^3 x - \sec x.$

13. $y = x^2 \sin x,$ $D^2 y = 4x \cos x - (x^2 - 2)\sin x.$

14. $y = \sin^3 x \cos x,$ $D^3 y = 6 - 60\sin^2 x + 64\sin^4 x.$

15. $y = \sin x \sinh x,$ $D^2 y = 2\cos x \cosh x.$

16. $y = \cos x \cosh x,$ $D^2 y = -2\sin x \sinh x.$

17. $y = \sin x \cosh x,$ $D^2 y = 2\cos x \sinh x.$

18. $y = \cos x \sinh x,$ $D^2 y = -2\sin x \cosh x.$

19. $y = \sin^{-1} x,$ $D^2 y = \dfrac{x}{(1-x^2)}.$

20. The first five derivatives of $\tan x$ are

$$1 + t^2, \quad 2t(1 + t^2), \quad 2(1 + 3t^2)(1 + t^2), \quad 8t(2 + 3t^2)(1 + t^2),$$
$$8(2 + 15t^2 + 15t^4)(1 + t^2),$$

where $t = \tan x$.

21. $y = x^3 \log \dfrac{x}{a}$, $\qquad\qquad D^4 y = \dfrac{6}{x}$.

22. $y = x^2 \log x$, $\qquad\qquad D^n y = (-)^{n-1} \dfrac{2(n-3)!}{x^{n-2}}$.

23. $y = x e^x$, $\qquad\qquad\qquad D^n y = (x + n)e^x$.

24. $y = x^2 e^x$, $\qquad\qquad\quad D^n y = \{x^2 + 2nx + n(n-1)\}e^x$.

25. By applying Leibnitz' Theorem to the differentiation of the identity

$$x^m \cdot x^n = x^{m+n},$$

prove that

$$m_r + \frac{r}{1} m_{r-1} n_1 + \frac{r(r-1)}{1 \cdot 2} m_{r-2} n_2 + \frac{r(r-1)(r-2)}{1 \cdot 2 \cdot 3} m_{r-3} n_3 + \ldots + n_r = (m+n)_r,$$

where $\qquad\qquad m_r = m(m-1)(m-2)\ldots(m-r+1)$.

26. By forming in different ways the nth derivative of x^{2n}, prove that

$$1 + \frac{n^2}{1^2} + \frac{n^2(n-1)^2}{1^2 \cdot 2^2} + \frac{n^2(n-1)^2(n-2)^2}{1^2 \cdot 2^2 \cdot 3^2} + \ldots = \frac{(2n)!}{(n!)^2}.$$

27. Prove that

$$D^n x^n (1-x)^n = n!(1-x)^n \left\{ 1 - \frac{n^2}{1^2} \frac{x}{1-x} + \frac{n^2(n-1)^2}{1^2 \cdot 2^2} \frac{x^2}{(1-x)^2} + \ldots \right\}.$$

28. If $\qquad\qquad\qquad y = \dfrac{1+x}{1-x}$,

$$D^n y^n = \frac{n!}{(1-x)^n} \left\{ 1 + \frac{n^2}{1^2} y + \frac{n^2(n-1)^2}{1^2 \cdot 2^2} y^2 + \ldots \right\}.$$

29. The equation $\qquad \dfrac{d^2 s}{dt^2} + k \dfrac{ds}{dt} + n^2 s = 0$

is satisfied by $\qquad\qquad s = A e^{-\frac{1}{2}kt} \cos(\sigma t + \epsilon)$,

for all values of A and ϵ, provided

$$\sigma^2 = n^2 - \tfrac{1}{4}k^2.$$

30. The equation $\qquad \dfrac{d^2 s}{dt^2} + 2n \dfrac{ds}{dt} + n^2 s = 0$

is satisfied by $\qquad\qquad s = (A + Bt)e^{-nt}$.

31. If
$$y = \{x + \sqrt{(1 + x^2)}\}^m,$$
prove that
$$(1 + x^2)\frac{d^2y}{dx^2} + x\frac{dy}{dx} - m^2y = 0.$$

32. If
$$x = \phi(t), \quad y = \chi(t),$$
prove that
$$\frac{d^2y}{dx^2} = \frac{\dfrac{dx}{dt}\dfrac{d^2y}{dt^2} - \dfrac{dy}{dt}\dfrac{d^2x}{dt^2}}{\left(\dfrac{dx}{dt}\right)^3}.$$

EXAMPLES. XXII.

1. Prove that, in a table of natural sines at intervals of 1′, the error of proportional parts never exceeds

·0000000106.

2. Shew that in a table of natural tangents the method of proportional parts fails for angles near 90°.

Also prove that the limit of error for angles near 45°, when the tangents are given at intervals of 1′, is

·0000000423.

3. Shew that in a table of log tangents the method of proportional parts fails both for angles near 0° and for angles near 90°.

Shew also that the maximum error involved in the method is least for angles near 45°.

4. Prove that the curve
$$y = \log x$$
is everywhere convex upwards.

5. Prove that the curve
$$y = x \log x$$
is everywhere concave upwards. Trace the curve.

6. Find the maximum ordinate, and the point of inflexion, of the curve
$$y = xe^{-x}.$$
Trace the curve.

> [The maximum ordinate corresponds to $x = 1$; the inflexion to $x = 2$.]

7. Shew that the curve $y = e^{-x^2}$ has inflexions at the points for which $x = \pm\dfrac{1}{\sqrt{2}}$; and trace it.

8. Find the maximum and minimum ordinates, and the inflexions, of the curve

$$y = xe^{-x^2}.$$

Trace the curve.

[The maximum and minimum ordinates are given by $x = \pm\sqrt{\frac{1}{2}}$; the inflexions by $x = 0$, $\pm\sqrt{\frac{3}{2}}$.]

9. A certain function $\phi(x)$ is constant and

$$= \tfrac{1}{2}(b^2 - a^2)$$

for $0 < x < a$; it

$$= \tfrac{1}{2}b^2 - \tfrac{1}{2}x^2 - \tfrac{1}{3}\frac{a^3}{x}$$

for $a < x < b$; and

$$= \tfrac{1}{3}\frac{b^3 - a^3}{x}$$

for $x > b$. Prove that $\phi(x)$ and $\phi'(x)$ are continuous, but that $\phi''(x)$ is discontinuous.

Trace the curve $y = \phi(x)$.

10. Shew that $\qquad y = x^2(3 - x)$

has an inflexion at the point $(1, 2)$. Trace the curve.

11. Shew that $\qquad y = x^2(1 - x^2)$

has inflexions at the points $\left(\pm\dfrac{1}{\sqrt{6}}, \dfrac{5}{36}\right)$. Trace the curve.

12. Find the points of inflexion of the curve

$$y = \frac{b^3}{a^2 + x^2}. \qquad\qquad [x = \pm a/\sqrt{3}.]$$

13. Shew that $\qquad y = \dfrac{a^2 x}{(x - a)^2}$

has a point of inflexion at $(-2a, -\tfrac{2}{9}a)$. Trace the curve.

14. Find the points of inflexion of the curve

$$y = \frac{x^3}{a^2 + x^2},$$

and trace the curve. $\qquad\qquad [x = 0, \ \pm a\sqrt{3}.]$

15. Shew that the curve

$$y = \frac{1 - x}{1 + x^2}$$

has three points of inflexion, and that they lie in a straight line. Trace the curve.

16. Prove that the equation

$$x^3 - 10x^2 + 15x - 6 = 0$$

has a triple root.

17. Prove that the equation

$$x^6 - 5x^5 + 5x^4 + 9x^3 - 14x^2 - 4x + 8 = 0$$

has a triple root; and find all the roots.

18. Find the maximum and minimum ordinates, and the points of inflexion on the curve

$$y = 1 - 3x^3 + 2x^4.$$

Trace the curve.

19. Find the points of inflexion on the curve

$$y = 2x^4 - 3x^3 - 3x^2 + 7x;$$

and trace the curve.

20. Find the maximum and minimum ordinates, and the point of inflexion, on the curve

$$y = 4x^3 + 9x^2 + 6x + 1.$$

Trace the curve.

21. Shew that the curves

$$y = x^3 + x^2 - x - 1,$$
$$y = 2\left(x^3 - x^2 + x - 1\right)$$

touch, and cross one another at the point of contact.

Trace the curves.

22. Determine the constants A, B, C so that the curve

$$y = Ax^3 + Bx^2 + Cx$$

may have a point of inflexion for $x = \frac{1}{2}$, be parallel to the axis of x for $x = -1$, and pass through the point $(1, 13)$.

23. Prove that the curve

$$y = \frac{x^2 - x + 1}{x^2 + x + 1}$$

has three real points of inflexion.

24. Find the turning points, and the points of inflexion, on the curve

$$y = 4 \sin x - \sin 2x.$$

Trace the curve.

25. If PN, $P'N'$ be two neighbouring ordinates of a curve $y = \phi(x)$, and if QH, any intermediate ordinate, meet the chord PP' in V, prove that

$$QV = \tfrac{1}{2}NH \cdot HN' \cdot \phi''(c),$$

ultimately, where c is the abscissa of some point between N and N'.

26. Shew that in the formula

$$\phi(a+h) = \phi(a) + h\phi'(a+\theta h)$$

of Art. 56, the limiting value of θ, when h is infinitely small, is in general $\frac{1}{2}$.

What is the geometrical meaning of this result?

27. Shew that the variation in the value of a function, in the neighbourhood of a maximum or minimum, is in general of the *second order* of small quantities.

28. Explain why the rate of a compensated chronometer, at any particular temperature, differs from the rate at the temperature of exact compensation by an amount proportional to the square of the difference of temperature.

29. Shew that, in a mathematical table calculated for equal intervals of the variable, the maximum error of interpolation by proportional parts, in any part of the table, is one-eighth of the 'second difference' (i.e. of the difference of the differences of successive entries).

30. The coordinates of three points P, Q, R on a curve are

$$(\cdot3250, \cdot8526), \quad (\cdot3500, \cdot8910), \quad (\cdot3750, \cdot9239),$$

respectively; find, approximately, the values of dy/dx and d^2y/dx^2 at Q.

31. Apply the formula (23) of Art. 70 to calculate $\log_{10} \cos 1°$ to six places of decimals.
$$[\bar{1}\cdot999934.]$$

CHAPTER VI

INTEGRATION

72. Nature of the Problem.

In the preceding chapters we have been occupied with the rate of variation of functions given *à priori*. The Integral Calculus, to which we now turn, is concerned with the inverse problem; viz. the rate of variation of a function being given, and the value of the function for some particular value of the independent variable being assigned, it is required to find the value of the function for any other assigned value of the independent variable. In symbols, it is required to solve the equation

$$\frac{dy}{dx} = \phi(x), \dots\dots\dots\dots\dots\dots\dots\dots(1)$$

where $\phi(x)$ is a given function of x, subject to the condition that for some specified value (a, say) of x, y shall have a given value (b).

For example, the law of velocity of a moving point being given, and the position of the point at the time t_0, it may be required to find its position at any other time t. This is equivalent to solving the equation

$$\frac{ds}{dt} = \phi(t), \dots\dots\dots\dots\dots\dots\dots\dots(2)$$

where $\phi(t)$ is a given function of t, subject to the condition that $s = s_0$ (say) for $t = t_0$.

If we can discover a continuous function $\psi(x)$ such that

$$\psi'(x) = \phi(x),$$

the equation (1) becomes

$$\frac{dy}{dx} = \frac{d}{dx}\psi(x). \dots\dots\dots\dots\dots\dots(3)$$

Hence if, as is the case in most practical applications of the subject, y be restricted to be continuous, we have, by Art. 56,

$$y = \psi(x) + C, \dots\dots\dots\dots\dots\dots\dots(4)$$

where C is a constant. The precise value of C is indeterminate, so far as the equation (1) is concerned; C is therefore called an 'arbitrary constant.' Its use is that it enables us to satisfy the remaining condition of the problem as above stated.

Thus if $y = b$ for $x = a$, we must have

$$b = \psi(a) + C,$$

whence

$$y - b = \psi(x) - \psi(a). \quad\quad\dots\dots\dots\dots\dots(5)$$

If, as in Art. 25, we use the symbol D for the operator d/dx, the equation (1) may be written

$$Dy = \phi(x), \quad\quad\dots\dots\dots\dots\dots(6)$$

and its solution may, consistently with the principles of algebraic notation, be written

$$y = D^{-1}\phi(x), \quad\quad\dots\dots\dots\dots(7)$$

the definition of the 'inverse' operator D^{-1} being that

$$D\{D^{-1}\phi(x)\} = \phi(x). \quad\quad\dots\dots\dots(8)$$

The function

$$D^{-1}\phi(x), \quad\quad\dots\dots\dots\dots\dots(9)$$

when it exists, is called the 'indefinite integral' of $\phi(x)$ with respect to x. It is more usually denoted by

$$\int \phi(x)\,dx. \quad\quad\dots\dots\dots\dots\dots(10)$$

The origin of this notation will be explained in the next chapter; in the meantime (10) is to be regarded as merely another way of writing (9).

The distinction between 'direct' and 'inverse' operations is one that occurs in many branches of Mathematics. A direct operation is one which can always be performed on any given function, according to definite rules, with an unambiguous result. An inverse operation is of the nature of a question: what function, operated on in a certain way, will produce an assigned result? To this question there may or may not be an answer, or there may be more than one answer (cf. Art. 16). In the case of the operator D^{-1} we have seen that if there is one answer, there are an infinite number, owing to the indeterminateness of the additive constant C. Whether there is, in every case, an answer is a matter yet to be investigated; but we may state, although this is rather more than we shall have occasion formally to prove, that every *continuous* function has an indefinite integral. In the rest of this chapter we shall be occupied with the problem of actually discovering indefinite integrals of various classes of mathematical functions.

Ex. Given that the velocity of a moving point is $u + gt$, we have

$$\frac{ds}{dt} = u + gt = \frac{d}{dt}(ut + \tfrac{1}{2}gt^2), \quad\quad\dots\dots\dots\dots(11)$$

whence

$$s = ut + \tfrac{1}{2}gt^2 + C. \quad\quad\dots\dots\dots\dots(12)$$

Determining C so that $s = s_0$ for $t = t_0$, we have

$$s - s_0 = u(t - t_0) + \tfrac{1}{2}g(t^2 - t_0^2). \quad\quad\dots\dots\dots\dots(13)$$

73. Standard Forms.

There are no infallible rules by which we can ascertain the indefinite integral

$$D^{-1}\phi(x) \text{ or } \int \phi(x)\,dx$$

of any given continuous function $\phi(x)$. As above stated, integration is an inverse process, in which we can only be guided by our recollections of the results of previous direct processes.

The integral, moreover, although in a certain sense it always exists, may not admit of being expressed (in a finite form) in terms of the functions, whether algebraic or transcendental, which are ordinarily employed in mathematics. The following are instances:

$$\int e^{x^2}dx, \quad \int \frac{\sin x}{x}\,dx, \quad \int \frac{dx}{\sqrt{(1+x^3)}},$$

and the list might easily be extended indefinitely.

The first step towards making a more or less systematic record of integrations is to write down a list of differentiations of various simple functions; each of these will, on inversion, furnish us with a result in indefinite integration. The arbitrary additive constant which always attaches to an indefinite integral need not be explicitly introduced, but its existence will occasionally be forced on the attention of the student by the fact of integrals of the same expression, arrived at in different ways, differing by a constant.

The student should make himself thoroughly familiar with the following results, which are fundamental:

$$\frac{d}{dx}\cdot x^n = nx^{n-1}, \qquad \int x^n dx = \frac{1}{n+1}x^{n+1}, \qquad (A)$$

$$[\text{except for } n = -1],$$

$$\frac{d}{dx}\cdot \log x = \frac{1}{x}, \qquad \int \frac{dx}{x} = \log x, \qquad (B)$$

$$\frac{d}{dx}\cdot e^{kx} = ke^{kx}, \qquad \int e^{kx} dx = \frac{1}{k}e^{kx}, \qquad (C)$$

$$\frac{d}{dx}\cdot \sin x = \cos x, \qquad \int \cos x\,dx = \sin x, \qquad (D)$$

$$\frac{d}{dx}\cdot \cos x = -\sin x, \qquad \int \sin x\,dx = -\cos x, \qquad (E)$$

$$\frac{d}{dx}\cdot \tan x = \sec^2 x, \qquad \int \sec^2 x\,dx = \tan x, \qquad (F)$$

$$\frac{d}{dx}.\cot x = -\operatorname{cosec}^2 x, \qquad \int \operatorname{cosec}^2 x\,dx = -\cot x, \qquad (G)$$

$$\frac{d}{dx}.\sin^{-1}\frac{x}{a} = \frac{1}{\sqrt{(a^2-x^2)}}, \qquad \int \frac{dx}{\sqrt{(a^2-x^2)}} = \sin^{-1}\frac{x}{a}, * \qquad (H)$$

$$\frac{d}{dx}.\tan^{-1}\frac{x}{a} = \frac{a}{a^2+x^2}, \qquad \int \frac{dx}{a^2+x^2} = \frac{1}{a}\tan^{-1}\frac{x}{a}, \qquad (I)$$

$$\frac{d}{dx}.\sinh x = \cosh x, \qquad \int \cosh x\,dx = \sinh x, \qquad (J)$$

$$\frac{d}{dx}.\cosh x = \sinh x, \qquad \int \sinh x\,dx = \cosh x, \qquad (K)$$

$$\frac{d}{dx}.\tanh x = \operatorname{sech}^2 x, \qquad \int \operatorname{sech}^2 x\,dx = \tanh x, \qquad (L)$$

$$\frac{d}{dx}.\coth x = -\operatorname{cosech}^2 x, \qquad \int \operatorname{cosech}^2 x\,dx = -\coth x, \qquad (M)$$

$$\frac{d}{dx}.\sinh^{-1}\frac{x}{a} = \frac{1}{\sqrt{(x^2+a^2)}}, \qquad \int \frac{dx}{\sqrt{(x^2+a^2)}} = \sinh^{-1}\frac{x}{a}$$
$$= \log\frac{x+\sqrt{(x^2+a^2)}}{a}, \qquad (N)$$

$$\frac{d}{dx}.\cosh^{-1}\frac{x}{a} = \frac{1}{\sqrt{(x^2-a^2)}}, \qquad \int \frac{dx}{\sqrt{(x^2-a^2)}} = \cosh^{-1}\frac{x}{a}$$
$$= \log\frac{x+\sqrt{(x^2-a^2)}}{a}, \dagger \qquad (O)$$

$$\frac{d}{dx}.\tanh^{-1}\frac{x}{a} = \frac{a}{a^2-x^2}, \qquad \int \frac{dx}{a^2-x^2} = \frac{1}{a}\tanh^{-1}\frac{x}{a}$$
$$[x^2 < a^2] \qquad\qquad = \frac{1}{2a}\log\frac{a+x}{a-x}, \qquad (P)$$

$$\frac{d}{dx}.\coth^{-1}\frac{x}{a} = -\frac{a}{x^2-a^2}, \qquad \int \frac{dx}{x^2-a^2} = -\frac{1}{a}\coth^{-1}\frac{x}{a}$$
$$[x^2 > a^2] \qquad\qquad = \frac{1}{2a}\log\frac{x-a}{x+a}. \qquad (Q)$$

A little care is necessary in the employment of some of these formulæ. In the first place, the sign of a in (H), (I), (N), (O), (P), (Q) is most conveniently taken to be positive; this is evidently always legitimate, since the square of a alone appears in the expression to be integrated.

* As to the question of sign, see Art. 33.

† As to the sign, see Art. 47.

Again, the formula (B) requires amendment when x is negative, since there is no logarithm of a negative quantity. Putting in this case $x = -x'$, and

$$y = \log x',$$

we have

$$\frac{dy}{dx} = -\frac{dy}{dx'} = -\frac{1}{x'} = \frac{1}{x}.$$

Hence

$$\int \frac{dx}{x} = \log x'.$$

The cases of x positive and x negative are both included in the formula

$$\int \frac{dx}{x} = \log |x|. \quad \dots\dots\dots\dots\dots\dots(B)$$

Again, the forms in (O) assume x to be positive. In (P) it is implied that $|x| < a$, and in (Q) that $|x| > a$.

74. Simple Extensions.

To extend the above results, we first notice that the addition of a constant to x makes no essential difference in the form of the result (cf. Art. 32, 1°).

Thus, obviously,

$$\int (x+a)^m \, dx = \frac{1}{m+1} (x+a)^{m+1}, \dots\dots\dots\dots\dots\dots(1)$$

$$\int \frac{dx}{x+a} = \log (x+a), \quad \dots\dots\dots\dots\dots\dots\dots(2)$$

$$\int \frac{dx}{\sqrt{(2ax - x^2)}} = \int \frac{dx}{\sqrt{\{a^2 - (x-a)^2\}}} = \sin^{-1} \frac{x-a}{a}, \dots\dots(3)$$

and so on. Some further illustrations occur in Arts. 75, 76.

Again, if x be multiplied by a factor k, the integral has the same form as before, except that it is divided by this factor (see Art. 32, 2°).

Thus

$$\int \sin kx \, dx = -\frac{1}{k} \cos kx, \quad \dots\dots\dots\dots\dots(4)$$

$$\int \frac{dx}{ax+b} = \frac{1}{a} \log (ax+b), \quad \dots\dots\dots\dots(5)$$

and so on.

Again, we have the theorems

$$\int C u \, dx = C \int u \, dx, \dots\dots\dots\dots\dots(6)$$

$$\int (u + v + w + \dots) \, dx = \int u \, dx + \int v \, dx + \int w \, dx + \dots ; \dots(7)$$

since, if we perform the operation d/dx on both sides we get in each case an identity, by Arts. 29, 30. It is assumed in (7) that the number of terms is finite.

Thus the indefinite integral of a rational integral function

$$A_0 x^m + A_1 x^{m-1} + \dots + A_{m-1} x + A_m, \dots\dots\dots(8)$$

is $$\frac{1}{m+1} A_0 x^{m+1} + \frac{1}{m} A_1 x^m + \dots + \tfrac{1}{2} A_{m-1} x^2 + A_m x. \dots(9)$$

Again, suppose we have a rational fraction of the form

$$\frac{F(x)}{x+a} \dots\dots\dots\dots\dots\dots(10)$$

By division this can be reduced to the sum of a rational integral function and a fraction

$$\frac{A}{x+a} \dots\dots\dots\dots\dots\dots(11)$$

The former part can be integrated as above, and the integral of (11) is

$$A \log (x + a). \dots\dots\dots\dots(12)$$

Ex. 1. $\int (x-1)^{\frac{3}{2}} \, dx = \dfrac{1}{\frac{3}{2}+1} (x-1)^{\frac{3}{2}+1} = \tfrac{2}{5} (x-1)^{\frac{5}{2}}.$

Ex. 2. $\int \dfrac{dx}{2x-1} = \tfrac{1}{2} \log (2x-1).$

Ex. 3. $\int \sin^2 x \, dx = \tfrac{1}{2} \int (1 - \cos 2x) \, dx = \tfrac{1}{2} x - \tfrac{1}{4} \sin 2x.$

Ex. 4. $\int \tan^2 x \, dx = \int (\sec^2 x - 1) \, dx = \tan x - x.$

Ex. 5. $\int \dfrac{x^3}{2x-1} \, dx = \int \left\{ \tfrac{1}{2} x^2 + \tfrac{1}{4} x + \tfrac{1}{8} + \dfrac{1}{8(2x-1)} \right\} dx$

$$= \tfrac{1}{6} x^3 + \tfrac{1}{8} x^2 + \tfrac{1}{8} x + \tfrac{1}{16} \log (2x-1).$$

75. Rational Fractions with a Quadratic Denominator.

We next shew how to integrate any expression of the form

$$\frac{F(x)}{x^2 + px + q}, \dots\dots\dots\dots\dots(1)$$

where $F(x)$ is rational and integral. If necessary, we first divide the numerator by the denominator until the remainder is of the form $ax + b$. We thus get the function (1) expressed as the sum of a rational integral function and a fraction

$$\frac{ax + b}{x^2 + px + q}. \qquad \dots\dots\dots\dots\dots\dots(2)$$

The former part can be integrated as in Art. 74; it remains only to consider the form (2).

We take first the case

$$\frac{1}{x^2 + px + q}. \qquad \dots\dots\dots\dots\dots\dots(3)$$

The form of the result will depend on whether $p^2 \gtreqless 4q$.

If $p^2 > 4q$, the quadratic expression can be resolved into real and distinct factors; thus

$$x^2 + px + q = (x - \alpha)(x - \beta). \qquad \dots\dots\dots\dots(4)$$

With a proper choice of the constants A, B we may then put

$$\frac{1}{(x - \alpha)(x - \beta)} = \frac{A}{x - \alpha} + \frac{B}{x - \beta}. \qquad \dots\dots\dots\dots(5)$$

For this will be an identity provided

$$1 = A(x - \beta) + B(x - \alpha), \qquad \dots\dots\dots\dots(6)$$

i.e. provided $\qquad A + B = 0, \quad A\beta + B\alpha = -1, \qquad \dots\dots\dots\dots(7)$

or $\qquad\qquad A = \frac{1}{\alpha - \beta}, \quad B = -\frac{1}{\alpha - \beta}. \qquad \dots\dots\dots\dots(8)$

Hence

$$\int \frac{dx}{(x - \alpha)(x - \beta)} = \frac{1}{\alpha - \beta} \left\{ \int \frac{dx}{x - \alpha} - \int \frac{dx}{x - \beta} \right\}$$

$$= \frac{1}{\alpha - \beta} \{ \log(x - \alpha) - \log(x - \beta) \}$$

$$= \frac{1}{\alpha - \beta} \log \frac{x - \alpha}{x - \beta}. \qquad \dots\dots\dots\dots\dots\dots(9)$$

When we have once learned that the two sides of (5) can be made identical, the values of A and B are most easily found as follows. We first multiply both sides by $x - \alpha$, and *afterwards* make $x = \alpha$; this gives A. Again, multiplying both sides by $x - \beta$, and then putting $x = \beta$, we find B. Hence the rule: To find A, omit the corresponding factor in the denominator of the expression which is to be resolved into partial fractions, and substitute α for x in the expression as thus modified. Similarly for B.

If $p^2 = 4q$, we have

$$x^2 + px + q = (x + \tfrac{1}{2}p)^2,$$

and

$$\int \frac{dx}{(x + \tfrac{1}{2}p)^2} = - \frac{1}{x + \tfrac{1}{2}p}. \quad\dots\dots\dots\dots(10)$$

If $p^2 < 4q$, we have

$$x^2 + px + q = (x + \tfrac{1}{2}p)^2 + (q - \tfrac{1}{4}p^2) = (x - \alpha)^2 + \beta^2,$$

where α, β are real, and β may be taken to be positive. Now

$$\int \frac{dx}{(x - \alpha)^2 + \beta^2} = \frac{1}{\beta} \tan^{-1} \frac{x - \alpha}{\beta}, \quad\dots\dots\dots(11)$$

by an obvious extension of Art. 73 (I).

The result when $p^2 > 4q$ can be put in a form analogous to (11). We may write

$$x^2 + px + q = (x + \tfrac{1}{2}p)^2 - (\tfrac{1}{4}p^2 - q) = (x - \alpha')^2 - \beta'^2, \quad\dots\dots(12)$$

where α', β' are real, and β' may be assumed positive. If $|x - \alpha'| < \beta'$, the formula is

$$\int \frac{dx}{\beta'^2 - (x - \alpha')^2} = \frac{1}{\beta} \tanh^{-1} \frac{x - \alpha'}{\beta}, \quad\dots\dots\dots(13)$$

by Art. 73 (P). If we put $\alpha' + \beta' = a$, $\alpha' - \beta' = \beta$, this is seen to be equivalent to (9); by Art. 46. If $|x - \alpha'| > \beta'$, the form is

$$\int \frac{dx}{(x - \alpha')^2 - \beta'^2} = - \frac{1}{\beta'} \coth^{-1} \frac{x - \alpha'}{\beta'}. \quad\dots\dots\dots(14)$$

Proceeding to the more general case (2), we observe that, by a proper choice of the constants λ, μ, we can make

$$ax + b = \lambda (2x + p) + \mu, \quad\dots\dots\dots\dots(15)$$

viz. we must have

$$\lambda = \tfrac{1}{2}a, \quad \mu = b - \tfrac{1}{2}pa. \quad\dots\dots\dots\dots(16)$$

Hence

$$\int \frac{ax + b}{x^2 + px + q} \, dx = \lambda \int \frac{2x + p}{x^2 + px + q} \, dx + \mu \int \frac{dx}{x^2 + px + q}. \quad\dots(17)$$

Of the two integrals on the right hand, the former is obviously equal to

$$\log (x^2 + px + q), \quad\dots\dots\dots\dots\dots(18)$$

and the latter has been dealt with above.

When the denominator can be resolved into real and distinct factors the integral on the left-hand side of (17) can be treated more simply by the method of 'partial fractions.' Thus, we have

$$\frac{ax + b}{(x - \alpha)(x - \beta)} = \frac{A}{x - \alpha} + \frac{B}{x - \beta}, \quad\dots\dots\dots(19)$$

provided $\qquad ax + b = A(x - \beta) + B(x - \alpha),$

i.e. provided $\qquad A + B = a, \quad A\beta + B\alpha = -b, \quad\ldots\ldots\ldots\ldots(20)$

or $\qquad A = \dfrac{a\alpha + b}{\alpha - \beta}, \quad B = \dfrac{a\beta + b}{\beta - \alpha}. \quad\ldots\ldots\ldots\ldots(21)$

It is unnecessary, however, to go through this work in every case, as the values of A, B can be found more simply by the artifice explained on p. 167.

The integration of (17) then gives

$$\int \frac{ax + b}{(x - \alpha)(x - \beta)}\, dx = A \log(x - \alpha) + B \log(x - \beta). \ldots(22)$$

Ex. 1. To find $\qquad \displaystyle\int \frac{dx}{2 - x - x^2}.$

Assuming $\qquad \dfrac{1}{(1 - x)(x + 2)} = \dfrac{A}{1 - x} + \dfrac{B}{x + 2},$

we find, by the method just referred to,

$$A = \tfrac{1}{3}, \quad B = \tfrac{1}{3}.$$

Hence $\qquad \displaystyle\int \frac{dx}{2 - x - x^2} = \tfrac{1}{3} - \log(1 - x) + \tfrac{1}{3}\log(x + 2)$

$$= \tfrac{1}{3}\log\frac{x + 2}{1 - x}.$$

Otherwise: $\quad \displaystyle\int \frac{dx}{2 - x - x^2} = \int \frac{dx}{\frac{9}{4} - (x + \frac{1}{2})^2} = \tfrac{2}{3}\tanh^{-1}\frac{x + \frac{1}{2}}{\frac{3}{2}}$

$$= \tfrac{2}{3}\tanh^{-1}\frac{2x + 1}{3}.$$

Ex. 2. $\quad \displaystyle\int \frac{dx}{1 - 4x + 4x^2} = \int \frac{dx}{(2x - 1)^2} = -\frac{1}{2}\frac{1}{2x - 1}.$

Ex. 3. $\quad \displaystyle\int \frac{dx}{1 - x + x^2} = \int \frac{dx}{(x - \frac{1}{2})^2 + \frac{3}{4}} = \frac{2}{\sqrt{3}}\tan^{-1}\frac{x - \frac{1}{2}}{\frac{1}{2}\sqrt{3}}$

$$= \frac{2}{\sqrt{3}}\tan^{-1}\frac{2x - 1}{\sqrt{3}}.$$

Ex. 4. $\quad \displaystyle\int \frac{1 - x}{1 - x + x^2}\, dx = \int \frac{\frac{1}{2} - \frac{1}{2}(2x - 1)}{1 - x + x^2}\, dx$

$$= \tfrac{1}{2}\int \frac{dx}{1 - x + x^2} - \tfrac{1}{2}\int \frac{2x - 1}{1 - x + x^2}\, dx$$

$$= \frac{1}{\sqrt{3}}\tan^{-1}\frac{2x - 1}{\sqrt{3}} - \tfrac{1}{2}\log(1 - x + x^2).$$

Ex. 5.   To integrate $\dfrac{2x-3}{(x-2)(x+1)}$.

Assuming that this $=\dfrac{A}{x-2}+\dfrac{B}{x+1}$,

we find $A=\tfrac{1}{3}, \quad B=\tfrac{5}{3}.$

The required integral is therefore

$$\tfrac{1}{3}\log(x-2)+\tfrac{5}{3}\log(x+1).$$

76.   Form $\dfrac{ax+b}{\sqrt{(Ax^2+Bx+C)}}$.

A somewhat similar treatment can be applied to functions of this type.

1°.   If A be positive, the form is equivalent to

$$\frac{ax+b}{\sqrt{(x^2+px+q)}}. \quad\dots\dots\dots\dots\dots(1)$$

Consider, in the first place, the form

$$\frac{1}{\sqrt{(x^2+px+q)}}. \quad\dots\dots\dots\dots\dots(2)$$

By completing the square, the expression under the root-sign may be put in one or other of the shapes

$$(x-\alpha)^2 \pm \beta^2.$$

Now, by Art. 73, (N), (O),

$$\int \frac{dx}{\sqrt{\{(x-\alpha)^2+\beta^2\}}} = \sinh^{-1}\frac{x-\alpha}{\beta}, \quad\dots\dots\dots(3)$$

and

$$\int \frac{dx}{\sqrt{\{(x-\alpha)^2-\beta^2\}}} = \cosh^{-1}\frac{x-\alpha}{\beta}. \quad\dots\dots\dots(4)$$

These functions have the alternative forms,

$$\log\frac{x-\alpha+\sqrt{\{(x-\alpha)^2\pm\beta^2\}}}{\beta},$$

or

$$\log\frac{x-\alpha+\sqrt{(x^2+px+q)}}{\beta}; \quad\dots\dots\dots\dots(5)$$

cf. Art. 46.

In the more general case (1), we assume

$$ax+b = \lambda(x+\tfrac{1}{2}p)+\mu, \quad\dots\dots\dots\dots(6)$$

which is satisfied by

$$\lambda = a, \quad \mu = b - \tfrac{1}{2}pa. \quad \text{......................(7)}$$

Hence

$$\int \frac{ax + b}{\sqrt{(x^2 + px + q)}}\, dx = \lambda \int \frac{x + \tfrac{1}{2}p}{\sqrt{(x^2 + px + q)}}\, dx + \mu \int \frac{dx}{\sqrt{(x^2 + px + q)}}.$$
$$\text{.........(8)}$$

The former of these two integrals is obviously equal to

$$\sqrt{(x^2 + px + q)},$$

and the latter has been dealt with above.

2°. We will next suppose that, in the form placed at the head of this Art., the coefficient A is negative. Without loss of generality we may put it $= -1$.

Consider, first, the function

$$\frac{1}{\sqrt{(q + px - x^2)}}. \quad \text{......................(9)}$$

Unless the quadratic expression be essentially negative, in which case the function would be imaginary for *all* real values of x, it can be put in the shape

$$\beta^2 - (x - \alpha)^2.$$

Now

$$\int \frac{dx}{\sqrt{\{\beta^2 - (x - \alpha)^2\}}} = \sin^{-1} \frac{x - \alpha}{\beta}. \quad \text{............(10)}$$

In the more general case of the function

$$\frac{ax + b}{\sqrt{(q + px - x^2)}}, \quad \text{....................(11)}$$

we assume

$$ax + b = \lambda\,(\tfrac{1}{2}p - x) + \mu, \quad \text{..................(12)}$$

or

$$\lambda = -a, \quad \mu = b + \tfrac{1}{2}pa. \quad \text{..................(13)}$$

Hence

$$\int \frac{ax + b}{\sqrt{(q + px - x^2)}}\, dx = \lambda \int \frac{\tfrac{1}{2}p - x}{\sqrt{(q + px - x^2)}}\, dx + \mu \int \frac{dx}{\sqrt{(q + px - x^2)}}.$$
$$\text{.........(14)}$$

The former of these two integrals is equal to

$$\sqrt{(q + px - x^2)},$$

and the latter has been treated above.

Ex. 1.

$$\int \frac{1+x}{\sqrt{(1-x-x^2)}}\, dx = \int \frac{(x+\frac{1}{2})+\frac{1}{2}}{\sqrt{(1-x-x^2)}}\, dx$$

$$= \int \frac{x+\frac{1}{2}}{\sqrt{(1-x-x^2)}}\, dx + \frac{1}{2} \int \frac{dx}{\sqrt{\{\frac{5}{4}-(x+\frac{1}{2})^2\}}}$$

$$= -\sqrt{(1-x-x^2)} + \frac{1}{2} \sin^{-1} \frac{x+\frac{1}{2}}{\frac{1}{2}\sqrt{5}}$$

$$= -\sqrt{(1-x-x^2)} + \frac{1}{2} \sin^{-1} \frac{2x+1}{\sqrt{5}}.$$

Ex. 2.

$$\int \frac{x+1}{\sqrt{(x^2+x+1)}}\, dx = \int \frac{(x+\frac{1}{2})+\frac{1}{2}}{\sqrt{(x^2+x+1)}}\, dx$$

$$= \int \frac{x+\frac{1}{2}}{\sqrt{(x^2+x+1)}}\, dx + \frac{1}{2} \int \frac{dx}{\sqrt{\{(x+\frac{1}{2})^2+\frac{3}{4}\}}}$$

$$= \sqrt{(x^2+x+1)} + \frac{1}{2} \sinh^{-1} \frac{x+\frac{1}{2}}{\frac{1}{2}\sqrt{3}}$$

$$= \sqrt{(x^2+x+1)} + \frac{1}{2} \sinh^{-1} \frac{2x+1}{\sqrt{3}}.$$

Ex. 3.
$$\int \sqrt{\left(\frac{1-x}{1+x}\right)}\, dx = \int \frac{1-x}{\sqrt{(1-x^2)}}\, dx$$

$$= \int \frac{dx}{\sqrt{(1-x^2)}} - \int \frac{x\, dx}{\sqrt{(1-x^2)}}$$

$$= \sin^{-1} x + \sqrt{(1-x^2)}.$$

77. Change of Variable.

There are two artifices of special use in integration; viz. the choice of a new independent variable, and the method of integration 'by parts.'

To change the variable in the integral

$$u = \int \phi(x)\, dx \dots\dots\dots\dots\dots\dots(1)$$

from x to t, where x is a given function of t, we have, by Art. 32,

$$\frac{du}{dt} = \frac{du}{dx}\frac{dx}{dt} = \phi(x)\frac{dx}{dt}, \dots\dots\dots\dots\dots(2)$$

and therefore, by the definition of the inverse symbol $\int$,

$$u = \int \phi(x)\frac{dx}{dt}\, dt.$$

Hence
$$\int \phi(x)\, dx = \int \phi(x)\, \frac{dx}{dt}\, dt. \quad \dots\dots\dots\dots(3)^*$$

Conversely, whenever a proposed integral is recognized to be of the form
$$\int \phi(u)\, \frac{du}{dx}\, dx, \quad \dots\dots\dots\dots\dots(4)$$
we may replace it by
$$\int \phi(u)\, du, \quad \dots\dots\dots\dots\dots\dots(5)$$
which is often easier to find.

The following are important cases:

1°.
$$\int \phi(x+a)\, dx = \int \phi(u)\, du, \quad \dots\dots\dots\dots(6)$$
where $u = x + a$.

2°.
$$\int \phi(kx)\, dx = \frac{1}{k}\int \phi(u)\, du, \quad \dots\dots\dots\dots(7)$$
where $u = kx$.

These results have already been employed in Art. 74.

3°.
$$\int \phi(x^2)\, x\, dx = \tfrac{1}{2}\int \phi(u)\, du, \quad \dots\dots\dots\dots(8)$$
where $u = x^2$.

The following are examples of (8).

Ex. 1.
$$\int \frac{dx}{x(1+x^2)} = \int \frac{x\, dx}{x^2(1+x^2)}$$
$$= \tfrac{1}{2}\int \frac{du}{u(1+u)} = \tfrac{1}{2}\int \left(\frac{1}{u} - \frac{1}{1+u}\right) du$$
$$= \tfrac{1}{2}\log \frac{u}{1+u} = \tfrac{1}{2}\log \frac{x^2}{1+x^2}$$
$$= \log \frac{x}{\sqrt{(1+x^2)}}.$$

Ex. 2.
$$\int \frac{x\, dx}{x^4-1} = \tfrac{1}{2}\int \frac{du}{u^2-1}$$
$$= \tfrac{1}{4}\int \left(\frac{1}{u-1} - \frac{1}{u+1}\right) du$$
$$= \tfrac{1}{4}\log \frac{u-1}{u+1}$$
$$= \tfrac{1}{4}\log \frac{x^2-1}{x^2+1}.$$

---

* Hence the rule: After the sign $\int$ replace dx by $\dfrac{dx}{dt}\, dt$.

Ex. 3.
$$\int \frac{x\,dx}{1+x^4} = \tfrac{1}{2}\int \frac{du}{1+u^2} = \tfrac{1}{2}\tan^{-1}u$$
$$= \tfrac{1}{2}\tan^{-1}x^2.$$

Ex. 4.
$$\int \frac{x\,dx}{\sqrt{(a^4-x^4)}} = \tfrac{1}{2}\int \frac{du}{\sqrt{(a^4-u^2)}} = \tfrac{1}{2}\sin^{-1}\frac{u}{a^2}$$
$$= \tfrac{1}{2}\sin^{-1}\frac{x^2}{a^2}.$$

The student will, after a little practice, find it easy to make such simple substitutions as the above mentally.

4°. Occasionally the integration of an algebraical function is facilitated by the substitution
$$x = 1/t, \qquad dx/dt = -1/t^2.*$$
Thus
$$\int \frac{dx}{x\sqrt{(a^2+x^2)}} = -\int \frac{dt}{t\sqrt{(a^2+t^{-2})}} = -\frac{1}{a}\int \frac{dt}{\sqrt{(t^2+a^{-2})}}$$
$$= -\frac{1}{a}\sinh^{-1}at$$
$$= -\frac{1}{a}\sinh^{-1}\frac{a}{x}$$
$$= \frac{1}{a}\log\frac{x}{a+\sqrt{(a^2+x^2)}}. \quad\ldots\ldots\ldots\ldots\ldots(9)$$

Similarly,
$$\int \frac{dx}{x\sqrt{(a^2-x^2)}} = -\frac{1}{a}\cosh^{-1}\frac{a}{x}$$
$$= \frac{1}{a}\log\frac{x}{a+\sqrt{(a^2-x^2)}}; \quad \ldots\ldots(10)$$

and
$$\int \frac{dx}{x\sqrt{(x^2-a^2)}} = -\frac{1}{a}\sin^{-1}\frac{a}{x}. \quad \ldots\ldots\ldots\ldots(11)$$

More generally, the integral
$$\int \frac{dx}{(x+a)\sqrt{(Ax^2+Bx+C)}} \quad\ldots\ldots\ldots\ldots\ldots(12)$$
is reduced by the substitution
$$x + a = 1/t$$
to one or other of the forms discussed in Art. 76.

---

* The substitution is equivalent to writing
$$-\frac{dt}{t} \quad \text{for} \quad \frac{dx}{x}.$$

Again, the substitution $x = 1/t$ gives

$$\int \frac{dx}{(a^2 + x^2)^{\frac{3}{2}}} = -\int \frac{dt}{t^2(a^2 + t^{-2})^{\frac{3}{2}}} = -\int \frac{t\,dt}{(1 + a^2 t^2)^{\frac{3}{2}}}$$

$$= \frac{1}{a^2} \frac{1}{(1 + a^2 t^2)^{\frac{1}{2}}}$$

$$= \frac{1}{a^2} \frac{x}{\sqrt{(a^2 + x^2)}}. \qquad\dots\dots\dots\dots(13)$$

Similarly

$$\int \frac{dx}{(a^2 - x^2)^{\frac{3}{2}}} = \frac{1}{a^2} \frac{x}{\sqrt{(a^2 - x^2)}}, \dots\dots\dots(14)$$

and

$$\int \frac{dx}{(x^2 - a^2)^{\frac{3}{2}}} = -\frac{1}{a^2} \frac{x}{\sqrt{(x^2 - a^2)}}. \qquad\dots\dots\dots(15)$$

The form

$$\int \frac{dx}{(Ax^2 + Bx + C)^{\frac{3}{2}}} \qquad\dots\dots\dots\dots(16)$$

can, by 'completing the square,' be brought under one or other of the preceding cases.

78. Integration of Trigonometrical Functions.

1°.

$$\int \tan x\,dx = \int \frac{\sin x}{\cos x}\,dx$$

$$= -\int \frac{d(\cos x)}{\cos x} = -\log \cos x$$

$$= \log \sec x. \dots\dots\dots\dots\dots(1)$$

Similarly

$$\int \cot x\,dx = \log \sin x. \dots\dots\dots\dots(2)$$

Again, by the same artifice,

$$\int \frac{\sin x}{\cos^2 x}\,dx = -\int \frac{d(\cos x)}{\cos^2 x}$$

$$= \frac{1}{\cos x} = \sec x. \dots\dots\dots\dots(3)$$

In a similar manner

$$\int \frac{\cos x}{\sin^2 x}\,dx = -\operatorname{cosec} x. \dots\dots\dots\dots(4)$$

Cf. Art. 31, 2°.

2°.
$$\int \frac{dx}{\sin x} = \int \frac{dx}{2 \sin \frac{1}{2} x \cos \frac{1}{2} x}$$

$$= \frac{1}{2} \int \frac{\sec^2 \frac{1}{2} x \, dx}{\tan \frac{1}{2} x} = \int \frac{d (\tan \frac{1}{2} x)}{\tan \frac{1}{2} x}$$

$$= \log \tan \frac{1}{2} x. \quad \dots\dots\dots\dots\dots\dots(5)$$

From this we deduce

$$\int \frac{dx}{\cos x} = \int \frac{dx}{\sin (\frac{1}{2} \pi + x)} = \log \tan (\tfrac{1}{4} \pi + \tfrac{1}{2} x). \dots\dots(6)$$

The formulæ (1) to (6) rank as standard results, and should be remembered.

3°.
$$\int \frac{dx}{a + b \cos x} = \int \frac{dx}{(a + b) \cos^2 \frac{1}{2} x + (a - b) \sin^2 \frac{1}{2} x}$$

$$= \int \frac{\sec^2 \frac{1}{2} x \, dx}{(a + b) + (a - b) \tan^2 \frac{1}{2} x}. \quad \dots\dots(7)$$

If we put $\tan \frac{1}{2} x = u$, this takes the shape

$$2 \int \frac{du}{(a + b) + (a - b) u^2}, \quad \dots\dots\dots\dots(8)$$

and so comes under one or other of the standard forms (I), (P), (Q) of Art. 73.

Similarly, with the same substitution,

$$\int \frac{dx}{a + b \sin x} = 2 \int \frac{du}{a + 2bu + au^2}. \quad \dots\dots\dots(9)$$

4°.
$$\int \frac{dx}{a^2 \cos^2 x + b^2 \sin^2 x} = \int \frac{\sec^2 x \, dx}{a^2 + b^2 \tan^2 x}. \quad \dots\dots(10)$$

If we put $\qquad \tan x = u,$

we get
$$\int \frac{du}{a^2 + b^2 u^2} = \frac{1}{b} \int \frac{d (bu)}{a^2 + (bu)^2} = \frac{1}{ab} \tan^{-1} \frac{bu}{a}$$

$$= \frac{1}{ab} \tan^{-1} \left(\frac{b}{a} \tan x \right). \quad \dots\dots\dots\dots(11)$$

The analogous results involving hyperbolic functions may be noted We easily find

$$\int \tanh x \, dx = \log \cosh x, \quad \int \coth x \, dx = \log \sinh x, \dots\dots\dots(12)$$

$$\int \frac{\sinh x}{\cosh^2 x} \, dx = - \operatorname{sech} x, \quad \int \frac{\cosh x}{\sinh^2 x} \int dx = - \operatorname{cosech} x, \quad \dots\dots(13)$$

$$\int \frac{dx}{\sinh x} = \log \tanh \tfrac{1}{2}x, \dots\dots\dots\dots\dots(14)$$

$$\int \frac{dx}{\cosh x} = 2\int \frac{e^x\, dx}{e^{2x}+1} = 2\tan^{-1} e^x. \dots\dots\dots(15)$$

Similarly the forms

$$\int \frac{dx}{a + b\cosh x}, \qquad \int \frac{dx}{a + b\sinh x}$$

can be integrated by the substitution $\tanh \tfrac{1}{2}x = u$.

79. Trigonometrical Substitutions.

The integration of an algebraic function involving the square root of a quadratic expression is often facilitated by the substitution of a trigonometrical or a hyperbolic function for the independent variable.

Thus : the occurrence of $\sqrt{(a^2 - x^2)}$ suggests the substitution

$$x = a\sin\theta, \ \text{ or } \ x = a\tanh u\,;$$

that of $\sqrt{(x^2 - a^2)}$ suggests

$$x = a\sec\theta, \ \text{ or } \ x = a\cosh u\,;$$

that of $\sqrt{(x^2 + a^2)}$ suggests

$$x = a\tan\theta, \ \text{ or } \ x = a\sinh u.$$

Ex. 1.   To find $\qquad \int \sqrt{(a^2 - x^2)}\, dx. \dots\dots\dots\dots\dots(1)$

Putting $\qquad\qquad x = a\sin\theta, \quad dx = a\cos\theta\, d\theta,$

we find $\qquad \int \sqrt{(a^2 - x^2)}\, dx = a^2 \int \cos^2\theta\, d\theta$

$$= \tfrac{1}{2}a^2\!\int (1 + \cos 2\theta)\, d\theta$$

$$= \tfrac{1}{2}a^2 (\theta + \tfrac{1}{2}\sin 2\theta)$$

$$= \tfrac{1}{2}a^2 \sin^{-1}\frac{x}{a} + \tfrac{1}{2}x \sqrt{(a^2 - x^2)}.\dots\dots\dots(2)$$

Ex. 2.   To find $\qquad \int \dfrac{\sqrt{(x^2 + a^2)}\, dx}{x^2}. \dots\dots\dots\dots\dots(3)$

Putting $\qquad\qquad x = a\sinh u, \quad dx = a\cosh u\, du,$

we obtain the form $\qquad \int \coth^2 u\, du,$

which $\qquad\qquad = \int(1 + \operatorname{cosech}^2 u)\, du = u - \coth u$

$$= \sinh^{-1}\frac{x}{a} - \frac{\sqrt{(x^2 + a^2)}}{x}. \dots\dots\dots\dots(4)$$

Ex. 3.   To find $\displaystyle\int \frac{dx}{(1-x)\sqrt{(1-x^2)}}$.

If we put $x = \cos\theta, \quad dx = -\sin\theta\, d\theta,$

the integral becomes

$$-\int \frac{d\theta}{1-\cos\theta} = -\tfrac{1}{2}\int \frac{d\theta}{\sin^2 \tfrac{1}{2}\theta} = \cot{}^1\theta$$

$$= \sqrt{\left(\frac{1+x}{1-x}\right)}.$$

80. Integration by Parts.

The second method referred to in Art. 77, viz. that of 'integration by parts,' consists in an inversion of the formula

$$\frac{d}{dx}(uv) = u\frac{dv}{dx} + v\frac{du}{dx}, \qquad \dots\dots\dots\dots(1)$$

given in Art. 30.   Integrating both sides, we find

$$uv = \int u\frac{dv}{dx}\, dx + \int v\frac{du}{dx}\, dx,$$

whence $\displaystyle\int u\frac{dv}{dx}\, dx = uv - \int v\frac{du}{dx}\, dx. \qquad \dots\dots\dots\dots(2)^*$

This gives the following rule:

If the expression to be integrated consists of two factors, one of which (dv/dx) is by itself immediately integrable, we may integrate as if the remaining factor (u) were constant, provided we subtract the integral of the product of the integrated factor (v) into the derivative (du/dx) of the other factor.

A very useful particular case is obtained by putting $v = x$, in (2).   Thus

$$\int u\, dx = xu - \int x\frac{du}{dx}\, dx. \dots\dots\dots\dots\dots(3)$$

The following are important applications of the method.

1°. $\displaystyle\int \log x\, dx = x\log x - \int x\cdot\frac{1}{x}\, dx$

$$= x\log x - x. \qquad \dots\dots\dots\dots\dots(4)$$

---

* If we write v for dv/dx, and therefore $D^{-1}v$ for v, this takes the form

$$D^{-1}(uv) = uD^{-1}v - D^{-1}(Du\cdot D^{-1}v).$$

2°.　To find　　　$\int \sqrt{(a^2 - x^2)}\, dx.$

Putting $u = \sqrt{(a^2 - x^2)}$ in (3), we have

$$\int \sqrt{(a^2 - x^2)}\, dx = x\sqrt{(a^2 - x^2)} + \int \frac{x^2\, dx}{\sqrt{(a^2 - x^2)}}. \quad \ldots\ldots(5)$$

But　　　$\int \sqrt{(a^2 - x^2)}\, dx = \int \frac{a^2 - x^2}{\sqrt{(a^2 - x^2)}}\, dx$

$$= a^2 \int \frac{dx}{\sqrt{(a^2 - x^2)}} - \int \frac{x^2\, dx}{\sqrt{(a^2 - x^2)}}$$

$$= a^2 \sin^{-1} \frac{x}{a} - \int \frac{x^2\, dx}{\sqrt{(a^2 - x^2)}}. \quad \ldots\ldots\ldots\ldots(6)$$

Adding to the former result, and dividing by 2, we find

$$\int \sqrt{(a^2 - x^2)}\, dx = \tfrac{1}{2} x\sqrt{(a^2 - x^2)} + \tfrac{1}{2} a^2 \sin^{-1} \frac{x}{a}; \quad \ldots\ldots(7)$$

cf. Art. 79, Ex. 1.

In exactly the same way we should find

$$\int \sqrt{(a^2 + x^2)}\, dx = \tfrac{1}{2} x\sqrt{(a^2 + x^2)} + \tfrac{1}{2} a^2 \sinh^{-1} \frac{x}{a}, \quad \ldots\ldots(8)$$

$$\int \sqrt{(x^2 - a^2)}\, dx = \tfrac{1}{2} x\sqrt{(x^2 - a^2)} - \tfrac{1}{2} a^2 \cosh^{-1} \frac{x}{a}. \quad \ldots\ldots(9)$$

3°.　To find the integrals

$$P = \int e^{ax} \cos \beta x\, dx, \quad Q = \int e^{ax} \sin \beta x\, dx. \ldots\ldots\ldots(10)$$

Putting　　　$u = \cos \beta x, \quad v = \frac{1}{a} e^{ax}$

in (2), we find

$$P = \frac{1}{a} e^{ax} \cos \beta x - \int \frac{1}{a} e^{ax} \cdot (-\beta \sin \beta x)\, dx$$

$$= \frac{1}{a} e^{ax} \cos \beta x + \frac{\beta}{a} Q. \ldots\ldots\ldots\ldots\ldots\ldots\ldots(11)$$

Similarly

$$Q = \frac{1}{a} e^{ax} \sin \beta x - \int \frac{1}{a} e^{ax} \cdot \beta \cos \beta x\, dx$$

$$= \frac{1}{a} e^{ax} \sin \beta x - \frac{\beta}{a} P. \quad \ldots\ldots\ldots\ldots\ldots\ldots(12)$$

Hence
$$\alpha P - \beta Q = e^{\alpha x} \cos \beta x, \\ \beta P + \alpha Q = e^{\alpha x} \sin \beta x, \Big\} \quad \dots\dots\dots\dots(13)$$

and therefore

$$\int e^{\alpha x} \cos \beta x \, dx = P = \frac{\beta \sin \beta x + \alpha \cos \beta x}{\alpha^2 + \beta^2} \, e^{\alpha x}, \\ \int e^{\alpha x} \sin \beta x \, dx = Q = \frac{\alpha \sin \beta x - \beta \cos \beta x}{\alpha^2 + \beta^2} \, e^{\alpha x}. \Bigg\} \quad \dots\dots(14)$$

81. Integration by Successive Reduction.

Sometimes, by an integration 'by parts,' or otherwise, one integral can be made to depend on another of simpler form.

1°. Let
$$u_n = \int x^n e^{\alpha x} dx. \quad \dots\dots\dots\dots\dots(1)$$

We have
$$u_n = \frac{1}{\alpha} e^{\alpha x} . x^n - \int \frac{1}{\alpha} e^{\alpha x} . n x^{n-1} dx$$

$$= \frac{1}{\alpha} x^n e^{\alpha x} - \frac{n}{\alpha} u_{n-1}. \quad \dots\dots\dots\dots\dots(2)$$

If n be a positive integer, we can by successive applications of this formula obtain u_n in terms of

$$u_0, = \int e^{\alpha x} dx, = \frac{1}{\alpha} e^{\alpha x}. \quad \dots\dots\dots\dots(3)$$

Ex. 1. Thus, if
$$u_n = \int x^n e^{-x} dx, \quad \dots\dots\dots\dots\dots(4)$$

we have
$$u_n = - x^n e^{-x} + n u_{n-1}. \quad \dots\dots\dots\dots(5)$$

For example,
$$u_3 = - x^3 e^{-x} + 3 u_2 = - x^3 e^{-x} + 3 \left(- x^2 e^{-x} + 2 u_1 \right)$$
$$= - x^3 e^{-x} - 3 x^2 e^{-x} + 6 \left(- x e^{-x} + u_0 \right),$$

or
$$\int x^3 e^{-x} dx = - \left(x^3 + 3 x^2 + 6 x + 6 \right) e^{-x}.$$

2°. Let
$$u_n = \int x^n \cos \beta x \, dx, \\ v_n = \int x^n \sin \beta x \, dx. \Big\} \quad \dots\dots\dots\dots(6)$$

We find
$$u_n = \frac{1}{\beta} \sin \beta x . x^n - \int \frac{1}{\beta} \sin \beta x . n x^{n-1} dx$$

$$= \frac{1}{\beta} \sin \beta x . x^n - \frac{n}{\beta} v_{n-1}, \quad \dots\dots\dots\dots(7)$$

and
$$v_n = - \frac{1}{\beta} \cos \beta x . x^n - \int \left(- \frac{1}{\beta} \cos \beta x \right) . n x^{n-1} dx$$

$$= - \frac{1}{\beta} \cos \beta x . x^n + \frac{n}{\beta} u_{n-1}. \quad \dots\dots\dots\dots(8)$$

If n is a positive integer, these formulæ enable us to express u_n and v_n in terms of either u_0 or v_0, which are known.

Ex. 2. Thus, if $\beta = 1$, we have

$$u_n = x^n \sin x - n v_{n-1}, \quad v_n = -x^n \cos x + n u_{n-1}. \quad \ldots\ldots\ldots(9)$$

For example,

$$u_3 = x^3 \sin x - 3 v_2 = x^3 \sin x - 3 \left(-x^2 \cos x + 2 u_1 \right)$$

$$= x^3 \sin x + 3 x^2 \cos x - 6 \left(x \sin x - v_0 \right),$$

or $\quad \int x^3 \cos x \, dx = (x^3 - 6x) \sin x + (3x^2 - 6) \cos x.$

3°. If $\qquad u_n = \int \tan^n \theta \, d\theta \quad \ldots\ldots\ldots\ldots\ldots \ldots\ldots\ldots(10)$

$$= \int \tan^{n-2} \theta \left(\sec^2 \theta - 1 \right) d\theta$$

$$= \int \tan^{n-2} \theta \, d \left(\tan \theta \right) - \int \tan^{n-2} \theta \, d\theta,$$

we have $\qquad u_n = \dfrac{1}{n-1} \tan^{n-1} \theta - u_{n-2}. \quad \ldots\ldots\ldots\ldots(11)$

Hence if n be a positive integer, u_n can be made to depend either on

$$u_1, \; = \int \tan \theta \, d\theta, \; = \log \sec \theta, \quad \ldots\ldots\ldots\ldots(12)$$

or on $\qquad u_0, \; = \int d\theta, \; = \theta, \quad \ldots\ldots\ldots\ldots\ldots\ldots(13)$

according as n is odd or even.

Similarly, if $\qquad v_n = \int \cot^n \theta \, d\theta, \quad \ldots\ldots\ldots\ldots(14)$

we find $\qquad v_n = -\dfrac{1}{n-1} \cot^{n-1} \theta - v_{n-2}. \quad \ldots\ldots\ldots\ldots(15)$

82. Reduction Formulæ, continued.

1°. Let $\qquad u_n = \int \cos^n \theta \, d\theta. \quad \ldots\ldots\ldots\ldots\ldots\ldots(1)$

We have

$$u_n = \int \cos^{n-1} \theta \, d \left(\sin \theta \right)$$

$$= \sin \theta \cos^{n-1} \theta - \int \sin \theta \, . \, (n-1) \cos^{n-2} \theta \, . \, (-\sin \theta) \, d\theta$$

$$= \sin \theta \cos^{n-1} \theta + (n-1) \int \left(1 - \cos^2 \theta \right) \cos^{n-2} \theta \, d\theta$$

$$= \sin \theta \cos^{n-1} \theta + (n-1) \left(u_{n-2} - u_n \right).$$

Transposing, and dividing by n, we find

$$u_n = \frac{1}{n} \sin \theta \cos^{n-1} \theta + \frac{n-1}{n} u_{n-2}. \quad \ldots\ldots\ldots\ldots(2)$$

By successive applications of this formula we reduce the index by 2 at each step; and finally, if n be a positive integer, the integral u_n is made to depend either on

$$u_1, \ = \int \cos \theta \, d\theta, \ = \sin \theta, \ \dots\dots\dots\dots(3)$$

or on

$$u_0, \ = \int d\theta, \ = \theta, \dots\dots\dots\dots\dots(4)$$

according as n is odd or even.

2°. By a similar process, if

$$v_n = \int \sin^n \theta \, d\theta, \ \dots\dots\dots\dots\dots(5)$$

we find

$$v_n = -\frac{1}{n} \cos \theta \sin^{n-1} \theta + \frac{n-1}{n} v_{n-2}. \ \dots\dots\dots(6)$$

In this way v_n, when n is a positive integer, is made to depend either on

$$v_1, \ = \int \sin \theta \, d\theta, \ = -\cos \theta, \ \dots\dots\dots\dots(7)$$

or on

$$v_0, \ = \int d\theta, \ = \theta. \ \dots\dots\dots\dots\dots(8)$$

3°. The same method can be applied to the more general form

$$u_{m,\,n} = \int \sin^m \theta \cos^n \theta \, d\theta. \ \dots\dots\dots\dots(9)$$

We have

$$u_{m,\,n} = \int \sin^m \theta \cos^{n-1} \theta \, d(\sin \theta)$$

$$= \frac{1}{m+1} \sin^{m+1} \theta \cos^{n-1} \theta$$

$$\qquad - \frac{1}{m+1} \int \sin^{m+1} \theta \,.\, (n-1) \cos^{n-2} \theta \,.\, (-\sin \theta) \, d\theta$$

$$= \frac{1}{m+1} \sin^{m+1} \theta \cos^{n-1} \theta$$

$$\qquad + \frac{n-1}{m+1} \int \sin^m \theta \cos^{n-2} \theta \, (1 - \cos^2 \theta) \, d\theta$$

$$= \frac{1}{m+1} \sin^{m+1} \cos^{n-1} \theta + \frac{n-1}{m+1} (u_{m,\,n-2} - u_{m,\,n}).$$

Clearing of fractions, transposing, and dividing by $m + n$, we obtain

$$u_{m,\,n} = \frac{1}{m+n} \sin^{m+1} \theta \cos^{n-1} \theta + \frac{n-1}{m+n} u_{m,\,n-2}. \ \dots(10)$$

In a similar manner we should find

$$u_{m,\,n} = -\frac{1}{m+n} \sin^{m-1} \theta \cos^{n+1} \theta + \frac{m-1}{m+n} u_{m-2,\,n}. \ \dots(11)$$

By successive applications of (10) and (11) we can reduce either index by 2 at each step, so that finally, if m, n are positive integers, the integral $u_{m,n}$ is made to depend on one or other of the following forms:

$$u_{1,1}, \; = \int \sin\theta \cos\theta\, d\theta, \; = \tfrac{1}{2}\sin^2\theta, \; \dots\dots\dots\dots(12)$$

$$u_{0,0}, \; = \int d\theta, \; = \theta, \; \dots\dots\dots\dots\dots\dots\dots(13)$$

$$u_{0,1}, \; = \int \cos\theta\, d\theta, \; = \sin\theta, \; \dots\dots\dots\dots\dots(14)$$

$$u_{1,0}, \; = \int \sin\theta\, d\theta, \; = -\cos\theta. \; \dots\dots\dots\dots(15)$$

The investigations of this section are chiefly important as leading to some simple and practically very useful results in definite integrals. See Art. 97.

83. Integration of Rational Fractions.

We return to the integration of algebraic functions. There are certain classes of such functions which can be treated by general methods.

We begin with the case of *rational* functions. A rational fraction

$$\frac{F(x)}{f(x)}, \quad \dots\dots\dots\dots\dots\dots\dots\dots(1)$$

in which the numerator is of lower dimensions than the denominator, is called a 'proper' fraction. Any rational fraction in which this condition is not fulfilled can by division be reduced to the sum of an integral function and a proper fraction; it will therefore be sufficient for us to consider the integration of proper fractions. Accordingly, if $f(x)$ be a polynomial of degree n, say

$$f(x) \equiv x^n + p_1 x^{n-1} + p_2 x^{n-2} + \dots + p_{n-1}x + p_n, \dots(2)$$

we shall suppose that $F(x)$ is at most of degree $n-1$.

To facilitate the integration, we resolve (1) into the sum of a series of 'partial fractions.' The possibility of this resolution depends on certain general theorems of Algebra, for the proof of which reference may be made to the special treatises on that subject*.

The student will find, however, that for such comparatively simple cases as are usually met with in practice, a mastery of the algebraical theory is not essential; since the results obtained by the rules to be given may be easily verified *à posteriori*.

---

* For the complete theory see Chrystal, *Algebra*, c. viii.

We will first suppose that the roots of the equation

$$f(x) = 0 \quad\text{............................}(3)$$

are all real and distinct, say they are α_1, α_2, ... α_n. The polynomial $f(x)$ then resolves into n distinct factors of the first degree, thus

$$f(x) \equiv (x - \alpha_1)(x - \alpha_2) \ldots (x - \alpha_n). \quad\text{............}(4)$$

There is no difficulty in shewing that in this case the fraction (1) can be resolved into the sum of n partial fractions whose denominators are the several factors of $f(x)$; thus

$$\frac{F(x)}{(x - \alpha_1)(x - \alpha_2)\ldots(x - \alpha_n)} = \frac{A_1}{x - \alpha_1} + \frac{A_2}{x - \alpha_2} + \ldots + \frac{A_n}{x - \alpha_n}, \ldots(5)$$

where A_1, A_2,..., A_n are certain constants. If we clear (5) of fractions, and then equate coefficients of x^{n-1}, x^{n-2}, ..., x^1, x^0 on the two sides we get exactly n conditions to determine the n constants; but it is not evident that the conditions in question are consistent and independent, and that the determination of A_1, A_2,..., A_n so as to satisfy (5) is therefore possible and unique. The two rational integral functions to be identified will, however, become equal for $x = \alpha_1$, $x = \alpha_2$, ..., $x = \alpha_n$, respectively, provided

$$\left.\begin{aligned}
A_1(\alpha_1 - \alpha_2)(\alpha_1 - \alpha_3) \ldots (\alpha_1 - \alpha_n) &= F(\alpha_1), \\
A_2(\alpha_2 - \alpha_1)(\alpha_2 - \alpha_3) \ldots (\alpha_2 - \alpha_n) &= F(\alpha_2), \\
\text{..} \\
A_n(\alpha_n - \alpha_1)(\alpha_n - \alpha_2) \ldots (\alpha_n - \alpha_{n-1}) &= F(\alpha_n),
\end{aligned}\right\} \quad\text{......}(6)*$$

or

$$A_1 = \frac{F(\alpha_1)}{f'(\alpha_1)}, \ A_2 = \frac{F(\alpha_2)}{f'(\alpha_2)}, \ldots, A_n = \frac{F(\alpha_n)}{f'(\alpha_n)} \quad\text{.........}(7)$$

Now two rational functions of degree $n - 1$ cannot be equal for more than $n - 1$ distinct values of x unless they are identical. Hence, with these values of the constants, (5) is an identity.

We then have

$$\int \frac{F(x)}{f(x)}\, dx = A_1 \log (x - \alpha_1) + A_2 \log (x - \alpha_2) + \ldots + A_n \log (x - \alpha_n).$$
$$\text{.........}(8)$$

Ex. 1.   To find $\qquad \displaystyle\int \frac{x^5\, dx}{x^4 - 5x^2 + 4} \cdot$ $\qquad\text{..............................}(9)$

We write

$$\frac{x^5}{x^4 - 5x^2 + 4} = x + \frac{5x^3 - 4x}{x^4 - 5x^2 + 4} = x + \frac{A}{x - 1} + \frac{B}{x + 1} + \frac{C}{x - 2} + \frac{D}{x + 2}. \ldots(10)$$

---

* This is an extension of Art. 75 (8).

If we clear of fractions and then equate coefficients, we get four linear equations to determine A, B, C, D. It is simpler, however, to use the method just explained. If we multiply the assumed identity by $x-1$, and afterwards put $x = 1$, we obtain the value of A; and similarly for the other coefficients. We thus find

$$\frac{x^5}{x^4 - 5x^2 + 4} = x - \frac{1}{6}\frac{1}{x-1} - \frac{1}{6}\frac{1}{x+1} + \frac{8}{3}\frac{1}{x-2} + \frac{8}{3}\frac{1}{x+2},\dots\dots(11)$$

a result which is easily verified. Hence the required integral is

$$\tfrac{1}{2}x^2 - \tfrac{1}{6}\log(x-1) - \tfrac{1}{6}\log(x+1) + \tfrac{8}{3}\log(x-2) + \tfrac{8}{3}\log(x+2)\dots(12)$$

Ex. 2.   To find $$\int \frac{dx}{x^2(1+x^2)}.\dots\dots\dots\dots\dots\dots(13)$$

Treating $1/x^2(1+x^2)$ as a function of x^2, we have

$$\frac{1}{x^2(1+x^2)} = \frac{1}{x^2} - \frac{1}{1+x^2},\dots\dots\dots\dots\dots(14)$$

whence $$\int \frac{dx}{x^2(1+x^2)} = -\frac{1}{x} - \tan^{-1}x.\dots\dots\dots(15)$$

84.   Case of Equal Roots.

If the roots of the equation $f(x) = 0$ are real but not all distinct, then, corresponding to an r-fold root β, we have a factor $(x-\beta)^r$ in $f(x)$. It is shewn, in the algebraic theory referred to, that the corresponding series of partial fractions, in the expansion of Art. 83 (1), is now

$$\frac{B_1}{x-\beta} + \frac{B_2}{(x-\beta)^2} + \dots + \frac{B_r}{(x-\beta)^r},\dots\dots\dots(1)$$

where $B_1, B_2, \dots B_r$ are r constants, to be determined by the method of equating coefficients, or otherwise.

The indefinite integral of the expression (1) is

$$B_1\log(x-\beta) - \frac{B_2}{x-\beta} - \frac{1}{2}\frac{B_3}{(x-\beta)^2} - \dots - \frac{1}{r-1}\frac{B_r}{(x-\beta)^{r-1}}.\dots(2)$$

Ex. 1.   To find $$\int \frac{dx}{x^2(1-x)}.\dots\dots\dots\dots\dots\dots\dots(3)$$

We assume $$\frac{1}{x^2(1-x)} = \frac{A}{x} + \frac{B}{x^2} + \frac{C}{1-x}.\dots\dots\dots\dots\dots(4)$$

If we multiply both sides by $1-x$, and then put $x = 1$, we get $C = 1$. Again, multiplying by x^2, and then putting $x = 0$, we find $B = 1$. The constant A remains to be found in some other way. If we multiply both sides of (4) by x, and then make $x \to \infty$, we find $A - C = 0$, whence $A = 1$. An equivalent method is to clear of fractions and equate the coefficients of x^2. Again, we might assign some other special value to x; for example, putting $x = -1$, we find

$$-A + B + \tfrac{1}{2}C = \tfrac{1}{2},$$

which, combined with the previous results, gives $A = 1$.

Hence
$$\int \frac{dx}{x^2 (1 - x)} = \int \left(\frac{1}{x} + \frac{1}{x^2} + \frac{1}{1 - x} \right) dx$$

$$= \log x - \frac{1}{x} - \log (1 - x) \quad \ldots\ldots\ldots (5)$$

Ex. 2. To find
$$\int \frac{2x + 1}{(x + 2)(x - 3)^2} dx. \quad \ldots\ldots\ldots\ldots\ldots (6)$$

Assume
$$\frac{2x + 1}{(x + 2)(x - 3)^2} = \frac{A}{x + 2} + \frac{B}{x - 3} + \frac{C}{(x - 3)^2}. \quad \ldots\ldots (7)$$

The short method of determining coefficients gives

$$A = \frac{-4 + 1}{(-2 - 3)^2} = -\frac{3}{25}, \quad C = \frac{6 + 1}{3 + 2} = \frac{7}{5}.$$

Also, multiplying by x, and then making $x \to \infty$, we find

$$A + B = 0, \quad \text{or} \quad B = \tfrac{3}{25}.$$

The integral is therefore

$$-\tfrac{3}{25} \log (x + 2) + \tfrac{3}{25} \log (x - 3) - \frac{7}{5 (x - 3)}. \quad \ldots\ldots (8)$$

Ex. 3. To find
$$\int \frac{x^3 \, dx}{(x^2 + 1)^2}. \quad \ldots\ldots\ldots\ldots\ldots\ldots (9)$$

We recall Art. 77, 3. Regarding $x^2/(x^2 + 1)^2$ as a function of x^2, we find (by inspection)

$$\frac{x^2}{(x^2 + 1)^2} = \frac{(x^2 + 1) - 1}{(x^2 + 1)^2} = \frac{1}{x^2 + 1} - \frac{1}{(x^2 + 1)^2}.$$

Hence
$$\int \frac{x^3 \, dx}{(x^2 + 1)^2} = \int \frac{x \, dx}{x^2 + 1} - \int \frac{x \, dx}{(x^2 + 1)^2}$$

$$= \tfrac{1}{2} \log (x^2 + 1) + \frac{1}{2 (x^2 + 1)}. \quad \ldots\ldots (10)$$

85. Case of Quadratic Factors.

The preceding methods are always applicable, but if some of the roots of $f(x) = 0$ are imaginary, the integral is obtained in the first instance in an imaginary form. If we wish to avoid the consideration of imaginary expressions, we may proceed as follows.

It is known from the Theory of Equations that a polynomial $f(x)$ whose coefficients are all real can be resolved into real factors of the first and second degrees. Then, in the resolution of the function

$$\frac{F(x)}{f(x)} \quad \dots\dots\dots\dots\dots\dots\dots\dots\dots(1)$$

into partial fractions, it may be shewn that we have

(a)　for each simple factor $x - \alpha$ which does not recur, a fraction of the form

$$\frac{A}{x - \alpha}; \quad \dots\dots\dots\dots\dots\dots\dots\dots(2)$$

(b)　for a simple factor $x - \beta$ which occurs r times, a series of r fractions, of the form

$$\frac{B_1}{x - \beta} + \frac{B_2}{(x - \beta)^2} + \dots + \frac{B_r}{(x - \beta)^r}; \quad \dots\dots\dots\dots(3)$$

(c)　for each quadratic factor $x^2 + px + q$ which does not recur, a fraction of the form

$$\frac{Cx + D}{x^2 + px + q}; \quad \dots\dots\dots\dots\dots\dots\dots(4)$$

(d)　for a quadratic factor $x^2 + px + q$ which occurs r times, a series of partial fractions, of the form

$$\frac{C_1 x + D_1}{x^2 + px + q} + \frac{C_2 x + D_2}{(x^2 + px + q)^2} + \dots + \frac{C_r x + D_r}{(x^2 + px + q)^r}. \quad \dots(5)$$

It is easily seen that in this way we have altogether just sufficient constants at our disposal to effect the identification of the function (1) with the complete system of partial fractions, by the method of equating coefficients.

*It only remains to shew how the indefinite integral of the partial fraction

$$\frac{C_s x + D_s}{(x^2 + px + q)^s}. \quad \dots\dots\dots\dots\dots\dots\dots(6)$$

can be found. The case $s = 1$ has been treated in Art. 74, and the general case can be reduced to this by a formula of reduction.

In the first place, we can find λ, μ so that

$$\frac{C_s x + D_s}{(x^2 + px + q)^s} = \lambda \frac{2x + p}{(x^2 + px + q)^s} + \frac{\mu}{(x^2 + px + q)^s}, \quad \dots\dots(7)$$

---

* The investigation which follows is given for the sake of completeness, but it is seldom required in practice. The student will lose little by postponing it. Another method of integrating expressions of the type (10) is indicated in Ex. 2, below.

viz. we have $\qquad\qquad \lambda = \tfrac{1}{2}C_s, \quad \mu = D_s - \tfrac{1}{2}pC_s.$ ....(8)

The integral of the first term on the right hand of (7) is

$$-\frac{\lambda}{s-1} \cdot \frac{1}{(x^2 + px + q)^{s-1}},\dots\dots\dots\dots\dots(9)$$

and it remains only to find

$$\int \frac{dx}{(x^2 + px + q)^s}, \quad \text{or} \quad \int \frac{dt}{(t^2 + c)^s}, \dots\dots\dots(10)$$

where $\qquad\qquad t = x + \tfrac{1}{2}p, \quad c = q - \tfrac{1}{4}p^2.$ ....(11)

Now, by differentiation, we find

$$\frac{d}{dt}\frac{t}{(t^2 + c)^{s-1}} = \frac{1}{(t^2 + c)^{s-1}} - (2s - 2)\frac{t^2}{(t^2 + c)^s}$$

$$= \frac{1}{(t^2 + c)^{s-1}} - (2s - 2)\frac{(t^2 + c) - c}{(t^2 + c)^s}$$

$$= -(2s - 3)\frac{1}{(t^2 + c)^{s-1}} + (2s - 2)\frac{c}{(t^2 + c)^s}. \dots\dots\dots(12)$$

Hence, integrating,

$$\frac{t}{(t^2 + c)^{s-1}} = -(2s - 3)\int \frac{dt}{(t^2 + c)^{s-1}} + (2s - 2)c\int \frac{dt}{(t^2 + c)^s},$$

or $\displaystyle \int \frac{dt}{(t^2 + c)^s} = \frac{1}{(2s - 2)c}\frac{t}{(t^2 + c)^{s-1}} + \frac{2s - 3}{2s - 2}\cdot\frac{1}{c}\int \frac{dt}{(t^2 + c)^{s-1}}.$ ....(13)

Returning to our previous notation, we have

$$\int \frac{dx}{(x^2 + px + q)^s} = \frac{1}{2(s - 1)(q - \tfrac{1}{4}p^2)}\frac{x + \tfrac{1}{2}p}{(x^2 + px + q)^{s-1}}$$

$$+ \frac{2s - 3}{2(s - 1)(q - \tfrac{1}{4}p^2)}\int \frac{dx}{(x^2 + px + q)^{s-1}},\dots\dots(14)$$

which is the formula of reduction required. By successive applications of this result, the integral (10) is made to depend ultimately on

$$\int \frac{dx}{x^2 + px + q},\dots\dots\dots\dots\dots\dots(15)$$

which is a known form (Art. 75).

Ex. 1.   To find $\displaystyle \int \frac{dx}{x^4 + x^2 + 1}.$ ....(16)

The denominator has here two quadratic factors, $x^2 + x + 1$ and $x^2 - x + 1$, which are not further resolvable. We therefore assume, in conformity with the above rule,

$$\frac{1}{x^4 + x^2 + 1} = \frac{Ax + B}{x^2 + x + 1} + \frac{Cx + D}{x^2 - x + 1},\dots\dots\dots(17)$$

or $\qquad 1 = (Ax + B)(x^2 - x + 1) + (Cx + D)(x^2 + x + 1).$

Equating coefficients of the several powers of x, we have

$$A + C = 0, \quad -A + C + B + D = 0,$$
$$A + C - B + D = 0, \qquad B + D = 1.$$

Hence $A = -C = \frac{1}{2}, \quad B = D = \frac{1}{2}.$ ....(18)

The integration can now be effected by the method of Art. 75. We have

$$\int \frac{dx}{x^4 + x^2 + 1} = \frac{1}{2} \int \frac{x+1}{x^2 + x + 1} \, dx - \frac{1}{2} \int \frac{x-1}{x^2 - x + 1} \, dx$$

$$= \frac{1}{4} \int \frac{(2x+1)+1}{x^2 + x + 1} \, dx - \frac{1}{4} \int \frac{(2x-1)-1}{x^2 - x + 1} \, dx$$

$$= \frac{1}{4} \log (x^2 + x + 1) - \frac{1}{4} \log (x^2 - x + 1)$$

$$+ \frac{1}{4} \int \frac{dx}{(x+\frac{1}{2})^2 + \frac{3}{4}} + \frac{1}{4} \int \frac{dx}{(x-\frac{1}{2})^2 + \frac{3}{4}}$$

$$= \frac{1}{4} \log \frac{x^2 + x + 1}{x^2 - x + 1} + \frac{1}{2\sqrt{3}} \left(\tan^{-1} \frac{2x+1}{\sqrt{3}} + \tan^{-1} \frac{2x-1}{\sqrt{3}} \right)$$

$$= \frac{1}{4} \log \frac{x^2 + x + 1}{x^2 - x + 1} + \frac{1}{2\sqrt{3}} \tan^{-1} \frac{\sqrt{3}x}{1 - x^2}. \quad \dots\dots\dots(19)$$

Ex. 2.   To find $\int \frac{dx}{(1+x^2)^2}.$ ....(20)

This comes under (14), but may be treated more simply as follows. If we put

$$x = \tan \theta,$$

we get

$$\int \frac{dx}{(1+x^2)^2} = \int \cos^2 \theta \, d\theta$$

$$= \frac{1}{2} \int (1 + \cos 2\theta) \, d\theta$$

$$= \frac{1}{2}\theta + \frac{1}{4} \sin 2\theta$$

$$= \frac{1}{2} \tan^{-1} x + \frac{1}{2} \frac{x}{1 + x^2}. \quad \dots\dots\dots\dots(21)$$

86. Integration of Irrational Functions.

The following are the leading results in this connection.

1°.   In the case of an algebraic function involving no irrationalities except fractional powers of the variable, we may put

$$x = t^m, \quad dx/dt = mt^{m-1}, \dots\dots\dots\dots\dots(1)$$

where m is the least common multiple of the denominators of the various fractional indices.   The problem is thus reduced to the integration of a rational function of t.

2°. Any rational function of x and X, where
$$X = \sqrt{(a + bx)}, \quad \dots\dots\dots\dots\dots\dots(2)$$
can be integrated by the substitution
$$a + bx = t^2, \quad dx/dt = 2t/b. \quad \dots\dots\dots\dots(3)$$

Thus $\displaystyle \int F(x, X)\, dx = \int F\left(\frac{t^2 - a}{b}, t\right) . \frac{2t\, dt}{b}, \quad \dots\dots\dots\dots(4)$

and the function of t which follows the integral-sign is now rational.

Ex. 1. To find $\displaystyle \int \frac{x\, dx}{\sqrt{x + 1}}.$

If we put $x = t^2$, this becomes
$$= 2 \int \frac{t^3\, dt}{t + 1} = 2 \int (t^2 - t + 1)\, dt - 2 \int \frac{dt}{t + 1}$$
$$= \tfrac{2}{3} t^3 - t^2 + 2t - 2 \log (t + 1)$$
$$= \tfrac{2}{3} x^{\frac{3}{2}} - x + 2x^{\frac{1}{2}} - 2 \log (x^{\frac{1}{2}} + 1).$$

Ex. 2. To find $\displaystyle \int \frac{dx}{(2 + x)\sqrt{(1 + x)}},$

put $\qquad\qquad 1 + x = t^2, \quad dx/dt = 2t.$

We obtain $\displaystyle \int \frac{2t\, dt}{(1 + t^2)\, t} = 2 \int \frac{dt}{1 + t^2}$
$$= 2 \tan^{-1} t = 2 \tan^{-1} \sqrt{(1 + x)}.$$

3°. If X stand for the square root of a quadratic expression, say
$$X = \sqrt{(ax^2 + bx + c)},$$
the problem of finding
$$\int F(x, X)\, dx, \quad \dots\dots\dots\dots\dots\dots(5)$$
where $F(x, X)$ is a rational function of x and X, can also be reduced to the integration of a rational function.

If a be positive, we may write
$$X = \sqrt{a} . \sqrt{(x^2 + px + q)}, \quad \dots\dots\dots\dots(6)$$
where $p = b/a,\ q = c/a$. Now assume
$$\sqrt{(x^2 + px + q)} = t - x,$$
whence $\qquad x = \dfrac{t^2 - q}{2t + p}, \quad \dfrac{dx}{dt} = \dfrac{2(t^2 + pt + q)}{(2t + p)^2} \quad \dots\dots\dots\dots(7)$

and $\qquad \sqrt{(x^2 + px + q)} = \dfrac{t^2 + pt + q}{2t + p}. \quad \dots\dots\dots\dots(8)$

It is evident that by these substitutions the problem is reduced to the integration of a rational function of t.

If the factors of $x^2 + px + q$ are real, say

$$x^2 + px + q = (x - a)(x - \beta), \quad \dots\dots\dots\dots\dots(9)$$

we may also make use of the substitution

$$x - \beta = (x - a) t^2, \quad \dots\dots\dots\dots\dots\dots(10)$$

whence

$$x = a + \frac{\beta - a}{1 - t^2}, \quad \frac{dx}{dt} = \frac{2(\beta - a)t}{(1 - t^2)^2} \quad \dots\dots\dots\dots(11)$$

and

$$\sqrt{(x^2 + px + q)} = (x - a)t = \frac{(\beta - a)t}{1 - t^2}. \quad \dots\dots\dots(12)$$

If a be negative, we may write

$$X = \sqrt{(-a)} \cdot \sqrt{(q + px - x^2)}, \quad \dots\dots\dots\dots(13)$$

where $p = -b/a$, $q = -c/a$. If the radical is to be real, the factors of $q + px - x^2$ must be real, for otherwise this expression would have the same sign for all values of x, and since it is obviously negative for sufficiently large values of x, it would always be negative. We have, then,

$$q + px - x^2 = (x - a)(\beta - x), \quad \dots\dots\dots\dots\dots(14)$$

where a, β are real. If we assume

$$\beta - x = (x - a) t^2, \quad \dots\dots\dots\dots\dots\dots(15)$$

we find

$$x = a + \frac{\beta - a}{1 + t^2}, \quad \frac{dx}{dt} = -\frac{2(\beta - a)t}{(1 + t^2)^2}, \quad \dots\dots\dots(16)$$

and

$$\sqrt{(q + px - x^2)} = (x - a)t = \frac{(\beta - a)t}{1 + t^2}. \quad \dots\dots\dots(17)$$

These substitutions evidently render

$$F(x, X)\frac{dx}{dt}$$

a rational function of t.

The above investigations are of some importance, as shewing that functions of the given forms can be integrated, and that the results will be of certain mathematical types; but the actual integration, in particular cases, can often be effected much more easily in other ways[*]. We have had instances of this fact in the course of the Chapter; and we add one or two further illustrations.

Ex. 3. By rationalizing the denominator, we have

$$\int \frac{dx}{\sqrt{(1 + x)} + \sqrt{x}} = \int \{\sqrt{(1 + x)} - \sqrt{x}\} \, dx$$

$$= \tfrac{2}{3}(1 + x)^{\frac{3}{2}} - \tfrac{2}{3}x^{\frac{3}{2}}.$$

[*] See especially the methods of Arts. 76, 77, 79.

Ex. 4.
$$\int \frac{dx}{x + \sqrt{(x^2-1)}} = \int \{x - \sqrt{(x^2-1)}\}\,dx$$
$$= \tfrac{1}{2}x^2 - \int \sqrt{(x^2-1)}\,dx$$
$$= \tfrac{1}{2}x^2 - \tfrac{1}{2}x\sqrt{(x^2-1)} + \tfrac{1}{2}\cosh^{-1}x.$$

Otherwise, putting $x = \cosh u$, the integral takes the form
$$\int \frac{\sinh u\,du}{\cosh u + \sinh u} = \int e^{-u}\sinh u\,du$$
$$= \tfrac{1}{2}\int (1 - e^{-2u})\,du = \tfrac{1}{2}u + \tfrac{1}{4}e^{-2u},$$

which may be easily shewn to differ from the former result only by an additive constant.

EXAMPLES. XXIII.

Find the indefinite integrals of the following expressions*:

1. $\dfrac{1}{1-x}, \quad \dfrac{1}{(1-x)^2}.$

2. $\dfrac{1}{2x-1}, \quad \dfrac{1}{(2x-1)^2}.$

3. $(x-1)^3, \quad \dfrac{1}{(x-1)^3}.$

4. $\sqrt{x}, \quad \dfrac{1}{\sqrt{x}}, \quad \dfrac{1+x}{\sqrt{x}}.$

5. $\dfrac{1}{\sqrt{(2+x)}}, \quad \dfrac{1}{\sqrt{(3-2x)}}.$

6. $\dfrac{1+x}{x}, \quad \dfrac{1+x}{x^2}.$

7. $\dfrac{1-2x}{3+x}, \quad \dfrac{2+x}{3-x}.$

8. $\left(x + \dfrac{1}{x}\right)^2, \quad \left(x + \dfrac{1}{x}\right)^3.$

9. $\dfrac{1+x}{1-x}, \quad \dfrac{1-x}{1+x}.$

10. $\dfrac{x^2}{1+x}, \quad \dfrac{x^2}{1-x}.$

11. $\dfrac{1-x^2}{1+x^2}, \quad \dfrac{1+x^2}{1-x^3}.$

12. $\cos^2 x, \quad \cot^2 x.$

13. $(\cos x - \sin x)^2.$

14. $\cosh^2 x, \quad \sinh^2 x.$

15. $\tanh^2 x, \quad \coth^2 x.$

16. $\dfrac{1-x^m}{1-x}, \quad \dfrac{1+x^{2m+1}}{1+x}.$

EXAMPLES. XXIV.

(Dynamical.)

1. A particle moves according to the law
$$\frac{ds}{dt} = u_0 - gt;$$
prove that the space described before it comes to rest is $u_0^2/2g$.

* The student should test the accuracy of the results by differentiation.

2. If a point start from rest at time $t = 0$ and move with a constant acceleration, and if v_1 be the velocity after any interval and $\bar{v}$ the mean velocity in this interval, then

$$\bar{v} = \tfrac{1}{2} v_1.$$

3. If, with the same notation, the acceleration vary as t^n, then

$$\bar{v} = \frac{1}{n+2}\, v_1.$$

4. A particle moves according to the law

$$\frac{ds}{dt} = v_0 \cos nt \,;$$

prove that the space described from time $t = 0$ until it first comes to rest is v_0/n.

5. If the velocity of a particle moving in a resisting medium be given by

$$\frac{ds}{dt} = v_0 e^{-kt},$$

prove that the particle never attains a distance v_0/k from its position when $t = 0$.

6. A particle moves according to the law

$$\frac{ds}{dt} = v_0 e^{-kt} \cos nt \,;$$

prove that the space described from time $t = 0$ until it first comes to rest is

$$\frac{n e^{-k\pi/2n} + k}{n^2 + k^2}\, v_0.$$

7. If the angular velocity of a body rotating about a fixed axis be given by

$$\frac{d\theta}{dt} = 2n \operatorname{sech} nt,$$

prove that

$$\theta = 4 \tan^{-1} e^{nt} - \pi,$$

supposing that θ vanishes for $t = 0$.

EXAMPLES. XXV.

(Quadratic Denominators.)

1. $\displaystyle \int \frac{1 + x}{1 + x^2}\, dx = \tan^{-1} x + \log \sqrt{(1 + x^2)}.$

2. $\displaystyle \int \frac{x^3}{1 + x^2}\, dx = \tfrac{1}{2} x^2 - \log \sqrt{(1 + x^2)}.$

3. $\int \dfrac{x^3}{1-x^2}\, dx = -\tfrac{1}{2}x^2 - \log \sqrt{(1-x^2)}.$

4. $\int \dfrac{dx}{1-2x+2x^2} = \tan^{-1}(2x-1).$

5. $\int \dfrac{dx}{1+3x+2x^2} = \log \dfrac{2x+1}{x+1}.$

6. $\int \dfrac{3x+7}{2x^2-3x+5}\, dx = \tfrac{3}{4}\log(2x^2-3x+5) + \dfrac{37}{2\sqrt{31}} \tan^{-1}\dfrac{4x-3}{\sqrt{31}}.$

7. $\int \dfrac{2x-1}{x^2+2x+3} = \log(x^2+2x+3) - \dfrac{3}{\sqrt{2}} \tan^{-1}\dfrac{x+1}{\sqrt{2}}.$

8. $\int \dfrac{x^2+x+1}{x^2-x+1}\, dx = x + \log(x^2-x+1) + \dfrac{2}{\sqrt{3}} \tan^{-1}\dfrac{2x-1}{\sqrt{3}}.$

9. $\int \dfrac{x+1}{(x-1)^2}\, dx = \log(x-1) - \dfrac{2}{x-1}.$

10. $\int \dfrac{x\, dx}{x^2+6x+8} = \log \dfrac{(x+4)^2}{x+2}.$

11. $\int \dfrac{x^2-1}{(x-2)(x-3)}\, dx = x - 3\log(x-2) + 8\log(x-3).$

12. $\int \dfrac{x^2+x+1}{x^2-x-1}\, dx = x + (\tfrac{3}{5}\sqrt{5}+1)\log(2x-\sqrt{5}-1)$
$\qquad\qquad\qquad\qquad\qquad - (\tfrac{3}{5}\sqrt{5}-1)\log(2x+\sqrt{5}-1).$

13. $\int \dfrac{1-x^{2m}}{1-x^2}\, dx = x + \tfrac{1}{3}x^3 + \tfrac{1}{5}x^5 + \ldots + \dfrac{1}{2m-1}x^{2m-1}.$

14. $\int \dfrac{1-(-)^m x^{2m}}{1+x^2}\, dx = x - \tfrac{1}{3}x^3 + \tfrac{1}{5}x^5 - \ldots + (-)^{m-1}\dfrac{1}{2m-1}x^{2m-1}.$

EXAMPLES. XXVI.

1. $\int \dfrac{dx}{\sqrt{(2-3x^2)}} = \dfrac{1}{\sqrt{3}} \sin^{-1}(\sqrt{\tfrac{3}{2}}x).$

2. $\int \dfrac{dx}{\sqrt{(2+3x^2)}} = \dfrac{1}{\sqrt{3}} \sinh^{-1}(\sqrt{\tfrac{3}{2}}x).$

3. $\int \dfrac{dx}{\sqrt{\{x(1-x)\}}} = \sin^{-1}(2x-1).$

4. $\int \dfrac{dx}{\sqrt{\{x(1+x)\}}} = \cosh^{-1}(2x+1).$

5. $\int \dfrac{dx}{\sqrt{\{x(x-1)\}}} = \cosh^{-1}(2x-1).$

6. $\int \dfrac{dx}{\sqrt{(1+2x+3x^2)}} = \dfrac{1}{\sqrt{3}} \sinh^{-1} \dfrac{3x+1}{\sqrt{2}}.$

7. $\int \dfrac{dx}{\sqrt{(1+2x-3x^2)}} = \dfrac{1}{\sqrt{3}} \sin^{-1} \dfrac{3x-1}{2}.$

8. $\int \dfrac{dx}{\sqrt{(2ax-x^2)}} = \cos^{-1}\left(1 - \dfrac{x}{a}\right).$

9. $\int \sqrt{\left(\dfrac{a-x}{x}\right)}\, dx = \sqrt{\{x(a-x)\}} + \tfrac{1}{2}a \sin^{-1} \dfrac{2x-a}{a}.$

10. $\int \sqrt{\left(\dfrac{1+x}{1-x}\right)}\, dx = \sin^{-1} x - \sqrt{(1-x^2)}.$

11. $\int \sqrt{\left(\dfrac{x+1}{x-1}\right)}\, dx = \sqrt{(x^2-1)} + \cosh^{-1} x.$

EXAMPLES. XXVII.

(Change of Variable.)

1. $\int \dfrac{x^2\, dx}{1-x^3} = \tfrac{1}{3} \log \dfrac{1}{1-x^3}.$

2. $\int \dfrac{x^2\, dx}{1-x^6} = \tfrac{1}{6} \log \dfrac{1+x^3}{1-x^3}, \qquad \int \dfrac{x^2\, dx}{1+x^6} = \tfrac{1}{3} \tan^{-1} x^3.$

3. $\int \dfrac{\log x}{x}\, dx = \tfrac{1}{2}(\log x)^2.$

4. $\int \sin x \cos x\, dx = \tfrac{1}{2} \sin^2 x.$

5. $\int \dfrac{\sin^{-1} x}{\sqrt{(1-x^2)}}\, dx = \tfrac{1}{2}(\sin^{-1} x)^2.$

6. $\int \dfrac{\cos x}{1+\sin x}\, dx = \log(1+\sin x),$

  $\int \dfrac{\sin x}{a+b \cos x}\, dx = -\dfrac{1}{b} \log(a+b \cos x).$

7. $\int \sin x \cos^3 x\, dx = -\tfrac{1}{4} \cos^4 x.$

8. $\int \dfrac{\sin x \cos x}{a \cos^2 x + b \sin^2 x}\, dx = \dfrac{1}{2(b-a)} \log(a \cos^2 x + b \sin^2 x).$

9. $\int \tan^3 x\, dx = \tfrac{1}{2} \tan^2 x + \log \cos x.$

10. $\int \dfrac{\sin x}{\cos^5 x}\, dx = \tfrac{1}{4} \sec^4 x.$

11. $\int \sec^4 x\, dx = \tan x + \tfrac{1}{3} \tan^3 x.$

12. $\int (\sec x + \tan x)\, dx = \log \dfrac{1}{1 - \sin x}$,

$\int (\sec x - \tan x)\, dx = \log (1 + \sin x)$.

13. $\int \dfrac{dx}{1 + \cos x} = \operatorname{cosec} x - \cot x,$ $\int \dfrac{dx}{1 - \cos x} = -\cot x - \operatorname{cosec} x.$

14. $\int \dfrac{dx}{1 + \sin x} = \tan x - \sec x,$ $\int \dfrac{dx}{1 - \sin x} = \tan x + \sec x.$

15. $\int \dfrac{dx}{\sin^2 x \cos^2 x} = \tan x - \cot x.$

16. $\int \dfrac{dx}{\sin x \cos^2 x} = \sec x + \log \tan \tfrac{1}{2} x.$

17. $\int \dfrac{dx}{\sin x \cos^3 x} = \tfrac{1}{2} \sec^2 x + \log \tan x.$

18. $\int \dfrac{dx}{1 + \tan x} = \tfrac{1}{2} x + \tfrac{1}{2} \log (\cos x + \sin x).$

19. $\int \dfrac{dx}{1 + \cos^2 x} = \dfrac{1}{\sqrt{2}} \tan^{-1} \left(\dfrac{1}{\sqrt{2}} \tan x \right).$

20. $\int x \sqrt{(a^2 + x^2)}\, dx = \tfrac{1}{3} (a^2 + x^2)^{\frac{3}{2}}.$

21. $\int \dfrac{dx}{x (x^n + 1)} = \dfrac{1}{n} \log \dfrac{x^n}{x^n + 1}.$

22.   Evaluate $\int \sqrt{(x^2 + a^2)}\, dx$ and $\int \sqrt{(x^2 - a^2)}\, dx$ by hyperbolic substitutions.

23. $\int \dfrac{dx}{x^2 \sqrt{(1 + x^2)}} = -\dfrac{\sqrt{(1 + x^2)}}{x}.$

24. $\int \dfrac{dx}{x^2 \sqrt{(a^2 - x^2)}} = -\dfrac{\sqrt{(a^2 - x^2)}}{a^2 x}.$

25. $\int \dfrac{x^3\, dx}{\sqrt{(a^2 + x^2)}} = \tfrac{1}{3} (x^2 - 2a^2) \sqrt{(a^2 + x^2)}.$

EXAMPLES. XXVIII.

(Integration by Parts.)

1. $\int x e^{x/a}\, dx = a (x - a) e^{x/a}.$

2. $\int x \log x\, dx = \tfrac{1}{2} x^2 (\log x - \tfrac{1}{2}).$

3. $\int x^m \log x\, dx = \dfrac{x^{m+1}}{m + 1} \left(\log x - \dfrac{1}{m + 1} \right).$

4. $\int x \sin x\, dx = -x \cos x + \sin x.$

5. $\int x \cos x\, dx = x \sin x + \cos x.$

6. $\int x \sin x \cos x \, dx = -\tfrac{1}{4}x \cos 2x + \tfrac{1}{8}\sin 2x.$

7. $\int \cos mx \cos nx \, dx = \dfrac{m \sin mx \cos nx - n \cos mx \sin nx}{m^2 - n^2}.$

8. $\int \sin mx \sin nx \, dx = \dfrac{n \sin mx \cos nx - m \cos mx \sin nx}{m^2 - n^2}.$

9. $\int \sin mx \cos nx \, dx = -\dfrac{m \cos mx \cos nx + n \sin mx \sin nx}{m^2 - n^2}.$

10. $\int \sin^{-1} x \, dx = x \sin^{-1} x + \sqrt{(1 - x^2)}.$

11. $\int \tan^{-1} x \, dx = x \tan^{-1} x - \log \sqrt{(1 + x^2)}.$

12. $\int \sec^{-1} x \, dx = x \sec^{-1} x - \cosh^{-1} x.$

13. $\int x \tan^{-1} x \, dx = \tfrac{1}{2}(1 + x^2) \tan^{-1} x - \tfrac{1}{2}x.$

14. $\int x \sec^2 x \, dx = x \tan x + \log \cos x.$

15. $\int \dfrac{x + \sin x}{1 + \cos x} \, dx = x \tan \tfrac{1}{2}x.$

16. $\int \dfrac{x \sin^{-1} x}{\sqrt{(1 - x^2)}} \, dx = - \sqrt{(1 - x^2)} \sin^{-1} x + x.$

17. $\int \cosh x \cos x \, dx = \tfrac{1}{2}(\sinh x \cos x + \cosh x \sin x).$

18. $\int \sinh x \sin x \, dx = \tfrac{1}{2}(\cosh x \sin x - \sinh x \cos x).$

19. $\int \cosh x \sin x \, dx = \tfrac{1}{2}(\sinh x \sin x - \cosh x \cos x).$

20. $\int \sinh x \cos x \, dx = \tfrac{1}{2}(\cosh x \cos x + \sinh x \sin x).$

21. $\int e^x \sin x \cos x \, dx = \tfrac{1}{10}(\sin 2x - 2 \cos 2x) e^x.$

22. $\int x^5 e^{-x} \, dx = -(x^5 + 5x^4 + 20x^3 + 60x^2 + 120x + 120) e^{-x}.$

23. $\int x^4 \sin x \, dx = -(x^4 - 12x^2 + 24) \cos x + (4x^3 - 24x) \sin x.$

24. If $\quad u_n = \int x^n \cosh x \, dx, \quad v_n = \int x^n \sinh x \, dx,$

prove that $\quad u_n = x^n \sinh x - n v_{n-1}, \quad v_n = x^n \cosh x - n u_{n-1}.$

Deduce the values of u_4 and v_4.

$$[u_4 = (x^4 + 12x^2 + 24) \sinh x - (4x^3 + 24x) \cosh x,$$
$$v_4 = (x^4 + 12x^2 + 24) \cosh x - (4x^3 + 24x) \sinh x.]$$

25. If u be a rational integral function of x, prove that

$$\int e^{ax} u \, dx = \frac{e^{ax}}{a}\left(1 - \frac{D}{a} + \frac{D^2}{a^2} - \ldots\right) u,$$

where $D = d/dx$.

26. Determine the coefficients A, B so that

$$\int \frac{dx}{(a + b \cos x)^2} = \frac{A \sin x}{a + b \cos x} + B \int \frac{dx}{a + b \cos x}.$$
$$[A = -b/(a^2 - b^2), \quad B = a/(a^2 - b^2).]$$

EXAMPLES. XXIX.

(Rational Fractions.)

1. $\int \dfrac{dx}{x\,(1-x^2)} = \log \dfrac{x}{\sqrt{(1-x^2)}}$.

2. $\int \dfrac{2x+3}{x\,(x-1)\,(x+2)}\,dx = \log \dfrac{(x-1)^{\frac{5}{3}}}{x^{\frac{3}{2}}\,(x+2)^{\frac{1}{6}}}$.

3. $\int \dfrac{x^2\,dx}{(x-1)\,(x-2)\,(x-3)}$
$$= \tfrac{1}{2}\log\,(x-1) - 4\log\,(x-2) + \tfrac{9}{2}\log\,(x-3).$$

4. $\int \dfrac{2x-3}{(x^2-1)\,(2x+3)}\,dx$
$$= \tfrac{5}{2}\log\,(x+1) - \tfrac{1}{10}\log\,(x-1) - \tfrac{12}{5}\log\,(2x+3).$$

5. $\int \dfrac{x\,dx}{x^4-x^2-2} = \tfrac{1}{6}\log\dfrac{x^2-2}{x^2+1}$.

6. $\int \dfrac{dx}{x^3+3x^2-4} = \dfrac{1}{3\,(x+2)} + \tfrac{1}{9}\log\dfrac{x-1}{x+2}$.

7. $\int \dfrac{x^2\,dx}{(x-a)\,(x-b)\,(x-c)} = \dfrac{a^2}{(a-b)\,(a-c)}\log\,(x-a)$
$$+ \dfrac{b^2}{(b-c)\,(b-a)}\log\,(x-b) + \dfrac{c^2}{(c-a)\,(c-b)}\log\,(x-c).$$

8. $\int \dfrac{dx}{(x^2+a^2)\,(x^2+b^2)} = \dfrac{1}{b^2-a^2}\left(\dfrac{1}{a}\tan^{-1}\dfrac{x}{a} - \dfrac{1}{b}\tan^{-1}\dfrac{x}{b}\right)$.

9. $\int \dfrac{x\,dx}{(x^2+a^2)\,(x^2+b^2)} = \dfrac{1}{2\,(b^2-a^2)}\log\dfrac{x^2+a^2}{x^2+b^2}$.

10. $\int \dfrac{x^2\,dx}{(x^2+a^2)\,(x^2+b^2)} = \dfrac{1}{a^2-b^2}\left(a\tan^{-1}\dfrac{x}{a} - b\tan^{-1}\dfrac{x}{b}\right)$.

11. $\int \dfrac{x^3\,dx}{(x^2+a^2)\,(x^2+b^2)} = \dfrac{1}{2\,(a^2-b^2)}\{a^2\log\,(x^2+a^2) - b^2\log\,(x^2+b^2)\}$.

12. $\int \dfrac{x\,dx}{(x+1)\,(x+2)^2} = -\dfrac{2}{x+2} + \log\dfrac{x+2}{x+1}$.

13. $\int \dfrac{x^2\,dx}{(x+1)\,(x+2)^2} = \dfrac{4}{x+2} + \log\,(x+1)$.

14. $\int \dfrac{dx}{x^3-x^2-x+1} = \tfrac{1}{4}\log\dfrac{x+1}{x-1} - \dfrac{1}{2}\dfrac{1}{x-1}$.

15. $\int \dfrac{dx}{(x^2-1)^2} = -\dfrac{1}{2}\dfrac{x}{x^2-1} + \tfrac{1}{4}\log\dfrac{x+1}{x-1}$.

16. $\int \dfrac{dx}{x^2 (1-x)^2} = \dfrac{2x-1}{x(1-x)} + 2 \log \dfrac{x}{1-x}.$

17. $\int \dfrac{x^2\,dx}{(x^2-1)^2} = -\dfrac{1}{2} \dfrac{x}{x^2-1} - \dfrac{1}{4} \log \dfrac{x+1}{x-1}.$

18. $\int \dfrac{dx}{x^3 (x^2-1)^2} = -\dfrac{3}{4} \log \dfrac{x-1}{x+1} - \dfrac{1}{x} - \dfrac{1}{2} \dfrac{x}{x^2-1}.$

19. $\int \dfrac{dx}{x^3 (x+1)} = -\dfrac{1}{2x^2} + \dfrac{1}{x} - \log (x+1) + \log x.$

20. $\int \dfrac{3x+2}{x(x+1)^3}\,dx = 2 \log \dfrac{x}{x+1} + \dfrac{2}{x+1} - \dfrac{1}{2(x+1)^2}.$

21. $\int \dfrac{dx}{(x^2-1)^3} = \dfrac{3}{16} \log \dfrac{x-1}{x+1} + \dfrac{3}{8} \dfrac{x}{x^2-1} - \dfrac{1}{4} \dfrac{x}{(x^2-1)^2}.$

22. $\int \dfrac{1+x}{x(1+x^2)}\,dx = \log \dfrac{x}{\sqrt{(1+x^2)}} + \tan^{-1} x.$

23. $\int \dfrac{dx}{1-x^4} = \dfrac{1}{4} \log \dfrac{1+x}{1-x} + \dfrac{1}{2} \tan^{-1} x.$

24. $\int \dfrac{x^2\,dx}{1-x^4} = \dfrac{1}{4} \log \dfrac{1+x}{1-x} - \dfrac{1}{2} \tan^{-1} x.$

25. $\int \dfrac{dx}{1+x^3} = \dfrac{1}{6} \log \dfrac{(1+x)^2}{1-x+x^2} + \dfrac{1}{\sqrt{3}} \tan^{-1} \dfrac{2x-1}{\sqrt{3}}.$

26. $\int \dfrac{x\,dx}{1+x^3} = \dfrac{1}{6} \log \dfrac{1-x+x^2}{(1+x)^2} + \dfrac{1}{\sqrt{3}} \tan^{-1} \dfrac{2x-1}{\sqrt{3}}.$

27. $\int \dfrac{dx}{(1+x)(1+x^2)} = \dfrac{1}{2} \log (1+x) - \dfrac{1}{4} \log (1+x^2) + \dfrac{1}{2} \tan^{-1} x.$

28. $\int \dfrac{x^2\,dx}{(1+x)(1+x^2)} = \dfrac{1}{2} \log (1+x) + \dfrac{1}{4} \log (1+x^2) - \dfrac{1}{2} \tan^{-1} x.$

29. $\int \dfrac{x^2\,dx}{x^4+x^2-2} = \dfrac{1}{6} \log \dfrac{x-1}{x+1} + \dfrac{\sqrt{2}}{3} \tan^{-1} \dfrac{x}{\sqrt{2}}.$

30. $\int \dfrac{x^3\,dx}{x^4+3x^2+2} = \log (x^2+2) - \dfrac{1}{2} \log (x^2+1).$

31. $\int \dfrac{x^2\,dx}{(x-1)^2(x^2+1)} = \dfrac{1}{2} \log (x-1) - \dfrac{1}{4} \log (x^2+1) - \dfrac{1}{2(x-1)}.$

32. $\int \dfrac{x^2-1}{x^4+x^2+1}\,dx = \dfrac{1}{2} \log \dfrac{x^2-x+1}{x^2+x+1}.$

33. $\int \dfrac{dx}{x^4(1+x^2)} = \tan^{-1} x + \dfrac{1}{x} - \dfrac{1}{3x^3}.$

34. $\int \dfrac{x^3\,dx}{(1+x^2)^3} = -\dfrac{1+2x^2}{4(1+x^2)^2}.$

35. $\int \dfrac{dx}{x^2 (1 + x^2)^2} = - \dfrac{2 + 3x^2}{2x (1 + x^2)} - \tfrac{3}{2} \tan^{-1} x.$

36. $\int \dfrac{dx}{1 + x^4} = \dfrac{1}{4 \sqrt{2}} \log \dfrac{1 + x \sqrt{2} + x^2}{1 - x \sqrt{2} + x^2} + \dfrac{1}{2 \sqrt{2}} \tan^{-1} \dfrac{x \sqrt{2}}{1 - x^2}.$

37. $\int \dfrac{x^2 dx}{1 + x^4} = \dfrac{1}{4 \sqrt{2}} \log \dfrac{1 - x \sqrt{2} + x^2}{1 + x \sqrt{2} + x^2} + \dfrac{1}{2 \sqrt{2}} \tan^{-1} \dfrac{x \sqrt{2}}{1 - x^2}.$

EXAMPLES. XXX.

(Irrational Functions.)

1. $\int x \sqrt{(1 + x)} \, dx = \tfrac{2}{5} (1 + x)^{\frac{5}{2}} - \tfrac{2}{3} (1 + x)^{\frac{3}{2}}.$

2. $\int \dfrac{dx}{\sqrt{(x + a)} + \sqrt{(x + b)}} = \dfrac{2}{3 (a - b)} \{(x + a)^{\frac{3}{2}} - (x + b)^{\frac{3}{2}}\}.$

3. $\int \dfrac{dx}{\sqrt{x} - 1} = 2 \sqrt{x} + 2 \log (1 - \sqrt{x}).$

4. $\int \dfrac{dx}{(1 - x) \sqrt{x}} = \log \dfrac{1 + \sqrt{x}}{1 - \sqrt{x}}.$

5. $\int \dfrac{dx}{(1 - x) \sqrt{(1 + x)}} = \sqrt{2} \tanh^{-1} \sqrt{\left(\dfrac{1 + x}{2}\right)}.$

6. $\int \dfrac{dx}{(1 + x) \sqrt{(1 - x)}} = - \sqrt{2} \tanh^{-1} \sqrt{\left(\dfrac{1 - x}{2}\right)}.$

7. $\int \dfrac{dx}{x + \sqrt{(x - 1)}} = \log \{x + \sqrt{(x - 1)}\} - \dfrac{2}{\sqrt{3}} \tan^{-1} \dfrac{2 \sqrt{(x - 1)} + 1}{\sqrt{3}}$

8. $\int \dfrac{\sqrt{x}}{x - 1} \, dx = 2 \sqrt{x} + \log \dfrac{\sqrt{x} - 1}{\sqrt{x} + 1}.$

9. $\int \dfrac{dx}{x \sqrt{(1 + x)}} = \log \dfrac{\sqrt{(1 + x)} - 1}{\sqrt{(1 + x)} + 1}.$

10. $\int \dfrac{dx}{x^2 \sqrt{(1 + x)}} = - \dfrac{\sqrt{(1 + x)}}{x} - \tfrac{1}{2} \log \dfrac{\sqrt{(1 + x)} - 1}{\sqrt{(1 + x)} + 1}.$

11. $\int \dfrac{x^2 dx}{\sqrt{(x^2 + 1)}} = \tfrac{1}{2} x \sqrt{(x^2 + 1)} - \tfrac{1}{2} \sinh^{-1} x.$

12. $\int \dfrac{x^2 dx}{(1 + x^2)^{\frac{3}{2}}} = - \dfrac{x}{\sqrt{(1 + x^2)}} + \sinh^{-1} x.$

13. $\int \dfrac{dx}{(1 + x^2) \sqrt{(1 - x^2)}} = \dfrac{1}{\sqrt{2}} \tan^{-1} \dfrac{x \sqrt{2}}{\sqrt{(1 - x^2)}}.$

14. $\int \dfrac{dx}{(1 - x^2) \sqrt{(1 + x^2)}} = \dfrac{1}{2 \sqrt{2}} \log \dfrac{x \sqrt{2} + \sqrt{(1 + x^2)}}{x \sqrt{2} - \sqrt{(1 + x^2)}}.$

CHAPTER VII

DEFINITE INTEGRALS

87. Introduction. Problem of Areas.

The problem of integration, in the sense now to be explained, is one of the oldest in Mathematics, but it was not till the time of Newton and Leibnitz that a general method of solution was evolved. We proceed in this Art. and the next to explain this method briefly, without special attention to logical details, taking the 'problem of areas' as a sufficiently typical case. In this way the essential principle will be easily apprehended. Afterwards, in Arts. 89–94, the question will be taken up *de novo* and discussed in a more general and more rigorous manner.

Suppose that it is required to find the area* included between a continuous curve

$$y = \phi(x), \quad \dots\dots\dots\dots\dots\dots\dots\dots\dots(1)$$

the axis of x, and two ordinates $x = a$, $x = b$. For definiteness we will suppose that y is positive over the range of x considered, and

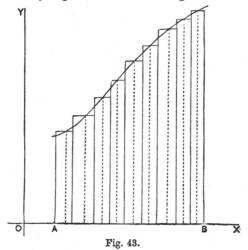

Fig. 43.

---

* The term 'area' is used, in this Art. and the next, in the ordinary intuitive sense. From the modern point of view the area of a curve needs definition; see Art. 99.

that $b > a$. We may divide this range $b - a$ into a series of subdivisions, $h_1, h_2, \ldots, h_n$, and erect on these as bases a series of rectangles whose altitudes $y_1, y_2, \ldots, y_n$ are ordinates of the curve at arbitrarily chosen points within the respective bases. The sum of the areas of the rectangles thus constructed may be regarded as an approximation to the area required. The result may, indeed, happen to be exact, but it is evident that the approximation will as a rule be better, the smaller the subdivisions, $h_1, h_2, \ldots, h_n$ are taken, their number being of course correspondingly increased. The limit to which the sum of the rectangles tends, when the subdivisions are infinitely small, is the area required.

Before the invention of the Calculus this procedure, or something equivalent to it, had to be carried out if possible for each curve separately, the methods employed being often highly ingenious. The following examples may serve as illustrations.

Ex. 1. To find the area included between the parabola $y = x^2$, the axis of x, and the ordinates $x = a$, $x = b$.

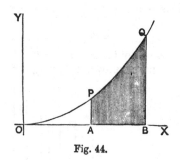

Fig. 44.

Putting
$$h_1 = h_2 = \ldots = h_n, = h,$$

$$y_1 = a^2, \quad y_2 = (a + h)^2, \quad y_3 = (a + 2h)^2, \quad \ldots\ldots \quad y_n = \{a + (n - 1) h\}^2,$$

we have to consider the sum

$$a^2 h + (a + h)^2 h + (a + 2h)^2 h + \ldots + \{a + (n - 1) h\}^2$$

$$= na^2 h + 2 \left\{1 + 2 + \ldots + (n - 1)\right\} ah^2 + \left\{1^2 + 2^2 + \ldots + (n - 1)^2\right\} h^3$$

$$= na^2 h + n (n - 1) ah^2 + \tfrac{1}{6} (n - 1) n (2n - 1) h^3$$

$$= a^2 (b - a) + \left(1 - \frac{1}{n}\right) a (b - a)^2 + \frac{1}{3} \left(1 - \frac{1}{n}\right) \left(1 - \frac{1}{2n}\right) (b - a)^3. \ldots (2)$$

The limiting value of this for $n \to \infty$ is

$$a^2 (b - a) + a (b - a)^2 + \tfrac{1}{3} (b - a)^3, \quad \text{or} \quad \tfrac{1}{3} (b^3 - a^3). \quad \ldots\ldots (3)$$

Ex. 2.   The general case of the curve

$$y = x^m, \quad\quad\quad\quad\quad\quad\quad\quad\quad\quad(4)$$

where m may have any integral or fractional value, positive or negative, except -1, may be treated as follows.

The abscissae of the dividing points of the range $b - a$ are taken in *geometric* instead of (as is more usual) arithmetic progression, viz. they are

$$a, \quad \mu a, \quad \mu^2 a, \quad ..., \quad \mu^n a,$$

where $\mu^n = b/a$.   The subdivisions are therefore

$$h_1 = (\mu - 1)\, a, \;\; h_2 = (\mu - 1)\, \mu a, \;\; h_3 = (\mu - 1)\, \mu^2 a, \;\; ..., \;\; h_n = (\mu - 1)\, \mu^{n-1} a.$$

The ordinates at the initial points of these are

$$a^m, \quad \mu^m a^m, \quad \mu^{2m} a^m, \quad ..., \quad \mu^{(n-1)m} a^m,$$

and the sum to be considered is therefore

$$(\mu - 1)\, a^{m+1} \left(1 + \mu^{m+1} + \mu^{2(m+1)} + ... + \mu^{(n-1)(m+1)} \right)$$

$$= (\mu - 1)\, a^{m+1} \frac{\mu^{n(m+1)} - 1}{\mu^{m+1} - 1} = \frac{\mu - 1}{\mu^{m+1} - 1} \left(b^{m+1} - a^{m+1} \right). \quad ...(5)$$

The subdivisions are made infinitely small by making μ tend to the limit 1, n becoming infinite.   Since

$$\lim_{\mu \to 1} \frac{\mu^{m+1} - 1}{\mu - 1} = m + 1, \quad\quad\quad\quad\quad(6)$$

by Art. 22, the result is

$$\frac{b^{m+1} - a^{m+1}}{m + 1}. \quad\quad\quad\quad\quad\quad(7)$$

If we put $m = 2$, we get the case of Ex. 1, above.

The above ingenious procedure is due to Wallis (1656).   It needs modification when $m = -1$.   In place of (5) we then have

$$n\,(\mu - 1) = n \left\{ \left(\frac{b}{a} \right)^{\frac{1}{n}} - 1 \right\}, \quad\quad\quad\quad\quad(8)$$

the limit of which when $n \to \infty$ is, by Art. 43 (9),

$$\log \frac{b}{a}. \quad\quad\quad\quad\quad\quad\quad\quad(9)$$

Ex. 3.   Let the curve be

$$y = \sin x, \quad\quad\quad\quad\quad\quad\quad\quad(10)$$

the range extending from $x = a$ to $x = \beta$.   Taking equal subdivisions

$$h = (\beta - a)/n, \quad\quad\quad\quad\quad\quad(11)$$

we consider the limit of the sum

$$\Sigma, = \{ \sin(a + \tfrac{1}{2}h) + \sin(a + \tfrac{3}{2}h) + ... + \sin(\beta - \tfrac{3}{2}h) + \sin(\beta - \tfrac{1}{2}h) \} h, \quad (12)$$

where the values of $\sin x$ at the middles of the respective intervals have been taken. Now

$$\frac{\sin \frac{1}{2}h}{\frac{1}{2}h} \cdot \Sigma = 2 \sin \tfrac{1}{2}h \sin (a + \tfrac{1}{2}h) + 2 \sin \tfrac{1}{2}h \sin (a + \tfrac{3}{2}h) + \dots$$

$$+ 2 \sin \tfrac{1}{2}h \sin (\beta - \tfrac{3}{2}h) + 2 \sin \tfrac{1}{2}h \sin (\beta - \tfrac{1}{2}h)$$

$$= \cos a \qquad\quad - \cos (a + h)$$

$$+ \cos (a + h) \quad - \cos (a + 2h)$$

$$+ \dots\dots\dots$$

$$+ \cos (\beta - 2h) - \cos (\beta - h)$$

$$+ \cos (\beta - h) \; - \cos \beta$$

$$= \cos a \qquad\quad - \cos \beta. \quad\dots\dots\dots\dots\dots\dots\dots(13)$$

Hence, proceeding to the limit $(h \to 0)$, the required area is

$$\cos a - \cos \beta. \quad\dots\dots\dots\dots\dots\dots\dots(14)$$

88. Connection with Inverse Differentiation.

Calculations of the above kind are now superseded by the rule of the Integral Calculus, to which we proceed.

If, keeping a fixed, we regard b as variable, the area considered in Art. 87 will be a function of b, which vanishes when $b = a$. When

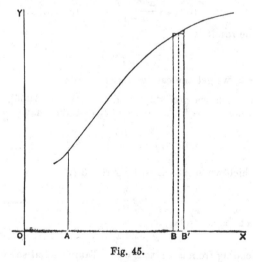

Fig. 45.

b receives an infinitesimal increment δb, the increment of the area will ultimately be equal to a rectangle of breadth δb, and height $\phi(b)$; see Fig. 45. Thus, if A be the area in question,

$$\delta A = \phi (b)\, \delta b, \quad\dots\dots\dots\dots\dots\dots(1)$$

or
$$\frac{dA}{db} = \phi\,(b). \qquad\qquad\text{.........................(2)}$$

Hence if $\psi\,(x)$ be a function such that $\psi'\,(x) = \phi\,(x)$, *i.e.* if $\psi\,(x)$ be the 'indefinite integral' of $\phi\,(x)$, we have

$$\frac{dA}{db} = \psi'\,(b). \qquad\qquad\text{.........................(3)}$$

It follows from Art. 56 that

$$A = \psi\,(b) + C, \qquad\qquad\text{.........................(4)}$$

where C is some constant; and since A must vanish for $b = a$, we must have $C = -\psi\,(a)$. Hence

$$A = \psi\,(b) - \psi\,(a). \qquad\qquad\text{.........................(5)}$$

The problem of finding the area is thus reduced to that of indefinite integration, which formed the subject of the preceding Chapter.

Ex. 1.　If $\phi\,(x) = x^m$, we have $\psi\,(x) = x^{m+1}/(m+1)$, and

$$A = \frac{b^{m+1} - a^{m+1}}{m+1}, \qquad\qquad\text{.........................(6)}$$

except when $m = -1$.

If $\phi\,(x) = 1/x$, we have $\psi\,(x) = \log x$, and

$$A = \log\frac{b}{a}. \qquad\qquad\text{.........................(7)}$$

Ex. 2.　If $\phi\,(x) = \sin x$, we have $\psi\,(x) = -\cos x$, and
$$A = -\cos\beta - (-\cos a) = \cos a - \cos\beta. \qquad\text{.........(8)}$$

The above results agree with those obtained, by much greater labour, in Art. 87.

89.　General Definition of an Integral.　Notation.

As the process of finding the limiting value of the sum of a series of infinitesimal quantities is one which has numerous applications in Geometry and Mechanics, we proceed to treat it in a more formal manner, attending at the same time to various points of theoretical importance which have hitherto been passed over.

Let $y, = \phi\,(x)$, be a function of x which is regarded as given (and therefore finite) for all values of x ranging from a to b, inclusively. Let the range $b - a$ be subdivided into a number of intervals

$$h_1, h_2, \ldots\ldots, h_n, \qquad\qquad\text{.........................(1)}$$

all of the same sign, so that

$$h_1 + h_2 + \ldots + h_n = b - a. \qquad\qquad\text{.................(2)}$$

Let y_1 be one of the values which y assumes in the interval h_1, y_2 one of the values which it assumes in the interval h_2, and so on; and let

$$\Sigma = y_1 h_1 + y_2 h_2 + \ldots + y_n h_n. \qquad \ldots\ldots\ldots\ldots(3)$$

The value of this sum will in general vary with the mode of subdivision of the range $b - a$, and with the choice of the values y_1, y_2, ..., y_n within the respective intervals (1). But if we introduce the condition that none of these intervals is to exceed some assigned magnitude k, then in certain cases, which include all the types of function ordinarily met with in the applications of the Calculus (and more), the value of Σ will tend, as k is diminished, to some definite limiting value S, in the sense that by taking k small enough we can ensure that Σ shall differ from S by less than any assigned magnitude, however small.

The sum which we have denoted by Σ is more fully expressed by

$$\Sigma_a^b y\, \delta x \quad \text{or} \quad \Sigma_a^b \phi(x)\, \delta x, \qquad \ldots\ldots\ldots\ldots\ldots\ldots(4)$$

δx standing for the increments h_1, h_2, ..., h_n of x. The limiting value (when it exists) to which this sum converges, as the increments δx are all indefinitely diminished, and their number in consequence indefinitely increased, is called the 'definite integral' of the function $\phi(x)$ between the limits a, and b*, and is denoted by

$$\int_a^b y\, dx \quad \text{or} \quad \int_a^b \phi(x)\, dx, \qquad \ldots\ldots\ldots\ldots\ldots(5)$$

the object of this notation being to recall the steps by which the limiting value was approached†.

Problems in which we require the limiting value of a sum of the type (3) occur in almost every branch of Mathematics. The area of a curve has already been referred to; other simple instances are: the length of a curved arc, regarded as the limit of an inscribed (or circumscribed) polygon, the volume of a solid of revolution, and so on. These will be considered more particularly in Chap. VIII.

Again, in Dynamics, the 'impulse' of a variable force, in any interval of time, is defined as the 'time-integral' of the force over that interval; viz. if F be the force, considered as a function of the time t, the impulse in the interval $t_1 - t_0$ is the limiting value of the sum

$$F_1 \tau_1 + F_2 \tau_2 + \ldots + F_n \tau_n, \qquad \ldots\ldots\ldots\ldots\ldots(6)$$

---

* It is a little unfortunate that the word 'limit' has to be used in several different senses. The word 'terminus' would perhaps be more appropriate in the present case.

† The symbol $\int$ is a specialized form of S, the sign of summation employed by the earlier analysts. The mode of indicating the range of integration was introduced by Fourier.

where $\tau_1, \tau_2, ..., \tau_n$ are subdivisions of the interval $t_1 - t_0$, such that

$$\tau_1 + \tau_2 + ... + \tau_n = t_1 - t_0, \quad\quad\quad\quad\quad......................(7)$$

whilst $F_1, F_2, ..., F_n$ denote values of the force in these respective intervals. Hence, in our present notation, the impulse is

$$\int_{t_0}^{t_1} F dt. \quad\quad\quad\quad\quad..............................(8)$$

Newton's Second Law of Motion asserts that the change of momentum of any mass (m) is equal to the impulse which it receives, or

$$mv_1 - mv_0 = \int_{t_0}^{t_1} F dt, \quad\quad\quad\quad......................(9)$$

where v_0, v_1 are the initial and final velocities.

Again, the work done by a variable force is defined as the space-integral of the force. If F denote the force, regarded now as a function of the position (s) of the body, the work done as s changes from s_0 to s_1 is

$$\int_{s_0}^{s_1} F ds. \quad\quad\quad\quad\quad..............................(10)$$

For example, the work done by unit mass of a gas as it expands from volume v_0 to volume v_1 is

$$\int_{v_0}^{v_1} p dv, \quad\quad\quad\quad\quad............................(11)$$

if p be the pressure when the volume is v. This is seen by supposing the gas to be enclosed, by a piston, in a cylinder of sectional area unity.

The graphical representation of the integral (10) or (11) is frequently employed in practice. Thus, in the case of (10), if a curve be constructed with s as abscissa and F as ordinate, the work is represented by the area included between the curve, the axis of s, and the ordinates corresponding to s_0 and s_1. This is the principle of Watt's indicator-diagram*.

90. Proof of Convergence.

Whenever the sum Σ has a definite limiting value, in the manner above explained, the function $\phi(x)$ is said to be 'integrable.' It may be shewn that every continuous function is integrable *in this sense*†, but as regards the formal proof we shall confine ourselves to the particular case where the range of the independent variable can be divided into a finite number of intervals within each of which the function either steadily increases or steadily decreases. This will be sufficient for all practical purposes.

Before, however, introducing any restriction (beyond that of finiteness) we may note that two fixed limits can be assigned

---

* See Maxwell, *Theory of Heat*, c. v.; Rankine, *The Steam-Engine*, Art. 43.
† It is not implied that a mathematical formula for the integral can be found.

between which Σ must necessarily lie. For if λ and μ be the lower and upper limits (Art. 17) of the values which the function $\phi(x)$ can assume in the interval $b - a$, it is evident that Σ will lie between

$$\lambda (h_1 + h_2 + \ldots + h_n), = \lambda (b - a),$$

and $$\mu (h_1 + h_2 + \ldots + h_n), = \mu (b - a).$$

We will now suppose, for definiteness, that $b > a$, and that $\phi(x)$ steadily increases as x increases from a to b. Consider any particular mode of subdivision

$$h_1, h_2, \ldots, h_n, \ldots\ldots\ldots\ldots\ldots\ldots\ldots(1)$$

of the range $b - a$, and let

$$\Sigma = y_1 h_1 + y_2 h_2 + \ldots + y_n h_n, \quad \ldots\ldots\ldots(2)$$

where, as in Art. 89, y_r denotes some value which the function assumes in the interval h_r.

Now if in (2) we replace $y_1, y_2, \ldots y_n$ by the values which the function has at the *beginnings* of the respective intervals, none of the terms will be increased; and if the resulting sum be denoted by Σ', we shall have

$$\Sigma' \not> \Sigma. \ldots\ldots\ldots\ldots\ldots\ldots\ldots(3)$$

Again, if we replace $y_1, y_2, \ldots y_n$ by the values which the function has at the *ends* of the respective intervals, none of the terms will be diminished; hence if the resulting sum be Σ'', we shall have

$$\Sigma'' \not< \Sigma. \quad \ldots\ldots\ldots\ldots\ldots\ldots\ldots(4)$$

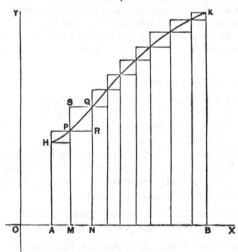

Fig. 46.

In Fig. 46 the quantity Σ' is represented by the sum of a series of rectangles such as PN, and Σ'' by the sum of a series of rectangles such as SN. Hence the difference $\Sigma'' - \Sigma'$ is represented by the sum of a series of rectangles such as SR. The sum of the altitudes of these latter rectangles is $KB - HA$, or $\phi(b) - \phi(a)$, and if k be the greatest of the bases, *i.e.* the greatest of the intervals (1), we shall have

$$\Sigma'' - \Sigma' \ngtr k\{\phi(b) - \phi(a)\}. \quad \ldots\ldots\ldots\ldots(5)$$

Now, considering all possible modes of subdivision of the range $b - a$, the sums Σ', being always less than $\mu(b - a)$, will have an upper limit, which we will denote by S', and the sums Σ'', being always greater than $\lambda(b - a)$, will have a lower limit, which we will denote by S'', and it is further evident that $S'' \nless S'$. It follows, from (5), that the difference $S'' - S'$ must lie between 0 and $k\{\phi(b) - \phi(a)\}$; and since, in this statement, k may be as small as we please, it appears that S' and S'' cannot but be equal. We will denote their common value by S.

Finally, it is evident that

$$|\Sigma - S| < \Sigma'' - \Sigma' < k\{\phi(b) - \phi(a)\}; \quad \ldots\ldots\ldots(6)$$

hence by taking k small enough we can ensure that $|\Sigma - S|$ shall be less than any assigned quantity, however small*.

A similar proof obviously applies if the function $\phi(x)$ steadily decreases throughout the range $b - a$.

It follows that the final result also holds when the range admits of being broken up into a finite number of smaller intervals within each of which the function either steadily increases or steadily decreases. See Fig. 47.

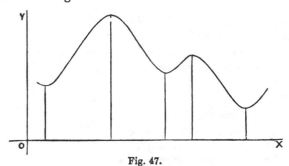

Fig. 47.

---

* The proof is a development of that given by Newton, *Principia*, lib. i., sect. i., lemma iii. (1687). It would be easy to eliminate all geometrical considerations and present the argument in a purely quantitative form.

It has been supposed that $b > a$. If $b < a$, the intervals h_1, h_2, ..., h_n will be negative, but the argument is substantially unaltered.

91.   Properties of $\int_a^b \phi(x)\,dx$.

1°.   If we compare the integrals

$$\int_a^b \phi(x)\,dx \quad \text{and} \quad \int_b^a \phi(x)\,dx,$$

we see that they may be regarded as limits of the same summation, with this difference, that in one case the increments h_1, h_2, ..., h_n of x, which make up the interval $b - a$ (or $a - b$) have the opposite sign to that which they have in the other.   Hence

$$\int_b^a \phi(x) = -\int_a^b \phi(x)\,dx. \quad ...(1)$$

2°.   Again, it follows from the definition that

$$\int_a^c \phi(x)\,dx = \int_a^b \phi(x)\,dx + \int_b^c \phi(x)\,dx.$$
$$.........(2)$$

This is illustrated graphically in Fig. 48.

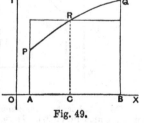

Fig. 48.

3°.   If λ, μ be the least and greatest values which $\phi(x)$ assumes as x ranges from a to b, the integral

$$\int_a^b \phi(x)\,dx,$$

being intermediate in value to $\lambda(b-a)$ and $\mu(b-a)$, must be equal to

$$\nu(b-a),$$

where ν is some quantity intermediate to λ, μ.

If, as we suppose, $\phi(x)$ is continuous, it assumes within the range $b - a$ *all* values intermediate to λ, μ.   Hence there must be some value (c) of x, between a and b, such that

$$\phi(c) = \nu.$$

In the graphical representation, Fig. 49, the area $PABQ$ is equal to a rectangle, on the base AB, whose altitude is equal to the ordinate at some point C of the range AB.

We may evidently write

$$c = a + \theta(b-a),$$

Fig. 49.

where θ is some quantity between 0 and 1. On this understanding,

$$\int_a^b \phi(x)\,dx = (b-a)\,\phi(a + \theta\,\overline{b-a}). \quad\ldots\ldots\ldots\ldots(3)$$

4°. More generally, if u, v, y be three functions such that for values of x ranging from a to b,

$$u > y > v, \quad\ldots\ldots\ldots\ldots\ldots\ldots\ldots\ldots\ldots(4)$$

then the integral $\qquad \int_a^b y\,dx \quad\ldots\ldots\ldots\ldots\ldots\ldots\ldots\ldots\ldots(5)$

will be intermediate in value to

$$\int_a^b u\,dx \quad\text{and}\quad \int_a^b v\,dx. \quad\ldots\ldots\ldots\ldots\ldots(6)$$

Suppose, first, that $b > a$. We have

$$\int_a^b u\,dx - \int_a^b y\,dx = \int_a^b (u - y)\,dx.$$

In virtue of (4), every term of the sum, of which the latter integral is the limit, will be positive. Hence

$$\int_a^b y\,dx < \int_a^b u\,dx. \quad\ldots\ldots\ldots\ldots\ldots\ldots(7)$$

Similarly $\qquad \int_a^b y\,dx > \int_a^b v\,dx. \quad\ldots\ldots\ldots\ldots\ldots\ldots(8)$

If $b < a$, the inequalities in (7) and (8) must be reversed.

92. Differentiation of a Definite Integral with respect to either Limit.

Let $\qquad I = \int_a^b \phi(x)\,dx. \quad\ldots\ldots\ldots\ldots\ldots\ldots(1)$

Evidently, I is a function of the 'limits of integration' a, b, and will in general vary when either of these varies. Regarding a as fixed, let us form the derived function of I with respect to the upper limit b. We have

$$I + \delta I = \int_a^{b+\delta b} \phi(x)\,dx$$

$$= \int_a^b \phi(x)\,dx + \int_b^{b+\delta b} \phi(x)\,dx, \quad\ldots\ldots\ldots\ldots(2)$$

by Art. 91, 2°. Hence

$$\delta I = \int_b^{b+\delta b} \phi(x)\,dx = \delta b\,.\,\phi(b + \theta\,\delta b), \quad\ldots\ldots\ldots\ldots(3)$$

by Art. 91, 3°. This shews that δI vanishes with δb, so that I is a continuous function of b. Also, since

$$\frac{\delta I}{\delta b} = \phi\,(b + \theta\,\delta b), \quad \dots\dots\dots\dots\dots\dots(4)$$

we have, on proceeding to the limit ($\delta b \to 0$),

$$\frac{dI}{db} = \phi\,(b). \quad \dots\dots\dots\dots\dots\dots(5)$$

In the same way, if we regard the upper limit b as fixed, and the lower limit a as variable, we find that I is a continuous function of a, and that

$$\frac{dI}{da} = -\,\phi\,(a). \quad \dots\dots\dots\dots\dots\dots(6)$$

93. Existence of an Indefinite Integral.

We can now shew that any function $\phi\,(x)$, having the character postulated in Art. 90, has an indefinite integral, *i.e.* there exists a definable (but not necessarily calculable) function $\psi\,(x)$ such that

$$\psi'\,(x) = \phi\,(x), \quad \dots\dots\dots\dots\dots\dots(1)$$

or
$$\psi\,(x) = D^{-1}\phi\,(x). \quad \dots\dots\dots\dots\dots\dots(2)$$

For if we write

$$\psi\,(\xi) = \int_a^\xi \phi\,(x)\,dx, \quad \dots\dots\dots\dots\dots\dots(3)$$

the expression on the right hand is, by Art. 90, a determinate function of ξ, and the investigation just given shews that it satisfies the condition

$$\psi'\,(\xi) = \phi\,(\xi). \quad \dots\dots\dots\dots\dots\dots(4)$$

The lower limit of integration in (3) is, from the present point of view, arbitrary, and the function $\psi\,(\xi)$ is therefore indeterminate to the extent of an additive constant. For, by Art. 91, 2°, the substitution of a' for a, as the lower limit in (3), is equivalent to the addition of

$$\int_{a'}^a \phi\,(x)\,dx$$

to the right-hand side. Cf. Art. 72.

94. Rule for calculating a Definite Integral.

Whenever the analytical form of a function $\psi\,(x)$, which has a given function $\phi\,(x)$ as its derivative, is known, the value of the definite integral

$$I = \int_a^b \phi\,(x)\,dx, \quad \dots\dots\dots\dots\dots\dots(1)$$

can be written down at once.　For, if we regard a as fixed, we have, by Art. 92,

$$\frac{dI}{db} = \phi(b)$$

$$= \psi'(b), \quad\dotfill(2)$$

by hypothesis.　It follows by Art. 56 that I and $\psi(b)$ can only differ by a 'constant,' *i.e.* a quantity independent of b; thus

$$\int_a^b \phi(x)\,dx = \psi(b) + C.\dotfill(3)$$

To find the value of C we may, since it does not vary with b, put $b = a$, whence

$$\psi(a) + C = \int_a^a \phi(x)\,dx = 0. \quad\dotfill(4)$$

Hence $C = -\psi(a)$, and

$$\int_a^b \phi(x)\,dx = \psi(b) - \psi(a). \quad\dotfill(5)$$

This is the fundamental proposition of the Integral Calculus. It reduces the problem of finding the definite integral of a given function $\phi(x)$ to the discovery of the inverse function $\psi(x)$, or $D^{-1}\phi(x)$. The reason why this inverse function is usually denoted by

$$\int \phi(x)\,dx \dotfill(6)$$

is now apparent.　The form (6) is simply an abbreviation for

$$\int_a^x \phi(x)\,dx, \quad\dotfill(7)$$

where a is arbitrary.　We have seen that a change in a is equivalent to the addition of a constant.

The notation $\qquad \left[\psi(b)\right]_a^b \quad\dotfill(8)$

is often used as an abbreviation for $\psi(b) - \psi(a)$.

Ex. 1.　To find $\qquad \displaystyle\int_a^b e^{kx}\,dx. \quad\dotfill(9)$

Here $\qquad \phi(x) = e^{kx}, \quad \psi(x) = \frac{1}{k}e^{kx},$

whence $\qquad \displaystyle\int_a^b e^{kx}\,dx = \frac{1}{k}(e^{kb} - e^{ka}). \quad\dotfill(10)$

Ex. 2.   To find $\int_0^{\frac{1}{2}\pi} \sin^2 x \, dx.$ ....(11)

Here $\phi(x) = \sin^2 x, \quad \psi(x) = \frac{1}{2}x - \frac{1}{4}\sin 2x.$

Hence $\int_0^{\frac{1}{2}\pi} \sin^2 x \, dx = \frac{1}{4}\pi.$ ....(12)

95.   Cases where the function $\phi(x)$, or the limits of integration, become infinite.

Before proceeding to further examples, it will be convenient to extend somewhat the definition of an integral given in Art. 89. It was there assumed that the limits of integration a, b were finite, and also that the function $\phi(x)$ was finite throughout the range $b - a$. We proceed to explain how, under certain conditions, these conditions may be relaxed.

1°.   Suppose $\phi(x)$ to be finite and continuous for all finite values of x, and consider the integral

$$\int_a^\omega \phi(x) \, dx, \qquad ....(1)$$

where $\omega > a$. If, as ω is increased indefinitely, the integral tends to a definite limiting value, this value is denoted by

$$\int_a^\infty \phi(x) \, dx. \qquad ....(2)$$

The integral (1) is then said to be 'convergent' for $\omega \to \infty$. As might be anticipated from the theory of infinite series (Art. 5) it is not a sufficient condition for convergence that

$$\lim_{x \to \infty} \phi(x) = 0 ; \qquad ....(3)$$

this condition is moreover not essential, for there may even be convergence when $\phi(x)$ has no definite limiting value for $x \to \infty$.

A similar definition of

$$\int_{-\infty}^b \phi(x) \, dx \qquad ....(4)$$

can obviously be framed.

2°.   Let $\phi(x)$ become infinite at or between the limits of integration.

It will be sufficient to consider the case where there is only one value of x for which $\phi(x) \to \infty$. The general case can be reduced to this by breaking up the range $b - a$ into smaller intervals*.

If $\phi(x)$ become infinite at the upper limit (only), we consider in the first place the integral

$$\int_a^{b-\epsilon} \phi(x)\,dx, \quad\dots\dots\dots\dots\dots(5)$$

where ϵ is positive. If, as ϵ is diminished indefinitely, this integral tends to a definite limiting value, this value is adopted as the definition of

$$\int_a^b \phi(x)\,dx.$$

A similar definition applies to the case where $\phi(x)$ becomes infinite at the lower limit a.

If $\phi(x)$ becomes infinite *between* the limits a, b, say for $x = c$, we consider the sum

$$\int_a^{c-\epsilon} \phi(x)\,dx + \int_{c+\epsilon'}^b \phi(x)\,dx. \quad\dots\dots\dots\dots(6)$$

If, with diminishing ϵ (and ϵ') each of these integrals tends to a finite limiting value, the sum of these values is adopted as the definition of

$$\int_a^b \phi(x)\,dx. \quad\dots\dots\dots\dots\dots(7)\dagger$$

The cases where $\phi(x)$ becomes infinite, or is discontinuous, at a finite number of isolated points, are dealt with by dividing the range into shorter intervals bounded by the points of discontinuity.

Ex. 1. $$\int_0^\omega e^{-ax}\,dx = \left[-\frac{1}{a} e^{-ax} \right]_0^\omega = \frac{1 - e^{-a\omega}}{a}. \quad\dots\dots\dots(8)$$

As ω increases this tends to the limit $1/a$. Hence we say that

$$\int_0^\infty e^{-ax}\,dx = \frac{1}{a}. \quad\dots\dots\dots\dots\dots(9)$$

---

* It being assumed that $\phi(x)$ becomes infinite only at a finite number of isolated points.

† Cases may arise in which each of the integrals

$$\int_a^{c-\epsilon} \phi(x)\,dx \text{ and } \int_{c+\epsilon'}^b \phi(x)\,dx$$

is ultimately infinite, whilst if *some special relation* be imposed on the ultimately vanishing quantities ϵ, ϵ', the infinite elements of the two integrals cancel in such a way that the sum remains finite. If the relation in question be $\epsilon' = \epsilon$, the result, when it exists, is called by Cauchy the 'principal value' of the integral (7).

Ex. 2.
$$\int_1^\omega \frac{dx}{x} = \Big[\log x\Big]_1^\omega = \log \omega. \ \dots \ \dots \dots \dots \dots (10)$$

This increases without limit with ω. Hence there is no limiting value for $\omega \to \infty$, although

$$\lim_{x \to \infty} \frac{1}{x} = 0. \ \dots\dots \ \dots\dots\dots\dots\dots (11)$$

Ex. 3.
$$\int_0^1 \frac{dx}{\sqrt{(1-x)}}. \ \dots\dots\dots\dots\dots\dots (12)$$

The function $1/\sqrt{(1-x)}$ becomes infinite for $x = 1$, but

$$\int_0^{1-\epsilon} \frac{dx}{\sqrt{(1-x)}} = \Big[-2\sqrt{(1-x)}\Big]_0^{1-\epsilon} = 2 - 2\sqrt{\epsilon}, \ \dots\dots (13)$$

and as ϵ is indefinitely diminished this tends to the limit 2. Hence

$$\int_0^1 \frac{dx}{\sqrt{(1-x)}} = 2. \ \dots\dots\dots\dots\dots (14)$$

Ex. 4.
$$\int_0^1 \log x\, dx. \ \dots\dots\dots\dots\dots (15)$$

We have

$$\int_\epsilon^1 \log x\, dx = \Big[x \log x - x\Big]_\epsilon^1. \ \dots\dots\dots\dots (16)$$

By Art. 43 (5) we have

$$\lim_{\epsilon \to 0} \epsilon \log \epsilon = 0.$$

Hence

$$\int_0^1 \log x\, dx = -1. \ \dots\dots\dots\dots\dots (17)$$

96. Applications of the Rule of Art. 94.

We give a few more typical examples of the evaluation of definite integrals.

Ex. 1.
$$\int_0^{\frac{1}{2}\pi} \sin x\, dx = \Big[-\cos x\Big]_0^{\frac{1}{2}\pi} = 1; \ \dots\dots\dots\dots (1)$$

$$\int_0^{\frac{1}{2}\pi} \cos x\, dx = \Big[\sin x\Big]_0^{\frac{1}{2}\pi} = 1; \ \dots\dots\dots\dots (2)$$

$$\int_0^{\frac{1}{2}\pi} \sin x \cos x\, dx = \Big[\tfrac{1}{2}\sin^2 x\Big]_0^{\frac{1}{2}\pi} = \tfrac{1}{2}. \ \dots\dots\dots (3)$$

Ex. 2. By Art. 80 we have

$$\int_0^\omega e^{-ax} \sin \beta x\, dx = -\left[\frac{a \sin \beta x + \beta \cos \beta x}{a^2 + \beta^2} e^{-ax}\right]_0^\omega$$

$$= \frac{\beta}{a^2 + \beta^2} - \frac{a \sin \beta\omega + \beta \cos \beta\omega}{a^2 + \beta^2} e^{-a\omega}.$$

If a be positive the last term tends, as ω is increased indefinitely, to the limiting value 0.   Hence

$$\int_0^\infty e^{-ax} \sin \beta x\, dx = \frac{\beta}{a^2 + \beta^2}. \qquad\qquad (4)$$

Similarly $$\int_0^\infty e^{-ax} \cos \beta x\, dx = \frac{a}{a^2 + \beta^2}. \qquad\qquad (5)$$

Ex. 3.   We have

$$\int_0^\omega \frac{dx}{1+x^2} = \left[\tan^{-1} x \right]_0^\omega = \tan^{-1} \omega - \tan^{-1} 0.$$

The function $\tan^{-1} x$ is many-valued (Art. 16), but it is immaterial which value we take, provided we suppose it to change continuously as x varies through the range of integration.   Hence if we take $\tan^{-1} 0 = 0$, we must understand by $\tan^{-1} \omega$ that value which increases continuously from 0 with ω.   As ω increases indefinitely, this value tends to the limit $\frac{1}{2}\pi$, so that

$$\int_0^\infty \frac{dx}{1+x^2} = \frac{1}{2}\pi. \qquad\qquad (6)$$

Ex. 4.   By Art. 78, 4° we have

$$\int_0^{\frac{1}{2}\pi} \frac{d\theta}{a \sin^2 \theta + \beta \cos^2 \theta} = \left[\frac{1}{\sqrt{(a\beta)}} \tan^{-1} \left(\sqrt{\frac{a}{\beta}} . \tan \theta \right) \right]_0^{\frac{1}{2}\pi}.$$

Now, as θ increases from 0 to $\frac{1}{2}\pi$, $\sqrt{(a/\beta)} . \tan \theta$ increases from 0 to ∞, and we may therefore suppose that $\tan^{-1} \{ \sqrt{(a/\beta)} . \tan \theta \}$ increases from 0 to $\frac{1}{2}\pi$.   Hence

$$\int_0^{\frac{1}{2}\pi} \frac{d\theta}{a \sin^2 \theta + \beta \cos^2 \theta} = \frac{\pi}{2\sqrt{(a\beta)}}. \qquad\qquad (7)$$

The student may have remarked in the course of the preceding Chapter that when an 'indefinite' integration is effected by a change of variable (Arts. 77, 79) the most troublesome part of the process consists often in the translation back to the original variable.   This part is, however, unnecessary when the object is merely to find the definite integral between given limits.   It is then sufficient to substitute the *altered* limits in the indefinite integral as first obtained.

Ex. 5.   To find $$\int_0^a \sqrt{(a^2 - x^2)}\, dx. \qquad\qquad (8)$$

We found (Art. 79), putting $x = a \sin \theta$, that

$$\int \sqrt{(a^2 - x^2)}\, dx = a^2 \int \cos^2 \theta\, d\theta = \tfrac{1}{2}a^2 (\theta + \tfrac{1}{2} \sin 2\theta).$$

Now, if θ increase from 0 to $\frac{1}{2}\pi$, x will increase from 0 to a.   Hence

$$\int_0^a \sqrt{(a^2 - x^2)}\, dx = \tfrac{1}{2}a^2 \left[\theta + \tfrac{1}{2} \sin 2\theta \right]_0^{\frac{1}{2}\pi} = \tfrac{1}{4}\pi a^2. \qquad\qquad (9)$$

97. Formulæ of Reduction.

The methods of Arts. 81, 82, when applied to the reduction of *definite* integrals, sometimes lead to specially simple results, owing to the vanishing of the integrated terms at both limits.

1°.   If
$$u_n = \int_0^{\frac{1}{2}\pi} \cos^n \theta \, d\theta, \ldots\ldots\ldots\ldots\ldots\ldots(1)$$

we have, by Art. 82 (2),

$$u_n = \left[\frac{1}{n} \sin \theta \cos^{n-1} \theta \right]_0^{\frac{1}{2}\pi} + \frac{n-1}{n} u_{n-2}. \ \ \ldots\ldots(2)$$

If $n > 1$, the first part vanishes, since

$$\sin 0 = 0, \quad \cos \tfrac{1}{2}\pi = 0.$$

Hence
$$\int_0^{\frac{1}{2}\pi} \cos^n \theta \, d\theta = \frac{n-1}{n} \int_0^{\frac{1}{2}\pi} \cos^{n-2} \theta \, d\theta. \ \ \ldots\ldots\ldots(3)$$

Similarly, from Art. 82 (6),

$$\int_0^{\frac{1}{2}\pi} \sin^n \theta \, d\theta = \frac{n-1}{n} \int_0^{\frac{1}{2}\pi} \sin^{n-2} \theta \, d\theta. \ \ \ldots\ldots\ldots(4)$$

If n be a positive integer, we can, by successive applications of (3), express

$$\int_0^{\frac{1}{2}\pi} \cos^n \theta \, d\theta$$

in terms of either

$$\int_0^{\frac{1}{2}\pi} \cos \theta \, d\theta, = 1, \ \text{ or } \int_0^{\frac{1}{2}\pi} d\theta, = \tfrac{1}{2}\pi, \ \ \ldots\ldots\ldots(5)$$

according as n is odd or even. In the same way

$$\int_0^{\frac{1}{2}\pi} \sin^n \theta \, d\theta$$

can be made to depend either on

$$\int_0^{\frac{1}{2}\pi} \sin \theta \, d\theta, = 1, \ \text{ or on } \int_0^{\frac{1}{2}\pi} d\theta, = \tfrac{1}{2}\pi. \ldots\ldots\ldots(6)$$

Ex. 1.
$$\int_0^{\frac{1}{2}\pi} \cos^5 \theta \, d\theta = \tfrac{4}{5} \int_0^{\frac{1}{2}\pi} \cos^3 \theta \, d\theta$$

$$= \tfrac{4}{5} \cdot \tfrac{2}{3} \int_0^{\frac{1}{2}\pi} \cos \theta \, d\theta = \tfrac{8}{15}.$$

After working out one or two examples in this way, the student will be able to supply the successive steps mentally, and write down at once the factors of the result; thus

$$\int_0^{\frac{1}{2}\pi} \cos^6\theta\, d\theta = \tfrac{5}{6} \cdot \tfrac{3}{4} \cdot \tfrac{1}{2} \cdot \tfrac{1}{2}\pi = \tfrac{5}{32}\pi.$$

The general values of the preceding integrals can be written down without difficulty. Thus, if n be odd, we have

$$\int_0^{\frac{1}{2}\pi} \cos^n\theta\, d\theta = \int_0^{\frac{1}{2}\pi} \sin^n\theta\, d\theta = \frac{(n-1)(n-3)\ldots 2}{n(n-2)\ldots 3}, \ldots(7)$$

whilst, if n be even,

$$\int_0^{\frac{1}{2}\pi} \cos^n\theta\, d\theta = \int_0^{\frac{1}{2}\pi} \sin^n\theta\, d\theta = \frac{(n-1)(n-3)\ldots 1}{n(n-2)\ldots 2} \cdot \frac{\pi}{2}. \quad (8)$$

Integrals of this type are of frequent occurrence in the physical applications of the Calculus.

2°. If $\qquad u_{m,n} = \int_0^{\frac{1}{2}\pi} \sin^m\theta \cos^n\theta\, d\theta, \ldots\ldots\ldots\ldots\ldots(9)$

we have, by Art. 82 (10),

$$u_{m,n} = \left[\frac{1}{m+n} \sin^{m+1}\theta \cos^{n-1}\theta \right]_0^{\frac{1}{2}\pi} + \frac{n-1}{m+n} u_{m,n-2}. \quad (10)$$

If $n > 1$, the expression in [ ] vanishes at both limits, and we have

$$\int_0^{\frac{1}{2}\pi} \sin^m\theta \cos^n\theta\, d\theta = \frac{n-1}{m+n} \int_0^{\frac{1}{2}\pi} \sin^m\theta \cos^{n-2}\theta\, d\theta. \ldots(11)$$

In the same way from Art. 82 (11) we obtain, if $m > 1$,

$$\int_0^{\frac{1}{2}\pi} \sin^m\theta \cos^n\theta\, d\theta = \frac{m-1}{m+n} \int_0^{\frac{1}{2}\pi} \sin^{m-2}\theta \cos^n\theta\, d\theta. \ldots(12)$$

By means of these formulæ, either index can be reduced by 2, and by repetitions of this process we can, if m, n be positive integers, make the integral (9) depend on one in which each index is 1 or 0. The result therefore finally involves one or other of the following forms:

$$\left. \begin{array}{ll} \int_0^{\frac{1}{2}\pi} \sin\theta \cos\theta\, d\theta, = \tfrac{1}{2}; & \int_0^{\frac{1}{2}\pi} d\theta, = \tfrac{1}{2}\pi; \\[2mm] \int_0^{\frac{1}{2}\pi} \sin\theta\, d\theta, = 1; & \int_0^{\frac{1}{2}\pi} \cos\theta\, d\theta, = 1. \end{array} \right\} \ldots\ldots\ldots(13)$$

Ex. 2. We have

$$\int_0^{\frac{1}{2}\pi} \sin^5 \theta \cos^3 \theta \, d\theta = \tfrac{4}{5} \int_0^{\frac{1}{2}\pi} \sin^3 \theta \cos^3 \theta \, d\theta = \tfrac{4}{5} \cdot \tfrac{2}{5} \int_0^{\frac{1}{2}\pi} \sin \theta \cos^3 \theta \, d\theta,$$

by (12).   Again, by (11),

$$\int_0^{\frac{1}{2}\pi} \sin \theta \cos^3 \theta \, d\theta = \tfrac{2}{4} \cdot \int_0^{\frac{1}{2}\pi} \sin \theta \cos \theta \, d\theta = \tfrac{2}{4} \cdot \tfrac{1}{2}$$

Hence $\int_0^{\frac{1}{2}\pi} \sin^5 \theta \cos^3 \theta \, d\theta = \tfrac{4}{5} \cdot \tfrac{2}{5} \cdot \tfrac{2}{4} \cdot \tfrac{1}{2} = \tfrac{1}{24}.$

After a little practice, the result can be written down immediately. Thus

$$\int_0^{\frac{1}{2}\pi} \sin^6 \theta \cos^2 \theta \, d\theta = \tfrac{5}{8} \cdot \tfrac{3}{6} \cdot \tfrac{1}{4} \cdot \tfrac{1}{2} \cdot \tfrac{1}{2}\pi = \tfrac{5}{256}\pi.$$

The formulæ (11) and (12), as well as (3) and (4), are often required in practice, and should be remembered.

Again, the algebraic integral

$$\int_0^1 x^m (1 - x)^n \, dx, \quad \dots\dots\dots\dots\dots (14)$$

is reduced by the substitution $x = \sin^2 \theta$ to the form

$$2 \int_0^{\frac{1}{2}\pi} \sin^{2m+1} \theta \cos^{2n+1} \theta \, d\theta, \quad \dots\dots\dots\dots (15)$$

and can therefore be evaluated by means of the formulæ given above, whenever $2m + 1$ and $2n + 1$ are positive integers or null.

Similarly, if we put $x = \sin \theta$, the integral

$$\int_0^1 x^m (1 - x^2)^n \, dx \quad \dots\dots\dots\dots\dots (16)$$

takes the form

$$\int_0^{\frac{1}{2}\pi} \sin^m \theta \cos^{2n+1} \theta \, d\theta. \quad \dots\dots\dots\dots (17)$$

Ex. 3. $\int_0^1 x^2 (1 - x)^{\frac{3}{2}} \, dx = 2 \int_0^{\frac{1}{2}\pi} \sin^5 \theta \cos^4 \theta \, d\theta$

$$= 2 \cdot \tfrac{4}{9} \cdot \tfrac{2}{7} \cdot \tfrac{3}{5} \cdot \tfrac{1}{3} \cdot 1 = \tfrac{16}{315}.$$

Ex. 4. $\int_0^1 x^2 (1 - x^2)^{\frac{3}{2}} = \int_0^{\frac{1}{2}\pi} \sin^2 \theta \cos^4 \theta \, d\theta$

$$= \tfrac{1}{6} \cdot \tfrac{3}{4} \cdot \tfrac{1}{2} \cdot \tfrac{1}{2}\pi = \tfrac{1}{32}\pi.$$

98. Related Integrals.

There are various theorems concerning definite integrals which follow almost intuitively from the definition of Art. 89.

For example,

$$\int_0^a \phi(x)\,dx = \int_0^a \phi(a-x)\,dx. \quad \dots\dots\dots\dots(1)$$

This is proved by writing

$$x = a - x', \quad dx = -dx',$$

the new limits of integration being $x' = a$, $x' = 0$, corresponding to $x = 0$, $x = a$, respectively. Thus

$$\int_0^a \phi(x)\,dx = -\int_a^0 \phi(a-x')\,dx' = \int_0^a \phi(a-x)\,dx,$$

the accent being dropped in the end, as no longer necessary.

This process is equivalent to transferring the origin to the point $x = a$, and reversing the direction of the axis of x. The areas represented by the integrals in (1) are thus seen to be identical.

An important case of (1) is

$$\int_0^{\frac{1}{2}\pi} f(\sin\theta)\,d\theta = \int_0^{\frac{1}{2}\pi} f(\cos\theta)\,d\theta. \quad \dots\dots\dots(2)$$

Ex. 1. Thus $\quad \displaystyle\int_0^{\frac{1}{2}\pi} \sin^2\theta\,d\theta = \int_0^{\frac{1}{2}\pi} \cos^2\theta\,d\theta.$

Hence each of these integrals

$$= \tfrac{1}{2}\int_0^{\frac{1}{2}\pi}(\sin^2\theta + \cos^2\theta)\,d\theta = \tfrac{1}{2}\int_0^{\frac{1}{2}\pi} d\theta = \tfrac{1}{4}\pi$$

Again, if $\phi(x)$ be an 'even' function of x, that is

$$\phi(-x) = \phi(x), \quad \dots\dots\dots\dots\dots\dots(3)$$

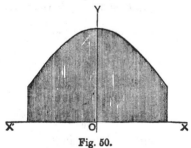

Fig. 50.

we have

$$\int_{-a}^{a} \phi\,(x)\,dx = 2\int_{0}^{a}\phi\,(x)\,dx, \quad \dots\dots\dots\dots(4)$$

the area represented by the former integral being obviously bisected by the axis of y.

On the other hand, if $\phi\,(x)$ be an 'odd' function of x, so that

$$\phi\,(-x) = -\,\phi\,(x), \dots\dots\dots\dots\dots\dots(5)$$

we have
$$\int_{-a}^{a}\phi\,(x)\,dx = 0, \quad \dots\dots\dots\dots\dots\dots(6)$$

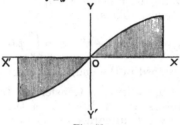

Fig. 51.

since in the sum, of which the definite integral is the limit (Art. 89), the element $\phi\,(x)\,\delta x$ is cancelled by the oppositely-signed element $\phi\,(-x)\,\delta x$.

Ex. 2.   We have

$$\int_{-\frac{1}{2}\pi}^{\frac{1}{2}\pi}\sin^{2}\theta\cos^{3}\theta\,d\theta = 2\int_{0}^{\frac{1}{2}\pi}\sin^{2}\theta\cos^{3}\theta\,d\theta = 2\cdot\tfrac{1}{5}\cdot\tfrac{2}{3}\cdot 1 = \tfrac{4}{15};$$

whilst
$$\int_{-\frac{1}{2}\pi}^{\frac{1}{2}\pi}\sin^{3}\theta\cos^{2}\theta\,d\theta = 0,$$

since $\sin^{3}\theta$ changes sign with θ.

For similar reasons, if

$$\phi\,(a-x) = \phi\,(x), \dots\dots\dots\dots\dots\dots(7)$$

we have
$$\int_{0}^{a}\phi\,(x)\,dx = 2\int_{0}^{\frac{1}{2}a}\phi\,(x)\,dx\,;\dots\dots\dots\dots(8)$$

whilst if
$$\phi\,(a-x) = -\,\phi\,(x), \quad \dots\dots\dots\dots\dots(9)$$

we have
$$\int_{0}^{a}\phi\,(x)\,dx = 0. \quad \dots\dots\dots\dots(10)$$

As a particular case of (8), we have

$$\int_{0}^{\pi}f\,(\sin\theta)\,d\theta = 2\int_{0}^{\frac{1}{2}\pi}f\,(\sin\theta)\,d\theta, \dots\dots\dots(11)$$

since
$$\sin\,(\pi-\theta) = \sin\theta.$$

Ex. 3.

$$\int_0^\pi \sin^3 \theta \cos^2 \theta \, d\theta = 2 \int_0^{\frac{1}{2}\pi} \sin^3 \theta \cos^2 \theta \, d\theta = 2 \cdot \tfrac{2}{5} \cdot \tfrac{1}{3} \cdot 1 = \tfrac{4}{15};$$

whilst $$\int_0^\pi \sin^2 \theta \cos^3 \theta \, d\theta = 0.$$

EXAMPLES. XXXI.

1. Prove by the method of Art. 87, Ex. 1 that

$$\int_a^b x^3 \, dx = \tfrac{1}{4}(b^4 - a^4).$$

2. Prove from first principles that

$$\int_a^\beta \cos x \, dx = \sin \beta - \sin \alpha.$$

3. Also that

$$\int_a^b e^{kx} \, dx = \frac{e^{kb} - e^{ka}}{k}.$$

4. Shew by graphical considerations that

$$\int_a^b k\phi(x) \, dx = k \int_a^b \phi(x) \, dx,$$

$$\int_a^b \phi(x) \, dx = \int_0^{b-a} \phi(x+a) \, dx,$$

$$\int_a^b \phi(kx) \, dx = \frac{1}{k} \int_{ka}^{kb} \phi(x) \, dx.$$

5. Prove that

$$\int_a^b \phi(x) \, dx = \int_a^b \phi(a+b-x) \, dx.$$

6. Prove that if n and p are positive integers

$$\lim_{n \to \infty} \left(\frac{1}{n} + \frac{1}{n+1} + \frac{1}{n+2} + \dots + \frac{1}{pn} \right) = \log p.$$

EXAMPLES. XXXII.

1. $\displaystyle\int_0^1 \sqrt{x} \, dx = \tfrac{2}{3}, \quad \int_0^1 \frac{dx}{\sqrt{x}} = 2, \quad \int_0^1 \frac{dx}{\sqrt{(3-2x)}} = \sqrt{3} - 1.$

2. $\displaystyle\int_0^1 \frac{dx}{\sqrt{(1-x^2)}} = \tfrac{1}{2}\pi, \quad \int_0^{1/\sqrt{3}} \frac{dx}{\sqrt{(2-3x^2)}} = \frac{\pi}{4\sqrt{3}}.$

3. $\displaystyle\int_0^\infty \frac{dx}{a^2 + b^2 x^2} = \frac{\pi}{2ab}.$

4. $\int_0^1 \dfrac{x\,dx}{\sqrt{(1-x^2)}} = 1,$ $\int_0^1 \dfrac{x\,dx}{\sqrt{(1+x^2)}} = \sqrt{2}-1.$

5. $\int_0^1 \dfrac{dx}{\sqrt{(x^2+1)}} = \log(1+\sqrt{2}),$ $\int_1^2 \dfrac{dx}{\sqrt{(x^2-1)}} = \log(2+\sqrt{3}).$

6. $\int_1^\infty \dfrac{dx}{x\,(1+x)} = \log 2,$ $\int_1^\infty \dfrac{dx}{x\,(1+x^2)} = \tfrac{1}{2}\log 2.$

7. $\int_0^1 \dfrac{dx}{x^2-x+1} = \dfrac{2\pi}{3\sqrt{3}}.$

8. $\int_0^1 \dfrac{1-x^2}{1+x^2}\,dx = \tfrac{1}{2}\pi - 1.$

9. $\int_0^\infty \dfrac{dx}{(x^2+a^2)\,(x^2+b^2)} = \dfrac{\pi}{2ab\,(a+b)}.$

10. $\int_0^\infty \dfrac{x^2\,dx}{(x^2+a^2)\,(x^2+b^2)} = \dfrac{\pi}{2\,(a+b)}.$

11. $\int_0^\infty \dfrac{x\,dx}{(x^2+a^2)\,(x^2+b^2)} = \dfrac{1}{a^2-b^2}\log\dfrac{a}{b}.$

12. $\int_1^\infty \dfrac{dx}{x^2\,(1+x)} = 1 - \log 2,$ $\int_1^\infty \dfrac{dx}{x\,(1+x)^2} = \log 2 - \tfrac{1}{2}.$

13. $\int_0^\infty \dfrac{dx}{(1+x^2)^2} = \tfrac{1}{4}\pi.$

14. $\int_1^\infty \dfrac{dx}{x\sqrt{(x^2-1)}} = \tfrac{1}{2}\pi,$ $\int_1^\infty \dfrac{dx}{x\sqrt{(x^2+1)}} = \log(1+\sqrt{2}).$

15. $\int_0^1 \dfrac{dx}{\sqrt{\{x\,(1-x)\}}} = \pi,$ $\int_0^1 \sqrt{\left(\dfrac{x}{1-x}\right)}\,dx = \tfrac{1}{2}\pi.$

16. $\int_{-1}^1 \sqrt{\left(\dfrac{1-x}{1+x}\right)}\,dx = \pi.$

17. $\int_a^\beta \dfrac{dx}{\sqrt{\{(x-a)\,(\beta-x)\}}} = \pi.$

18. $\int_a^b \sqrt{\left(\dfrac{b-x}{x-a}\right)}\,dx = \int_a^b \sqrt{\left(\dfrac{x-a}{b-x}\right)}\,dx = \tfrac{1}{2}\pi\,(b-a).$

19. $\int_0^{\frac{1}{4}\pi} \sec^2\theta\,d\theta = 1,$ $\int_0^{\frac{1}{4}\pi} \tan^2\theta\,d\theta = 1 - \tfrac{1}{4}\pi.$

20. $\int_0^{\frac{1}{4}\pi} \sin 2\theta\,d\theta = 1,$ $\int_0^{\frac{1}{2}\pi} \cos 2\theta\,d\theta = 0.$

21. $\int_0^{\frac{1}{4}\pi} \dfrac{\cos\theta}{1+\sin^2\theta}\,d\theta = \int_0^{\frac{1}{4}\pi} \dfrac{\sin\theta}{1+\cos^2\theta}\,d\theta = \tfrac{1}{4}\pi.$

22. $\int_0^{\frac{1}{4}\pi} \sec^4\theta\,d\theta = \tfrac{3}{4},$ $\int_0^{\frac{1}{4}\pi} \tan^4\theta\,d\theta = \tfrac{1}{4}\pi - \tfrac{2}{3}.$

23. $\int_0^\pi \dfrac{d\theta}{1 + e\cos\theta} = \dfrac{\pi}{\sqrt{(1 - e^2)}}, \quad [e < 1].$

24. $\int_0^{\frac{1}{2}\pi} \dfrac{d\theta}{1 + \cos\theta} = \int_0^{\frac{1}{2}\pi} \dfrac{d\theta}{1 + \sin\theta} = 1.$

25. $\int_{-\frac{1}{4}\pi}^{\frac{1}{4}\pi} \tan\theta\, d\theta = 0, \qquad \int_{-\frac{1}{4}\pi}^{\frac{1}{4}\pi} \sec\theta\, d\theta = \log(3 + 2\sqrt{2}).$

26. $\int_0^{\frac{1}{2}\pi} (\sec\theta - \tan\theta)\, d\theta = \log 2.$

27. $\int_0^\pi \dfrac{d\theta}{a^2 - 2ab\cos\theta + b^2} = \dfrac{\pi}{|a^2 - b^2|}.$

28. $\int_0^1 x^n \log x\, dx = -\dfrac{1}{(n+1)^2}.$

29. $\int_0^1 \sin^{-1}x\, dx = \frac{1}{2}\pi - 1, \qquad \int_0^1 \tan^{-1}x\, dx = \frac{1}{4}\pi - \frac{1}{2}\log 2.$

30. $\int_0^{\frac{1}{2}\pi} \theta\sin\theta\, d\theta = 1, \qquad \int_0^{\frac{1}{2}\pi} \theta\cos\theta\, d\theta = \frac{1}{2}\pi - 1.$

31. $\int_0^{\frac{1}{2}\pi} \theta^2\sin\theta\, d\theta = \pi - 2, \qquad \int_0^{\frac{1}{2}\pi} \theta^2\cos\theta\, d\theta = \frac{1}{4}\pi^2 - 2.$

32. $\int_0^\pi \theta(\pi - \theta)\sin\theta\, d\theta = 4, \qquad \int_0^\pi \theta(\pi - \theta)\cos\theta\, d\theta = 0.$

33. $\int_{-1}^1 (1 - x^2)\cos\beta x\, dx = \dfrac{4}{\beta^3}(\sin\beta - \beta\cos\beta).$

34. $\int_0^\infty \dfrac{x\tan^{-1}x}{(1 + x^2)^2}\, dx = \frac{1}{8}\pi.$

35. $\int_{-1}^1 \dfrac{\sqrt{(1 - x^2)}}{a - x}\, dx = \pi\{a - \sqrt{(a^2 - 1)}\}, \text{ provided } a > 1.$

36. $\int_0^{\frac{1}{4}\pi} \theta\sec^2\theta\, d\theta = \frac{1}{4}\pi - \frac{1}{2}\log 2.$

37. $\int_0^\infty e^{-x}\cos(x + \frac{1}{4}\pi)\, dx = 0.$

38. $\int_{-\infty}^\infty \dfrac{du}{\cosh u} = \pi.$

39. $\int_0^\infty \dfrac{dx}{a^2 e^x + b^2 e^{-x}} = \dfrac{1}{ab}\tan^{-1}\dfrac{b}{a}.$

40. $\int_0^\infty \dfrac{dx}{a^2\cosh^2 x + b^2\sinh^2 x} = \dfrac{1}{ab}\tan^{-1}\dfrac{b}{a}.$

41. $\displaystyle\int_0^\infty \frac{a\,dx}{a^2 + \sinh^2 x} = \frac{\cos^{-1} a}{\sqrt{(1 - a^2)}}$, or $\displaystyle\frac{\cosh^{-1} a}{\sqrt{(a^2 - 1)}}$, according as $a^2 \lessgtr 1$.

42. $\displaystyle\int_0^{\frac12\pi} \frac{d\theta}{\cosh^2 a - \cos^2 \theta} = \frac{\pi}{\sinh 2a}$, $\displaystyle\int_0^{\frac12\pi} \frac{\cos^2 \theta\,d\theta}{\cosh^2 a - \cos^2 \theta} = \frac{\pi e^{-a}}{2\sinh a}$.

EXAMPLES. XXXIII.

(Formulæ of Reduction, &c.)

1.　Write down the values of the following integrals:

(1)　$\displaystyle\int_0^{\frac12\pi} \sin^3 \theta\,d\theta$,　　$\displaystyle\int_0^{\frac12\pi} \cos^4 \theta\,d\theta$,　　$\displaystyle\int_0^{\frac12\pi} \sin^7 \theta\,d\theta$.

(2)　$\displaystyle\int_0^{\frac12\pi} \sin^2 \theta \cos \theta\,d\theta$,　　$\displaystyle\int_0^{\frac12\pi} \sin^3 \theta \cos^3 \theta\,d\theta$,　　$\displaystyle\int_0^{\frac12\pi} \sin^5 \theta \cos^6 \theta\,d\theta$.

(3)　$\displaystyle\int_0^\pi \sin^5 \theta\,d\theta$,　　$\displaystyle\int_0^\pi \cos^5 \theta\,d\theta$,　　$\displaystyle\int_0^\pi \sin^3 \theta \cos^4 \theta\,d\theta$.

(4)　$\displaystyle\int_{-\frac12\pi}^{\frac12\pi} \sin^4 \theta\,d\theta$,　　$\displaystyle\int_{-\frac12\pi}^{\frac12\pi} \sin^5 \theta\,d\theta$,　　$\displaystyle\int_{-\frac12\pi}^{\frac12\pi} \cos^5 \theta\,d\theta$,

$$\int_{-\frac12\pi}^{\frac12\pi} \sin^3 \theta \cos^3 \theta\,d\theta.$$

2.　Prove from first principles that

$$\int_0^1 x^m (1 - x)^n \,dx = \int_0^1 x^n (1 - x)^m \,dx.$$

Prove that the common value is

$$\frac{m!\,n!}{(m + n + 1)!}.$$

3.　Prove that

$$\int_{-a}^a \phi(x^2)\,dx = 2\int_0^a \phi(x^2)\,dx, \qquad \int_{-a}^a \phi(x^2)\,x\,dx = 0.$$

4.　Prove that

$$\int_{-a}^a \phi(x)\,dx = \int_0^a \{\phi(x) + \phi(-x)\}\,dx,$$

and 　　　　　　$$\int_{-a}^a \{\phi(x) - \phi(-x)\}\,dx = 0.$$

5. If
$$u_n = \int_0^{\frac{1}{4}\pi} \tan^n \theta \, d\theta,$$

prove that
$$u_n = \frac{1}{n-1} - u_{n-2}.$$

6. $\int_0^1 x^3 (1-x^2)^{\frac{5}{2}} \, dx = \frac{2}{63}.$

7. $\int_0^1 x^3 (1-x)^3 \, dx = \frac{1}{140}.$

8. $\int_0^1 x^{\frac{3}{2}} (1-x)^{\frac{3}{2}} \, dx = \frac{3}{128} \pi.$

9. $\int_0^1 \frac{x^5 \, dx}{\sqrt{(1-x^2)}} = \frac{8}{15}, \qquad \int_0^1 \frac{x^6 \, dx}{\sqrt{(1-x^2)}} = \frac{5}{32} \pi.$

10. $\int_{-\frac{1}{2}\pi}^{\frac{1}{2}\pi} \theta \sin \theta \cos \theta \, d\theta = \frac{1}{4}\pi, \qquad \int_{-\frac{1}{2}\pi}^{\frac{1}{2}\pi} \theta \sin^2 \theta \, d\theta = 0.$

11. $\int_0^\pi \theta \sin \theta \cos^2 \theta \, d\theta = \frac{1}{3}\pi, \qquad \int_0^\pi \theta \sin^2 \theta \cos \theta \, d\theta = -\frac{4}{9}.$

12. $\int_0^{\frac{1}{4}\pi} \frac{\sin 3\theta}{\sin \theta} \, d\theta = \frac{1}{2}\pi, \qquad \int_0^{\frac{1}{4}\pi} \frac{\sin 4\theta}{\sin \theta} \, d\theta = \frac{3}{4}, \qquad \int_0^{\frac{1}{4}\pi} \frac{\sin 5\theta}{\sin \theta} \, d\theta = \frac{1}{2}\pi.$

13. $\int_0^\pi (a \sin \theta + b \cos \theta)^5 \, d\theta = \frac{16}{15} a^5 + \frac{8}{3} a^3 b^2 + 2ab^4.$

14. $\int_0^{\frac{1}{4}\pi} \frac{\cos x - \sin x}{1 + \sin x \cos x} \, dx = 0.$

15. Prove that
$$\int_0^{\frac{1}{2}\pi} \frac{\cos^2 \theta \, d\theta}{a^2 \cos^2 \theta + b^2 \sin^2 \theta} = \frac{\pi}{2a \, (a+b)},$$
$$\int_0^{\frac{1}{2}\pi} \frac{\sin^2 \theta \, d\theta}{a^2 \cos^2 \theta + b^2 \sin^2 \theta} = \frac{\pi}{2b \, (a+b)}.$$

16. Prove that
$$\int_{-1}^1 (1+x)^m (1-x)^n \, dx = 2^{m+n+2} \int_0^{\frac{1}{2}\pi} \sin^{2n+1} \theta \cos^{2m+1} \theta \, d\theta.$$

17. Prove that
$$\int_0^\infty \frac{dx}{(1+x^2)^n} = \frac{2n-3}{2n-2} \int_0^\infty \frac{dx}{(1+x^2)^{n-1}}.$$

[Put $x = \tan \theta$.]

18. Prove that if n be a positive integer
$$\int_0^\infty \frac{dx}{(1+x^2)^{n+\frac{1}{2}}} = \frac{2 . 4 . 6 \dots (2n-2)}{1 . 3 . 5 \dots (2n-1)}.$$

19.   Prove that, if $n > 1$,

$$\int_0^\infty \frac{dx}{\{x + \sqrt{(x^2 + 1)}\}^n} = \frac{n}{n^2 - 1}.$$

[Put $x = \sinh u$.]

20.   Prove that

$$\int_0^\infty \frac{du}{\cosh^n u} = \frac{n - 2}{n - 1} \int_0^\infty \frac{du}{\cosh^{n-2} u}.$$

[Put $\cosh u = \sec \theta$]

21.   If

$$u_n = \int_0^{\frac{1}{2}\pi} \theta \cos^n \theta \, d\theta,$$

prove that

$$u_n = -\frac{1}{n^2} + \frac{n - 1}{n} u_{n-2}.$$

Prove that

$$u_3 = \cdot 2694 \ldots .$$

22.   If

$$u_n = \int_0^{2a} x^n \sqrt{(2ax - x^2)} \, dx,$$

prove that

$$u_n = \frac{2n + 1}{n + 2} a u_{n-1}.$$

Find, geometrically or otherwise, the value of u_0, and deduce the values of u_1, u_2.

EXAMPLES.   XXXIV.

1.   By considering the value of

$$\int_0^1 (1 - x^2)^n \, dx,$$

prove that if n be a positive integer

$$1 - \frac{n}{1 \cdot 3} + \frac{n(n-1)}{1 \cdot 2 \cdot 5} - \frac{n(n-1)(n-2)}{1 \cdot 2 \cdot 3 \cdot 7} + \ldots = \frac{2 \cdot 4 \cdot 6 \ldots 2n}{3 \cdot 5 \cdot 7 \ldots (2n+1)}.$$

2.   If

$$(1 + x)^n = 1 + p_1 x + p_2 x^2 + \ldots + p_n x^n,$$

prove that

$$p_1 - \tfrac{1}{2} p_2 + \tfrac{1}{3} p_3 - \ldots + (-)^{n-1} \frac{p_n}{n} = 1 + \tfrac{1}{2} + \tfrac{1}{3} + \ldots + \frac{1}{n}.$$

3.   Prove that when a is large the sum to infinity of the series

$$\frac{1}{a^2} + \frac{1}{a^2 + 1^2} + \frac{1}{a^2 + 2^2} + \ldots$$

is $\tfrac{1}{2}\pi/a$, approximately.

4.   Prove that

$$\int_0^\pi \frac{x \sin x}{1 + \cos^2 x} \, dx = \tfrac{1}{4}\pi^2.$$

5. Prove from first principles that

$$\int_0^{\frac{1}{2}\pi} \sin^{n+1} \theta \, d\theta < \int_0^{\frac{1}{2}\pi} \sin^n \theta \, d\theta.$$

Hence shew that $\frac{1}{2}\pi$ lies between

$$\frac{2 \cdot 2 \cdot 4 \cdot 4 \cdot 6 \cdot 6 \ldots 2n \cdot 2n}{1 \cdot 3 \cdot 3 \cdot 5 \cdot 5 \cdot 7 \ldots (2n-1)(2n+1)}$$

and the fraction obtained by omitting the last factors in the numerator and denominator. (Wallis.)

6. Prove from first principles that

$$\int_0^{\frac{1}{2}\pi} \tan^{n+1} \theta \, d\theta < \int_0^{\frac{1}{2}\pi} \tan^n \theta \, d\theta.$$

Hence, using the result of Ex. XXXIII, 5, shew that

$$\int_0^{\frac{1}{2}\pi} \tan^n \theta \, d\theta$$

lies between $\dfrac{1}{2(n-1)}$ and $\dfrac{1}{2(n+1)}$.

7. Shew that

$$\int_0^\infty \sin \theta \, d\theta \quad \text{and} \quad \int_0^\infty \cos \theta \, d\theta$$

are indeterminate.

8. Shew from graphical considerations that

$$\int_0^\infty \frac{\sin \theta}{\theta} \, d\theta$$

is finite and determinate.

9. Prove that if $\phi(x)$ be finite and continuous for values of x ranging from 0 to a, except for $x = 0$, when it becomes infinite, the integral

$$\int_0^a \phi(x) \, dx$$

will be finite, provided a positive quantity m can be found, less than unity, and such that

$$\lim_{x \to 0} x^m \phi(x)$$

is finite. [Put $x = t^n$.]

10. If $\phi(x)$ be finite and continuous for all values of x, the integral

$$\int_0^\infty \phi(x) \, dx$$

will be finite, provided a quantity m can be found, greater than unity, and such that

$$\lim_{x \to \infty} x^m \phi(x)$$

is finite. [Put $x = t^{-n}$.]

11. Prove that

$$\int_0^\infty \cos x^2\, dx \ \text{ and } \ \int_0^\infty \sin x^2\, dx$$

are finite and determinate.

12. Prove that

$$\int_0^\infty x e^{-x^2}\, dx = \tfrac{1}{2}, \qquad \int_{-\infty}^\infty x e^{-x^2}\, dx = 0.$$

13. Prove that the integral

$$\int_0^\infty x^n e^{-x^2}\, dx$$

(where $n > -1$) is finite and determinate.

14. Prove that, if $n > 1$,

$$\int_0^\infty x^n e^{-x^2}\, dx = \tfrac{1}{2}\,(n-1) \int_0^\infty x^{n-2} e^{-x^2}\, dx.$$

Hence shew that, if n be a positive integer,

$$\int_0^\infty x^{2n+1} e^{-x^2}\, dx = \tfrac{1}{2} \cdot n\,!.$$

15. If
$$u_n = \int_0^\infty x^n e^{-ax}\, dx,$$

prove that
$$u_n = \frac{n}{a}\, u_{n-1},$$

n being positive.

Hence shew that, if n be integral,

$$\int_0^\infty x^n e^{-ax}\, dx = \frac{n\,!}{a^{n+1}}.$$

16. Tables of the elliptic integral

$$\int_0^\phi \frac{d\theta}{\sqrt{(1 - k^2 \sin^2 \theta)}}$$

have been constructed for values of ϕ at intervals of one minute of angle. Find a formula for the difference of successive entries in a given part of the table.

For example, if $k = \tfrac{1}{2}$, $\phi = 60°$, prove that the difference will be ·000323, approximately.

17. If $f(x)$ and $\phi(x)$ be finite and continuous, and if $\phi(x)$ retain the same sign, throughout the interval from $x = a$ to $x = b$, then

$$\int_a^b f(x)\, \phi(x)\, dx = f[a + \theta\,(b - a)] \int_a^b \phi(x)\, dx,$$

where $1 > \theta > 0$.

18. Shew how it follows from the equality

$$\int_1^x \frac{dx}{x} = \log x$$

that the sum of n terms of the harmonic series

$$1 + \tfrac{1}{2} + \tfrac{1}{3} + \dots$$

lies between $\log(n+1)$ and $1 + \log n$.

Shew that the sum of a million terms of this series lies between 13·8 and 14·8.

19. Shew from graphical considerations that if $f(x)$ steadily diminishes, as x increases from 0 to ∞, the series

$$f(1) + f(2) + f(3) + \dots$$

is convergent, and that its sum lies between I and $I + f(1)$, provided the integral

$$I, = \int_1^\infty f(x)\, dx,$$

be finite.

Apply this to the series

$$\frac{1}{(n+1)^2} + \frac{1}{(n+2)^2} + \frac{1}{(n+3)^2} + \dots.$$

EXAMPLES. XXXV.

1. Prove that if the pressure (p) and volume (v) of a gas be connected by the relation

$$pv = \text{const.},$$

the work done in expanding from volume v_0 to volume v_1 is

$$p_0 v_0 \log \frac{v_1}{v_0}.$$

2. Prove also that if the relation be

$$pv^\gamma = \text{const.},$$

the work done is

$$\frac{1}{\gamma - 1}(p_0 v_0 - p_1 v_1).$$

3. If the tension of an elastic string vary as the increase over the natural length, prove that the work done in stretching the string from one length to another is the same as if the tension had been constant and equal to half the sum of the initial and final tensions.

4. Prove that the work done by gravity on a pound of matter, as it is brought from an infinite distance to the surface of the Earth, is n foot-lbs., where n is the number of feet in the Earth's radius. [Assume that the force varies inversely as the square of the distance from the Earth's centre.]

CHAPTER VIII

GEOMETRICAL APPLICATIONS

99. Definition of an Area.

In Euclid's Elements a system of propositions is developed by means of which we are able to give a precise meaning to the term 'area,' as applied to any figure bounded wholly by straight lines. In particular it is shewn that a rectangle can be constructed equal to the given figure, and having any given base, say the (arbitrarily chosen) unit of length. The 'area' of the figure in question is then measured by the ratio of this rectangle to the square on the unit length.

This process obviously does not apply to a figure bounded, in whole or in part, by *curved* lines, and we require therefore a definition of what is to be understood by the 'area' in such a case. To supply this, we imagine two rectilinear figures to be constructed, one including, and the other included by, the given curved figure. There is an upper limit to the area of the inscribed figure, and a lower limit to that of the circumscribed figure, and these limits can be proved to be identical. The common limiting value is adopted, *by definition*, as the measure of the 'area' of the given curvilinear figure.

Thus, in the case of a circle, if, in Fig. 17, p. 31, PQ be the side of an inscribed polygon, the area of the polygon will be $\frac{1}{2}\Sigma(ON.PQ)$. Now ON is less than the radius, and $\Sigma(PQ)$ is less than the perimeter, of the circle. Hence the upper limit to the area of an inscribed polygon cannot exceed $\frac{1}{2}a \times 2\pi a$, or πa^2, where a is the radius. Similarly we may shew that the lower limit to the area of a circumscribed polygon cannot be less than πa^2. Moreover, the difference between the area of an inscribed polygon, and that of the corresponding circumscribed polygon, is represented by $\Sigma(PN.NT)$, and is therefore less than $\Sigma(PN).\epsilon$, where ϵ is the greatest value of NT. Since this can be made as small as we please, the upper and lower limits aforesaid must be equal, and each is therefore equal to πa^2.

In the same way we may prove that the area of any sector of a circle of radius a is $\frac{1}{2}a^2\theta$, where θ is the angle of the sector.

100. Formula for an Area, in Cartesian Coordinates.

If the equation of a curve in rectangular coordinates be

$$y = \phi(x), \quad \dots\dots\dots\dots\dots\dots\dots(1)$$

the area included between the curve, the axis of x, and the ordinates $x = a$, $x = b$, is

$$\int_a^b \phi(x)\, dx \text{ or } \int_a^b y\, dx, \quad \dots\dots\dots\dots\dots(2)$$

it being assumed that $\phi(x)$ is a function of the type contemplated in Art. 90.

This follows at once from the definition of the preceding Art. and the investigation of Art. 90.

If the axes of coordinates be oblique, making (say) an angle ω with one another, the elementary rectangles $y\,\delta x$ which occur in the sum, of which the area is the limit, are replaced by elementary parallelograms $y\,\delta x \sin\omega$; the area included between the curve, the axis of x, and two bounding ordinates is therefore given by

$$\sin\omega \int y\, dx \quad \dots\dots\dots\dots\dots\dots\dots(3)$$

taken between the proper limits.

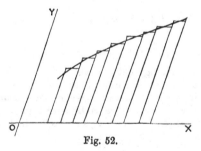

Fig. 52.

Ex. 1. The area of a quadrant of the ellipse

$$\frac{x^2}{a^2} + \frac{y^2}{b^2} = 1 \quad \dots\dots\dots\dots\dots\dots\dots(4)$$

is given by

$$\int_0^a y\, dx = \frac{b}{a} \int_0^a \sqrt{(a^2 - x^2)}\, dx.$$

The value of the definite integral was found in Art. 96 to be $\frac{1}{4}\pi a^2$. Hence the whole area of the ellipse is πab.

Ex. 2. In the rectangular hyperbola

$$x^2 - y^2 = 1, \quad \dots\dots\dots\dots\dots\dots\dots(5)$$

we may put, for the positive branch,

$$x = \cosh u, \quad y = \sinh u, \quad \dots\dots\dots\dots(6)$$

since these satisfy (5), and give the required range of values of x and y.

The area included between the curve, the axis of x, and the ordinate defined by the variable u, is therefore

$$\int_1^x y\,dx = \int_0^u \sinh^2 u\,du = \tfrac{1}{2}\int_0^u (\cosh 2u - 1)\,du$$

$$= \tfrac{1}{4}\sinh 2u - \tfrac{1}{2}u. \quad \dots\dots\dots\dots(7)$$

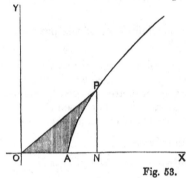
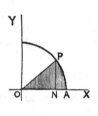

Fig. 53.

This gives the area PAN in the left-hand figure.   Hence the area of the hyperbolic sector AOP is

$$\tfrac{1}{2}PN \cdot ON - \text{area } PAN = \tfrac{1}{2}u. \quad \dots\dots\dots\dots(8)$$

We have here an analogy between the 'amplitude' (u) of the hyperbolic functions $\cosh u$, $\sinh u$, &c., and the amplitude (θ) of the circular functions $\cos \theta$, $\sin \theta$, &c.; viz. the independent variable in each case represents twice the sectorial area AOP corresponding to the point P whose coordinates are $(\cosh u, \sinh u)$, or $(\cos \theta, \sin \theta)$, respectively.

In the case of the general hyperbola

$$\frac{x^2}{a^2} - \frac{y^2}{b^2} = 1, \quad \dots\dots\dots\dots(9)$$

the coordinates of any point on the positive branch may be represented by

$$x = a\cosh u, \quad y = b\sinh u, \quad \dots\dots\dots\dots(10)$$

and the sectorial area is $\tfrac{1}{2}ab\,u$.

Ex. 3.   The equation of a parabola, referred to any diameter and the tangent at its extremity, is

$$y^2 = 4a'x. \quad \dots\dots\dots\dots(11)$$

The area of the segment cut off by the chord $x = a$ is therefore

$$2 \sin \omega \int_0^a y \, dx = 4a'^{\frac{1}{2}} \sin \omega \int_0^a x^{\frac{1}{2}} \, dx = \tfrac{8}{3} a'^{\frac{1}{2}} a^{\frac{3}{2}} \sin \omega$$

$$= \tfrac{4}{3} a \beta \sin \omega, \quad \dots\dots\dots\dots\dots\dots\dots\dots\dots(12)$$

if 2β be the length of the chord.

Hence the area of any segment of a parabola is two-thirds the rectangle contained by the intercept (a) of the chord on its diameter and the projection ($2\beta \sin \omega$) of the chord on the directrix.

101.   On the Sign to be attributed to an Area.

It was tacitly implied in Art. 100 that $b > a$, and that the ordinate $\phi(x)$ is positive throughout the range of integration.   If we drop these restrictions, it is easily seen that the integral

$$\int_a^b \phi(x) \, dx \quad \dots\dots\dots\dots\dots\dots\dots\dots(1)$$

is equal to $\pm S$, where S is the area included between the curve, the axis of x, and the extreme ordinates; the sign being $+$ or $-$ according as the area in question lies to the right or left of the curve, supposed described in the direction from P to Q, where

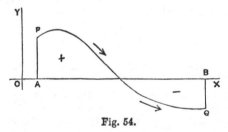

Fig. 54.

PA, QB are the ordinates corresponding to $x=a$, $x=b$, respectively*. If the curve cuts the axis of x between A and B, the integral gives the excess (positive or negative) of the area which lies to the right over that which lies to the left.

Even with these generalizations, the formula

$$\int_a^b \phi(x) \, dx = \pm S \quad \dots\dots\dots\dots\dots\dots(2)$$

still applies in strictness only when there is a unique value of y, or $\phi(x)$, for each value of x within the range $b - a$.   If however

---

* It is assumed here that the axes of x and y have the relative directions shewn in the figures.   In the opposite case, the words 'right' and 'left' must be interchanged.

we replace x, as independent variable, by a quantity t such that, as t increases, the corresponding point P moves in a continuous manner along the curve*, the formula

$$\int_{t_0}^{t_1} y \frac{dx}{dt}\, dt \quad \dots\dots\dots\dots\dots\dots(3)$$

will give in a generalized sense the area included between the

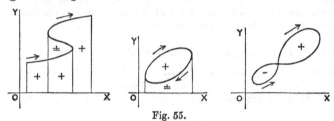

Fig. 55.

curve, the axis of x, and the ordinates of the points P_0, P_1 for which $t = t_0$, $t = t_1$, respectively, viz. it will give the excess of those portions of the area swept over by the ordinate y as it moves to the right over those swept over as it moves to the left, or *vice versâ*, according as y is positive or negative.

If, for a certain value of t, P return to its former position, having described a closed curve, the integral

$$\int y \frac{dx}{dt}\, dt, \quad \dots\dots\dots\dots\dots\dots(4)$$

taken between proper limits of t, will give the area included by the curve, with the sign + or −, according as the area lies to the right or left of P, when this point describes the curve in accordance with the variation of t†. If the curve cut itself, the formula (4) gives the excess of those portions of the included area which lie to the right over those which lie to the left.   (See Fig. 55.)

It is sometimes convenient, in finding the area of a curve, to use y as independent variable, instead of x.   The area included between the curve, the axis of y, and the lines $y = h$, $y = k$, is evidently given, with the same kind of qualification as before, by

$$\int_{h}^{k} x\, dy. \quad \dots\dots\dots\dots\dots\dots(5)$$

---

* For instance, we may take as the new variable the arc s of the curve, measured from some fixed point on it.

† Thus, in the indicator-diagrams referred to on p. 207, the area enclosed by the curve gives the excess of the work done by the steam on the piston during the forward stroke over the work done by the piston in expelling the steam during the back stroke, and so represents the net energy communicated to the piston in a complete stroke.

The more general formula, analogous to (3), is

$$\int_{t_0}^{t_1} x \frac{dy}{dt} dt, \qquad \dotfill (6)$$

but it will be found on examination that the rule of signs must be reversed.

Allowing for this, we have the expression

$$\tfrac{1}{2} \int \left(x \frac{dy}{dt} - y \frac{dx}{dt} \right) dt \dotfill (7)$$

for the area of a closed curve, the limits of t being such that the point (x, y) returns to its initial position. The rule of signs is now that the expression (7) is positive when the area lies to the left of a point describing the curve in the direction in which t increases.

102. Areas referred to Polar Coordinates.

If the equation of a curve in polar coordinates be

$$r = \phi(\theta), \qquad \dotfill (1)$$

the area included between the curve and any two radii vectores $\theta = \alpha, \theta = \beta$ is given by the formula

$$\tfrac{1}{2} \int_{\alpha}^{\beta} r^2 d\theta \text{ or } \tfrac{1}{2} \int_{\alpha}^{\beta} \{\phi(\theta)\}^2 d\theta \dots (2)$$

For we can construct, in the manner indicated by the figure, an including area S, and an included area S', each built up of sectors of circles. The area of any one of these sectors is equal to $\tfrac{1}{2} r^2 \delta\theta$, where r is its radius, and $\delta\theta$ its angle, and the sum of either series of sectors is therefore given by a series of the type

$$\tfrac{1}{2} \Sigma_{\alpha}^{\beta} r^2 \delta\theta \dotfill (3)$$

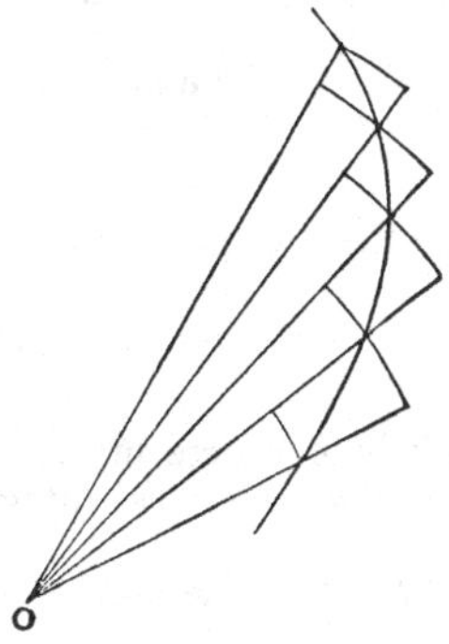

Fig. 56.

Hence either series has the unique limit denoted by (2).

It is here assumed that $\beta > \alpha$ and that each radius vector through the origin intersects the arc considered in one point only. If however we introduce a new independent variable t, such that, as t increases, the corresponding point P moves in a continuous manner along the curve, the expression

$$\tfrac{1}{2} \int_{t_0}^{t_1} r^2 \frac{d\theta}{dt} dt \qquad \dotfill (4)$$

will give the net area swept over by the radius vector as t varies from t_0 to t_1, i.e. the (positive or negative) excess of those parts which are swept over in the direction of θ increasing over those swept over in the contrary direction. Moreover if, as t increases, P at length returns to its original position, having described a closed curve, the expression

$$\tfrac{1}{2}\int r^2 \frac{d\theta}{dt}\,dt, \qquad \dots\dots\dots\dots\dots(5)$$

taken between suitable limits of t, gives in a generalized sense the area enclosed by the curve; viz., it represents the excess of that part of the area which lies to the left of P (as it describes the curve in accordance with the variation of t) over that part which lies to the right. Cf. Art. 101.

It may be noted that the formula (5) is equivalent to (7) of Art. 101. If the coordinates of two adjacent points P, Q be (x, y) and $(x+\delta x, y+\delta y)$, respectively, the area of the elementary triangle OPQ is, subject to a convention as to sign,

$$\tfrac{1}{2}(x\,\delta y - y\,\delta x)$$

by a formula of Analytical Geometry. The same thing is denoted in our present notation by $\tfrac{1}{2}r^2\delta\theta$.

Ex. 1. The area of the circle

$$r = 2a\sin\theta \qquad \dots\dots\dots\dots\dots\dots(6)$$

(see Fig. 38, p. 124) is given by

$$\tfrac{1}{2}\int_0^\pi r^2 d\theta = 2a^2\int_0^\pi \sin^2\theta\,d\theta = \pi a^2. \qquad \dots\dots\dots\dots(7)$$

This is of course merely a verification, or rather a new evaluation of the trigonometrical integral.

Ex. 2. The area of a sector of the parabola

$$r = \frac{2a}{1+\cos\theta} \qquad \dots\dots\dots\dots\dots\dots(8)$$

included between two focal radii is

$$\begin{aligned}
\tfrac{1}{2}\int r^2 d\theta &= \tfrac{1}{2}a^2\int\sec^4\tfrac{1}{2}\theta\,d\theta \\
&= \tfrac{1}{2}a^2\int(1+\tan^2\tfrac{1}{2}\theta)\sec^2\tfrac{1}{2}\theta\,d\theta \\
&= \tfrac{1}{2}a^2[\tan\tfrac{1}{2}\theta + \tfrac{1}{3}\tan^3\tfrac{1}{2}\theta], \qquad \dots\dots\dots\dots(9)
\end{aligned}$$

taken between proper limits. If the limits be $-\tfrac{1}{2}\pi$ and $\tfrac{1}{2}\pi$, we get the area of the segment cut off by the latus rectum, viz. $\tfrac{4}{3}a^2$.

Ex. 3.   The equation of an ellipse in polar coordinates, the centre being pole, is

$$\frac{1}{r^2} = \frac{\cos^2 \theta}{a^2} + \frac{\sin^2 \theta}{b^2}. \quad\quad\quad\quad\quad\quad(10)$$

Hence the area is

$$\frac{1}{2}\int_0^{2\pi} \frac{a^2 b^2}{a^2 \sin^2 \theta + b^2 \cos^2 \theta}\, d\theta = 2a^2 b^2 \int_0^{\frac{1}{2}\pi} \frac{d\theta}{a^2 \sin^2 \theta + b^2 \cos^2 \theta}. \quad (11)$$

The value of the latter integral has been found in Art. 96 to be $\frac{1}{2}\pi/(ab)$. Hence the required area is πab.

103.   Area swept over by a Moving Line.

The area swept over by a moving line, of constant or variable length, may be calculated as follows.

Let PQ, $P'Q'$ be two consecutive positions of the line, and let their directions meet in C.   Let R, R' be the middle points of PQ, $P'Q'$, and let RS be an arc of a circle with centre C.   Then if the angle PCP' be denoted by $\delta\theta$, we have, ultimately,

$$\begin{aligned}
\text{area } PQQ'P' &= \triangle QCQ' - \triangle PCP'\\
&= \tfrac{1}{2}CQ^2 . \delta\theta - \tfrac{1}{2}CP^2 . \delta\theta\\
&= PQ . \tfrac{1}{2}(CP + CQ)\,\delta\theta\\
&= PQ . CR . \delta\theta = PQ . RS.
\end{aligned}$$

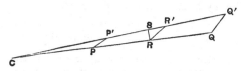

Fig. 57.

Hence, if we denote the length PQ by u, and the elementary displacement of R, estimated in the direction perpendicular to the moving line, by $\delta\sigma$, the area swept over may be represented by

$$\int u\,d\sigma. \quad\quad\quad\quad\quad\quad(1)$$

It will be noticed that the formulæ of Arts. 100, 102 are particular cases of this result.   Thus in the case of Art. 100 (3) we have $u = y$, $\delta\sigma = \delta x . \sin \omega$.

It is tacitly assumed, in the foregoing proof, that the areas are swept over always in the same direction.   It is easy to see, however, that the formula (1) will apply without any such restriction, provided areas be reckoned positive or negative according as they are swept over towards the side of the line PQ on which $\delta\sigma$ is reckoned positive, or the reverse.   For example, the area swept over by a straight line whose middle point is fixed is on this reckoning zero.

We will suppose, for definiteness, that $\delta\sigma$ is positive when the motion of R is to the *left* of PQ as regards a spectator looking along the straight line in the direction from P to Q. If PQ return finally to its original position, its extremities P, Q having described closed curves, the integral (1) will, on the above convention, represent the excess of the area enclosed by the path of Q over that enclosed by the path of P, provided the signs attributed to these areas be in accordance with the rule of Art. 102.

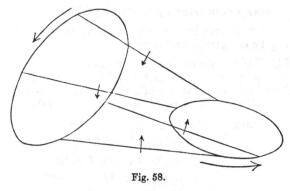

Fig. 58.

104. Theory of Amsler's Planimeter.

A 'planimeter' is an instrument by which the area of any figure drawn on paper is measured mechanically.

Many such instruments have been devised*, but the simplest and most popular is the one invented by Amsler, of Schaffhausen, in 1854. This consists of two bars OP, PQ, freely jointed at P, the former of which can rotate about a fixed point at O. If a tracing point attached

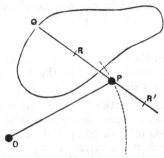

Fig. 59.

* See Henrici, 'Report on Planimeters,' *Brit. Ass. Rep.*, 1894, p. 496.

to the bar PQ at Q be carried round any closed curve, P will oscillate to and fro along an arc of a circle, describing (as it were) a contour of zero area. Hence by the theorem stated at the end of Art. 103, the area of the curve described by Q will be equal to

$$\int l\,d\sigma, \quad\dots\dots\dots\dots\dots\dots\dots\dots\dots\dots(1)$$

where l is the length PQ, and $\int d\sigma$ represents the integral motion of R, the middle point of PQ, estimated always in the direction perpendicular to PQ*.

Now if, as is generally the case in the actual use of the instrument, PQ return to its original position without making a complete revolution, the *integral* motion of R at right angles to PQ is the same as that of any other point R' in the line PQ. For if $\delta\sigma$, $\delta\sigma'$ be corresponding elements of the paths of R, R', estimated as aforesaid, we have

$$\delta\sigma - \delta\sigma' = R'R\,.\,\delta\theta,$$

where $\delta\theta$ is the angle between the consecutive positions of $R'R$. Hence

$$\int d\sigma = \int d\sigma' + R'R \int d\theta$$
$$= \int d\sigma', \quad\dots\dots\dots\dots\dots\dots\dots\dots\dots\dots(2)$$

since, under the circumstances supposed, we have $\int d\theta = 0$.

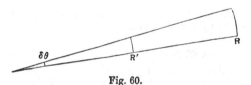

Fig. 60.

In the instrument, as actually constructed, the integral motion normal to the bar of a point R' in QP produced backwards, is recorded by means of a small wheel having its axis in the direction PQ. As Q describes any curve, the wheel partly rolls and partly slides over the plane of the paper on which the curve is drawn, and the rotation of the wheel is in exact proportion to the displacement of R' perpendicular to its axis. The wheel is graduated, and has a fixed index for the record of partial revolutions; the whole revolutions are recorded by a dial and counter.

There is also an arrangement for varying the length PQ; this merely alters the *scale* of the record.

A more compact proof can be given analytically. Taking rectangular axes through O, let θ and ϕ be the angles which OP and PQ make with the positive direction of the axis of x. Putting

$$OP = a, \quad PQ = l,$$

---

* This is of course *not* in general the same thing as the length of the path of R.

the coordinates of Q are

$$x = a \cos \theta + l \cos \phi, \quad y = a \sin \theta + l \sin \phi. \quad \ldots\ldots\ldots\ldots(3)$$

Hence

$$\begin{aligned}
x \, \delta y - y \, \delta x &= (a \cos \theta + l \cos \phi) \, (a \cos \theta \, \delta\theta + l \cos \phi \, \delta\phi) \\
&\quad + (a \sin \theta + l \sin \phi) \, (a \sin \theta \, \delta\theta + l \sin \phi \, \delta\phi) \\
&= a^2 \, \delta\theta + l^2 \, \delta\phi + al \cos (\phi - \theta) \, (\delta\theta + \delta\phi) \\
&= \delta \{ a^2 \theta + l^2 \phi + al \sin (\phi - \theta) \} \\
&\qquad + 2al \cos (\phi - \theta) \, \delta\theta. \quad \ldots\ldots\ldots\ldots\ldots\ldots(4)
\end{aligned}$$

An infinitesimal displacement $\delta\sigma$ of any point R in PQ, estimated at right angles to PQ, may be regarded as made up of the displacement relative to P, and the resolved displacement of P. Hence, if $PR = b$,

$$\delta\sigma = b \, \delta\phi + a \cos (\phi - \theta) \, \delta\theta. \quad \ldots\ldots\ldots\ldots(5)$$

Hence

$$\begin{aligned}
\tfrac{1}{2} (x \, \delta y - y \, \delta x) &= \tfrac{1}{2}\delta \{ a^2 \theta + l(l - b) \, \phi + al \sin (\phi - \theta) \} \\
&\qquad + l \, \delta\sigma. \quad \ldots\ldots\ldots\ldots\ldots\ldots(6)
\end{aligned}$$

It follows that, if Q describes a complete circuit such that θ and ϕ return to their initial values,

$$\tfrac{1}{2} \int (x \, dy - y \, dx) = \int l \, d\sigma. \quad \ldots\ldots\ldots\ldots(7)$$

The expression on the left hand is then equal to the area included within the circuit, by Art. 101 (7).

105. Volumes of Solids.

It is impossible to give a general definition of the 'volume,' even of a solid bounded wholly by plane faces, without introducing, in one form or another, the notion of a 'limiting value.'

It may, indeed, be proved by Euclidean methods that two rectangular parallelepipeds are to one another in the ratio compounded of the ratios, each to each, of three concurrent edges of the one to three concurrent edges of the other; and, more generally, that two *prisms* are to one another in the ratio compounded of the ratio of their altitudes and the ratio of their bases. In this way we may define the ratio of any prism to that of the unit cube.

But it is not in general possible to dissect a given polyhedron into a finite number of *prisms*. The simplest general mode of dissection is into *pyramids* having a common vertex at some internal point O, and the faces of the polyhedron as their bases. And the volume of a pyramid cannot be compared with that of a prism without having recourse to the notion of a limiting value. A triangular prism may, indeed, be dissected into three pyramids of equal altitude standing on equal bases (Euc. XII. 7); but we cannot prove these pyramids equal to one another except by a process which involves the consideration of infinitesimal elements (Euc. XII. 5).

A general definition of the volume of a solid bounded by any surfaces, plane or curved, may be framed similar to that of the area of a plane figure (Art. 99). It is always possible to construct two figures, built up of prisms, such that one figure includes, and the other is included by, the given solid, and that the difference between their volumes admits of being made as small as we please. The limiting value to which the volume of either of these figures tends, as the difference between them is indefinitely diminished, is adopted as the definition of the 'volume' of the given solid. We may easily satisfy ourselves, as before, that this limiting value is unique.

The volume of any cylinder (right or oblique), with plane parallel ends, is equal to the product of the area of either end into the perpendicular distance between the two ends. For we may construct an including, and also an included, prismatic figure, whose bases are polygons respectively including, and included by, the base of the cylinder. The above statement is true of each of these figures, and therefore in the limit it is true of the cylinder. Thus the volume of a circular cylinder is $\pi a^2 h$, where a is the radius of the base, and h is the altitude.

Having found the volume of a cylinder with parallel plane ends, we are at liberty, if we find it convenient, to use such cylinders, in place of prisms, to build up the accessory figures employed in the general definition given above. The limit finally obtained in either way must evidently be the same.

106.   General expression for the Volume of any Solid.

The axis of x being drawn in any convenient direction, let the area of the section of the solid by a plane perpendicular to this axis, at a distance x from the origin, be $f(x)$. If we draw a system of planes perpendicular to x, at intervals δx, it is evident that the required volume will be the limit of the sum

$$\Sigma f(x)\,\delta x, \quad\dots\dots\dots\dots\dots\dots(1)$$

since each element of this sum represents the volume of a cylinder of height δx and base $f(x)$. Hence the volume will be given by

$$\int f(x)\,dx, \quad\dots\dots\dots\dots\dots\dots(2)$$

taken between suitable limits of x.

Ex. 1.   Thus, in the case of a cone (or a pyramid), right or oblique, on any base, we take the origin O at the vertex, and the axis of x perpendicular to the base.   If A be the area of the base, and h the altitude, the area of a section at a distance x from O will be

$$f(x) = \left(\frac{x}{h}\right)^2 . A, \quad \dots\dots\dots\dots\dots\dots(3)$$

since similar areas are proportional to the squares on corresponding lines.   Hence the volume, being equal to

$$\frac{1}{h^2} A \int_0^h x^2 dx, \quad \text{or} \quad \tfrac{1}{3}hA, \quad \dots\dots\dots\dots\dots(4)$$

is one-third the altitude into the area of the base.

Ex. 2.   The volume of a tetrahedron is

$$\tfrac{1}{6}haa' \sin a, \quad \dots\dots\dots\dots\dots\dots\dots(5)$$

where a, a' are any pair of opposite edges, h their shortest distance, and a the angle between their directions.

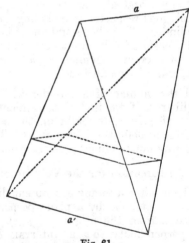

Fig. 61.

Divide the tetrahedron into laminæ by planes parallel to the edges a, a', and therefore perpendicular to the shortest distance h.   It is evident, on reference to Fig. 61, that the section made by a plane of the system at a distance x from the edge a is a parallelogram whose sides are

$$\frac{x}{h} . a' \quad \text{and} \quad \frac{h-x}{h} a,$$

and whose area is therefore

$$\frac{aa'}{h^2} x (h - x) \sin a.$$

Hence the volume $\quad = \dfrac{aa'}{h^2} \sin a \displaystyle\int_0^h x (h - x) \, dx,$

which reduces to the value given above.

107. Solids of Revolution.

Let the equation of the generating curve be

$$y = \phi (x), \quad \dots\dots\dots\dots\dots\dots\dots(1)$$

the axis of x being that of symmetry, and let the solid be bounded by plane ends perpendicular to x. In this case, the area $f(x)$, being that of a circle of radius y, is πy^2. Hence the required volume is

$$\int \pi y^2 \, dx, \quad \dots\dots\dots\dots\dots\dots\dots(2)$$

taken between proper limits. Each element of the sum, of which this integral is the limit, represents, in fact, the volume of a circular plate of thickness δx and area πy^2.

Ex. 1. The equation of a circle, referred to a point on its circumference as origin, is

$$y^2 = x (2a - x). \quad \dots\dots\dots\dots\dots\dots(3)$$

Hence the volume of a segment of a sphere, of height h, is

$$\pi \int_0^h x (2a - x) \, dx = \pi \left[ax^2 - \tfrac{1}{3} x^3 \right]_0^h$$
$$= \pi h^2 (a - \tfrac{1}{3} h), \quad \dots\dots\dots\dots(4,$$

a being the radius of the sphere. For the complete sphere, we have $h = 2a$, and the volume is $\tfrac{4}{3} \pi a^3$, or two-thirds the volume ($\pi a^2 \times 2a$) of the circumscribed circular cylinder.

Ex. 2. The volume of a segment, of height h, of the paraboloid generated by the revolution of the curve

$$y^2 = 4ax \quad \dots\dots\dots\dots\dots\dots\dots(5)$$

about the axis of x, is

$$\pi \int_0^h y^2 \, dx = 4\pi a \left[\tfrac{1}{2} x^2 \right]_0^h = 2\pi a h^2. \quad \dots\dots\dots\dots(6)$$

If b be the radius of the base, we have $b^2 = 4ah$. Hence the volume is $\tfrac{1}{2} \pi b^2 . h$, or one-half that of the cylinder of the same height on the same base.

Ex. 3.   To find the volume of the 'anchor-ring,' or 'tore,' generated by the revolution of the circle

$$x^2 + (y - a)^2 = b^2, \quad \dots\dots\dots\dots\dots(7)$$

where $a > b$, about the axis of x.   See Fig. 66, p. 256.

For each value of x between $\pm b$ we have two values of y, say y_1, y_2; viz.

$$y_1 = a + \sqrt{(b^2 - x^2)}, \quad y_2 = a - \sqrt{(b^2 - x^2)}. \quad \dots\dots\dots(8)$$

The area of a section of the ring by a plane perpendicular to x is therefore

$$\pi y_1^2 - \pi y_2^2 = 4\pi a \sqrt{(b^2 - x^2)} \quad \dots\dots\dots\dots(9)$$

and the required volume is

$$4\pi a \int_{-b}^{b} \sqrt{(b^2 - x^2)}\, dx = 2\pi^2 a b^2 ; \quad \dots\dots\dots\dots(10)$$

by Art. 96, Ex. 5.

This is the same as the volume of a cylinder whose section (πb^2) is equal to that of the ring, and whose length $(2\pi a)$ is equal to the circumference of the circle described by the centre of the generating circle.

108.   Some related Cases.

We give some further examples of the general formula (2) of Art. 106.

Ex. 1.   The section of the elliptic paraboloid

$$2x = \frac{y^2}{p} + \frac{z^2}{q} \quad \dots\dots\dots\dots\dots\dots(1)$$

by a plane $x = $ const. is an ellipse of semi-axes $\sqrt{(2px)}$ and $\sqrt{(2qx)}$, and therefore of area $2\pi\sqrt{(pq)}\, x$.   Hence the volume of the segment cut off by the plane $x = h$ is

$$2\pi \sqrt{(pq)} \int_{0}^{h} x\, dx = \pi \sqrt{(pq)}\, h^2 \dots\dots\dots\dots(2)$$

This is one-half the volume of a cylinder of the same height h on the same elliptic base.

Ex. 2.   In the ellipsoid

$$\frac{x^2}{a^2} + \frac{y^2}{b^2} + \frac{z^2}{c^2} = 1, \quad \dots\dots\dots\dots\dots(3)$$

the section by a plane $x = $ const. is an ellipse of semi-axes

$$b \sqrt{\left(1 - \frac{x^2}{a^2}\right)} \text{ and } c \sqrt{\left(1 - \frac{x^2}{a^2}\right)}, \quad \dots\dots\dots\dots(4)$$

and therefore of area $\pi bc \left(1 - \frac{x^2}{a^2}\right).$ $\dots\dots\dots\dots(5)$

The volume included between any two planes perpendicular to x therefore

$$= \pi bc \int \left(1 - \frac{x^2}{a^2} \right) dx, \quad\ldots\ldots\ldots\ldots\ldots\ldots\ldots(6)$$

taken between the proper limits of x. For the whole volume the limits of x are $\pm a$, and the result is $\frac{4}{3}\pi abc$.

109. Simpson's Rule.

Most of the preceding results are virtually included in a general formula applicable to all cases where the area of the section by a plane perpendicular to x is a quadratic function of x.

The volume included between two parallel planes can then be simply expressed in terms of the areas of the sections made by these planes, of the section half-way between them, and of the interval $(2h)$ between the two extreme planes.

Since the form of a quadratic function is not altered by the addition of a constant to x, we may conveniently take the origin in the middle section. Putting

$$f(x) = A + Bx + Cx^2, \quad\ldots\ldots\ldots\ldots\ldots(1)$$

we have $$\int_{-h}^{h} f(x)\, dx = 2Ah + \tfrac{2}{3}Ch^3. \quad\ldots\ldots\ldots\ldots(2)$$

Denoting the areas of the sections $x = -h$, $x = 0$, $x = h$ by S_1, S_2, S_3, respectively, we have

$$A - Bh + Ch^2 = S_1, \quad A = S_2, \quad A + Bh + Ch^2 = S_3, \ \ldots(3)$$

whence $$\int_{-h}^{h} f(x)\ dx = \tfrac{1}{3}h\,(S_1 + 4S_2 + S_3), \quad\ldots\ldots\ldots\ldots(4)$$

which gives the rule referred to. We may interpret this as expressing that the 'mean' section is

$$\tfrac{1}{6}\,(S_1 + 4S_2 + S_3). \quad\ldots\ldots\ldots\ldots\ldots(5)$$

It is easily seen that the addition of a term Dx^3 in (1) would make no difference to the form of (4). The result is thus extended to the case where $f(x)$ is of the *third* degree.

The formula (1) is obviously applicable to the case of a cone, pyramid, or sphere, and also to the case of a paraboloid, ellipsoid, or hyperboloid, provided the bounding sections be perpendicular to a principal axis. The student who is familiar with the theory of surfaces of the second degree will easily convince himself, moreover, that the latter condition is not essential.

Another case coming under the present rule is that of a solid bounded by two parallel plane polygonal faces and by plane lateral faces which are triangles or trapeziums. We may even include the case where some or all of the lateral faces are curved surfaces (hyperbolic paraboloids) generated by straight lines moving parallel to the planes of the polygons, and each intersecting two straight lines each of which joins a vertex of one polygon to a vertex of the other (see Fig. 63).

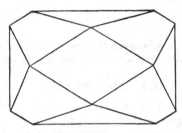

Fig. 62.

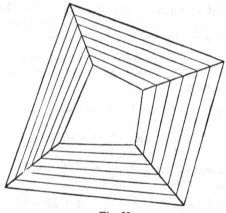

Fig. 63.

And since the number of sides in each polygonal face may be increased indefinitely, the rule will also apply to a solid bounded by any two plane parallel faces and by a curved surface generated by a straight line which meets always the perimeters of those faces.

Ex. 1.   To find the volume of a frustum of a right circular cone.

If a, b be the radii of the two plane ends, that of the middle section will be $\frac{1}{2}(a+b)$.   Hence

$$S_1 = \pi a^2, \quad S_2 = \tfrac{1}{4}\pi (a+b)^2, \quad S_3 = \pi b^2.$$

The volume is therefore

$$\tfrac{1}{6}\pi h\,(a^2 + ab + b^2), \quad\quad\dots\dots\dots\dots\dots\dots\dots(6)$$

if h be the height of the frustum.

Ex. 2.   A cylindrical hole of radius b is bored centrally through a solid sphere of radius a; to find the volume which remains.

Here $\qquad S_1 = 0, \quad S_2 = \pi \left(a^2 - b^2\right), \quad S_3 = 0.$

The mean section is therefore $\frac{2}{3}\pi \left(a^2 - b^2\right)$. The length of the hole is $2\sqrt{(a^2 - b^2)}$. The required volume is therefore

$$\frac{4}{3}\pi \left(a^2 - b^2\right)^{\frac{3}{2}}. \qquad \dots\dots\dots\dots\dots\dots(7)$$

110. Rectification of Curved Lines.

The perimeter of a *rectilinear* figure is the length obtained by placing end to end in succession, in a straight line, lengths equal to the respective sides of the figure.

But since a *curved* line, however short, cannot be superposed on any portion of a straight line, we require some definition of what is to be understood by the 'length' of a curve. The definition usually adopted is that it is the limit to which the perimeter of an inscribed polygon tends as the lengths of the sides are indefinitely diminished. It is assumed that the gradient of the curve is continuous, except possibly at isolated points; *i.e.* that the mutual inclination of the tangents at two adjacent points P, Q can be made as small as we please by taking Q sufficiently near to P. It will appear that, under proper conditions, the above limit is unique; and it can also be shewn that it coincides with the corresponding limit for a circumscribed polygon.

If (x, y) and $(x + \delta x, y + \delta y)$ be the rectangular coordinates of two adjacent points P, Q on a curve, the length of the chord PQ is

$$\sqrt{\{(\delta x)^2 + (\delta y)^2\}}.$$

It has been shewn, in Art. 56, that if y and dy/dx be finite and continuous, the ratio $\delta y/\delta x$ is equal to the value of the derived function dy/dx for *some* point of the curve between P and Q. Hence with a properly chosen value of dy/dx, we have

$$PQ = \sqrt{\left\{1 + \left(\frac{dy}{dx}\right)^2\right\}}\, \delta x.$$

The limiting value of the perimeter of the inscribed polygon is therefore

$$\int \sqrt{\left\{1 + \left(\frac{dy}{dx}\right)^2\right\}}\, dx, \qquad \dots\dots\dots\dots\dots(1)$$

taken between proper limits of x. The fact that this limiting value is unique has been established in Art. 90.

If x be regarded as a function of y, the corresponding formula is

$$\int \sqrt{\left\{1 + \left(\frac{dx}{dy}\right)^2\right\}}\, dy. \qquad \dots\dots\dots\dots\dots(2)$$

If we denote by s the arc of the curve measured from some arbitrary point (x_0), and if as in Art. 60, we put $dy/dx = \tan \psi$, the formula (1) becomes

$$s = \int_{x_0}^{x} \sec \psi \, dx. \quad \dots\dots\dots\dots\dots(3)$$

There is of course a similar transformation of the formula (2).

Ex. 1.   In the catenary

$$y = c \cosh \frac{x}{c}, \quad \dots\dots\dots\dots\dots\dots\dots(4)$$

we have $\quad \displaystyle\int \sqrt{\left\{ 1 + \left(\frac{dy}{dx}\right)^2 \right\}} \, dx = \int \sqrt{\left(1 + \sinh^2 \frac{x}{c}\right)} \, dx$

$$= \int \cosh \frac{x}{c} \, dx = c \sinh \frac{x}{c}.$$

Since this vanishes with x, the arc (s) measured from the lowest point is given by

$$s = c \sinh \frac{x}{c}. \quad \dots\dots\dots\dots\dots\dots(5)$$

Ex. 2.   In the parabola

$$y^2 = 4ax, \quad \dots\dots\dots\dots\dots\dots \dots\dots(6)$$

we have $\quad \displaystyle\int \sqrt{\left\{ 1 + \left(\frac{dy}{dx}\right)^2 \right\}} \, dx = \int \sqrt{\left(\frac{x+a}{x}\right)} \, dx. \quad \dots\dots\dots(7)$

This may be integrated by the method of Art. 76, first rationalizing the numerator, or we may put

$$x = a \sinh^2 u,$$

and obtain $\quad \displaystyle 2a \int \cosh^2 u \, du = a \int (1 + \cosh 2u) \, du$

$$= a \left(u + \tfrac{1}{2} \sinh 2u\right). \quad \dots\dots\dots\dots(8)$$

Since u vanishes with x, this gives the length of the arc measured from the vertex.

For example, at the end of the latus-rectum we have

$$\sinh u = 1, \quad \cosh u = \sqrt{2}, \quad u = \log (1 + \sqrt{2}),$$

whence we find that the length of the arc up to this point is

$$a \left\{\log (1 + \sqrt{2}) + \sqrt{2}\right\} = 2 \cdot 296a.$$

111.   Generalized Formulæ.

It is a consequence of the definition above given that any infinitely small arc PQ of a curve is ultimately in a ratio of equality to the chord PQ.

For a formal proof of this theorem, which is a generalization of that of Art. 22, 2°, let PP_1, P_1P_2, ..., $P_{n-1}Q$ be the sides of an open polygon inscribed in the arc, and let ϵ_1, ϵ_2, ..., ϵ_n be the angles (positive or negative) which they respectively make with the chord PQ. It is obvious that

$$PQ < PP_1 + P_1P_2 + ... + P_{n-1}Q. \qquad (1)$$

On the other hand

$$PQ = PP_1 \cos \epsilon_1 + P_1P_2 \cos \epsilon_2 + ... + P_{n-1}Q \cos \epsilon_n$$

$$> (PP_1 + P_1P_2 + ... + P_{n-1}Q) \cos \epsilon, \qquad (2)$$

where ϵ is the greatest in absolute value of the angles ϵ_1, ϵ_2, ... ϵ_n. The ratio

$$\frac{PQ}{PP_1 + P_1P_2 + ... + P_{n-1}Q} \qquad (3)$$

therefore lies between 1 and $\cos \epsilon$. Since the chords PP_1, P_1P_2, ..., $P_{n-1}Q$, as well as PQ, are parallel to tangents to the arcs at points between their respective extremities, it follows from the continuity of the gradient that $|\epsilon|$ can be made as small as we please by taking Q sufficiently near to P. The limiting value of the ratio (3) is therefore unity.

This may be verified immediately by differentiating the formula (3) of the preceding Art. with respect to the upper limit (x) of the integral. We thus find

$$\lim \frac{\delta s}{\delta x} = \sec \psi. \qquad (4)$$

Since, when Q is taken infinitely near to P, $\sec \psi$ is the limiting value of the ratio of the chord PQ to δx, we have, ultimately,

$$\lim \frac{\delta s}{PQ} = 1. \qquad (5)$$

The above principle leads to several important formulæ. In the first place, if the coordinates x, y of any point P on the curve be regarded as functions of the arc s, we have, in Fig. 19, p. 45,

$$\cos QPR = \frac{PR}{PQ} = \frac{\delta x}{\delta s} \cdot \frac{\delta s}{PQ}, \qquad \sin QPR = \frac{QR}{PQ} = \frac{\delta y}{\delta s} \cdot \frac{\delta s}{PQ},$$

and therefore $\qquad \cos \psi = \frac{dx}{ds}, \qquad \sin \psi = \frac{dy}{ds}. \qquad (6)$

It follows that

$$\left(\frac{dx}{ds}\right)^2 + \left(\frac{dy}{ds}\right)^2 = 1. \qquad (7)$$

Again, if x, y be functions of any other variable t, we have

$$PQ = \sqrt{\left\{\left(\frac{\delta x}{\delta t}\right)^2 + \left(\frac{\delta y}{\delta t}\right)^2\right\}}\,\delta t,$$

and therefore $\quad \lim \dfrac{PQ}{\delta t} = \sqrt{\left\{\left(\frac{dx}{dt}\right)^2 + \left(\frac{dy}{dt}\right)^2\right\}}$,

or

$$\frac{ds}{dt} = \sqrt{\left\{\left(\frac{dx}{dt}\right)^2 + \left(\frac{dy}{dt}\right)^2\right\}}. \quad\quad\quad\quad\quad (8)$$

Hence

$$s = \int \sqrt{\left\{\left(\frac{dx}{dt}\right)^2 + \left(\frac{dy}{dt}\right)^2\right\}}\, dt. \quad\quad\quad (9)$$

This may be regarded as a generalization of Art. 110 (1). The formula referred to was obtained on the supposition that there is only one value of y for each value of x, within the arc considered. The result (9) is free from this restriction; all that is essential is that as t increases, the point P should describe the curve continuously.

In the same way, the formulæ (6) may be taken to apply to any rectifiable curve, provided we understand by ψ the angle which the tangent, drawn in the direction of s increasing, makes with the positive direction of the axis of x.

The formulæ (8) and (9) have an obvious interpretation in Dynamics. If x, y be the rectangular coordinates of a moving point, regarded as functions of the time t, then dx/dt and dy/dt are the component velocities parallel to the coordinate axes, and if v be the actual velocity, we have

$$v = \sqrt{\left\{\left(\frac{dx}{dt}\right)^2 + \left(\frac{dy}{dt}\right)^2\right\}}. \quad\quad\quad\quad (10)$$

The formulæ (8), (9) are thus equivalent to

$$\frac{ds}{dt} = v, \quad s = \int v\, dt. \quad\quad\quad\quad\quad (11)$$

Ex. In the ellipse

$$x = a \sin\phi, \quad y = b \cos\phi \quad\quad\quad\quad\quad (12)$$

we have $\quad \left(\dfrac{ds}{d\phi}\right)^2 = \left(\dfrac{dx}{d\phi}\right)^2 + \left(\dfrac{dy}{d\phi}\right)^2 = a^2 \cos^2\phi + b^2 \sin^2\phi$

$$= a^2 (1 - e^2 \sin^2\phi),$$

where e is the eccentricity. Hence the arc s, measured from the extremity of the minor axis, is

$$s = a \int_0^\phi \sqrt{(1 - e^2 \sin^2\phi)}\, d\phi. \quad\quad\quad\quad (13)$$

This cannot be expressed (in a finite form) in terms of the ordinary functions of Mathematics. The integral is called an 'elliptic integral of the second kind,' and is denoted by $E (e, \phi)$. It may be regarded as a known function, having been tabulated by Legendre[*]. The whole perimeter of the ellipse is expressed by

$$4a \int_0^{\frac{1}{2}\pi} \sqrt{(1 - e^2 \sin^2 \phi)}\, d\phi. \qquad \dots\dots\dots\dots(14)$$

The integral in this expression is denoted by $E (e, \frac{1}{2}\pi)$, or more shortly by $E_1 (e)$. It is called a 'complete' elliptic integral of the second kind[†]. The quantity e is called the 'modulus' of the integral.

For the calculation of the integral (14) by means of a series see Art. 180.

112. Arcs referred to Polar Coordinates.

Let OP, OP' be two consecutive radii of the curve, and let PN be drawn perpendicular to OP'. If we write

$$OP = r, \quad OP' = r + \delta r, \quad \angle POP' = \delta\theta,$$

then, as in Art. 63, PN will differ from $r\delta\theta$, and NP' from δr, by quantities which are infinitely small in comparison with PN, NP' respectively. Hence PP', or $\sqrt{(PN^2 + NP'^2)}$, is ultimately in a ratio of equality to

$$\sqrt{\{r^2 (\delta\theta)^2 + (\delta r)^2\}}.$$

It follows that, if θ be the independent variable,

$$\frac{ds}{d\theta} = \lim \frac{PP'}{\delta\theta} = \sqrt{\left\{r^2 + \left(\frac{dr}{d\theta}\right)^2\right\}}, \qquad \dots\dots\dots(1)$$

and therefore

$$s = \int \sqrt{\left\{r^2 + \left(\frac{dr}{d\theta}\right)^2\right\}}\, d\theta, \qquad \dots\dots\dots\dots(2)$$

provided the integral be taken between the appropriate limits of θ.

If r and θ be given as functions of an independent variable t, we have

$$\frac{ds}{dt} = \lim \frac{PP'}{\delta t} = \sqrt{\left\{\left(\frac{dr}{dt}\right)^2 + r^2 \left(\frac{d\theta}{dt}\right)^2\right\}}, \qquad \dots\dots(3)$$

---

[*] *Traité des Fonctions Elliptiques* (1826).
[†] The elliptic integral of the 'first kind' is

$$\int_0^\phi \frac{d\phi}{\sqrt{(1 - e^2 \sin^2 \phi)}}$$

and is denoted by $F(e, \phi)$. The corresponding 'complete' integral (with $\frac{1}{2}\pi$ as the upper limit) is denoted by $F_1(e)$.

and therefore

$$s = \int \sqrt{\left\{ \left(\frac{dr}{dt}\right)^2 + r^2 \left(\frac{d\theta}{dt}\right)^2 \right\}} \, dt, \quad \ldots \ldots \ldots \ldots (4)$$

which includes (2) as a particular case.

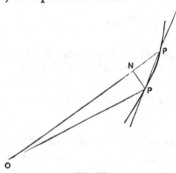

Fig. 64.

Again, if r, θ be regarded as functions of the arc s, and if ϕ denote the angle which the tangent-line, drawn in the direction of s increasing, makes with the positive direction of the radius vector, we have

$$\cos NP'P = \frac{NP'}{PP'}, \qquad \sin NP'P = \frac{PN}{PP'},$$

and therefore, in the limit,

$$\cos \phi = \frac{dr}{ds}, \qquad \sin \phi = r \frac{d\theta}{ds}. \quad \ldots \ldots \ldots \ldots (5)$$

These results, again, have a dynamical illustration. If v denote the velocity of a moving point, the component velocities along the transverse to the radius are

$$v \cos \phi = \frac{dr}{ds} \frac{ds}{dt} = \frac{dr}{dt},$$

and

$$v \sin \phi = \frac{rd\theta}{ds} \frac{ds}{dt} = \frac{rd\theta}{dt}, \quad \left. \right\} \quad \ldots \ldots \ldots \ldots \ldots (6)$$

respectively; and the formula (4) is equivalent to

$$s = \int v \, dt, \quad \ldots \ldots \ldots \ldots \ldots \ldots \ldots \ldots (7)$$

as before.

113. Areas of Surfaces of Revolution.

To frame a general definition of the area of a curved surface, and to prove that the area so defined has a determinate value, is a matter of some nicety. We shall here only consider the case of a surface of revolution limited (if at all) by planes perpendicular to the axis.

We begin with the circular cylinder. The curved surface may be defined as the limiting value of the sum of the areas of the lateral faces of an inscribed prism. These faces have all the same length, and their sum is equal to this common length multiplied by the perimeter of the cross-section of the prism. In the limit this perimeter becomes the perimeter of the cylinder. Hence the curved surface of a right cylinder of radius a and height h is $2\pi ah$.

Take next the surface of a right cone included between two planes perpendicular to the axis. We can inscribe in this a frustum of a pyramid, whose bases are similar and similarly situated regular polygons, inscribed in the two bounding circles. The curved surface in question may be defined as the limit of the lateral area of this frustum. This area is made up of a number of trapeziums having a common altitude, viz. the perpendicular distance between their parallel sides, and is therefore equal to this common altitude multiplied by the arithmetic mean of the perimeters of the two polygons. In the limit, these perimeters become the circumferences of the bounding circles, and the common altitude becomes the length of a generating line of the cone included between these circles. In other words, the curved surface generated by the revolution of a straight line PQ about any axis in the same plane with it is equal to PQ multiplied by the arithmetic mean of the circumferences of the circles described by P and Q. This is the same as the product of PQ into the circumference of the circle described by its middle point.

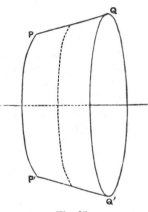

Fig. 65.

Next consider the surface generated by the revolution of any arc of a curve

$$y = \phi(x), \quad \dots\dots\dots\dots\dots\dots\dots(1)$$

about the axis of x. Taking any number of points in this arc, and joining them by straight lines, we obtain an open polygon; the curved surface is then defined as the limiting value to which the sum of the areas described by the sides of the polygon tends, when the lengths of the sides are indefinitely diminished. Hence if PQ be the chord of any element δs of the generating curve, and

y the ordinate of the middle point of PQ, the curved surface is the limiting value of the sum $\Sigma\,(2\pi y\,.\,PQ)$. Ultimately, PQ is in a ratio of equality to δs, and y may be taken to be the ordinate of the curve; the surface is then equal to the limiting value of $2\pi\Sigma\,(y\,.\,\delta s)$, that is, to

$$2\pi \int y\,ds, \dots\dots\dots\dots\dots\dots\dots(2)$$

taken over the proper range of s.

Ex. 1. In the case of the sphere the coordinates of any point of the generating curve may be written

$$x = a\cos\theta, \quad y = a\sin\theta, \dots\dots\dots\dots(3)$$

whence

$$ds/d\theta = a. \dots\dots\dots\dots\dots\dots(4)$$

Hence the surface of a zone bounded by planes perpendicular to x is

$$2\pi a^2 \int_{\theta_1}^{\theta_2} \sin\theta\,d\theta = 2\pi a^2\,(\cos\theta_1 - \cos\theta_2)$$

$$= 2\pi a\,(x_1 - x_2) \dots\dots\dots\dots\dots(5)$$

where the suffixes refer to the bounding circles. Hence the zone is equal in area to the corresponding zone of a circumscribing cylinder having its axis perpendicular to the planes of the bounding circles. In particular, the whole surface of the sphere is $2\pi a\,.\,2a$, or $4\pi a^2$.

Ex. 2. To find the surface of the ring generated by the revolution of a circle of radius b about a line in its plane at a distance a from its centre, we may put

$$x = b\sin\theta, \quad y = a - b\cos\theta, \quad ds/d\theta = b, \dots\dots\dots(6)$$

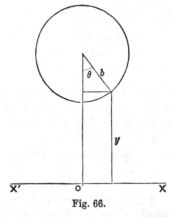

Fig. 66.

and obtain

$$2\pi \int y\,ds = 2\pi b \int (a - b\cos\theta)\,d\theta. \dots\dots\dots\dots(7)$$

The limits of θ being 0 and 2π, the result is $2\pi b \times 2\pi a$, which is equal to the curved surface of a right cylinder of radius b and length $(2\pi a)$

equal to the circumference described by the centre of the generating circle.

Ex. 3. To find the surface generated by the revolution of the ellipse

$$x = a \sin \phi, \quad y = b \cos \phi \quad \dots\dots\dots\dots\dots(8)$$

about the *major* axis, we have

$$2\pi \int y\, ds = 2\pi \int y\, \frac{ds}{d\phi}\, d\phi$$

$$= 2\pi ab \int \sqrt{(1 - e^2 \sin^2 \phi)}\, d\,(\sin \phi), \dots\dots\dots\dots(9)$$

by Art. 111. If we put $\quad e \sin \phi = \sin \theta, \dots\dots\dots\dots\dots(10)$
this

$$= \frac{2\pi ab}{e} \int \cos^2 \theta\, d\theta = \frac{\pi ab}{e}\, [\theta + \sin \theta \cos \theta]. \dots\dots\dots(11)$$

To find the whole surface we take this between the limits $\phi = \mp \frac{1}{2}\pi$, or $\theta = \mp \sin^{-1} e$. The result is

$$\frac{2\pi ab}{e} \{\sin^{-1} e + e \sqrt{(1 - e^2)}\},$$

or $\qquad\qquad 2\pi b^2 + 2\pi ab\, \dfrac{\sin^{-1} e}{e}. \dots\dots\dots\dots\dots(12)$

By a similar process we find, for the surface generated by the revolution of the ellipse about its *minor* axis, the value

$$\frac{2\pi ab}{e'} \{\sinh^{-1} e' + e' \sqrt{(1 + e'^2)}\},$$

where $e' = \sqrt{(a^2 - b^2)}/b$, or

$$2\pi a^2 + 2\pi ab\, \frac{\sinh^{-1} e'}{e'}. \dots\dots\dots\dots(13)$$

This may also be put into the form

$$2\pi a^2 + \pi b^2 \cdot \frac{1}{e} \log \frac{1 + e}{1 - e}. \dots\dots\dots\dots(14)$$

114. Approximate Integration.

Various methods have been devised for finding an approximate value of a definite integral, when the indefinite integral of the function involved cannot be obtained. For brevity of statement, we will consider the problem in its geometrical form; viz. it is required to find an approximate value of the area included between a given curve, the axis of x, and two given ordinates.

The methods referred to all consist in substituting for the actual curve another which shall follow the same course more or less closely, whilst it is represented by a function of an easily integrable character.

The simplest, and roughest, mode is to draw n equidistant ordinates of the curve, and to join their extremities by straight lines. The required area is thus replaced by the sum of a series of trapeziums. If h be the distance between consecutive ordinates, and $y_1, y_2, \ldots, y_n$ the lengths of the ordinates, the sum of the trapeziums is

$$\tfrac{1}{2}(y_1 + y_2)\, h + \tfrac{1}{2}(y_2 + y_3)\, h + \ldots + \tfrac{1}{2}(y_{n-2} + y_{n-1})\, h + \tfrac{1}{2}(y_{n-1} + y_n)\, h$$

$$= (\tfrac{1}{2}y_1 + y_2 + y_3 + \ldots + y_{n-2} + y_{n-1} + \tfrac{1}{2}y_n)\, h;\ \ldots(1)$$

that is, we add to the arithmetic mean of the first and last ordinates the sum of the intervening ordinates, and multiply the result by the common interval h.

The value thus obtained will obviously be in excess if the curve is convex to the axis of x, and in defect in the opposite case.

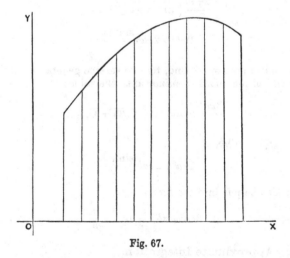

Fig. 67.

Another method, originally given by Newton and Cotes[*], is to assume for y a rational integral expression of degree $n - 1$, thus

$$y = A_0 + A_1 x + A_2 x^2 + \ldots + A_{n-1} x^{n-1}, \ldots\ldots\ldots(2)$$

[*] See the latter's tract *De Methodo Differentiali*, printed as a supplement to the *Harmonia Mensurarum*, Cambridge, 1722.

and to determine the coefficients $A_1, A_2, \dots A_{n-1}$ so that, for the n equidistant values of x, y shall have the prescribed values $y_1, y_2, \dots, y_n$. The area is then given by

$$\int y\, dx, = A_0 x + \tfrac{1}{2} A_1 x^2 + \tfrac{1}{3} A_2 x^3 + \dots + \frac{1}{n} A_{n-1} x^n, \quad \dots(3)$$

taken between proper limits of x.

Thus, in the case of *three* equidistant ordinates, taking the origin at the foot of the middle ordinate, we assume

$$y = A_0 + A_1 \frac{x}{h} + A_2 \left(\frac{x}{h}\right)^2, \quad \dots\dots\dots\dots\dots(4)$$

with the conditions that

$$\begin{aligned} y = \quad & y_1, \quad y_2, \quad y_3, \\ \text{for} \qquad x = & -h, \quad 0, \quad h, \end{aligned} \Bigg\} \dots\dots\dots\dots\dots(5)$$

respectively. These give

$$A_0 - A_1 + A_2 = y_1, \quad A_0 = y_2, \quad A_0 + A_1 + A_2 = y_3, \quad \dots(6)$$

so that

$$A_0 = y_2, \quad A_1 = \tfrac{1}{2}(y_3 - y_1), \quad A_2 = \tfrac{1}{2}(y_1 + y_3 - 2y_2). \quad \dots\dots(7)$$

Hence

$$\int_{-h}^{h} y\, dx = 2(A_0 + \tfrac{1}{3} A_2) h$$

$$= \tfrac{1}{3}(y_1 + 4y_2 + y_3) h. \dots\dots\dots\dots(8)$$

Cf. Art. 109.

The method here employed is equivalent to replacing the actual curve by an arc of a parabola having its axis vertical; and the result represents the difference between the trapezium

$$\tfrac{1}{2}(y_1 + y_3) \cdot 2h$$

and the parabolic segment

$$\tfrac{2}{3} \cdot \{\tfrac{1}{2}(y_1 + y_3) - y_2\} \cdot 2h;$$

see Art. 100, Ex. 3.

In the case of *four* equidistant ordinates a similar process leads to the formula

$$\tfrac{3}{8}(y_1 + 3y_2 + 3y_3 + y_4) h, \quad \dots\dots\dots\dots(9)$$

whilst for *five* ordinates we get

$$\tfrac{2}{45}(7y_1 + 32y_2 + 12y_3 + 32y_4 + 7y_5) h. \quad \dots\dots(10)$$

With an increasing number of ordinates the coefficients in this method become more and more unwieldy*. A simple, but generally less accurate, rule was devised by Simpson†. Taking an odd number of ordinates, the areas between alternate ordinates, beginning with the first, are calculated from the 'parabolic' formula (8), and the results added. We thus obtain

$$\tfrac{1}{3}\,\{y_1 + 4y_2 + y_3$$
$$+ y_3 + 4y_4 + y_5$$
$$+ y_5 + 4y_6 + y_7$$
$$+ \ldots\ldots\ldots\ldots$$
$$+ y_{2n-1} + 4y_{2n} + y_{2n+1}\}\,h$$
$$= \tfrac{1}{3}\,\{(y_1 + y_{2n+1}) + 2\,(y_3 + y_5 + \ldots + y_{2n-1}) + 4\,(y_2 + y_4 + \ldots + y_{2n})\}\,h.$$
$$\ldots\ldots\ldots(11)$$

That is, we take the sum of the first and last ordinates, twice the sum of the intervening odd ordinates, and four times the sum of the even ordinates, and multiply one-third the aggregate thus obtained by the common interval h.

Ex. To calculate the value of π from the formula

$$\tfrac{1}{4}\pi = \int_0^1 \frac{dx}{1 + x^2}. \quad\ldots\ldots\ldots\ldots\ldots\ldots(12)$$

Dividing the range into 10 equal intervals, so that $h = \cdot1$, we find

$$y_1 = 1, \qquad y_2 = \cdot9900990, \qquad y_3 = \cdot9615385,$$
$$y_4 = \cdot9174312, \qquad y_5 = \cdot8620690,$$
$$y_6 = \cdot8000000, \qquad y_7 = \cdot7352941,$$
$$y_8 = \cdot6711409, \qquad y_9 = \cdot6097561.$$
$$y_{11} = \cdot5, \qquad y_{10} = \cdot5524862,$$

Hence
$$y_1 + y_{11} = 1\cdot5,$$
$$y_3 + y_5 + y_7 + y_9 = 3\cdot1686577,$$
$$y_2 + y_4 + y_6 + y_8 + y_{10} = 3\cdot9311573.$$

The formula (11) then gives

$$\tfrac{1}{4}\pi = \tfrac{1}{30}\,(1\cdot5 + 6\cdot3373154 + 15\cdot7246292)$$
$$= \cdot78539815,$$

---

* The coefficients for the cases $n = 3, 4, 5, \ldots 11$, were calculated by Cotes; see also Bertrand, *Calcul Intégral*, Art. 363.

† *Mathematical Dissertations* (1743).

whence, retaining only seven figures,
$$\pi = 3\cdot141593.$$
This is correct to the last figure.

The formula (1) would have given
$$\tfrac{1}{4}\pi = \cdot78498150, \quad \pi = 3\cdot139926,$$
which is too small by about one part in 2000.

115. Mean Values.

Let $y_1, y_2, \ldots\ldots y_n$ be the values of a function $\phi(x)$ corresponding to n equidistant values of x distributed over the range $b - a$, say to the values of x which mark the middle points of the n equal intervals (h) into which this range may be subdivided. The limiting value to which the arithmetic mean

$$\frac{1}{n}(y_1 + y_2 + \ldots + y_n) \ldots\ldots\ldots\ldots\ldots\ldots(1)$$

tends, as n is indefinitely increased, is called the 'mean value' of the function over the range $b - a$.

Since $h = (b-a)/n$, the expression (1) may be written
$$\frac{y_1 h + y_2 h + \ldots + y_n h}{b - a},$$
and the limiting value of this for $n \to \infty$, $h \to 0$, is

$$\frac{\int_a^b \phi(x)\,dx}{b - a}. \ldots\ldots\ldots\ldots\ldots\ldots\ldots(2)$$

In the geometrical representation the mean value is the altitude of the rectangle on base $b - a$ whose area is equal to that included between the curve $y = \phi(x)$, the extreme ordinates, and the axis of x. See Fig. 49, p. 210.

The theorem of Art. 91, 3° may now be stated as follows: The mean value of a continuous function, over any range of the independent variable, is equal to the value of the function for some value of the independent variable within the range.

The various formulæ of Art. 114 may be interpreted as giving approximate expressions for the mean value of a function, over a given range, in terms of a series of values of the function taken at equidistant intervals covering the range. For example, in terms of three and of four such values, the mean values, as given by Cotes' method, are

$$\tfrac{1}{6}(y_1 + 4y_2 + y_3) \quad \text{and} \quad \tfrac{1}{8}(y_1 + 3y_2 + 3y_3 + y_4),$$
respectively.

In applying the conception of a mean value it is essential to have a clear understanding as to what is the independent variable to which (in the first instance) equal increments are given. Thus, in the case of a particle descending with a constant acceleration g, the mean value of the velocity in any interval of *time* t_1 from rest is

$$\frac{1}{t_1} \int_0^{t_1} v \, dt = \frac{1}{t_1} \int_0^{t_1} gt \, dt = \tfrac{1}{2} g t_1 ;$$

i.e. it is *one-half* the final velocity. But if we seek the mean velocity for equal infinitesimal increments of the *space* (s), we have, since $v^2 = 2gs$,

$$\frac{1}{s_1} \int_0^{s_1} v \, ds = \frac{(2g)^{\frac{1}{2}}}{s_1} \int_0^{s_1} s^{\frac{1}{2}} \, ds = \tfrac{2}{3} (2g s_1)^{\frac{1}{2}} ;$$

i.e. it is *two-thirds* the final velocity.

Ex. 1. The mean value of $\sin \theta$ for equidistant intervals of θ ranging from 0 to π is

$$\frac{1}{\pi} \int_0^{\pi} \sin \theta \, d\theta = \frac{2}{\pi} = \cdot 6366. \quad \dots \dots \dots \dots \dots (3)$$

Hence the mean value of the ordinates of a semicircle of radius a, drawn through equidistant points of the *arc*, is $\cdot 6366a$.

If the ordinates had been drawn through equidistant points on the *diameter*, the mean value would have been

$$\frac{1}{2a} \int_{-a}^{a} \sqrt{(a^2 - x^2)} \, dx = \tfrac{1}{2} a \int_{-\frac{1}{2}\pi}^{\frac{1}{2}\pi} \cos^2 \theta \, d\theta = \tfrac{1}{4} \pi a, \quad \dots \dots (4)$$

or $\cdot 7854a$. It is easily seen *à priori* why this latter mean should be the greater.

Ex. 2. A disk has the form of a very flat ellipsoid of revolution. To find the ratio of its mean thickness to the thickness at the centre.

If a be the radius, the ratio of the thickness at a distance r from the centre to that at the centre is

$$\sqrt{\left(1 - \frac{r^2}{a^2}\right)}. \quad \dots \dots \dots \dots \dots \dots \dots (5)$$

The required ratio is therefore

$$\frac{1}{\pi a^2} \int_0^a \sqrt{\left(1 - \frac{r^2}{a^2}\right)} \, 2\pi r \, dr = \tfrac{2}{3}. \quad \dots \dots \dots \dots (6)$$

Ex. 3. If the density ρ of a sphere be a function of the distance r from the centre, taking as the element of volume

$$\delta V = \delta \left(\tfrac{4}{3} \pi r^3\right) = 4 \pi r^2 \delta r,$$

the mean density is

$$\bar{\rho} = 4\pi \int_0^a \rho r^2 dr \div \tfrac{4}{3}\pi a^3 = \frac{3}{a^3} \int_0^a \rho r^2 dr, \quad \ldots\ldots\ldots\ldots(7)$$

if a be the external radius.

Thus, if $\rho \propto r^n$, the mean density is $3/(n+3)$ of the density at the surface.

Again, assuming that in the Earth

$$\rho = \rho_0 \left(1 - k\frac{r^2}{a^2} \right), \quad \ldots\ldots\ldots\ldots\ldots\ldots(8)$$

we find
$$\bar{\rho} = \rho_0 (1 - \tfrac{3}{5}k) = \tfrac{1}{5}(2\rho_0 + 3\rho_1), \quad \ldots\ldots\ldots\ldots(9)$$

where ρ_1 is the density at the surface $(r = a)$. If the above law of density be applicable to the case of the Earth, then since $\bar{\rho} = 2\rho_1$, roughly, we infer that $\rho_0 = \tfrac{7}{2}\rho_1$, or the density at the centre is $3\tfrac{1}{2}$ times the density at the surface.

116. Mean Centres of Geometrical Figures.

The 'mean centre' (G, say) of a system of geometrical points

$$(x_1, y_1), \quad (x_2, y_2), \ldots, \quad (x_n, y_n) \quad \ldots\ldots\ldots\ldots(1)$$

may be defined as the point whose coordinates are

$$\left. \begin{aligned} \bar{x} &= \frac{1}{n}(x_1 + x_2 + \ldots + x_n) = \frac{\Sigma(x)}{n}, \\ \bar{y} &= \frac{1}{n}(y_1 + y_2 + \ldots + y_n) = \frac{\Sigma(y)}{n}. \end{aligned} \right\} \quad \ldots\ldots\ldots(2)$$

Since these relations are linear, and since transformations of Cartesian coordinates are effected by linear formulæ, it easily follows that the distance of G from *any* line is equal to the arithmetic mean of the distances of the given points from the line, these distances being taken of course with the proper signs according as they lie on one side of the line or the other.

There is, in like manner, a mean centre of a plane curve, or of a plane area, whose distance from any line in the plane is the mean (in the sense of Art. 115) of the distances of the infinitesimal elements of the curve, or of the area, from the line.

Thus, for a curve we have

$$\bar{x} = \lim \frac{\Sigma(x\,\delta s)}{\Sigma(\delta s)}, \quad \bar{y} = \lim \frac{\Sigma(y\,\delta s)}{\Sigma(\delta s)}; \quad \ldots\ldots\ldots\ldots(3)$$

and for an area

$$\bar{x} = \lim \frac{\Sigma\,(x\,\delta A)}{\Sigma\,(\delta A)}, \quad \bar{y} = \lim \frac{\Sigma\,(y\,\delta A)}{\Sigma\,(\delta A)}, \quad \dots\dots(4)$$

where δA is an element of area. In the limit the summations take the form of integrals.

Ex. 1. In the case of a circular arc, if the origin be taken at the centre, and the axis of x along the medial line, we have $\bar{y} = 0$, by symmetry. Writing $x = a\cos\theta$, $\delta s = a\,\delta\theta$,

$$\bar{x} = \frac{1}{2a\alpha} \int_{-\alpha}^{\alpha} a\cos\theta\,d\theta = \frac{\sin\alpha}{\alpha} . a, \dots\dots\dots\dots(5)$$

if 2α denote the angle which the whole arc subtends at the centre.

As α increases from an infinitely small value to π, $\bar{x}$ decreases from a to 0. For the semicircle, we have $\alpha = \frac{1}{2}\pi$, and

$$\bar{x} = \frac{2}{\pi}\,a = \cdot 637\,a.$$

Ex. 2. For the area of a segment of the parabola

$$y^2 = 4ax \dots\dots\dots\dots\dots\dots(6)$$

bounded by the double ordinate $x = h$, we have

$$\bar{x} = \frac{\int_0^h xy\,dx}{\int_0^h y\,dx} = \frac{\int_0^h x^{\frac{3}{2}}\,dx}{\int_0^h x^{\frac{1}{2}}\,dx} = \tfrac{3}{5}h. \dots\dots\dots\dots(7)$$

The notion of the mean centre can obviously be extended to three-dimensional figures, distances from a line being now replaced by distances from a plane. Thus in the case of a surface, we have

$$\bar{x} = \lim \frac{\Sigma\,(x\,\delta S)}{\Sigma\,(\delta S)}, \quad \bar{y} = \lim \frac{\Sigma\,(y\,\delta S)}{\Sigma\,(\delta S)}, \quad \bar{z} = \lim \frac{\Sigma\,(z\,\delta S)}{\Sigma\,(\delta S)}, \quad \dots(8)$$

where δS denotes an element of the surface. Similarly, for a volume,

$$\bar{x} = \lim \frac{\Sigma\,(x\,\delta V)}{\Sigma\,(\delta V)}, \quad \bar{y} = \lim \frac{\Sigma\,(y\,\delta V)}{\Sigma\,(\delta V)}, \quad \bar{z} = \lim \frac{\Sigma\,(z\,\delta V)}{\Sigma\,(\delta V)}, \dots(9)$$

where δV is an element of volume.

In the case of a surface, or a solid, of revolution the mean centre is evidently on the axis of symmetry, and if we take this as axis of x, we have only to calculate the value of $\bar{x}$. If y be the ordinate of the generating curve, we put, in (8), $\delta S = 2\pi y\,\delta s$, this being

(Art. 113) the area of an annular element of surface, whose points are all at the same distance from the plane $x = 0$.   Hence

$$\bar{x} = \frac{\int x \cdot 2\pi y\, ds}{\int 2\pi y\, ds} = \frac{\int xy\, ds}{\int y\, ds}. \qquad \text{...................(10)}$$

Similarly, in (9) we put $\delta V = \pi y^2 \delta x$, and obtain

$$\bar{x} = \frac{\int x \cdot \pi y^2\, dx}{\int \pi y^2\, dx} = \frac{\int xy^2\, dx}{\int y^2\, dx}. \qquad \text{...................(11)}$$

Ex. 3.   For a zone of a spherical surface, putting

$$x = a\cos\theta, \quad y = a\sin\theta, \quad \delta s = a\delta\theta, \quad \text{...............(12)}$$

we have
$$\bar{x} = a\,\frac{\displaystyle\int_a^\beta \cos\theta\sin\theta\, d\theta}{\displaystyle\int_a^\beta \sin\theta\, d\theta} = \tfrac{1}{2}a\,\frac{\cos^2 a - \cos^2 \beta}{\cos a - \cos \beta}$$

$$= \tfrac{1}{2}a\,(\cos a + \cos \beta) = \tfrac{1}{2}\,(x_1 + x_2), \quad \text{...................(13)}$$

if a, β be the limits of θ, and x_1, x_2 the abscissæ of the bounding circles. Hence the mean centre of the zone is on the axis, half-way between the planes of the bounding circles.

For example, the mean centre of a hemispherical surface bisects the axial radius.

These results might also have been inferred immediately from the equality of area of corresponding zones on the sphere and on an enveloping cylinder (Art. 113, Ex. 1).

Ex. 4.   In the case of a solid circular cone, the origin being at the vertex, the section varies as x^2, so that

$$\bar{x} = \frac{\displaystyle\int_0^h x^3\, dx}{\displaystyle\int_0^h x^2\, dx} = \tfrac{3}{4}h, \qquad \text{.......................(14)}$$

if h be the altitude.

Ex. 5.   For the segment of an elliptic paraboloid

$$2x = \frac{y^2}{p} + \frac{z^2}{q} \qquad \text{.......... ....(15)}$$

cut off by a plane $x = h$, since the section varies as x, as in Art. 108, Ex. 1, we have

$$\bar{x} = \frac{\displaystyle\int_0^h x^2\, dx}{\displaystyle\int_0^h x\, dx} = \tfrac{2}{3}h. \qquad \text{...... ....(16)}$$

Ex. 6. For a hemisphere of radius a, putting $y^2 = a^2 - x^2$, we have

$$\bar{x} = \frac{\int_0^a x\,(a^2 - x^2)\,dx}{\int_0^a (a^2 - x^2)\,dx} = \tfrac{3}{8}a. \quad \dots\dots\dots\dots\dots(17)$$

The same formula gives the position of the mean centre of the half of the ellipsoid

$$\frac{x^2}{a^2} + \frac{y^2}{b^2} + \frac{z^2}{c^2} = 1, \quad \dots\dots\dots\dots\dots\dots(18)$$

which lies on the positive side of the plane yz, since $f(x)$ in this case also varies as $a^2 - x^2$. See Art. 108, Ex. 2.

117. Theorems of Pappus.

1°. If an arc of a plane curve revolve about an axis in its plane, not intersecting it, the *surface* generated is equal to the length of the *arc* multiplied by the length of the path of its mean centre.

Let the axis of x coincide with the axis of rotation, and let y be the ordinate of the generating curve. The surface generated in a complete revolution is, by Art. 113, equal to

$$2\pi \int y\,ds,$$

the integration extending over the arc. But if $\bar{y}$ refer to the mean centre of the arc, we have

$$\bar{y} = \frac{\int y\,ds}{\int ds},$$

by Art. 116. Hence

$$2\pi \int y\,ds = 2\pi\bar{y} \times \int ds, \quad \dots\dots\dots\dots\dots(1)$$

which is the theorem.

2°. If a plane area revolve about an axis in its plane, not intersecting it, the *volume* generated is equal to the *area* multiplied by the length of the path of its mean centre.

If δA be an element of the area, the volume generated in a complete revolution is

$$\lim \Sigma\,(2\pi y \,.\, \delta A).$$

But if $\bar{y}$ refer to the centre of mass of the area, we have

$$\bar{y} = \lim \frac{\Sigma\,(y\,\delta A)}{\Sigma\,(\delta A)},$$

by Art. 116.   Hence

$$\lim \Sigma \left(2\pi y \,.\, \delta A \right) = 2\pi \bar{y} \times \lim \Sigma \left(\delta A \right), \ldots\ldots\ldots\ldots(2)$$

which is the theorem *.

The revolutions have been taken to be complete, but the restriction is obviously unessential.

The theorems may be used, conversely, to find the mean centre of a plane arc, or of a plane area, when the surface, or the volume, generated by its revolution is known independently.   See Ex. 3, below.

Ex. 1.   The ring generated by the revolution of a circle of radius b about a line in its own plane at a distance a from its centre.

The surface is $2\pi b \times 2\pi a, \,=4\pi^2 ab$;

and the volume is $\pi b^2 \times 2\pi a, \,=2\pi^2 ab^2.$

Cf. Art. 107, Ex. 3, and Art. 113, Ex. 2.

Ex. 2.   A segment of the parabola $y^2 = 4ax$, bounded by the double ordinate $x = h$, revolves about this ordinate.

If $2k$ be the length of the double ordinate, the area of the segment is $\frac{4}{3} hk$, by Art. 100 ; and the distance of its mean centre from the ordinate is $\frac{2}{5} h$, by Art. 116, Ex. 2.   Hence the volume generated is

$$\tfrac{4}{3} hk \times \tfrac{4}{5} \pi h = \tfrac{16}{15} \pi h^2 k.$$

Ex. 3.   For a semicircular *arc* revolving about the diameter joining its extremities, we have

$$\pi a \times 2\pi \bar{y} = 4\pi a^2,$$

whence $$\bar{y} = \frac{2}{\pi} a.$$

Again, for a semicircular *area* revolving about its bounding diameter,

$$\tfrac{1}{2} \pi a^2 \times 2\pi \bar{y} = \tfrac{4}{3} \pi a^3,$$

whence $$\bar{y} = \frac{4}{3\pi} a.$$

A similar calculation leads to a simple formula for the volume of a prism or a cylinder (of any form of cross-section) bounded by plane ends.

In the first place we will suppose that one of the ends, which we will call the base, is perpendicular to the length.   Let P be any point of the base, and let z be the length of the ordinate PP'

---

* These theorems are contained in a treatise on Mechanics by Pappus, who flourished at Alexandria about A.D. 300.   They were given as new by Guldinus, *de centro gravitatis* (1635—1642).   (Ball, *History of Mathematics*.)

drawn parallel to the length, to meet the opposite end in P', and let $\bar{z}$ be the ordinate of the centre of the oblique end. If δA, $\delta A'$ be corresponding elements of area at P and P', we have

$$\bar{z} = \lim \frac{\Sigma(z.\delta A')}{\Sigma(\delta A')} = \lim \frac{\Sigma(z.\delta A)}{\Sigma(\delta A)},$$

since δA, being the orthogonal projection of $\delta A'$, is in a constant ratio to it. Hence the volume of the solid

$$= \Sigma(z.\delta A) = \bar{z} \times \Sigma(\delta A); \quad \dots\dots\dots\dots\dots(3)$$

that is, it is equal to the area of the base multiplied by the ordinate of the mean centre of the opposite face. It is easily seen that this is the same as the ordinate drawn through the mean centre of the base.

A prism or a cylinder with both ends oblique may be regarded as the sum or as the difference of two prisms or cylinders each having one end perpendicular to the length. We infer that in all cases the volume is equal to the area of the cross-section multiplied by the distance between the mean centres of the two ends.

Ex. 4. The volume of the wedge-shaped solid cut off from a right circular cylinder by a plane through the centre of the base, making an angle a with the plane of the base, is

$$\tfrac{1}{2}\pi a^2 \times \frac{4}{3\pi} a \tan a = \tfrac{2}{3}a^3 \tan a;$$

cf. Art. 118, Ex. 1.

The theorems of Pappus may be generalized in various ways; but it may be sufficient here to state the following extension of the second theorem.

If a plane area, constant or continuously variable, move about in any manner in space, but so that consecutive positions of the plane do not intersect within the area, the volume generated is equal to

$$\int S d\sigma, \quad \dots\dots\dots\dots\dots\dots\dots(4)$$

where S is the area, and $d\sigma$ is the projection of an element of the locus of the mean centre of the area on the normal to the plane. If ds denote an element of this locus, and θ the angle between ds and the normal to the plane, the formula may also be written

$$\int S \cos \theta\, ds. \quad \dots\dots\dots\dots\dots\dots(5)$$

This theorem is the three-dimensional analogue of the proposition of Art. 103, relating to the area swept over by a moving line. It is a simple corollary from the theorem above proved.

118. Multiple Integrals.

This book deals mainly with functions of a single variable, and therefore, as regards the Integral Calculus, with problems which depend, or can be made to depend, upon a single integration. Multiple integrals occur however so frequently, as a matter of notation, in the physical applications of the subject, that it may be useful to give here a few explanations concerning them. We shall pass very lightly over theoretical points; what is wanting in this respect may be supplied by a proper adaptation of the method of Art. 90.

Let z be a continuous and single-valued function of the independent variables x, y; say

$$z = \phi(x, y). \quad\dots\dots\dots\dots\dots\dots\dots\dots(1)$$

This may be interpreted, geometrically, as the equation of a surface (Art. 34). Take any finite region S in the plane xy, and let a cylindrical surface be generated by a straight line which meets always the perimeter of S, and is parallel to the axis of z. We consider the volume (V) included between this cylinder, the plane xy, and the surface (1). See Fig. 68.

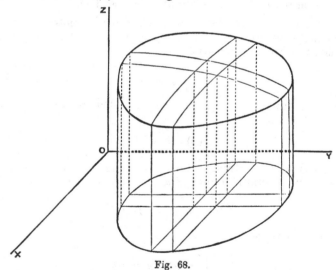

Fig. 68.

If the region S be divided into elements of area δA_1, δA_2, δA_3, ..., and if z_1, z_2, z_3, ... be ordinates of the surface (1) at arbitrarily chosen points within these elements, then, the coordinate axes being supposed rectangular, the sum

$$z_1 \delta A_1 + z_2 \delta A_2 + z_3 \delta A_3 + \dots \quad\dots\dots\dots\dots\dots(2)$$

will give us the total volume of a system of prisms, of altitudes $z_1, z_2, z_3, \ldots$, standing on the bases $\delta A_1, \delta A_2, \delta A_3, \ldots$. And if the function $\phi(x, y)$ be subject to certain generally satisfied conditions*, the above sum will, when the dimensions of $\delta A_1, \delta A_2, \delta A_3, \ldots$ are taken infinitely small, tend to a unique limiting value, viz. the aforesaid volume V.

If the subdivision of the region S be made by lines drawn parallel to the axes of x and y, the elements $\delta A_1, \delta A_2, \delta A_3, \ldots$ are rectangular areas of the type $\delta x \delta y$, and the sum (2) may be denoted by

$$\Sigma\Sigma z \delta x \delta y, \quad\ldots\ldots\ldots\ldots\ldots\ldots\ldots(3)$$

where the sign Σ is duplicated because the summation is in two dimensions. The limiting value of this sum is denoted by

$$\iint z\,dx\,dy, \quad\ldots\ldots\ldots\ldots\ldots\ldots\ldots(4)$$

and we have the formula

$$V = \iint \phi(x, y)\,dx\,dy. \quad\ldots\ldots\ldots\ldots\ldots\ldots(5)$$

The expression on the right hand is called a 'double integral'; it is of course not determinate unless the range of the variables x, y, as limited by the boundary of S, be specified.

The volume V may, however, be obtained in another way. If $f(x)$ denote the area of a section by a plane parallel to yz, whose abscissa is x, we have, by Art. 106,

$$V = \int_a^b f(x)\,dx, \quad\ldots\ldots\ldots\ldots\ldots\ldots(6)$$

where a, b are the extreme values of x belonging to the area S. But, by Art. 100,

$$f(x) = \int_\alpha^\beta z\,dy, \quad\ldots\ldots\ldots\ldots\ldots\ldots(7)$$

where α, β are the extreme values of y in the section $f(x)$, and are therefore in general functions of x. Hence we have

$$V = \int_a^b \left\{ \int_\alpha^\beta \phi(x, y)\,dy \right\} dx, \quad\ldots\ldots\ldots\ldots(8)$$

or, as it is more usually written,

$$V = \int_a^b \int_\alpha^\beta \phi(x, y)\,dx\,dy. \quad\ldots\ldots\ldots\ldots\ldots(9)\dagger$$

---

* The already stipulated condition of continuity is sufficient, but the proof is simplified if we introduce the additional condition that $\phi(x, y)$ shall have only a finite number of maxima and minima within any finite area of the plane xy. Cf. Art. 90.

$\dagger$ The first $\int$ refers to the dx, and the second to the dy. There is not absolute uniformity of usage, however, on this point.

If the limits of *both* integrations be constants, *i.e.* if the region S take the form of a rectangle having its sides parallel to x and y, the volume V is expressed also by

$$\int_a^\beta \left\{ \int_a^b \phi(x, y)\, dx \right\} dy, \quad\dots\dots\dots\dots(10)$$

and we may assert that

$$\int_a^b \int_a^\beta \phi(x, y)\, dx\, dy = \int_a^\beta \int_a^b \phi(x, y)\, dy\, dx. \dots\dots\dots(11)$$

This is illustrated by Fig. 23, p. 63. In other cases the limits of the respective integrations require to be adjusted when we invert the order.

The above explanations have been clothed in a geometrical form, but this is not of the essence of the matter. The same principles are involved, for example, in the calculation of the mass of a plane lamina, having given the density at any point (x, y) of it, and in many other physical problems.

Another mode of decomposition of the area S is often useful. Taking polar coordinates r, θ in the plane xy, we may divide the area into quasi-rectangular elements by means of concentric circles and radii. The area of any one of these elements may be denoted by $r\delta\theta \cdot \delta r$, if r be the arithmetic mean of the radii of the two curved sides. The formula (8) is then replaced by

$$V = \iint z r\, d\theta\, dr, \dots\dots\dots\dots\dots\dots(12)$$

where z is supposed given as a function of r and θ.

After what precedes, the meaning of a 'triple-integral,'

$$\iiint \phi(x, y, z)\, dx\, dy\, dz, \quad\dots\dots\dots\dots(13)$$

will not require much development. If a finite region R be divided into rectangular elements $\delta x \delta y \delta z$ by planes parallel to the coordinate axes, and if we multiply the volume of each of these elements by the value of the function $\phi(x, y, z)$ at some arbitrarily chosen point of it, the expression (13) is used to denote the limiting value to which (under certain conditions) the sum of the products tends when the dimensions of the elements are infinitely small. The same limiting value may be obtained by a succession of three simple integrations, thus

$$\int_a^b \left\{ \int_a^\beta \left(\int_a^b \phi(x, y, z)\, dz \right) dy \right\} dx, \quad\dots\dots\dots(14)$$

where the integration is with respect to z between the limits a and b, which are in general functions of x and y, then with respect to

y between the limits α and β, which are in general functions of x, and finally with respect to x between the limits a and b. If the limits of integration be all constants, they are unchanged when the order of integration is varied.

As an example, consider the determination of the mass of a solid, whose density is a function of x, y, z.

Ex. 1. To find the volume of the wedge included between the plane $z = 0$, the cylinder

$$x^2 + y^2 = a^2, \qquad \ldots\ldots\ldots\ldots\ldots\ldots\ldots(15)$$

and the part of the plane $z = \tan \alpha$ for which z is positive.

We have $\qquad \iint z\,dx\,dy = \tan \alpha \int_0^a \int_{-\sqrt{(a^2-x^2)}}^{\sqrt{(a^2-x^2)}} x\,dx\,dy. \quad \ldots\ldots\ldots\ldots(16)$

The integration with respect to y gives

$$\left[xy \right]_{-\sqrt{(a^2-x^2)}}^{\sqrt{(a^2-x^2)}} = 2x \sqrt{(a^2 - x^2)}.$$

We then have

$$\int_0^a 2x \sqrt{(a^2 - x^2)}\, dx = \left[-\tfrac{2}{3} (a^2 - x^2)^{\frac{3}{2}} \right]_0^a = \tfrac{2}{3} a^3.$$

The required volume is therefore

$$\tfrac{2}{3} a^3 \tan \alpha. \qquad \ldots\ldots\ldots\ldots\ldots\ldots(17)$$

Ex. 2. To find the volume included within the sphere

$$x^2 + y^2 + z^2 = a^2, \qquad \ldots\ldots\ldots\ldots\ldots(18)$$

by the cylinder $\qquad x^2 + y^2 = ax. \qquad \ldots\ldots\ldots\ldots\ldots\ldots(19)$

(The cylinder has a radius half that of the sphere, and its axis bisects at right angles a radius of the sphere.)

If we introduce polar coordinates in the plane xy, the equation (19) takes the form

$$r = a \cos \theta, \qquad \ldots\ldots\ldots\ldots\ldots\ldots(20)$$

and (18) gives $\qquad z = \pm \sqrt{(a^2 - r^2)}. \qquad \ldots\ldots\ldots\ldots\ldots\ldots(21)$

The required volume is therefore given by

$$2 \int_{-\frac{1}{2}\pi}^{\frac{1}{2}\pi} \int_0^{a \cos \theta} z r\,d\theta\,dr = 4 \int_0^{\frac{1}{2}\pi} \int_0^{a \cos \theta} \sqrt{(a^2 - r^2)}\, r\,d\theta\,dr. \quad \ldots(22)$$

Now

$$\int_0^{a \cos \theta} \sqrt{(a^2 - r^2)}\, r\,dr = \left[-\tfrac{1}{3} (a^2 - r^2)^{\frac{3}{2}} \right]_0^{a \cos \theta} = \tfrac{1}{3} a^3 (1 - \sin^3 \theta),$$

and $\qquad \int_0^{\frac{1}{2}\pi} (1 - \sin^3 \theta)\, d\theta = \tfrac{1}{2}\pi - \tfrac{2}{3}.$

The final result is $\qquad \tfrac{2}{3} \pi a^3 - \tfrac{8}{9} a^3. \ldots\ldots\ldots\ldots\ldots\ldots\ldots(23)$

EXAMPLES*. XXXVI.

(Areas.)

1. If a curve be such that

$$y^m \propto x^n,$$

the rectangle enclosed between the coordinate axes and lines drawn parallel to them through any point on the curve is divided by the curve into two portions whose areas are as $m : n$.

2. The area included between the axis of x and one semi-undulation of the curve

$$y = b \sin x/a$$

is $2ab$.

3. The area included between the catenary

$$y = c \cosh x/c,$$

the axis of x, and the lines $x = 0$, $x = x_1$, is

$$c^2 \sinh x_1/c.$$

4. The curve $\qquad a^2 y = x^2 (x + a)$

includes, with the axis of x, an area $\frac{1}{12} a^2$.

5. The areas included between the axis of x and successive semi-undulations of the curve

$$y = e^{-ax} \sin \beta x$$

form a descending geometric series, the common ratio being $e^{-\pi a/\beta}$.

6. The area included between the axis of x and the parabola

$$cy = (x - a) (x - b)$$

is $\frac{1}{6} (a - b)^2/c$.

7. Find the area included between the two curves

$$2y^2 - 3y = x - 1, \quad y^2 - 2y = x - 3. \qquad \left[\tfrac{9}{2}.\right]$$

8. Find the area included between the two parabolas

$$y = 3x^2 - x - 3, \quad y = -2x^2 + 4x + 7. \qquad \left[\tfrac{4\cdot 6}{2}.\right]$$

9. Find the area of the segment of the parabola

$$y = x^2 - 7x + 9$$

cut off by the straight line

$$y = 3 - 2x. \qquad \left[\tfrac{1}{6}.\right]$$

10. The area included between the two parabolas

$$y^2 = 4a (x + a), \quad y^2 = 4b (b - x)$$

is $\frac{8}{3} (a + b) \sqrt{(ab)}$.

---

* Some further Examples for practice will be found at the end of the Chapter (IX) on 'Special Curves.'

11. Prove that the whole area (when finite) included between the axis of x and the curve

$$y = \frac{1}{a} \, \phi \left(\frac{x}{a} \right)$$

is independent of the value of a.

12. The area included between the positive branch of the curve

$$y = b \tanh x/a,$$

its asymptote, and the axis of y, is $ab \log 2$.   (See Fig. 26, p. 81.)

13. The area common to the two ellipses

$$\frac{x^2}{a^2} + \frac{y^2}{b^2} = 1, \quad \frac{x^2}{b^2} + \frac{y^2}{a^2} = 1$$

is $4ab \tan^{-1} b/a$.

14. The area included between the coordinate axes and the parabola

$$\left(\frac{x}{a} \right)^{\frac{1}{2}} + \left(\frac{y}{b} \right)^{\frac{1}{2}} = 1$$

is $\frac{1}{6} ab \sin \omega$, where ω is the inclination of the axes.

$$[\text{Put } x = a \sin^4 \theta, \quad y = b \cos^4 \theta.]$$

15. The area between the parabola

$$2cy = x^2 + a^2$$

and the two tangents drawn to it from the origin is $\frac{1}{3} a^3/c$.

16. The area common to the two parabolas

$$y^2 = 4ax, \quad x^2 = 4ay$$

is $\frac{16}{3} a^2$.

17. Prove by integration that the area of an ellipse is

$$\pi a \beta \sin \omega,$$

where a, β are the lengths of any pair of conjugate semi-diameters, and ω is the angle between these.

18. The formula (Art. 80 (2)) for integration by parts may be written

$$\int u dv = uv - \int v du \, ;$$

interpret this geometrically in terms of areas.

19. A curve AB is traced on a lamina which turns in its own plane about a fixed point O through an angle θ.   Prove that the area swept over by the curve is

$$\tfrac{1}{2} \, (OA^2 \sim OB^2) \, \theta.$$

20. Trace the curve

$$r = 3 + 2 \cos \theta,$$

and find its area. $[11\pi.]$

21. Find the area of that portion of the curve

$$r = 2a (1 + \cos \theta)$$

which lies outside the parabola

$$r = \frac{2a}{1 + \cos \theta}. \qquad [3\pi a^2 + \tfrac{16}{3} a^2.]$$

22. Prove by transformation to polar coordinates that the area of the ellipse

$$Ax^2 + 2Hxy + By^2 = 1$$

is $\pi / \sqrt{(AB - H^2)}$.

23. A weightless string of length l, attached to a fixed point O, passes through a small ring which can slide along a horizontal rod AB in the same vertical plane with O, and the lower portion hangs vertically, carrying a small weight P. Find the locus of P, and prove that the area between this locus and AB is

$$l \sqrt{(l^2 - h^2)} - h^2 \cosh^{-1} l/h,$$

where h is the depth of AB below O.

24. Prove directly from geometrical considerations that the area included between two focal radii of a parabola and the curve is half that included between the curve, the corresponding perpendiculars on the directrix, and the directrix.

25. What is indicated by the record of the wheel in Amsler's Planimeter when the bar PQ (Fig. 59) makes a complete revolution whilst the point Q traces out the closed curve?

26. In a certain form of planimeter the arm carrying the tracing point is pivoted at the other end on a vertical axis carried by a small waggon which can roll (without slipping) backwards and forwards over the paper, and has a recording wheel attached to it, to measure the rolling. Prove that when the tracing point describes a closed curve, the record gives the area, on a certain scale.

27. If S_A, S_B, S_C be the areas of the closed curves described by three points A, B, C on a bar which moves in one plane, and returns to its original position without performing a complete revolution, prove that

$$BC . S_A + CA . S_B + AB . S_C = 0,$$

where the lines BC, CA, AB have signs attributed to them according to their directions, and the signs of S_A, S_B, S_C are determined by the rule of Art. 101.

28. If P be a point on a bar AB which moves in one plane, and returns to its original position after accomplishing one revolution, prove that

$$S_P = \frac{aS_B + bS_A}{a + b} - \pi ab,$$

where $a = AP$, $b = PB$, and the meanings of S_A, S_B, S_P are as in the preceding question.

Hence shew that if the extremities A, B of the bar move on a closed oval curve

$$S_A - S_P = \pi ab.$$ (Holditch.)

29. If a straight line AB of constant length move with its extremities on two fixed intersecting straight lines, any point P on it describes an ellipse of area $\pi \cdot AP \cdot PB$.

EXAMPLES*. XXXVII.

(Volumes.)

1. The volume generated by the revolution of one semi-undulation of the curve

$$y = b \sin x/a$$

about the axis of x is one-half that of the circumscribing cylinder.

2. The volume of a frustum of any cone, with parallel ends, is

$$\tfrac{1}{3} h \{A_1 + \sqrt{(A_1 A_2)} + A_2\},$$

where A_1, A_2 are the areas of the two ends, and h is the perpendicular distance between them.

3. In the solid generated by the revolution of the rectangular hyperbola

$$x^2 - y^2 = a^2$$

about the axis of x, the volume of a segment of height a, measured from the vertex, is equal to that of a sphere of radius a.

4. The volume of a segment of a sphere bounded by two parallel planes at a distance h apart exceeds that of a cylinder of height h and sectional area equal to the arithmetic mean of the areas of the plane ends, by the volume of a sphere of diameter h.

5. A plane is drawn parallel to the base of a hemisphere of radius a at a distance $2a \sin 10°$ from the base. Prove that it bisects the volume of the hemisphere.

6. The portion of a solid sphere of radius a which is included within a spherical surface of radius b $(< 2a)$, having its centre on the surface of the sphere, is removed. Prove that the volume of the cavity is less than that of a hemisphere of radius b by $\tfrac{1}{4}\pi b^4/a$.

7. The volume generated by the revolution about the axis of x of the area included between that axis and the parabola

$$cy = (x - a)(x - b)$$

is $$\tfrac{1}{30}\pi (a - b)^5/c^2.$$

* See the footnote on p. 273.

8. If a segment of a parabola revolve about the ordinate, the volume generated is $\frac{8}{15}$ of that of the circumscribing cylinder.

9. The volume of the solid generated by the revolution of a parabola about the tangent at the vertex is $\frac{1}{5}$ that of the circumscribing cylinder.

10. The segment of the parabola $y^2 = 4ax$ which is cut off by the latus-rectum revolves about the directrix. Prove that the volume of the annular solid generated is $\frac{152}{15}\pi a^3$.

11. The segment cut off from the curve

$$ay^2 = x^3$$

by the chord $x = h$ revolves about the axis of x. Prove that the volume generated is one-fourth that of a cylinder of height h on the same base.

12. The volume of a frustum of a triangular prism cut off by any two planes is

$$\tfrac{1}{3}\left(h_1 + h_2 + h_3\right)A,$$

where h_1, h_2, h_3 are the lengths of the three parallel edges, and A is the area of the section perpendicular to these edges.

13. If b be the radius of the middle section of a cask, and a the radius of either end, prove that the volume of the cask is

$$\tfrac{1}{15}\pi\left(3a^2 + 4ab + 8b^2\right)h,$$

where h is the length, it being assumed that the generating curve is an arc of a parabola.

14. An arc of a circle revolves about its chord; prove that the volume of the solid generated is

$$\tfrac{4}{3}\pi a^3 \sin a + \tfrac{2}{3}\pi a^3 \sin a \cos^2 a - 2\pi a^3 a \cos a,$$

where a is the radius, and $2a$ is the angular measure of the arc.

15. The figure bounded by a quadrant of a circle of radius a, and the tangents at its extremities, revolves about one of these tangents; prove that the volume of the solid thus generated is

$$\left(\frac{5}{3} - \frac{\pi}{2}\right)\pi a^3.$$

16. The volume enclosed by two right circular cylinders of equal radius a, whose axes intersect at right angles, is $\frac{16}{3}a^3$.

If the axes intersect at an angle a, the volume is $\frac{16}{3}a^3 \operatorname{cosec} a$.

17. If the hyperbola

$$\frac{x^2}{a^2} - \frac{y^2}{b^2} = 1$$

revolve about the axis of x, the volume included between the surface thus generated, the cone generated by the asymptotes, and two planes perpendicular to x, at a distance h apart, is equal to that of a circular cylinder of height h and radius b.

18. A right circular cone of semi-angle a has its vertex on the surface of a sphere of radius a, and its axis passes through the centre. Prove that the volume of the portion of the sphere which is exterior to the cone is $\frac{4}{3}\pi a^3 \cos^4 a$.

19. If in Simpson's method (Art. 109) of estimating the volume included between two parallel sections S_1, S_3 the intermediate section S_2 is at unequal distances h, k from S_1, S_3, respectively, the formula is

$$\frac{h+k}{6hk}\{(2h-k)\,kS_1 + (h+k)^2\,S_2 + (2k-h)\,hS_3\}.$$

EXAMPLES*. XXXVIII.

(Curved Lines and Surfaces.)

1. The length of a complete undulation of the curve of sines

$$y = b \sin x/a$$

is equal to the perimeter of an ellipse whose semi-axes are $\sqrt{(a^2 + b^2)}$ and a.

2. Prove the following formula for the length of the perpendicular (p) from the origin on any tangent to a curve :

$$p = x\frac{dy}{ds} - y\frac{dx}{ds}.$$

Also prove that the orthogonal projection of the radius vector on the tangent is

$$x\frac{dx}{ds} + y\frac{dy}{ds} \quad\text{or}\quad r\frac{dr}{ds}.$$

3. The surface generated by the revolution about the directrix of an arc of the catenary

$$y = c \cosh x/c,$$

commencing at the vertex, is

$$\pi\,(cx + ys),$$

where x, y, s refer to the extremity of the arc.

4. The curved surface cut off from a paraboloid of revolution by a plane perpendicular to the axis is

$$\tfrac{1}{6}\frac{\pi b}{h^2}\{(4h^2 + b^2)^{\frac{3}{2}} - b^3\},$$

where h is the length of the axis, and b the radius of the bounding circle.

5. The curved surface generated by the revolution about the axis of x of the portion of the parabola $y^2 = 4ax$ included between the origin and the ordinate $x = 3a$ is $\tfrac{56}{3}\pi a^2$.

---

* See the footnote on p. 273.

6. The segment of a parabola included between the vertex and the latus-rectum revolves about the axis; prove that the curved surface of the figure generated is 1·219 times the area of its base.

7. A circular arc revolves about its chord; prove that the surface generated is

$$4\pi a^2 (\sin a - a \cos a),$$

where a is the radius, and $2a$ the angular measure of the arc.

8. A quadrant of a circle of radius a revolves about the tangent at one extremity; prove that the area of the curved surface generated is $\pi (\pi - 2) a^2$.

9. A variable sphere of radius r is described with its centre on the surface of a fixed sphere of radius a; prove that the area of its surface intercepted by the fixed sphere is a maximum when $r = \frac{4}{3} a$.

10. A tangent cone is drawn to a sphere, and with the vertex of the cone as centre two spherical surfaces are described cutting both the sphere and the cone. Prove that the areas of the zones intercepted on the sphere and on the cone are equal.

EXAMPLES. XXXIX.

(Approximate Quadrature. Mean Values.)

1. Apply Simpson's rule to calculate $\log_e 2$ from the formula

$$\log_e 2 = \int_0^1 \frac{dx}{1 + x}.$$

[The correct value is $\log_e 2 = ·693147....$]

2. Calculate the value of π from the formula

$$\tfrac{1}{6}\pi = \int_0^{\frac{1}{2}} \frac{dx}{\sqrt{(1 - x^2)}}.$$

3. If in Simpson's method with three ordinates (Art. 114) the middle ordinate y_2 is at unequal distances h, k from y_1, y_3, respectively, the formula is

$$\tfrac{1}{6} (h + k) (y_1 + 4y_2 + y_3) + \tfrac{1}{6} (h^2 - k^2) \left(\frac{y_1 - y_2}{h} - \frac{y_2 - y_3}{k} \right).$$

4. The mean of the squares on the diameters of an ellipse, drawn at equal angular intervals, is equal to the rectangle contained by the major and minor axes.

5. A point is taken at random on a straight line of length a; prove that the mean area of the rectangle contained by the two segments is $\frac{1}{6} a^2$, and that the mean value of the sum of the squares on the two segments is $\frac{2}{3} a^2$.

6. If a point move with constant acceleration, the mean square of the velocities at equal infinitely small intervals of time is equal to

$$\tfrac{1}{3} (v_0^2 + v_0 v_1 + v_1^2),$$

where v_0, v_1 are the initial and final velocities.

7. Prove that in simple-harmonic motion the mean kinetic energy is one-half the maximum kinetic energy.

8. The mean horizontal range of a particle projected with given velocity, but arbitrary elevation, is ·6366 of the maximum range.

9. The mean of the focal radii of an ellipse, drawn at equal angular intervals, is equal to the semi-minor axis.

10. The mean distance of points on the curved surface of a hemisphere from the plane of the base is one-half the radius.

11. The mean distance of points of a hemispherical surface of radius a from the pole of the hemisphere is ·9429a.

12. Find the mean values of the reciprocals of the distances of the points of a circular area of radius a from the centre, and from a point of the circumference.   [$2/a$, $4/\pi a$.]

13. A rod has the form of a very elongate prolate ellipsoid of revolution, prove that its mean sectional area is two-thirds that at the centre.

14. The surface-density on an electrified circular disk of radius a varies as $(a^2 - r^2)^{-\frac{1}{2}}$, where r denotes distance from the centre. Find the ratio of the average density to the density at the centre.

15. If the orbits of comets were uniformly distributed through space, their mean inclination to the ecliptic would be equal to the radian (57·296°).

16. The mean distance of the points of a spherical surface of radius a from a point P at a distance c from the centre is

$$c + \tfrac{1}{3} \frac{a^2}{c} \quad \text{or} \quad a + \tfrac{1}{3} \frac{c^2}{a},$$

according as P is external or internal.

17. The mean distance of points on the circumference of a circle of radius a from a fixed point on the circumference is $1 \cdot 273a$.

18. The mean distance of points within a circular area of radius a from a fixed point on the circumference is $1 \cdot 132a$.

19. The mean distance of points within a sphere from a given point on the surface is $\tfrac{6}{5} a$.

20. If the density at a distance r from the centre of the Earth be given by the formula

$$\rho = \rho_0 \frac{\sin kr}{kr},$$

where k is a constant, prove that the mean density is

$$3\rho_0 \frac{\sin ka - ka \cos ka}{k^3 a^3},$$

a denoting the Earth's radius.

21. If, in a spherical mass whose density ρ is a function of the distance (r) from the centre, D denote the mean density of the matter included within a concentric sphere of radius r, then

$$\rho = D + \tfrac{1}{3} r \frac{dD}{dr}.$$

EXAMPLES. XL.

(Mean Centres.)

1. Prove by integration that the mean centre of a trapezium divides the line joining the middle points of the parallel sides in the ratio $2a + b : a + 2b$, where a, b are the lengths of the parallel sides.

2. The mean centre of the area included between one semi-undulation of the curve

$$y = b \sin x/a$$

and the axis of x is at a distance $\tfrac{1}{8}\pi b$ from this axis.

3. The mean centre of the area included between the curve

$$y = \frac{a^3}{a^2 + x^2}$$

and the axis of x is at the point $(0, \tfrac{1}{4} a)$.

4. Prove that the mean centre of the area of the circular spandril formed by a quadrant of a circle and the tangents at its extremities is at a distance $\cdot 2234 a$ from either tangent, a being the radius.

5. The mean centre of the area included between the coordinate axes and the parabola

$$\left(\frac{x}{a}\right)^{\frac{1}{2}} + \left(\frac{y}{b}\right)^{\frac{1}{2}} = 1$$

is at the point $(\tfrac{1}{5} a, \tfrac{1}{5} b)$.   [Put $x = a \sin^4 \theta$, $y = b \cos^4 \theta$.]

6. The distances from the centre of a sphere of radius a of the mean centres of the two segments into which it is divided by a plane at a distance c from the centre of figure are

$$\frac{3}{4} \frac{(a \pm c)^2}{2a \pm c}.$$

7. The figure formed by a quadrant of a circle of radius a and the tangents at its extremities revolves about one of these tangents; prove that the distance of the mean centre of the solid thus generated from the vertex is $\cdot 869a$.

8. A solid ogival shot has the form produced by rotating a portion APN of a parabolic area, where A is the vertex, and PN an ordinate, about PN; prove that the mean centre divides the axis in the ratio $5:11$.

9. AP is an arc of a parabola beginning at the vertex, and PN is a perpendicular on the tangent at the vertex; prove that the mean centre of the solid generated by the revolution of the figure APN about AN is at a distance from A equal to $\frac{5}{6}AN$.

10. The mean centre of the volume included between two equal circular cylinders, whose axes meet at right angles, and the plane of these axes, is at a distance from this plane equal to $\frac{3}{8}$ of the common radius.

11. A quadrant of a circle revolves about the tangent at one extremity; prove that the distance of the mean centre of the curved surface generated, from the vertex, is $\cdot 876a$.

12. The mean centre of either half of the surface of an anchorring cut off by the equatorial plane is at a distance $2b/\pi$ from this plane, where b is the radius of the generating circle.

13. The mean centre of either half of the volume of an anchorring cut off by the equatorial plane is at a distance $4b/3\pi$ from the plane, where b is the radius of the generating circle.

14. If the ellipse $\qquad \dfrac{x^2}{a^2} + \dfrac{y^2}{b^2} = 1$

revolve about the axis of x, the mean centre of the curved surface generated by either of the two halves into which the curve is divided by the axis of y is at a distance

$$\frac{2}{3} \cdot \frac{a^2 + ab + b^2}{a + b} \cdot \frac{a}{b + a\,(\sin^{-1} e)/e}$$

from the centre, where e is the eccentricity, it being supposed that $b < a$.

Obtain the corresponding result when $b > a$.

15. Apply the theorems of Pappus to find the volume and the curved surface of a right circular cone, and of a frustum of such a cone.

16. A groove of semicircular section, of radius b, is cut round a cylinder of radius a; prove that the volume removed is

$$\pi^2 ab^2 - \tfrac{4}{3}\pi b^3.$$

Also that the surface of the groove is

$$2\pi^2 ab - 4\pi b^2.$$

17. A screw-thread of rectangular section is cut on a cylinder of radius R. Prove that the volume of one turn of the thread is $2\pi abR + \pi ab^2$, where a, b are the sides of the rectangle, a being that side which is at right angles to the surface of the cylinder.

18. If a straight line be drawn through the mean centre of the perimeter of a closed curve, the surfaces generated by the revolution about this line of the two portions into which the perimeter is divided will be equal.

19. If an area A, revolving about the axes of x and y, generates volumes U and V, respectively, find the area generated when it revolves about the line

$$x \cos a + y \sin a = p,$$

assuming that this line does not cut the area.

EXAMPLES. XLI.

(Multiple Integrals.)

1. Find the values of the integrals

$$\iint (x^2 + y^2)\, dx\, dy, \quad \iint \sqrt{\left(1 - \frac{x^2}{a^2} - \frac{y^2}{b^2}\right)}\, dx\, dy,$$

taken over the area of the ellipse

$$\frac{x^2}{a^2} + \frac{y^2}{b^2} = 1.$$

2. Prove that the volume enclosed by the cylinders

$$x^2 + y^2 = 2ax, \quad z^2 = 2ax$$

is $\frac{128}{15}a^3$.

3. A sphere of radius $2a$ is described with its centre on the surface of a cylinder of radius a; prove that the area of that portion of the surface of the cylinder which is within the sphere is $16a^2$.

4. The volume included between the elliptic paraboloid

$$2z = \frac{x^2}{p} + \frac{y^2}{q},$$

the cylinder $x^2 + y^2 = a^2$, and the plane $z = 0$, is

$$\frac{\pi a^4 (p + q)}{8pq}.$$

CHAPTER IX

SPECIAL CURVES

119. Algebraic Curves with an Axis of Symmetry.

The method of tracing algebraic curves of the type

$$y = f(x), \quad \dots\dots\dots\dots\dots\dots(1)$$

where $f(x)$ is a rational function, including the determination of asymptotes, maximum and minimum ordinates, and points of inflexion, has been illustrated in various parts of this book; see Arts. 13, 14, 51, 67.

The study of algebraic curves in general is beyond our limits, but a little space may be devoted to the discussion of curves of the type

$$y^2 = f(x). \quad \dots\dots\dots\dots\dots\dots(2)$$

Two points of novelty here present themselves. Since the equation gives two equal, but oppositely-signed, values of y for every value of x, the curve will be symmetrical with respect to the axis of x; also since y^2 must be positive, there can be no real part of the curve within those ranges of x (if any) for which $f(x)$ is negative.

Thus if $f(x)$ contain a *simple* factor $x - x_1$, so that the equation is of the form

$$y^2 = (x - x_1)\,\phi\,(x), \quad \dots\dots\dots\dots\dots(3)$$

the right-hand member will change sign as x passes through the value x_1. Hence on one side of the point $(x_1, 0)$ the ordinate is imaginary.

Also, we have, at this point,

$$\left(\frac{\delta y}{\delta x}\right)^2 = \frac{y^2}{(x - x_1)^2} = \frac{\phi\,(x)}{x - x_1},$$

and therefore, $dy/dx = \infty$. The tangent is therefore perpendicular to Ox.

If, on the other hand, $f(x)$ contain a *double* factor, say

$$y^2 = (x - x_1)^2\,\phi\,(x), \quad \dots\dots\dots\dots\dots(4)$$

the right-hand side does not change sign as x passes through the value x_1. Hence the ordinate is real on both sides of the point $(x_1, 0)$, or imaginary on both sides. In the former case we have two branches of the curve intersecting at an angle and forming what is called a 'node'; in the latter case $(x_1, 0)$ is an isolated or 'conjugate' point on the locus. The directions of the tangent-lines at the node are given by

$$\left(\frac{dy}{dx}\right)^2 = \lim_{x \to x_1} \frac{y^2}{(x - x_1)^2} = \phi(x_1).$$

If $f(x)$ contain a *triple* factor, say

$$y^2 = (x - x_1)^3 \phi(x), \quad \dots\dots\dots\dots\dots\dots(5)$$

the right-hand side changes sign at the point $(x_1, 0)$; the curve is therefore imaginary on one side of this point. Also since dy/dx here $= 0$, the curve touches the axis of x.

We proceed to some examples, beginning with cases where $f(x)$ is integral as well as rational.

Ex. 1.   In the cases where $f(x)$ is of the first or second degree, say

$$y^2 = Ax + B, \quad y^2 = Ax^2 + Bx + C, \quad \dots\dots\dots\dots(6)$$

the curve is a conic having the axis of x as a principal axis.

Ex. 2.   The cubic curves

$$y^2 = Ax^3 + Bx^2 + Cx + D \quad \dots\dots\dots\dots\dots(7)$$

include some interesting varieties.

(*a*)   If the linear factors of the right-hand side be real and distinct, we may write

$$ay^2 = (x - a)(x - \beta)(x - \gamma), \quad \dots\dots\dots\dots\dots(8)$$

and there is no loss of generality in supposing that a is positive and $a < \beta < \gamma$. The ordinates are then imaginary for $x < a$, and for $\beta < x < \gamma$. Between $(a, 0)$ and $(\beta, 0)$ there is a maximum value of y^2. The curve consists therefore of a closed oval, and of an infinite branch. For large values of x we have

$$\frac{y^2}{x^2} = \frac{x}{a}\left(1 - \frac{a}{x}\right)\left(1 - \frac{\beta}{x}\right)\left(1 - \frac{\gamma}{x}\right),$$

so that the curve tends to become more and more nearly perpendicular to the axis of x.

(*b*)   If the expression on the right-hand of (7) has only one real factor, we may write

$$ay^2 = (x - a)(x^2 + px + q), \quad \dots\dots\dots\dots\dots(9)$$

where $p^2 < 4q$. The curve then only meets the axis of x once.

(c) The transition from the form (8) to the form (9) may be imagined to take place in two ways. In the first way, the intermediate critical case is marked by the coalescence of the two greater of the quantities a, β, γ, so that

$$ay^2 = (x - a)(x - \beta)^2. \quad \dotfill (10)$$

Here y is imaginary for $x < a$, and real for $x > a$, but vanishes for $x = \beta$ The point $(\beta, 0)$ is here a node; it may be regarded as due to the union of the oval in the former case with the infinite branch.

(d) If, however, the two smaller of the quantities a, β, γ coalesce, so that

$$ay^2 = (x - a)^2(x - \gamma), \quad \dotfill (11)$$

y will be imaginary for $x < \gamma$, except for $x = a$, when it vanishes. The point $(a, 0)$ is therefore an isolated point; it may be regarded as due to the evanescence of the oval in the first case.

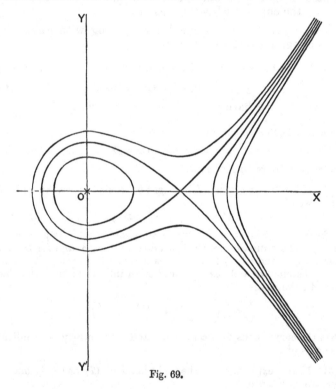

Fig. 69.

All these cases are illustrated in Fig. 69. Beginning on the right we have a curve of the type (9), consisting of a single infinite branch.

Next to it comes the case of an infinite branch associated with an isolated point (at O), the equation being of the type (11). Next in order comes an infinite branch, and with it an oval surrounding the point O; the equation is now of the type (8). In the next stage, the oval and the infinite branch have united to form a node with a loop, the corresponding type of equation being (10). Finally, we have a single branch passing outside the loop in the last case; the equation is again of the type (9)*.

(e) In the very special case where all three quantities a, β, γ, in (8), coincide, so that

$$ay^2 = (x - a)^3, \quad \dots\dots\dots \quad \dots\dots\dots\dots(12)$$

the curve is known as the 'semi-cubical parabola.' It has a 'cusp' at $(a, 0)$; this may be regarded as an extreme form of a node, due to the evanescence of the loop. See Fig. 70, where $a = 0$.

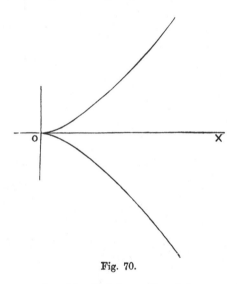

Fig. 70.

If, in the equation (2), $f(x)$ be rational but not integral, the real roots (if any) of the denominator will give asymptotes parallel to y, provided that, for values of x differing infinitely little from these roots, y^2 be positive.

---

* The curves in the figure have been traced from the equation

$$y^2 = \tfrac{1}{4}(x^3 - 3x^2 + C),$$

where $C = -2, 0, 2, 4, 6$. The relation between them is most easily conceived by regarding them as successive contour-lines of a surface (Art. 34), as in the neighbourhood of a pinnacle on a mountain side.

Ex. 3. $$\frac{y^2}{a^2} = \frac{x-a}{x}. \quad \dots\dots\dots\dots\dots\dots(13)$$

The axis of y is an asymptote. Also, for large values of x we have $y = \pm a$, nearly. There is no real part of the curve between $x = 0$ and $x = a$. See Fig. 71.

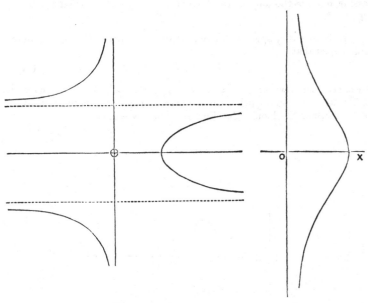

Fig. 71.   Fig. 72.

Ex. 4. $$\frac{y^2}{a^2} = \frac{a-x}{x}. \quad \dots\dots\dots\dots\dots\dots(14)$$

Here y is imaginary for x negative, and for $x > a$. See Fig. 72. The curve is known as the 'witch' of Agnesi.

Ex. 5. $$y^2 = x^2 \frac{a+x}{b-x}. \quad \dots\dots\dots\dots\dots\dots(15)$$

There is a node at the origin, and the curve cuts the axis of x again at $(-a, 0)$. For $x > b$, and $x < -a$, y is imaginary. The line $x = b$ is an asymptote. See Fig. 73.

Ex. 6. $$y^2 = \frac{x^3}{b-x}. \quad \dots\dots\dots\dots\dots\dots(16)$$

This is obtained by putting $a = 0$ in (15). The loop now shrinks into a cusp; see Fig. 74. The curve is known as the 'cissoid.'

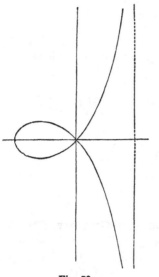

Fig. 73.   Fig. 74.

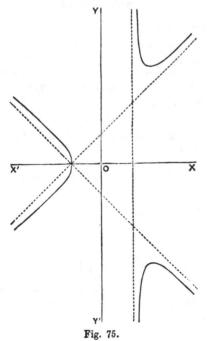

Fig. 75.

Ex. 7.
$$y^2 = x^2 \frac{x+a}{x-a}. \quad \dots\dots\dots\dots\dots(17)$$

Since y is imaginary for $a > x > -a$, except for $x = 0$, the origin is an isolated point. To find the oblique asymptotes we have

$$\frac{y}{x} = \pm \left(\frac{1 + \dfrac{a}{x}}{1 - \dfrac{a}{x}} \right)^{\frac{1}{2}} = \pm \left(1 + \frac{a}{x} \right) \left(1 - \frac{a^2}{x^2} \right)^{-\frac{1}{2}}$$

$$= \pm \left(1 + \frac{a}{x} + \tfrac{1}{2} \frac{a^2}{x^2} + \dots \right). \quad \dots\dots\dots\dots(18)$$

Hence the lines
$$y = \pm (x + a) \quad \dots\dots\dots\dots\dots(19)$$

are asymptotes. See Fig. 75.

120. Transcendental Curves; Catenary, Tractrix.

We proceed to the discussion of some important curves, mainly transcendental, which are most conveniently defined by equations of the type already referred to in Art. 61, viz.

$$x = \phi(t), \quad y = \chi(t), \quad \dots\dots\dots\dots(1)$$

where t is a variable parameter.

The 'catenary' is the curve in which a uniform chain hangs freely under gravity. It appears from elementary statical principles that if s be the arc of the curve measured from the lowest point (A) up to any point P, and ψ the inclination to the horizontal of the tangent at P, then

$$s = a \tan \psi, \quad \dots\dots\dots\dots\dots(2)$$

where a is a constant. Hence if x, y be horizontal and vertical coordinates, we have

$$\left. \begin{aligned} \frac{dx}{d\psi} &= \frac{dx}{ds} \frac{ds}{d\psi} = \cos \psi \,.\, a \sec^2 \psi = a \sec \psi, \\ \frac{dy}{d\psi} &= \frac{dy}{ds} \frac{ds}{d\psi} = \sin \psi \,.\, a \sec^2 \psi = a \tan \psi \sec \psi. \end{aligned} \right\} \quad \dots(3)$$

Integrating, we find

$$x = a \log \tan (\tfrac{1}{4}\pi + \tfrac{1}{2}\psi), \quad y = a \sec \psi. \quad \dots\dots(4)$$

The omission of the additive constants merely amounts to a special choice of the origin, which was so far undetermined. Since the formulæ (4) make $x = 0$, $y = a$ for $\psi = 0$, it appears that the origin is at a distance a vertically beneath A.

From (4) the Cartesian equation can be deduced without difficulty.   We have

$$\frac{x}{a} = \log \tan \left(\tfrac{1}{4}\pi + \tfrac{1}{2}\psi\right) = \log \left(\sec \psi + \tan \psi\right),$$

whence $\qquad\qquad \sec \psi + \tan \psi = e^{x/a},$

and therefore $\qquad \sec \psi - \tan \psi = e^{-x/a}.$ $\qquad$ ....(5)

Hence, by addition and subtraction,

$$y = a \sec \psi = a \cosh \frac{x}{a},$$

$$s = a \tan \psi = a \sinh \frac{x}{a}.$$

................(6)

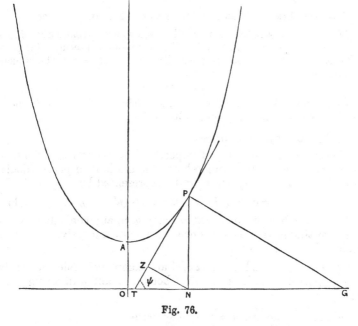

Fig. 76.

Some further properties follow easily from a figure. If PN be the ordinate, PT the tangent, PG the normal, NZ the perpendicular from the foot of the ordinate on the tangent, we have

$$NZ = y \cos \psi = a, \quad PZ = a \tan \psi = s.$$

Since PZ is equal to the arc of the catenary, it is easily seen that the consecutive position of Z is in ZN; in other words, ZN is a tangent to the locus of Z. Hence this locus possesses the property that its tangent ZN is of constant length. The curve thus characterized is called the 'tractrix,' from the fact that it is the path of a heavy particle dragged along a rough horizontal plane by a string, the other end (N) of which is made to describe a straight line (OX).

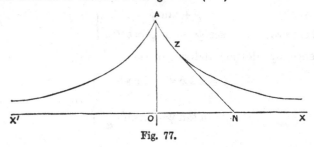

Fig. 77.

The curve has a cusp at A, and the axis of x is an asymptote.

Many properties of the tractrix follow immediately from the constancy (in length) of the tangent. For example, since two consecutive tangents make an angle $\delta\psi$ with one another, the area swept over by the tangent is given by

$$\tfrac{1}{2}\int a^2 d\psi,$$

taken between the proper limits. The whole area between the curve and its asymptote is thus found to be $\tfrac{1}{2}\pi a^2$.

121. Lissajous' Curves.

These curves, which are of importance in Acoustics, result from the composition of two simple-harmonic motions in perpendicular directions. They may therefore be represented by

$$x = a \cos (nt + \epsilon), \quad y = b \cos (n't + \epsilon'), \quad \ldots\ldots\ldots\ldots(1)$$

and it is further obvious that we may give any convenient value to *one* of the quantities ϵ, ϵ', since this amounts merely to a special choice of the origin of t.

When the periods $2\pi/n$, $2\pi/n'$ are commensurable, we can by elimination of t obtain the relation between x and y in an algebraic form.

Ex. 1. In the case $n' = n$, we may write

$$x = a \cos (nt + \epsilon), \quad y = b \cos nt, \quad \ldots\ldots\ldots\ldots\ldots(2)$$

whence $\quad \dfrac{x}{a} - \dfrac{y}{b} \cos \epsilon = -\sin nt \sin \epsilon, \quad \dfrac{y}{b} \sin \epsilon = \cos nt \sin \epsilon.$

Squaring, and adding, we find

$$\frac{x^2}{a^2} - \frac{2xy}{ab} \cos \epsilon + \frac{y^2}{b^2} = \sin^2 \epsilon. \quad \ldots\ldots\ldots\ldots\ldots(3)$$

This represents an ellipse. In the special case of $\epsilon = 0$ or $\epsilon = \pi$, the ellipse degenerates into a straight line

$$\frac{x}{a} \mp \frac{y}{b} = 0. \qquad\dots\dots\dots\dots\dots\dots\dots(4)$$

If the equality of periods be not quite exact, the figure described may be regarded as an ellipse which gradually changes its form owing to a continuous variation of the relative phase (ϵ) of the two component motions.

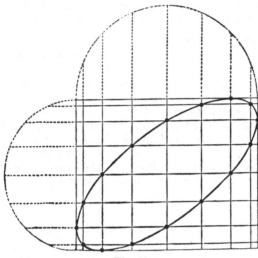

Fig. 78.

When the ellipse (3) is referred to its principal axes, the coordinates of the moving point take the forms

$$x = a \cos(nt + \epsilon), \quad y = b \sin(nt + \epsilon). \qquad\dots\dots\dots\dots(5)$$

We identify $nt + \epsilon$ with the 'eccentric angle'; and since this increases uniformly with the time it appears that the point (x, y) moves like the orthogonal projection of a point describing a circle of radius a with a constant velocity na. Since in the transition from the circle to the ellipse any infinitely small chord is altered in the same ratio as the radius parallel to it, we see that in the elliptic motion the velocity at any point P will be $n \cdot CD$, where CD is the semi-diameter conjugate to CP, C being the centre.

The type of motion here considered is called 'elliptic harmonic.'

Ex. 2.　If $n' = 2n$, we may write

$$x = a \cos nt, \quad y = b \cos(2nt + \epsilon). \qquad\dots\dots\dots\dots\dots(6)$$

Here y goes through its period twice as fast as x, and the point $(0, -b \cos \epsilon)$ is passed through twice as nt increases by 2π. The curve therefore consists in general of two loops.

For $\epsilon = \pm \frac{1}{2}\pi$, the curve is symmetrical with respect to both axes, the algebraic equation being

$$\frac{y^2}{b^2} = 4\frac{x^2}{a^2}\left(1 - \frac{x^2}{a^2}\right). \qquad\qquad\dots\dots\dots\dots\dots(7)$$

When $\epsilon = 0$, or π, the curve degenerates into an arc of a parabola, viz.

$$\frac{y}{b} = \pm\left(2\frac{x^2}{a^2} - 1\right). \qquad\qquad\dots\dots\dots\dots\dots(8)$$

When the relation of the periods is not quite exact the curve oscillates between these two parabolic arcs as extreme forms*.

122. The Cycloid.

The 'cycloid' is the curve traced by a point on the circumference of a circle which rolls in contact with a fixed straight line. It evidently consists of an endless succession of exactly congruent portions, each of which represents a complete revolution of the circle. The points (such as A in the figure) where the curve is furthest from the fixed straight line or 'base' (BD) are called 'vertices'; the points (D) half-way between successive vertices, where the curve meets the base, are the 'cusps.' A line (AB) through a vertex and perpendicular to the base is called an 'axis' of the curve. It is evidently a line of symmetry.

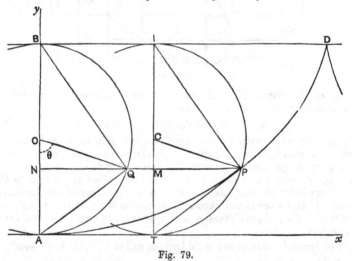

Fig. 79.

* A method of constructing Lissajous' curves is indicated in Fig. 78, where the vertical and horizontal lines, being drawn through equidistant points on the respective auxiliary circles, mark out equal intervals of time.

There are numerous optical and mechanical contrivances for producing the curves. For a description of these, and for specimens of the curves described, we must refer to books on experimental Acoustics.

It is convenient to employ the circle described on an axis AB as diameter as a circle of reference. Let IPT be any other position of the rolling circle, I the point of contact with the base, C the centre, T the opposite extremity of the diameter through I, and let P be the position of the tracing-point. Draw PMN parallel to the base, meeting TI and AB in M and N respectively, and the circle of reference in Q. If AT, AB be taken as axes of x and y, the coordinates of P will be

$$x = NP = BI + MP, \quad y = AN = CT - CM.$$

Let a be the radius of the rolling circle, and θ the angle (PCT) through which it turns as the tracing point travels from A to P. We have, then, $BI = a\theta$, $PM = a \sin \theta$, $CM = a \cos \theta$, and therefore

$$\begin{aligned} x &= a\,(\theta + \sin \theta), \\ y &= a\,(1 - \cos \theta). \end{aligned} \Big\} \qquad \dots\dots\dots\dots\dots(1)$$

From these equations all the properties of the curve can be deduced. Thus if ψ denote the inclination of the tangent to AT, or of the normal to BA, we have

$$\tan \psi = \frac{dy}{dx} = \frac{dy}{d\theta} \Big/ \frac{dx}{d\theta} = \frac{\sin \theta}{1 + \cos \theta} = \tan \tfrac{1}{2}\theta,$$

whence
$$\psi = \tfrac{1}{2}\theta. \dots\dots\dots\dots\dots\dots\dots(2)$$

Since the angle TIP is one-half of TCP, it follows that IP is the normal, and PT the tangent, to the curve at P. Cf. Art. 146, below.

Again, to find the arc (s) of the curve, we have

$$\left(\frac{dx}{d\theta}\right)^2 + \left(\frac{dy}{d\theta}\right)^2 = a^2 \{(1 + \cos \theta)^2 + \sin^2 \theta\} = 4a^2 \cos^2 \tfrac{1}{2}\theta$$

whence, by Art. 111,
$$s = 2a \int \cos \tfrac{1}{2}\theta\, d\theta = 4a \sin \tfrac{1}{2}\theta,$$

or, in terms of ψ, $\qquad s = 4a \sin \psi, \dots\dots\dots\dots\dots\dots(3)$

no additive constant being required, if the origin of s be at A. This relation is important in Dynamics.

Since $TP = TI \sin \psi$, we have
$$\text{arc } AP = 2TP = 2 \text{ chord } AQ. \quad \dots\dots\dots\dots(4)$$

In particular, the length of the arc from one cusp to the next is $8a$.

If we put $\qquad y' = IM = a\,(1 + \cos \theta), \quad \dots\dots\dots\dots(5)$

the area included between the curve and the base is given by

$$\int y'\, dx = a^2 \int (1 + \cos \theta)^2\, d\theta = 4a^2 \int \cos^4 \tfrac{1}{2}\theta\, d\theta = 8a^2 \int \cos^4 \psi\, d\psi.$$

Taking this between the limits $\mp \frac{1}{2}\pi$, we find that the area included between the base and one arch of the curve is three times the area of the generating circle.

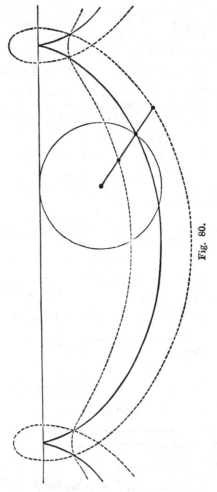

Fig. 80.

The curve traced by *any* point fixed relatively to a circle which rolls on a fixed straight line is called a 'trochoid.'

If, in Fig. 79, the tracing point be in the radius CP, at a distance k from the centre, its coordinates will be

$$x = a\theta + k\sin\theta, \quad y = a - k\cos\theta. \quad \dots\dots\dots\dots\dots(6)$$

When $k > a$ we have loops, which in the particular case ($k = a$) of the cycloid degenerate into cusps. When $k < a$, the curve does not meet the base. Fig. 80 shews the cases $k = \frac{1}{2}a$, $k = a$, $k = \frac{3}{2}a$.

It is readily proved from (6) that the normal at any point of the trochoid passes through the corresponding position of the point of contact of the rolling circle. Cf. Art. 146.

123. Epicycloids and Hypocycloids.

The path traced out by a point on the circumference of a circle which rolls in contact with a fixed circle is called an 'epicycloid' or a 'hypocycloid' according as the rolling circle is outside or inside the fixed circle*. Those epicycloids in which the rolling circle surrounds the fixed circle may be referred to, when a distinction is desired, as 'pericycloids.'

Let O be the centre of the fixed circle, C that of the rolling circle in any position, I the point of contact, P the tracing point; and suppose that, initially, the other extremity P' of the diameter PCP' was in contact with A. We take as our standard case that in which each circle is external to the other. Let

$$OA = a, \quad CP = b, \quad \angle IOA = \theta, \quad \angle ICP' = \phi.$$

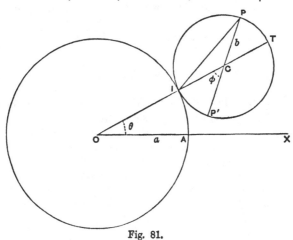

Fig. 81.

The inclination of CP to OA will be $\theta + \phi$. Hence if we take O as origin of rectangular coordinates, and OA as axis of x, we find, by orthogonal projections, that the coordinates of P are

$$\left. \begin{array}{l} x = (a + b) \cos \theta + b \cos (\theta + \phi), \\ y = (a + b) \sin \theta + b \sin (\theta + \phi), \end{array} \right\} \quad \dots\dots\dots\dots(1)$$

* This is the definition as improved by Proctor in his *Treatise on the Cycloid, etc.* (1878).

or, since

$$a\theta = \text{arc } AI = \text{arc } P'I = b\phi, \quad \dots\dots\dots\dots(2)$$

$$\left.\begin{aligned} x &= (a+b)\cos\theta + b\cos\frac{a+b}{b}\theta, \\ y &= (a+b)\sin\theta + b\sin\frac{a+b}{b}\theta. \end{aligned}\right\} \quad \dots\dots\dots(3)$$

The epicycloid traced out by P' is found by changing the sign of b in the *coefficient* of the second terms; viz. we have

$$\left.\begin{aligned} x &= (a+b)\cos\theta - b\cos\frac{a+b}{b}\theta, \\ y &= (a+b)\sin\theta - b\sin\frac{a+b}{b}\theta. \end{aligned}\right\} \quad \dots\dots\dots\dots(4)$$

This has a cusp at A.

In the above standard case the circles lie on opposite sides of the tangent at I. If they lie on the same side, as in the 'peri-cycloids' and 'hypocycloids,' we have merely to reverse the sign of b throughout, the formulæ corresponding to (3) being then

$$\left.\begin{aligned} x &= (a-b)\cos\theta - b\cos\frac{a-b}{b}\theta, \\ y &= (a-b)\sin\theta + b\sin\frac{a-b}{b}\theta. \end{aligned}\right\} \quad \dots\dots\dots(5)$$

The verification is left to the reader; see Fig. 82. In the hypo-cycloids we have $a > b$, in the pericycloids $a < b$.

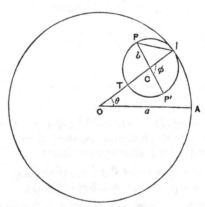

Fig. 82.

Similarly, for the locus of P' we obtain

$$x = (a - b) \cos \theta + b \cos \frac{a - b}{b} \theta, \left.\vphantom{\begin{array}{c} \\ \\ \end{array}}\right\}$$
$$y = (a - b) \sin \theta - b \sin \frac{a - b}{b} \theta. \left.\vphantom{\begin{array}{c} \\ \\ \end{array}}\right\} \quad \dots\dots\dots\dots(6)$$

To find the tangent at any point of an epicycloid, we have from (1), since $d\phi/d\theta = a/b$,

$$\frac{dy}{dx} = -\frac{\cos \theta + \cos (\theta + \phi)}{\sin \theta + \sin (\theta + \phi)} = -\cot (\theta + \tfrac{1}{2} \phi). \quad \dots\dots(7)$$

On reference to Fig. 81, we see that $\theta + \tfrac{1}{2} \phi$ is the inclination of IP to OA. Hence IP is normal to the epicycloid at P. A similar result can be deduced, of course, for the pericycloids and hypocycloids, from the equations (5). Cf. Art. 146.

Again, from (1),

$$\left(\frac{dx}{d\phi}\right)^2 + \left(\frac{dy}{d\phi}\right)^2 = \frac{(a+b)^2 b^2}{a^2} (2 + 2 \cos \phi) = \frac{4 (a+b)^2 b^2}{a^2} \cos^2 \tfrac{1}{2} \phi,$$

or

$$\frac{ds}{d\phi} = \frac{2 (a + b) b}{a} \cos \tfrac{1}{2}\phi. \quad \dots\dots\dots\dots(8)$$

Hence

$$s = \frac{4 (a + b) b}{a} \sin \tfrac{1}{2}\phi, \quad \dots\dots\dots\dots(9)$$

no additive constant being necessary, if $s = 0$ for $\phi = 0$.

If we denote by ψ the inclination of the normal IP to OA, we have

$$\psi = \theta + \tfrac{1}{2} \phi = \frac{a + 2b}{2a} \phi, \quad \dots\dots\dots\dots(10)$$

and therefore

$$s = \frac{4 (a + b) b}{a} \sin \frac{a}{a + 2b} \psi. \quad \dots\dots\dots(11)$$

The formula (9) has a simple interpretation. It appears from Fig. 81 that $TP = 2b \sin \tfrac{1}{2} \phi$, whence

$$s = 2 \frac{a + b}{a} \times \text{chord } TP. \quad \dots\dots\dots\dots(12)*$$

In particular, the length of the curve from one cusp to the next is $8 (a + b) b/a$.

The corresponding results for the pericycloids and hypocycloids are easily inferred by changing the sign of b.

* Newton, *Principia*, lib. i., prop. xlix.

The curve traced out by a point of the rolling circle which is not on the circumference is called an 'epitrochoid,' or a 'hypotrochoid,' as the case may be.   If k denote the distance of the tracing point from the centre of the rolling circle, the expressions for the coordinates x, y in the various cases are obtained by writing k for b in the coefficients (only) of the second terms.

124. Special Cases.

1°. If the radius of the *fixed* circle be infinitely great we fall back on the case of the cycloid.   The corresponding equations are easily deduced from Art. 123 (1), writing $x + a$ for x, $a\theta = b\phi$, and (finally) $\theta = 0$.

2°. Again, making the radius of the *rolling* circle infinite, we get the path described by a point of a straight line which rolls on a fixed circle. The curve thus defined is called the 'involute of the circle'; see Art. 144. The equations may be obtained as limiting forms of Art. 123 (4), or they may be written down at once from a figure.   We find

$$x = a \cos \theta + a\theta \sin \theta, \atop y = a \sin \theta - a\theta \cos \theta. \Big\} \quad\dots\dots\dots\dots\dots(1)$$

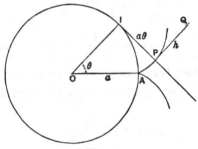

Fig. 83.

The corresponding trochoidal curve is

$$x = (a + h) \cos \theta + a\theta \sin \theta, \atop y = (a + h) \sin \theta + a\theta \cos \theta, \Big\} \quad\dots\dots\dots\dots(2)$$

where $h = PQ$ in the figure, Q being the tracing point. The particular case of $h = -a$ gives the 'spiral of Archimedes,' see Art. 126.

3°. If the radii a, b be commensurable, then after some exact number of revolutions the tracing point will have returned to its original position, and its subsequent course will be a repetition of the previous path. In such cases the curve is algebraic, since the trigonometrical functions can be eliminated between the expressions for x and y. Sometimes the equation is more conveniently expressed in polar coordinates.

Figs. 84, 85, 86 shew the epi- and hypo-cycloids in which the ratio of the radius of the rolling circle to that of the fixed circle has the values $1, \frac{1}{2}, \frac{1}{3}$, respectively.

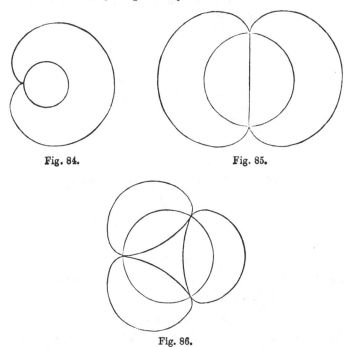

Fig. 84.　　　　　　　　　　Fig. 85.

Fig. 86.

We proceed to notice in detail one or two of the cases which have specially important properties.

Ex. 1.　The 'cardioid.'

If in Art. 123 (3) we put $b = a$, we get

$$x = 2a \cos \theta + a \cos 2\theta, \quad y = 2a \sin \theta + a \sin 2\theta,$$

whence　　$x + a = 2a(1 + \cos \theta) \cos \theta, \quad y = 2a(1 + \cos \theta) \sin \theta. \quad \ldots\ldots(3)$

This shews that the radius vector drawn from the point $(-a, 0)$ as pole is given by

$$r = 2a(1 + \cos \theta). \ldots \ldots\ldots\ldots\ldots\ldots\ldots\ldots(4)$$

This is otherwise evident from Fig. 87, p. 302, where

$$A'P = 2A'N = 2(OI + A'M).$$

The corresponding trochoids are given by

$$x = 2a \cos \theta + k \cos 2\theta, \quad y = 2a \sin \theta + k \sin 2\theta.$$

Referred to the point $(-k, 0)$ as pole these formulæ are equivalent to

$$r = 2 (a + k \cos \theta), \dots\dots\dots\dots\dots\dots(5)$$

which is the polar equation of the 'limaçon' (Art. 127). This equation, again, is easily obtained geometrically.

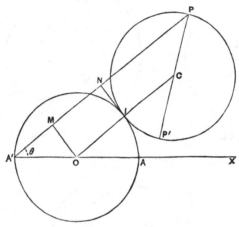

Fig. 87.

Ex. 2.   A circle rolls inside another of twice its radius.

If in Art. 123 (6) we put $b = \frac{1}{2}a$, we get

$$x = a \cos \theta, \quad y = 0 ; \dots\dots\dots\dots\dots(6)$$

i.e. the tracing point on the circumference of the rolling circle traces out a diameter of the fixed circle.

Again, the corresponding trochoidal curve is given by

$$x = (b + k) \cos \theta, \quad y = (b - k) \sin \theta, \dots\dots\dots(7)$$

and is therefore an ellipse of semi-axes $b \pm k$.   Moreover if the rolling circle have a constant angular velocity, the motion of the tracing point is elliptic-harmonic.

These results also follow easily from geometrical considerations. The rolling circle passes always through the centre O of the fixed circle; also, if P be the point of the rolling circle which initially coincides with A, the arc IP is equal to the arc IA. Hence, since the radii are as $1 : 2$, the angle which the arc IP subtends at the circumference of its circle must be equal to the angle which the arc IA subtends at the centre of *its* circle; that is, OP and OA coincide in direction, and P describes the fixed diameter OA. Again, since the angle POP' is a right angle, the other extremity of the diameter PP' of the rolling circle describes the diameter of the fixed circle which is perpendicular to OA. Hence PP' is a line of constant length whose extremities move on two fixed straight lines at right angles to one another. It is known that under these circumstances any other point on PP' describes an ellipse. Cf. Art. 145, Ex. 1.

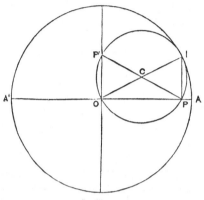

Fig. 88.

Ex. 3.   A circle rolls on the outside of a fixed circle of one-half the radius, which it encloses.

The formulæ (5) of Art. 123 give, for $b = 2a$,

$$x = - a \cos \theta - 2a \cos \tfrac{1}{2}\theta, \quad y = - a \sin \theta - 2a \sin \tfrac{1}{2}\theta,$$

or $x - a = - 2a \left(1 + \cos \tfrac{1}{2}\theta\right) \cos \tfrac{1}{2}\theta, \quad y = - 2a \left(1 + \cos \tfrac{1}{2}\theta\right) \sin \tfrac{1}{2}\theta$....(8)

If we put $\theta' = \tfrac{1}{2}\theta + \pi,$

it appears that the pericycloid, referred to the point $(a, 0)$ as pole, has the equation

$$r = 2a \left(1 - \cos \theta'\right), \quad \dots\dots\dots\dots\dots\dots(9)$$

and is therefore a cardioid.

The connection between this result and that of Ex. 1, above, will appear in Art. 150.

Ex. 4.   The 'four-cusped hypocycloid.'

If in Art. 123 (6) we put $b = \tfrac{1}{4}a$, we get

$$\left.\begin{array}{l} x = \tfrac{3}{4}a \cos \theta + \tfrac{1}{4}a \cos 3\theta = a \cos^3 \theta, \\ y = \tfrac{3}{4}a \sin \theta - \tfrac{1}{4}a \sin 3\theta = a \sin^3 \theta, \end{array}\right\} \quad \dots\dots\dots(10)$$

from which the curve is easily traced.   The Cartesian form is

$$x^{\frac{2}{3}} + y^{\frac{2}{3}} = a^{\frac{2}{3}}. \quad \dots\dots\dots \dots\dots\dots(11)$$

This curve is sometimes called the 'astroid.'   It possesses the property that the length of the tangent intercepted between the co-ordinate axes is constant.   If, in Fig. 89, P be the tracing point, TP is the tangent, and it is easily seen that the angle CTP is double the angle AOC. Hence $HK = 2OT = a$. See also Fig. 125, Art. 145.

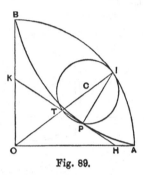

Fig. 89.

125. Superposition of Circular Motions. Epicyclics.

The cycloidal and trochoidal curves discussed in Arts. 122—124 present themselves in another manner, as the paths of points whose motion is compounded of two uniform circular motions.

If an arm OQ revolve about a fixed point O with constant angular velocity n, its projections on rectangular axes through O may be taken to be

$$x = c \cos nt, \quad y = c \sin nt, \quad \dots\dots\dots\dots(1)$$

where $c = OQ$, provided the origin of t be suitably chosen. If another arm OQ' revolve about O with constant angular velocity n', starting simultaneously with OQ from coincidence with the axis of x, the projections of OQ' will be

$$x = c' \cos n't, \quad y = c' \sin n't, \quad \dots\dots\dots\dots(2)$$

where $c' = OQ'$. If we complete the parallelogram $OQPQ'$, the vector OP will represent the geometric sum of OQ and OQ', and the coordinates of P will be

$$x = c \cos nt + c' \cos n't, \quad y = c \sin nt + c' \sin n't \dots\dots(3)^*$$

Since QP is always equal and parallel to OQ', the path of P is that of a point describing uniformly a circular orbit relatively to a point Q which itself has a uniform circular motion about O. Curves described in this manner are called 'epicyclics.' If the angular velocities n, n' have the same sign, the epicyclic is

---

* If the parallelogram $OQPQ'$ consist of four jointed rods, and if OQ, OQ' be made to revolve at the proper rates about O, the distance of P from any fixed line through O will represent the sum of two simple-harmonic motions of periods $2\pi/n$, $2\pi/n'$. This is the principle of Lord Kelvin's 'tidal clock,' which performs mechanically the superposition of the solar and lunar tides.

said to be 'direct'; if they have opposite signs it is said to be 'retrograde.'

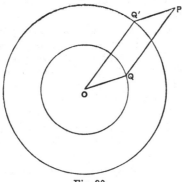

Fig. 90.

Figs. 91—94 shew some specimens of direct and retrograde epicyclics*.

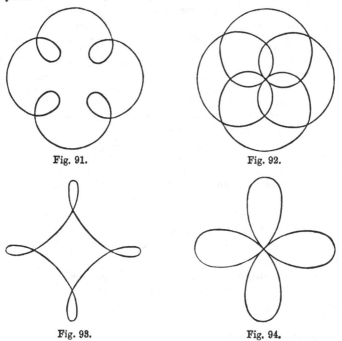

Fig. 91.   Fig. 92.

Fig. 93.   Fig. 94.

* The variety of such figures is of course endless. Epicyclics are easily described mechanically with a lathe; a number of very interesting diagrams obtained in this manner are reproduced in Proctor's treatise cited on p. 297 *ante*.

Since the path of P may be equally well defined as that of a point moving in a circular orbit relatively to a point Q' which itself is in uniform circular motion about O, we see that every epicyclic can be generated in two distinct ways.

It is evident that every epi- or hypo-cycloid, and (more generally) every epi- and hypo-trochoid, is an epicyclic, since if the rolling circle has a constant angular velocity its centre C will describe a circle uniformly about O, the centre of the fixed circle, whilst the radius CP which contains the tracing point P has a uniform rotation about C. See Figs. 81, 82.

Conversely, it may be proved that every epicyclic is either an epi- or a hypo-trochoid: more particularly that every direct epicyclic is an epitrochoid, and every retrograde epicyclic is a hypo-trochoid. This may be shewn by a comparison of (3), above, with the results of Art. 123. A simple geometrical proof will be given later (Art. 150), in connection with the theory of the 'instantaneous centre.'

The connection of the direct and retrograde epicyclics in Figs. 91—9; with the four-cusped epi- and hypo-cycloids will be apparent.

Epicyclics played a great part in ancient Astronomy. If we ignore the eccentricities and inclinations of the planetary orbits, the Sun may be regarded as describing a circle round the Earth, and any other planet describes a circle in the same plane about the Sun. The path of the planet relatively to the Earth is therefore an epicyclic. This was the accepted view of the matter from the time of Ptolemy down to the sixteenth century, when it was gradually superseded by the simpler mode of describing the phenomena discovered by Copernicus.

The relative orbits of the planets have loops, as in Fig. 91. This accounts for the 'stationary points' and 'retrograde motions,' which were in fact the occasion of Ptolemy's invention of epicyclics.

The orbit of the moon relatively to the sun, on the other hand, though an epicyclic, has no loops; it is, moreover, everywhere concave inwards.

Ex. 1. If the angular velocities of the component circular motions are equal and opposite $(n' = -n)$, we have

$$x = (c + c') \cos nt, \quad y = (c - c') \sin nt, \dots\dots\dots\dots(4)$$

so that the resultant motion is elliptic-harmonic. In the particular case of $c' = c$, the ellipse degenerates into a straight line.

This example is of importance in Physical Optics.

Ex. 2. The special form assumed by an epicyclic when $c' = c$ may be noticed.

The equations (3) then become

$$x = 2c \cos \tfrac{1}{2}(n + n')\, t \,.\, \cos \tfrac{1}{2}(n - n')\, t, \\ y = 2c \sin \tfrac{1}{2}(n + n')\, t \,.\, \cos \tfrac{1}{2}(n - n')\, t, \quad\Big\} \dots\dots\dots(5)$$

or $$x = r \cos \theta, \quad y = r \sin \theta, \dots\dots\dots\dots\dots\dots(6)$$

where $$\theta = \tfrac{1}{2}(n + n')t, \quad r = 2c \cos \frac{n - n'}{n + n'}\theta. \quad\dots\dots\dots(7)$$

The polar equation of the curve is therefore of the form

$$r = a \cos m\theta, \dots\dots\dots\dots\dots\dots(8)$$

where $m \lessgtr 1$, according as the epicyclic is direct or retrograde.

Figs. 92, 94 on p. 305 correspond to the cases $m = \tfrac{2}{3}$, $m = 2$, respectively.

126. Curves referred to Polar Coordinates. The Spirals.

There are several curves of interest whose equations are most conveniently expressed in polar coordinates. We begin with the 'spirals.'

1°. The 'equiangular spiral' is defined by the property that the curve makes a constant angle with the radius vector.

Denoting this angle by α, we have, by Art. 63,

$$\frac{dr}{d\theta} = r \cot \alpha, \dots\dots\dots\dots\dots\dots(1)$$

the solution of which is (Art. 38)

$$r = ae^{\theta \cot \alpha}. \quad\dots\dots\dots\dots\dots\dots(2)$$

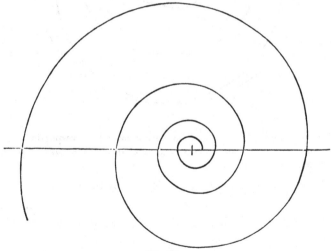

Fig. 95.

As θ ranges from $-\infty$ to $+\infty$, r ranges from 0 to ∞. See Fig. 95.

Since by Art. 112, we have $dr/ds = \cos a$, it appears that the length of the curve, between the radii r_1, r_2, is

$$\int_{r_1}^{r_2} \frac{ds}{dr}\, dr = (r_2 - r_1)\sec a. \quad \dots\dots\dots\dots\dots(3)$$

2°. The 'spiral of Archimedes' is the curve described by a point which travels along a straight line with constant velocity, whilst the line rotates with constant angular velocity about a fixed point in it.

In symbols, $r = ut, \quad \theta = nt,$

whence $r = a\theta, \quad\dots\dots\dots\dots\dots\dots\dots\dots\dots(4)$

if $a = u/n.$

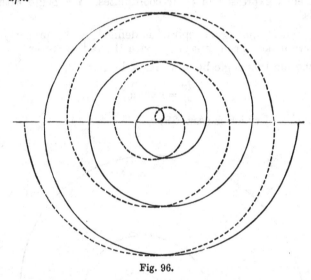

Fig. 96.

Fig. 96 shews the curve. The dotted branch corresponds to negative values of θ. Another mode of generation of this curve has been explained in Art. 124.

3°. The 'reciprocal spiral' is defined by the equation

$$r = a/\theta. \quad \dots\dots\dots\dots\dots\dots\dots(5)$$

If y be the ordinate drawn to the initial line, we have

$$y = r \sin \theta = a\frac{\sin \theta}{\theta}.$$

As θ approaches the value zero, r becomes infinite, but y approaches the finite limit a. Hence the line $y = a$ is an asymptote.

The dotted part of the curve in Fig. 97 corresponds to negative values of θ.

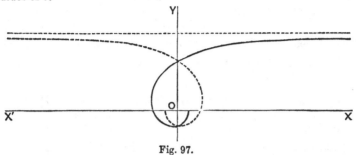

Fig. 97.

127. The Limaçon, and Cardioid.

If a point O on the circumference of a fixed circle of radius $\frac{1}{2}a$ be taken as pole, and the diameter through O as initial line, the radius vector of any point Q on the circumference is given by

$$r = a \cos \theta. \dots\dots\dots\dots\dots\dots\dots\dots\dots(1)$$

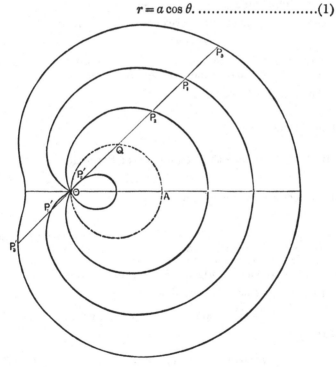

Fig. 98.

If on this radius we take two points P, P' at equal constant distances c from Q, the locus of these points is called a 'limaçon.' Its equation is evidently

$$r = a \cos \theta + c. \quad \ldots\ldots\ldots\ldots\ldots(2)$$

This includes the paths both of P and of P', if θ range from 0 to 2π.

If $c < a$, the curve passes through O when

$$\theta = \cos^{-1}(-c/a),$$

and forms a loop. See the curve traced by the points P_2, P_2' in Fig. 98. If $c > a$, r cannot vanish; see the curve traced by P_3, P_3' in the figure.

In the critical case of $c = a$, the loop shrinks into a cusp. The locus is now called a 'cardioid' or heart-shaped curve. Its equation is

$$r = a(1 + \cos \theta). \quad \ldots\ldots\ldots \ldots\ldots(3)$$

See the curve traced by P_1, P_1' in the figure. Also Fig. 84, p. 301.

128. The Curves $r^n = a^n \cos n\theta$.

A number of important curves are included in the type

$$r^n = a^n \cos n\theta. \quad \ldots\ldots\ldots\ldots\ldots(1)$$

The curves corresponding to equal, but oppositely-signed values of n, are 'inverse' to one another; see Art. 130.

Thus if $n = \pm 1$, we have the circle

$$r = a \cos \theta \quad \ldots\ldots\ldots\ldots\ldots(2)$$

and the straight line

$$r \cos \theta = a. \quad \ldots\ldots\ldots\ldots\ldots(3)$$

If $n = \pm 2$ we have the 'lemniscate of Bernoulli'

$$r^2 = a^2 \cos 2\theta, \quad \ldots\ldots\ldots\ldots\ldots(4)$$

and the rectangular hyperbola

$$r^2 \cos 2\theta = a^2. \quad \ldots\ldots\ldots\ldots\ldots(5)$$

The equation (4) makes r real for values of θ between $\pm \frac{1}{4}\pi$, imaginary for values between $\frac{1}{4}\pi$ and $\frac{3}{4}\pi$, and so on. Also r^2 is a maximum for $\theta = 0$, $\theta = \pi$, etc. It follows that the lemniscate consists of two loops, with a node at the origin. See Fig. 106, p. 321.

If $n = \pm \frac{1}{2}$, we have the cardioid

$$r^{\frac{1}{2}} = a^{\frac{1}{2}} \cos \tfrac{1}{2}\theta, \quad \text{or} \quad r = \tfrac{1}{2}a(1 + \cos \theta), \quad \ldots\ldots(6)$$

and the parabola

$$r^{\frac{1}{2}} \cos \tfrac{1}{2}\theta = a^{\frac{1}{2}}, \quad \text{or} \quad r = \frac{2a}{1 + \cos \theta}. \quad \ldots\ldots\ldots(7)$$

If we differentiate (1) logarithmically, we find, if ϕ denote the angle between the tangent and the radius vector,

$$\cot \phi = \frac{1}{r}\frac{dr}{d\theta} = -\tan n\theta, \quad\dotfill(8)$$

or
$$\phi = \tfrac{1}{2}\pi + n\theta. \quad\dotfill(9)$$

The student should examine the meaning of this result in the various special cases mentioned above.

129. The Tangential-Polar Equation.

If p be the perpendicular from the origin on any tangent, and r the radius vector of the point of contact, p will in general be a function of r. The equation expressing this relation is called the 'tangential-polar' equation of the curve.

If the ordinary polar equation be given, the tangential-polar equation is to be found by eliminating θ and ϕ between the formulæ

$$p = r \sin \phi, \quad \frac{1}{r}\frac{dr}{d\theta} = \cot \phi \quad\dotfill(1)$$

(for which see Art. 63) and the given equation.

From (1) we obtain

$$\frac{1}{p^2} = \frac{1}{r^2}(1 + \cot^2 \phi) = \frac{1}{r^2} + \frac{1}{r^4}\left(\frac{dr}{d\theta}\right)^2. \quad\dotfill(2)$$

It is occasionally convenient to employ the reciprocal of the radius vector instead of the radius itself. If we write

$$u = \frac{1}{r}, \quad \text{we have} \quad \frac{du}{d\theta} = -\frac{1}{r^2}\frac{dr}{d\theta}, \quad\dotfill(3)$$

and the formula (2) takes the shape

$$\frac{1}{p^2} = u^2 + \left(\frac{du}{d\theta}\right)^2. \quad\dotfill(4)$$

It is important, with a view to some applications in Dynamics, to notice that if the tangential-polar equation be given, say

$$p = f(r), \quad\dotfill(5)$$

the curve is determinate save as to orientation. For we have

$$r\frac{d\theta}{dr} = \tan \phi = \frac{p}{\sqrt{(r^2 - p^2)}}, \quad\dotfill(6)$$

whence
$$\theta - a = \int \frac{p\,dr}{r\sqrt{(r^2 - p^2)}}. \quad\dotfill(7)$$

A variation of the additive constant a has merely the effect of turning the curve bodily through an angle about O.

Ex. 1.   In the equiangular spiral, we have

$$p = r \sin \alpha. \quad \dots\dots\dots\dots\dots\dots\dots(8)$$

Ex. 2.   In the circle $r = 2a \sin \theta \quad \dots\dots\dots\dots\dots\dots\dots(9)$

we have, as in Art. 63, $\phi = \theta$, and therefore $p/r = r/2a$, or

$$p = r^2/2a. \quad \dots\dots\dots\dots\dots\dots\dots(10)$$

Ex. 3.   In the parabola

$$r = a \sec^2 \tfrac{1}{2}\theta, \quad \dots\dots\dots\dots\dots\dots(11)$$

where the focus is the pole, we find $\phi = \tfrac{1}{2}\pi - \tfrac{1}{2}\theta$, $p = r \cos \tfrac{1}{2}\theta$, whence

$$p^2 = ar. \quad \dots\dots\dots\dots\dots\dots(12)$$

This is a well-known property of the curve.

This example, like the preceding, is included in a general result embracing all curves of the type

$$r^n = a^n \cos n\theta. \quad \dots\dots\dots\dots\dots(13)$$

By Art. 128 (9) we have

$$p = r \sin \phi = r \cos n\theta, \quad \dots\dots\dots\dots\dots(14)$$

whence, eliminating θ,

$$p = r^{n+1}/a^n. \quad \dots\dots\dots\dots\dots\dots(15)$$

Thus in the case of the cardioid $(n = \tfrac{1}{2})$, we have

$$p^2 = r^3/a. \quad \dots\dots\dots\dots\dots\dots(16)$$

Ex. 4.   The tangential-polar equations of the central conics may be given here, as they are sometimes employed in Dynamics, although the proofs do not require the use of the Calculus.

First let the origin be at the centre. The Cartesian equation of the conic being

$$\frac{x^2}{a^2} \pm \frac{y^2}{b^2} = 1, \quad \dots\dots\dots\dots\dots\dots(17)$$

we have, if β be the conjugate semi-diameter,

$$p\beta = ab, \quad \text{and} \quad \beta^2 \pm r^2 = b^2 \pm a^2, \quad \dots\dots\dots\dots(18)$$

by known properties of central conics.   Hence

$$\frac{a^2 b^2}{p^2} = b^2 \pm a^2 \mp r^2. \quad \dots\dots\dots\dots\dots(19)$$

In the particular case of the rectangular hyperbola we have

$$pr = a^2, \quad \dots\dots\dots\dots\dots\dots\dots(20)$$

since $\beta = r$.   This is also obtained by making $n = -2$ in (15) above.

Ex. 5.   Again, taking a focus as pole, let us denote the perpendicular and radius vector corresponding to the other focus by p' and r'.   Since the tangent makes equal angles with the two focal radii, we have $p/r = p'/r'$, and therefore

$$\frac{p^2}{r^2} = \frac{pp'}{rr'}.$$

Now $pp' = b^2$, and, in the ellipse, $r + r' = 2a$. Hence, for this curve,
$$p^2/r = b^2/(2a - r),$$
or, if l denote the semi-latus-rectum (b^2/a),

$$\frac{l}{p^2} = \frac{2}{r} - \frac{1}{a}. \qquad\qquad\dots\dots\dots(21)$$

In the hyperbola, we find

$$\frac{l}{p^2} = \pm\frac{2}{r} + \frac{1}{a}, \qquad\qquad\dots\dots\dots(22)$$

the upper sign relating to the branch nearest to the origin, the lower to the further branch.

Ex. 6.   To find the curve in which

$$p = r^3/a^2. \qquad\qquad\dots\dots\dots(23)$$

Substituting in (6), and integrating, we find

$$\theta - a = \int \frac{r\,dr}{\sqrt{(a^4 - r^4)}} = \tfrac{1}{2}\sin^{-1}\frac{r^2}{a^2},$$

or
$$r^2 = a^2 \sin 2\,(\theta - a),\dots\dots\dots(24)$$

a lemniscate.

130.   Related Curves.   Inversion.

There are various geometrical theories in which one curve is associated with another connected with it by a definite relation.

A simple instance is that of 'inversion.'

If from a fixed origin O we draw a radius vector OP to any given curve, and in OP take a point P' such that

$$OP \cdot OP' = k^2, \qquad\qquad\dots\dots\dots(1)$$

where k is a given constant, the locus of P' is said to be the 'inverse' of that of P. The point O is called the 'centre,' and k^2 is called the 'constant,' of inversion.

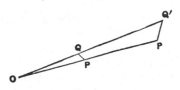

Fig. 99.

A curve and its inverse make supplementary angles with the radius vector. For if P, Q be the consecutive points of a curve, and P', Q' the corresponding points on the inverse curve, we have $OP \cdot OP' = OQ \cdot OQ'$, and therefore

$$OP : OQ = OQ' : OP'. \quad\ldots\ldots\ldots\ldots\ldots\ldots(2)$$

Hence the triangles POQ, $Q'OP'$ are similar, and the angles OPQ, $OP'Q'$ are supplementary. In the limit, when Q is infinitely close to P, these are the angles which the respective tangents make with the radius vector.

It follows that if two curves intersect, the respective inverse curves will intersect at the same angle. In particular, orthogonal curves invert into orthogonal curves.

It is proved in books on elementary Geometry that the inverse of a circle is a circle, except in the particular case where the centre of inversion is on the circumference, when the inverse locus becomes a straight line.

There are various devices by which the inverse of a given curve can be traced mechanically.

1°. *Peaucellier's Linkage.*

This consists of a rhombus $PAQB$ formed of four rods freely jointed at their extremities, and of two equal bars connecting two opposite corners A, B to a fixed pivot at O.

It is evident that, whatever shape and position the linkage assumes, the points P, Q will always be in a straight line with O. If N be the intersection of the diagonals of the rhombus, we have

$$OP \cdot OQ = ON^2 \sim PN^2 = OA^2 \sim AP^2 = \text{const.} \quad\ldots\ldots(1)$$

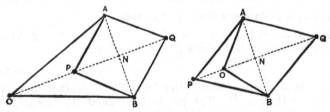

Fig. 100.

Hence if P (or Q) be made to describe any curve, Q (or P) will describe the inverse curve with respect to O.

In particular if, by a link, P be pivoted to a fixed point S, such that $SO = SP$, the locus of P is a circle through O, and consequently the locus of Q will be a straight line perpendicular to OS. This gives an exact solution of the important mechanical

problem of converting circular into rectilinear motion by means
of link-work.

2°. *Hart's Linkage.*

This consists of a 'crossed parallelogram' $ABCD$ formed of

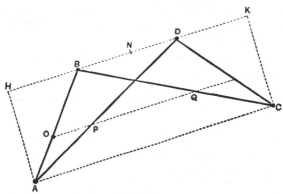

Fig. 101.

four rods jointed at their extremities, the alternate sides being
equal. A point O in one side AB is made a fixed pivot, and P, Q
are points in AD and BC such that

$$AP : PD = CQ : QB = AO : OB, = m : n, \text{ say.}$$

Evidently O, P, Q will lie in a straight line parallel to AC and
BD. If H, K be the orthogonal projections of A, C on BD, and
N be the middle point of BD, we have

$$AC . BD = 2NH . 2NB = DH^2 - BH^2 = AD^2 - AB^2.$$

Now $OP : BD = AO : AB = m : m + n,$

and $OQ : AC = BO : AB = n \ : m + n.$

Hence $OP . OQ = \dfrac{mn}{(m+n)^2} (AD^2 - AB^2) = \text{const.}$ ......(2)

Hence P and Q describe inverse curves with respect to O.

As before, by connecting P to a fixed pivot S by a link PS
equal to SO, we can convert circular into rectilinear motion.

131. Pedal Curves. Reciprocal Polars.

If a perpendicular OZ be drawn from a fixed point O to the
tangent to a curve, the locus of the foot Z of this perpendicular
is called the 'pedal' of the original curve with respect to the
origin O.

Thus: the pedal of a parabola with respect to the focus is the tangent at the vertex. The pedal of an ellipse or hyperbola with respect to either focus is the 'auxiliary circle.'

If $OZ = p$, and if ψ be the angle which OZ makes with any fixed straight line, then p, ψ may be taken to be the polar coordinates of Z with respect to O as pole. Hence if the relation between p and ψ can be found, the polar equation of the pedal can be at once written down.

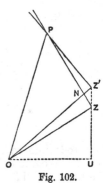

Fig. 102.

The angle which the tangent makes with the radius vector at corresponding points is the same for a curve and its pedal. For let OZ, OZ' be the perpendiculars from O on two consecutive tangents PZ, PZ', and let OU be drawn perpendicular to ZZ' produced. The points Z, Z' lie on the circle described on OP as diameter. Hence the exterior angle OZU of the quadrilateral $OZZ'P$ is equal to the interior and opposite OPZ'. In the limit these are the angles which OZ and OP make with the tangent to the pedal, and with the tangent to the original curve, respectively.

Also, by similar triangles, we have

$$OU : OZ = OZ' : OP. \dots\dots\dots\dots(1)$$

Hence if r be the radius vector of the original curve, p the perpendicular from O on the tangent, and p' the perpendicular from O on the tangent to the pedal, we have, ultimately,

$$p'/p = p/r, \text{ or } p' = p^2/r. \dots\dots\dots\dots(2)$$

Again, if OZ' meet PZ in N, we may write

$$OZ = p, \quad OZ' = p + \delta p, \quad \angle ZOZ' = \angle ZPZ' = \delta\psi.$$

Neglecting small quantities of the second order, we have

$$\delta p = NZ' = PZ'. \delta\psi.$$

Hence, proceeding to the limit, when PZ' coincides with PZ, we obtain an expression for the projection of the radius vector on the tangent to a curve, viz.

$$PZ = \frac{dp}{d\psi}. \dots\dots\dots\dots(3)$$

This result enables us easily to solve the problem of 'negative pedals,' viz. to find the curve having a given pedal. Taking O as origin, and the initial line of ψ as axis of x, the coordinates of the point of contact P are given by

$$x = OZ \cos \psi - ZP \sin \psi, \quad y = OZ \sin \psi + ZP \cos \psi,$$

or

$$\left. \begin{array}{l} x = p \cos \psi - \dfrac{dp}{d\psi} \sin \psi, \\[2mm] y = p \sin \psi + \dfrac{dp}{d\psi} \cos \psi. \end{array} \right\} \quad \dots\dots\dots\dots(4)$$

Ex. 1.   If the origin be at the centre of the conic

$$\frac{x^2}{a^2} \pm \frac{y^2}{b^2} = 1, \quad \dots\dots\dots\dots\dots\dots\dots\dots(5)$$

and ψ be the angle which p makes with Ox, it is shewn in books on Conic Sections that

$$p^2 = a^2 \cos^2 \psi \pm b^2 \sin^2 \psi. \quad \dots\dots\dots\dots\dots(6)$$

Hence the polar equation of the pedal is

$$r^2 = a^2 \cos^2 \theta \pm b^2 \sin^2 \theta. \quad \dots\dots\dots\dots\dots(7)$$

In the case of the rectangular hyperbola

$$x^2 - y^2 = a^2 \quad \dots\dots\dots\dots\dots\dots \dots(8)$$

the pedal is the lemniscate

$$r^2 = a^2 \cos 2\theta. \quad \dots\dots\dots\dots\dots\dots(9)$$

Ex. 2.   In the case of a circle of radius a, the pole O being at a distance c from the centre C, and the line OC being the origin of ψ, we have at once from a figure

$$p = a + c \cos \psi. \quad \dots\dots\dots\dots\dots\dots(10)$$

Hence the pedal is the limaçon

$$r = a + c \cos \theta. \quad \dots\dots\dots\dots\dots\dots(11)$$

If O be on the circumference, we have $c = a$, and the pedal is the cardioid

$$r = a\,(1 + \cos \theta). \quad \dots\dots\dots\dots\dots\dots(12)$$

Ex. 3.   To find the curve whose pedal is the cardioid

$$r = a\,(1 + \cos \theta). \quad \dots\dots\dots\dots\dots\dots(13)$$

Writing

$$p = a\,(1 + \cos \psi), \quad \dots\dots\dots\dots\dots\dots(14)$$

the formulæ (4) make

$$x = a \cos \psi + a, \quad y = a \sin \psi,$$

whence

$$(x - a)^2 + y^2 = a^2, \quad \dots\dots\dots\dots\dots\dots(15)$$

a circle through the origin.

The locus of the pole of the tangent to a curve S, with respect to a fixed conic Σ, is called the 'reciprocal polar' of S. It is proved in books on Conics that if S' be the locus of the poles of the tangents to S then S is the locus of the poles of the tangents to S'. This explains the use of the word 'reciprocal.'

We shall here only notice the case where the fixed conic Σ is a circle. If O be the centre of this circle, and k denote its radius, the pole P' of any tangent to the curve S is found by drawing OZ

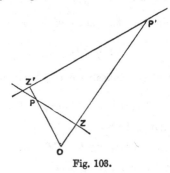

Fig. 103.

perpendicular to this tangent, and by taking in OZ a point P' such that

$$OZ \cdot OP' = k^2. \quad\quad\quad\quad\quad (16)$$

Hence the reciprocal polar is in this case the *inverse of the pedal* of the given curve, with respect to the point O.

By the reciprocal property above cited, the original curve must be the inverse of the pedal of the locus of P'. This is easily verified; for if P be the point of contact of the tangent to the original curve, and if OP meet the tangent to the locus of P' in Z', the angles $OP'Z'$ and OPZ will be equal. Hence $OZ'P'$ is a right angle, and Z' traces out the pedal of P'. And, since $PZP'Z'$ is a cyclic quadrilateral, we have

$$OP \cdot OZ' = OZ \cdot OP' = k^2. \quad\quad\quad\quad (17)$$

Hence P describes the inverse of the locus of Z'.

Ex. 4. The reciprocal polar of a circle with respect to any origin is a conic having the origin as focus.

As in Ex. 2, the formula for the pedal of the circle is

$$p = a + c \cos \psi. \quad\quad\quad\quad\quad (18)$$

Writing θ for ψ, and k^2/r for p, we get the equation of the reciprocal polar in the form

$$\frac{k^2}{r} = a + c \cos \theta, \quad\quad\quad\quad\quad (19)$$

which represents a conic, having its focus at the origin, of eccentricity c/a. Hence the conic is an ellipse, parabola, or hyperbola, according as the origin is inside, on, or outside the circle.

Ex. 5. The pedal of the conic

$$\frac{x^2}{a^2} \pm \frac{y^2}{b^2} = 1, \quad \text{...........................(20)}$$

with respect to the centre, is given by

$$p^2 = a^2 \cos^2 \psi \pm b^2 \sin^2 \psi. \quad \text{...................(21)}$$

Hence the reciprocal polar is

$$\frac{k^4}{r^2} = a^2 \cos^2 \theta \pm b^2 \sin^2 \theta, \quad \text{...................(22)}$$

or $\qquad\qquad a^2 x^2 \pm b^2 y^2 = k^4, \quad \text{................ ......(23)}$

a concentric conic.

132. Bipolar Coordinates.

A curve may be defined by a relation between the distances (r, r') of any point P on it from two fixed points, or foci, S, S'; thus

$$f(r, r') = 0. \quad \text{...........................(1)}$$

If we denote the angles PSS', $PS'S$ by θ, θ', respectively, and the angles which the radii r, r' make with the tangent by ϕ, ϕ', we have, as in Art. 112,

$$\left.\begin{array}{ll} \dfrac{dr}{ds} = \cos \phi, & \dfrac{dr'}{ds} = \cos \phi', \\[2mm] r\dfrac{d\theta}{ds} = \sin \phi, & r'\dfrac{d\theta'}{ds} = \sin \phi'. \end{array}\right\} \quad \text{...........(2)}$$

We have, in addition, the relations

$$r \sin \theta = r' \sin \theta', \quad r \cos \theta + r' \cos \theta' = 2c, \quad \text{......(3)}$$

where $c = \frac{1}{2} SS'$.

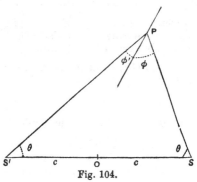

Fig. 104.

Ex. 1.   In the ellipse we have

$$r + r' = 2a, \quad\dotfill(4)$$

and therefore

$$\frac{dr}{ds} + \frac{dr'}{ds} = 0,$$

that is,

$$\cos \phi + \cos \phi' = 0, \quad \text{or} \quad \phi' = \pi - \phi. \quad\dotfill(5)$$

The focal distances therefore make supplementary angles with the curve.

Similarly, in the hyperbola

$$r \sim r' = 2a, \quad\dotfill(6)$$

we find

$$\cos \phi = \cos \phi', \quad\dotfill(7)$$

or, the focal distances make equal angles with the curve on opposite sides.

Ex. 2.   To find the form which a reflecting or refracting surface must have in order that incident rays whose directions pass through a fixed point S may be reflected or refracted in directions passing through a fixed point S'.

The case of *reflection* is merely the converse of Ex. 1.   The surface must have the form generated by the revolution of an ellipse or hyperbola about the line joining the foci (S, S').

In the case of *refraction*, we have, if μ and μ' be the refractive indices of the two media,

$$\mu \sin \chi = \mu' \sin \chi', \quad\dotfill(8)$$

where

$$\chi = \pm (\tfrac{1}{2}\pi - \phi), \quad \chi' = \pm (\tfrac{1}{2}\pi - \phi'). \quad\dotfill(9)$$

Hence

$$\mu \cos \phi \pm \mu' \cos \phi' = 0, \quad\dotfill(10)$$

or

$$\frac{d}{ds} (\mu r \pm \mu' r') = 0. \quad\dotfill(11)$$

Integrating, we have

$$\mu r \pm \mu' r' = \text{const.} \quad\dotfill(12)$$

These curves, in which the sum (or difference) of given multiples of the two radii is constant, are called 'Cartesian ovals,' after Descartes, by whom the optical problem was first discussed.

When the lower sign in (12) is taken, the family includes the *circle*

$$r/r' = \mu/\mu'. \quad\dotfill(13)$$

See Fig. 105.

Ex. 3.   The 'ovals of Cassini' are defined by

$$rr' = k^2, \quad\dotfill(14)$$

k being a given constant.   Since for a point P in SS' the greatest value of rr' is c^2, it follows that the curve will consist of two detached ovals surrounding S, S', respectively, or of a single oval embracing both points, according as $k \lessgtr c$.

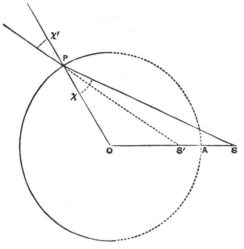

Fig. 105.

In the critical case of $k = c$ the curve is known as the 'lemniscate of Bernoulli'; this presents itself in various mathematical problems. If O, the middle point of $S'S$, be taken as pole, and OS as initial line, of a system of coordinates r_1, θ_1, we have

$$r^2 = r_1^2 + c^2 - 2cr_1 \cos \theta_1, \quad r'^2 = r_1^2 + c^2 + 2cr_1 \cos \theta_1;$$

the equation of the lemniscate is therefore

$$(r_1^2 + c^2)^2 - 4c^2 r_1^2 \cos^2 \theta_1 = c^4,$$

which reduces to

$$r_1^2 = 2c^2 \cos 2\theta_1. \quad \dots\dots\dots\dots\dots\dots(15)$$

Cf. Art. 128.

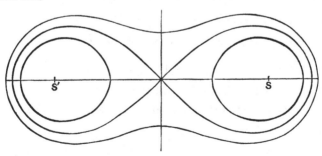

Fig. 106.

Ex. 4.   The magnetic curves.

If S, S' be the N. and S. poles of a magnet, the forces at any point P may be represented by μ/r^2 along SP, and μ/r'^2 along PS'. A 'line of force' is a line drawn from point to point always in the direction of the resultant force. Expressing that the total force at right angles to the line is zero, we have

$$\frac{\mu}{r^2} \sin \phi + \frac{\mu}{r'^2} \sin \phi' = 0,$$

or
$$\frac{1}{r} \frac{d\theta}{ds} + \frac{1}{r'} \frac{d\theta'}{ds} = 0. \quad \ldots\ldots\ldots\ldots\ldots\ldots(16)$$

Hence, since $r \sin \theta = r' \sin \theta'$, we have

$$\sin \theta \frac{d\theta}{ds} + \sin \theta' \frac{d\theta'}{ds} = 0,$$

or
$$\cos \theta + \cos \theta' = \text{const.} \quad \ldots\ldots\ldots\ldots\ldots(17)$$

An 'equipotential line' is a line such that no work is done on a magnetic pole describing it. Expressing that the total force in the direction of the line is zero, we find

$$\frac{\mu}{r^2} \cos \phi - \frac{\mu}{r'^2} \cos \phi' = 0,$$

or
$$\frac{1}{r^2} \frac{dr}{ds} - \frac{1}{r'^2} \frac{dr'}{ds} = 0, \quad \ldots\ldots\ldots\ldots\ldots\ldots(18)$$

whence
$$\frac{1}{r} - \frac{1}{r'} = \text{const.} \quad \ldots\ldots\ldots\ldots\ldots(19)$$

The equipotential lines will necessarily cut the lines of force at right angles.

EXAMPLES.   XLII.

(Algebraic Curves.)

1.   Trace the curves
$$y^2 = 4x(1-x), \quad y^2 = x^2 + x + 1.$$

2.   Trace the curve
$$ay^2 = x^2(a-x),$$
and shew that it forms a loop of area $\frac{8}{15}a^2$.

Find where the breadth of the loop is greatest. $[x = \frac{2}{3}a.]$

3.   Trace the curve
$$a^2 y^2 = x^2(a^2 - x^2),$$
and shew that it forms two loops, each of area $\frac{2}{3}a^2$.

4. Trace the curves
$$y^2 = x (x^2 - 1), \quad y^2 = x^3 (1 - x).$$

5. Trace the curve
$$a^4 y^2 = x^4 (a^2 - x^2),$$
and shew that it encloses an area $\frac{1}{4} \pi a^2$.

6. Trace the curve
$$a^2 y^4 = x^4 (a^2 - x^2),$$
and shew that it encloses an area $\frac{8}{5} a^2$.

7. The length of an arc of the curve
$$a y^2 = x^3$$
(Fig. 70), from the vertex to the point whose abscissa is x, is
$$\frac{1}{27 \sqrt{a}} (9x + 4a)^{\frac{3}{2}} - \frac{8}{27} a.$$

8. The mean centre of the area included between the curve $ay^2 = x^3$ and the line $x = h$ is at the point $(\frac{5}{7} h, \ 0)$.

9. If the curve $ay^2 = x^3$ revolve about the axis of x, the volume included between the surface generated, and any plane perpendicular to the axis, is one-fourth that of a cylinder of the same length on the same circular base.

10. Trace the curves
$$y^2 = \frac{1}{x}, \quad y^2 = \frac{1}{x (1 - x)}.$$

11. The area included between the curve
$$\frac{y^2}{a^2} = \frac{a - x}{x}$$
(Fig. 72) and its asymptote is πa^2.

If the same curve revolve about its asymptote, the volume of the solid generated is $\frac{1}{2} \pi^2 a^3$.

12. Trace the curves
$$y^2 = \frac{x^2 + 1}{x}, \quad y^2 = \frac{x^2 - 1}{x}.$$

13. Trace the curves
$$y^2 = \frac{x^2 + 1}{x^2}, \quad y^2 = \frac{x^2 - 1}{x^2}.$$

14. Trace the curves
$$y^2 = \frac{x}{1 + x^2}, \quad y^2 = \frac{x}{1 - x^2}.$$

Determine the maximum and minimum ordinates (if any), and the points of inflexion.

15. The area included between the curve

$$y^2 = \frac{x^3}{a - x}$$

(Fig. 74) and its asymptote is $\frac{3}{4}\pi a^2$.

If the same curve revolve about its asymptote the volume of the solid generated is $\frac{1}{4}\pi^2 a^3$.

16. Trace the curve

$$y^2 = \frac{a^2 x^2}{a^2 + x^2},$$

and shew that the area included between its two branches and either asymptote is $2a^2$.

17. Shew that the area included between the curve

$$y^2 = x^2 \frac{a + x}{a - x}$$

(Fig. 73) and its asymptote is $\frac{1}{2}(\pi + 4) a^2$.

18. Trace the curve

$$y^2 = \frac{x^4}{a^2 - x^2},$$

and shew that the area included between the curve and either asymptote is $\frac{1}{2}\pi a^2$.

19. Trace the curve

$$y^2 = x^2 \frac{a^2 - x^2}{a^2 + x^2},$$

and shew that it forms a loop of area $\frac{1}{2}(\pi - 2) a^2$.

20. Trace the curve

$$y^2 = \frac{x^2}{a^2}(2a - x)(x - a)$$

and shew that it encloses an area $\frac{3}{8}\pi a^2$.

21. Trace the curve

$$y^2 = \frac{x}{a}(x - b)^2 + c^2.$$

22. Trace the curve

$$x = t - t^3, \quad y = 1 - t^4,$$

for real values of t; and prove that it forms a loop of area $\frac{16}{35}$.

EXAMPLES. XLIII.

(Catenary, Cycloid, etc.)

1. Prove that, in the catenary $y = c \cosh x/c$,

$$s^2 = y^2 - c^2.$$

2. Prove that the catenary is the only curve in which the perpendicular from the foot of the ordinate on the tangent is of constant length.

3. Of all the catenaries which pass through two given points at the same level, and have their axes vertical, shew that there is one in which the depth of the directrix below the given points is a minimum.

Also prove that in this catenary the tangents at the given points meet on the directrix.

If $2b$ be the distance between the given points, the depth of the directrix is $b \sinh u$, the arc of the curve is $b (\sinh u)/u$, and the inclination to the horizontal at the given points is $\cos^{-1} (\operatorname{sech} u)$, where u is the positive root of $u \tanh u = 1$.

4. The coordinates of any point on the tractrix may be expressed in the forms

$$x = a (u - \tanh u), \quad y = a \operatorname{sech} u,$$

where u is a variable parameter.

5. Prove that, in the tractrix,

$$y = ae^{-s/a},$$

the arc s being measured from the cusp.

6. The volume of the solid generated by the revolution of the tractrix about its asymptote is $\frac{2}{3}\pi a^3$.

The surface of the same solid is $4\pi a^2$.

7. If the coordinates of a moving point be

$$x = a \cosh nt, \quad y = b \sinh nt,$$

where t is the time, the path is a hyperbola, and the velocity varies as the length of the semi-conjugate diameter measured up to its intersection with the conjugate hyperbola.

Also shew that the area swept over by the radius vector increases uniformly with the time.

8. The area of either loop of the Lissajous' curve

$$x = a \sin 2 (nt - \epsilon), \quad y = b \cos nt$$

is $\frac{4}{3}ab \cos 2\epsilon$.

9. Prove that the Lissajous' curve

$$x = a \cos nt, \quad y = b \cos 3nt$$

consists of part of the curve

$$\frac{y}{b} = \frac{x}{a}\left(4\,\frac{x^2}{a^2} - 3\right),$$

Trace this curve.

10. If, in the cycloid, the rolling circle has a constant angular velocity, the velocity of the tracing point P is proportional to the normal IP (see Fig. 79).

11. The volume generated by the revolution of a cycloid about its base is $5\pi^2a^3$, if a be the radius of the generating circle.

The surface of the same solid is $\frac{64}{3}\pi a^2$.

12. The portion of a cycloid between two consecutive cusps revolves about the tangent at the vertex; prove that the area of the surface generated is $\frac{32}{3}\pi a^2$.

Also prove that the volume included by the above surface and the planes of the circles described by the cusps is π^2a^3.

13. The volume generated by the revolution of a cycloid about its axis is $\frac{1}{6}(9\pi^2 - 16)\,\pi a^3$.

14. The surface of the same solid is $\frac{8}{3}(3\pi - 4)\,\pi a^2$.

15. The mean centre of the arc of a cycloid, from cusp to cusp, is at a distance $\frac{4}{3}a$ from the base.

16. The mean centre of the area included between a cycloid and its base is at a distance $\frac{5}{6}a$ from the base.

17. Prove that, in the curve

$$x^{\frac{2}{3}} + y^{\frac{2}{3}} = a^{\frac{2}{3}},$$

the intercepts made by the tangent at any point on the coordinate axes are $a^{\frac{2}{3}}x^{\frac{1}{3}}$, $a^{\frac{2}{3}}y^{\frac{1}{3}}$, respectively.

Hence verify that the length of the tangent intercepted by the axes is constant.

18. Prove, from the equations

$$x = a\cos^3\theta, \quad y = a\sin^3\theta,$$

that, in the astroid,

$$\frac{ds}{d\theta} = 3a\sin\theta\cos\theta,$$

and thence that the whole length of the curve is $6a$.

19. Prove that the area of the astroid is $\frac{3}{8}\pi a^2$.

20. The volume generated by the revolution of the astroid about the line joining two opposite cusps is $\frac{32}{105}\pi a^3$.

21. The length of a quadrant of the curve

$$x = a \cos^3 \theta, \quad y = b \sin^3 \theta,$$

is $$(a^2 + ab + b^2)/(a + b).$$

The area enclosed by the same curve is $\frac{3}{8}\pi ab$.

22. The whole perimeter of an n-cusped epi- or hypo-cycloid is

$$\frac{8(n \pm 1)}{n}\, a,$$

where a is the radius of the fixed circle.

23. Sketch the curve obtained by compounding two uniform circular motions when the radii of the circles are equal, but the periods slightly different, (i) when the rotations are in the same direction, and (ii) when they are in opposite directions.

24. Prove that in an epicyclic the tangent line cannot pass through the centre unless $nc < n'c'$, where c is the greater of the two quantities c, c'. (Art. 125.)

25. Prove that the length of a complete undulation of the trochoid

$$x = a\theta + k \sin \theta, \quad y = a - k \cos \theta$$

is equal to the perimeter of an ellipse whose semi-axes are $a + k, a - k$.

EXAMPLES. XLIV.

(Polar Coordinates.)

1. Prove that all equiangular spirals of the same angle are identically equal.

2. Prove that in an equiangular spiral of angle a the area swept over by the radius vector (r) is

$$\tfrac{1}{4}(r_2^2 - r_1^2) \tan a,$$

where r_1, r_2 are the extreme values of r.

3. Prove that in the spiral of Archimedes the angle (ϕ) between the tangent and the radius vector is given by

$$\cos \phi = \frac{a}{\sqrt{(a^2 + r^2)}}.$$

4. Prove that in the reciprocal spiral the area swept over by the radius increases proportionally to the radius.

5.   Shew that all chords drawn through the pole of a cardioid are of the same length.

Does the same hold of the limaçon ?

6.   The area of the cardioid

$$r = a \left(1 + \cos \theta\right)$$

is $\frac{3}{2}\pi a^2$.

7.   Trace the curve

$$r = a + 2a \cos \theta,$$

and prove that the area of the inner loop is $\cdot5435a^2$.

8.   Prove that, in the cardioid,

$$\frac{ds}{d\theta} = 2a \cos \tfrac{1}{2}\theta,$$

and thence that the whole perimeter is $8a$.

9.   The volume generated by the revolution of the cardioid about its axis is $\frac{8}{3}\pi a^3$.

10.   Prove that, in the cardioid, the maximum breadth (perpendicular to the axis) is $\frac{3}{2}\sqrt{3}a$, and that the double tangent cuts the axis at a distance $\frac{1}{4}a$ from the pole.

11.   Find the maximum ordinate, and the minimum abscissa, in the limaçon

$$r = a \cos \theta + c.$$

12.   The area of the limaçon

$$r = a \cos \theta + c,$$

when $c > a$, is $\pi \left(c^2 + \tfrac{1}{2}a^2\right)$.

13.   Prove geometrically that if two straight lines, touching two fixed circles, make a constant angle with one another, their intersection traces out a limaçon.

14.   The whole area of the lemniscate

$$r^2 = a^2 \cos 2\theta$$

is a^2.

15.   The perimeter of either loop of the same curve is

$$2a \int_0^{\frac{1}{4}\pi} \frac{d\theta}{\sqrt{(1 - 2 \sin^2 \theta)}}.$$

Prove that, in the notation of elliptic integrals (Art. 111), this is equal to

$$\sqrt{2}aF_1\left(\frac{1}{\sqrt{2}}\right).$$

16. The mean centre of the area of either loop of the lemniscate is at a distance $\frac{1}{8}\sqrt{2\pi}a$ from the pole.

17. Shew that the area included by one loop of the epicyclic

$$r = a \sin m\theta$$

is $\pi a^2/4m$.

18. Trace the curve $\qquad r^2 = a^2 \cos \theta$.

19. Prove the following properties of the 'solid of greatest attraction' (viz. the figure generated by the revolution of the curve $r^2 = a^2 \cos \theta$ about the initial line):

(1) The volume is $\frac{4}{15}\pi a^3$;

(2) The greatest breadth is $1\cdot2408a$, at a distance $\cdot4389a$ from the pole;

(3) The mean centre of the volume is at a distance $\frac{15}{32}a$ from the pole.

20. If the 'polar subtangent' of a curve be defined to be the length intercepted by the tangent on a perpendicular drawn to the radius vector from the pole, prove that it is equal to $r^2d\theta/dr$.

Prove that in the reciprocal spiral the polar subtangent is constant.

21. The tangential-polar equation of the involute of a circle of radius a is

$$p^2 = r^2 - a^2,$$

the centre being pole.

22. Shew that in the spiral of Archimedes (Fig. 111)

$$p^2 = \frac{r^4}{a^2 + r^2}.$$

23. Shew that in the reciprocal spiral (Fig. 97)

$$\frac{1}{p^2} = \frac{1}{r^2} + \frac{1}{a^2}.$$

24. Shew that in the curve

$$r = \frac{a}{\cos m\theta},$$

$$\frac{1}{p^2} = \frac{1 - m^2}{r^2} + \frac{m^2}{a^2}.$$

25. Shew that in the curves

$$r = \frac{a}{\cosh m\theta}, \quad r = \frac{a}{\sinh m\theta},$$

$$\frac{1}{p^2} = \frac{1 + m^2}{r^2} \mp \frac{m^2}{a^2},$$

respectively.

26. Prove that, in the epicycloid (Art. 123),

$$r^2 = a^2 + \frac{4\,(a+b)\,b}{(a+2b)^2}\,p^2.$$

What is the corresponding formula for the hypocycloid?

27. Find the Cartesian equation of the curve such that

$$p = a \sin \psi \cos \psi. \qquad\qquad [x^{\frac{2}{3}} + y^{\frac{2}{3}} = a^{\frac{2}{3}}.]$$

28. Find the polar equation of the curve in which

$$p^2 = \frac{4r^6}{a^4 + 3r^4}.$$

29. Prove the formula

$$s = \int \frac{r\,dr}{\sqrt{(r^2 - p^2)}}$$

for the arc of a curve whose tangential-polar equation is given.

30. Prove the formula

$$p\,ds = r^2\,d\theta,$$

and give its geometrical interpretation.

Hence shew that if the area swept over by the radius vector of a moving point increase uniformly with the time, the velocity will vary inversely as the perpendicular from the origin on the tangent to the path.

EXAMPLES. XLV.

(Related Curves.   Bipolar Coordinates.)

1. The inverse of an equiangular spiral with respect to the pole is an equal spiral.

2. The inverse of a hyperbola with respect to the centre has a node at the centre.

3. The inverse of a rectangular hyperbola with respect to the centre is a lemniscate of Bernoulli.

4. Prove by means of the polar equations that the inverse of a straight line is a circle through the pole of inversion, and conversely.

5. Prove by means of the polar equation that the inverse of a circle is a circle.

6. The inverse of a parabola with respect to the focus is a cardioid.

The inverse of any conic with respect to a focus is a limaçon.

7. Prove that the inverse of the ellipse

$$\frac{x^2}{a^2} + \frac{y^2}{b^2} = 1$$

with respect to the centre is the curve

$$(x^2 + y^2)^2 = k^4 \left(\frac{x^2}{a^2} + \frac{y^2}{b^2}\right).$$

Also shew that the curve, where it cuts the axis of y, will be concave or convex to the origin according as $b^2 \lessgtr 2a^2$.

8. From the fact that the cardioid is the inverse of a parabola with respect to the focus, or otherwise, prove that the normals at the extremities of any chord through the cusp are at right angles, and that the line joining their intersection to the cusp is perpendicular to the chord.

9. Prove by inversion, or otherwise, that the cardioids

$$r = a\,(1 + \cos\theta), \quad r = b\,(1 - \cos\theta)$$

cut one another at right angles.

10. If ds, ds' be corresponding elements of a curve and its inverse,

$$ds : ds' = r^2 : k^2 = k^2 : r'^2,$$

where r, r' are the radii.

11. The pedal of a parabola with respect to its vertex is the cissoid (Art. 119 (16)).

12. If two tangents to a curve make a constant angle with one another, the locus of their intersection (P) touches the circle through P and the two points of contact.

13. Prove that the area of a pedal curve is given by the formula

$$\tfrac{1}{2}\int p^2 d\psi.$$

14. Prove that the arc of a pedal curve is expressed by

$$\int r\,d\psi.$$

15. The area of the pedal of an ellipse, the centre being pole, is

$$\tfrac{1}{2}\pi\,(a^2 + b^2),$$

where a, b are the semi-axes.

16. The pedal of the hyperbola

$$\frac{x^2}{a^2} - \frac{y^2}{b^2} = 1$$

with respect to the centre consists of two loops, each of area

$$\tfrac{1}{2}ab + \tfrac{1}{2}\,(a^2 - b^2)\tan^{-1}\frac{a}{b}.$$

17. If p_0, p_1 be perpendiculars on the tangent to a curve from the origin of (rectangular) coordinates, and from the point (x_1, y_1) respectively, prove that

$$p_1 = p_0 - x_1\cos\psi - y_1\sin\psi,$$

where ψ is the inclination of the perpendiculars to the axis of x.

18. If A_0, A_1 be the areas of the pedals of a closed oval curve with respect to the origin O and with respect to the point (x_1, y_1), both these points being within the curve, prove that

$$A_1 = A_0 - x_1 \int_0^{2\pi} p_0 \cos \psi \, d\psi - y_1 \int_0^{2\pi} p_0 \sin \psi \, d\psi + \tfrac{1}{2}\pi (x_1^2 + y_1^2).$$

19. Prove that the locus of a point such that the pedal of a given closed oval curve with respect to it as pole has a given constant area is a circle; and that the circles corresponding to different values of the constant are concentric.

Also that, if O be the common centre, the area of the pedal with respect to any other point P exceeds the area of the pedal with respect to O by the area of the circle whose radius is OP.

20. The negative pedal of the parabola

$$y^2 = 4ax$$

with respect to the vertex is the curve

$$27ay^2 = (x - 4a)^3.$$

21. In what case is

$$p = a \cos \psi \, ?$$

22. Prove that the curve for which

$$p = a \sin \psi \cos \psi$$

is the astroid

$$x^{\frac{2}{3}} + y^{\frac{2}{3}} = a^{\frac{2}{3}}.$$

23. State what property follows by differentiation with respect to the arc (s) from the equation

$$r^2 + r'^2 = k^2,$$

and verify the result geometrically.

24. Prove the following construction for the normal at any point P of a Cassini's oval: In PS, PS' take points Q, Q', respectively, such that $PQ = PS'$, and $PQ' = PS$; the line joining P to the middle point of QQ' is the required normal.

25. A system of parallel rays is to be reflected so as to pass through a fixed point; prove that the reflecting curve must be a parabola.

26. A system of parallel rays is to be refracted so that their directions pass through a fixed point; prove that the refracting curve must be a conic, and that the eccentricity of the conic will be equal to the ratio of the refractive indices.

27. Prove that the equation of a Cartesian oval, referred to either focus as pole, is of the form

$$r^2 - 2 (a + b \cos \theta) r + c^2 = 0.$$

28. Prove that a Cartesian oval is necessarily closed, if we except the case where the curve is a branch of a hyperbola.

CHAPTER X

CURVATURE

133. Measure of Curvature.

As regards the applications of the Calculus to the theory of plane curves we have so far been concerned chiefly with the direction of the tangent at various points. We have not considered specially the manner in which this direction varies from point to point.

The subject of curvature, to which we now proceed, can be treated from several independent stand-points, and although all the methods lead to identically the same formulæ, it is important for the student to observe that they are in their foundations logically distinct.

In the first of these methods*, we begin by defining the 'total' or 'integral' curvature of an arc of a curve as the angle ($\delta\psi$) through which the tangent turns as the point of contact travels from one end of the arc to the other.

The 'mean curvature' of the arc is defined as the ratio of the total curvature to the length (δs) of the arc; it is therefore equal to

$$\frac{\delta\psi}{\delta s}.$$

The 'curvature at a point' P of a curve is defined as the mean curvature of an infinitely small arc terminated by that point. In conformity with the previous notation it is denoted by

$$\frac{d\psi}{ds}. \quad \dots\dots\dots\dots\dots\dots\dots\dots(1)$$

In a circle of radius R we have $\delta s = R\delta\psi$, and therefore

$$\frac{d\psi}{ds} = \frac{1}{R},$$

i.e. the curvature of a circle is measured by the reciprocal of its radius. Hence, if ρ be the radius of the circle which has the same curvature as the given curve at the point P, we have

$$\rho = \frac{ds}{d\psi}. \quad \dots\dots\dots\dots\dots\dots(2)$$

* Other methods are explained in Arts. 136, 137.

A circle of this radius, having the same tangent at P, and its concavity turned the same way, as in the given curve, is called the 'circle of curvature,' its radius is called the 'radius of curvature,' and its centre the 'centre of curvature.'

The length intercepted by this circle on a straight line drawn through P in any specified direction is called the 'chord of curvature' in that direction. If θ be the angle which the direction makes with the normal, the length (q) of the chord is given by

$$q = 2\rho \cos \theta. \quad\dots\dots\dots\dots\dots\dots\dots(3)$$

If ξ. η be the rectangular coordinates of the centre of curvature, we have by orthogonal projections

$$\xi = x - \rho \sin \psi, \quad \eta = y + \rho \cos \psi, \quad\dots\dots\dots(4)$$

provided the zero of ψ be when the tangent is parallel to the axis of x.

The centre of curvature is the intersection of two consecutive normals to the given curve. For if PC, $P'C$ be the normals at two consecutive points, including an angle $\delta\psi$, and if δs be the arc PP', then drawing the chord PP' we have (see Fig. 107)

$$\frac{CP}{PP'} = \frac{\sin CP'P}{\sin \delta\psi},$$

or $\qquad CP = \sin CP'P . \dfrac{PP'}{\delta s} . \dfrac{\delta\psi}{\sin \delta\psi} . \dfrac{\delta s}{\delta\psi}.$

When P' is taken infinitely near to P, the limiting value of each factor on the right hand, except the last, is unity. Hence, ultimately, $CP = ds/d\psi = \rho$.

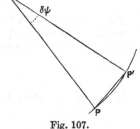

In modern geometry a curve is regarded as generated in a two-fold way, first as the locus of a point, and secondly as the envelope of a straight line (see Art. 141). Considering any continuous succession of these associated elements, the straight line is at any instant rotating about the point, and the point is travelling along the straight line; and

Fig. 107.

the curvature $d\psi/ds$ expresses the relation between these two motions.

If at any point the curvature is zero, the rotation of the tangent is momentarily arrested, and we have what is called a 'stationary tangent.' The simplest instance of this is at a point of inflexion (Art. 67), where the direction of the rotation of the tangent is reversed after the stoppage.

If at any point the radius of curvature $(ds/d\psi)$ vanishes, the motion of the point along the line is momentarily arrested, and we have a 'stationary point.' The simplest instance of this is at a 'cusp' such as we have met with in Figs. 70, 74, 79, 83, etc. The direction of motion of the point is in such cases reversed after the stoppage. In the examples of Art. 119 a cusp was regarded as due to the evanescence of a loop: this shews in another way why the radius of curvature should vanish there.

The consideration of curvature is of importance in numerous dynamical and physical problems. For example, in Dynamics, if the force acting on a moving particle be resolved into two components, along the tangent and normal to the path, respectively, the former component affects the velocity, and the latter the direction of motion. If from a fixed origin we draw a vector OV to represent the velocity at any instant, the polar coordinates of V may be taken to be v, ψ, where $v = ds/dt$. Hence the radial and transverse velocities of V will (see Art. 112 (6)) be

$$\frac{dv}{dt} \text{ and } v\frac{d\psi}{dt}, \quad \dots\dots\dots\dots\dots\dots\dots(5)$$

respectively. These are the rates of change of the velocity estimated in the direction of the tangent and normal to the path of the particle. Since

$$v\frac{d\psi}{dt} = v\frac{d\psi}{ds}\frac{ds}{dt} = \frac{v^2}{\rho}, \quad \dots\dots\dots\dots\dots\dots(6)$$

the latter component is equal to the product of the curvature into the square of the velocity.

134. Intrinsic Equation of a Curve.

The formula

$$\rho = \frac{ds}{d\psi} \quad \dots\dots\dots\dots\dots\dots\dots\dots(1)$$

is of course most immediately applicable when the relation between s and ψ for the curve in question is given in the form

$$s = f(\psi). \quad \dots\dots\dots\dots\dots\dots(2)$$

This is called the 'intrinsic' equation of the curve, for the reason that its form does not depend materially on space-elements extraneous to the curve. The only arbitrary elements are the origin of s and the origin of ψ, and a change in either of these merely adds a constant to the corresponding variable.

If the intrinsic equation be not known, we may employ one or other of the formulæ of Art. 135; or we may, in particular cases have recourse to special artifices. See Exs. 4, 5, below.

Ex. 1. In the catenary we have

$$s = a \tan \psi, \quad \dots\dots\dots\dots\dots\dots\dots\dots(3)$$

whence
$$\rho = a \sec^2 \psi = y \sec \psi, \quad \dots\dots\dots\dots\dots\dots(4)$$

the notation being as in Art. 120. On reference to the figure there given it appears that the radius of curvature is equal to the normal PG.

Ex. 2. In the cycloid (Art. 122) we have

$$s = 4a \sin \psi, \quad \dots\dots\dots\dots\dots\dots\dots(5)$$

and therefore
$$\rho = 4a \cos \psi. \quad \dots\dots\dots\dots\dots\dots(6)$$

Hence in Fig. 79, p. 294, we have $\rho = 2PI$, or the radius of curvature is double the normal.

Ex. 3. Again, in the epicycloid we have (Art. 123 (11))

$$s = \frac{4 (a + b) b}{a} \sin \frac{a}{a + 2b} \psi, \quad \dots\dots\dots\dots(7)$$

and therefore

$$\rho = \frac{4 (a + b) b}{a + 2b} \cos \frac{a}{a + 2b} \psi = \frac{4 (a + b) b}{a + 2b} \cos \tfrac{1}{2}\phi. \quad \dots\dots(8)$$

Hence, on reference to Fig. 81, p. 297, it appears that

$$\rho = \frac{2 (a + b)}{a + 2b} PI, \quad \dots\dots\dots\dots\dots\dots(9)$$

where PI is the length of the normal between the tracing point and the fixed circle.

Ex. 4. In the parabola $y^2 = 4ax$ we have

$$y = 2a \cot \psi, \quad \dots\dots\dots \dots\dots\dots\dots(10)$$

whence
$$\sin \psi = \frac{dy}{ds} = -\frac{2a}{\sin^2 \psi} \frac{d\psi}{ds},$$

or
$$\rho = -\frac{2a}{\sin^3 \psi}, \quad \dots\dots\dots\dots\dots(11)$$

the negative sign indicating that ψ diminishes as s increases.

Ex. 5. If the ellipse

$$x = a \cos \phi, \ y = b \sin \phi \quad \dots\dots\dots\dots\dots(12)$$

be supposed derived by orthogonal projection from the circle

$$x = a \cos \phi, \ y = a \sin \phi, \quad \dots\dots\dots\dots\dots(13)$$

we have
$$\frac{ds}{d\phi} = \beta, \quad \dots\dots\dots\dots\dots\dots\dots(14)$$

where β is the conjugate semi-diameter. For the element of arc is altered from $a\delta\phi$ to δs, and the parallel radius from a to β. Also since $\frac{1}{2}\beta^2\delta\psi$ and $\frac{1}{2}a^2\delta\phi$ represent corresponding elements of area, we have

$$\beta^2\delta\psi = \frac{b}{a} \times a^2\delta\phi,$$

or

$$\frac{d\phi}{d\psi} = \frac{\beta^2}{ab}. \qquad \dots\dots\dots\dots\dots\dots\dots(15)$$

Hence

$$\rho = \frac{ds}{d\psi} = \frac{ds}{d\phi}\frac{d\phi}{d\psi} = \frac{\beta^3}{ab}. \qquad \dots\dots\dots\dots\dots(16)$$

If p be the perpendicular from the centre on the tangent-line, we have $p\beta = ab$, so that our result may also be written

$$\rho = \frac{\beta^2}{p}, \quad \text{or} \quad \rho = \frac{a^2 b^2}{p^3}. \qquad \dots\dots\dots\dots\dots(17)$$

Since $\quad p^2 = a^2 \cos^2\psi + b^2 \sin^2\psi = a^2 (1 - e^2 \sin^2\psi),$

the last form is equivalent to

$$\rho = a\, \frac{1 - e^2}{(1 - e^2 \sin^2\psi)^{\frac{3}{2}}}. \qquad \dots\dots\dots\dots(18)$$

This formula leads to an important result in Geodesy. The figure of the Earth being taken to be an ellipsoid of revolution, the expression for the radius of curvature in terms of the latitude ψ is, if we neglect e^4,

$$\frac{ds}{d\psi} = a\, (1 - e^2 + \tfrac{3}{2}e^2 \sin^2\psi) = a\, (1 - \tfrac{1}{2}\epsilon - \tfrac{3}{2}\epsilon \cos 2\psi), \quad \dots\dots(19)$$

where $\epsilon = (a - b)/a = \tfrac{1}{2}e^2$; that is, ϵ denotes the 'ellipticity' of the meridian. Integrating (19) we find, for the length of an arc of the meridian, from the equator to latitude ψ,

$$s = a\, (1 - \tfrac{1}{2}\epsilon)\, \psi - \tfrac{3}{4}a\epsilon \sin 2\psi. \qquad \dots\dots\dots\dots(20)$$

Ex. 6. In the equiangular spiral (Art. 126), we have

$$\psi = \theta + a, \qquad \dots\dots\dots\dots\dots\dots\dots\dots(21)$$

whence $\quad d\psi/ds = d\theta/ds = (\sin a)/r,$

or

$$\rho = \frac{r}{\sin a}. \qquad \dots\dots\dots\dots\dots\dots\dots\dots(22)$$

Hence the radius of curvature subtends a right angle at the origin.

135. Formulæ for the Radius of Curvature.

The expression $d\psi/ds$ for the curvature is easily translated into a variety of other forms.

1°. In rectangular Cartesian coordinates, we have

$$\tan\psi = \frac{dy}{dx}, \qquad \dots\dots\dots\dots\dots\dots(1)$$

and therefore

$$\sec^2 \psi \frac{d\psi}{ds} = \frac{d}{ds}\left(\frac{dy}{dx}\right) = \frac{d^2y}{dx^2}\frac{dx}{ds} = \cos \psi \frac{d^2y}{dx^2},$$

whence

$$\frac{1}{\rho} = \frac{\dfrac{d^2y}{dx^2}}{\left\{1 + \left(\dfrac{dy}{dx}\right)^2\right\}^{\frac{3}{2}}}. \qquad\qquad\ldots\ldots\ldots\ldots(2)$$

This form shews, again, that the curvature vanishes at a point of inflexion, where $d^2y/dx^2 = 0$ (Art. 67).

When dy/dx is a small quantity the formula (2) gives, approximately,

$$\frac{1}{\rho} = \frac{d^2y}{dx^2}, \qquad\qquad\ldots\ldots\ldots\ldots\ldots\ldots\ldots(3)$$

the proportional error being of the *second* order. This formula is an obvious transcript of $d\psi/ds$, since when ψ is small we may write $dy/dx \,(= \tan \psi)$ for ψ, and d/dx for d/ds. It has important practical applications, *e.g.* in the theory of flexure of bars.

2°. It was proved in Art. 131 that the projection (t) of the radius on the tangent is given by

$$t = \frac{dp}{d\psi}. \qquad\qquad\ldots\ldots\ldots\ldots\ldots\ldots\ldots(4)$$

If OU, OU' be the perpendiculars from the origin on two consecutive normals PC, $P'C$, and if OU' meet PC in N, we have, ultimately,

$$OU' - OU = U'N = CN\delta\psi, \quad \text{or} \quad \delta t = CN\delta\psi.$$

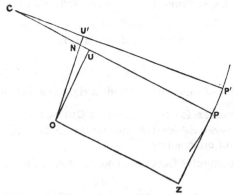

Fig. 108.

The limiting value of CU or CN is therefore $dt/d\psi$, whence

$$\rho = CP = OZ + CU = p + \frac{dt}{d\psi} = p + \frac{d^2p}{d\psi^2}. \qquad \ldots\ldots (5)$$

3°. With the notation of Art. 112 we have

$$\frac{t}{r} = \cos\phi = \frac{dr}{ds}. \qquad \ldots\ldots\ldots\ldots\ldots\ldots(6)$$

Since
$$\frac{ds}{d\psi} = \frac{ds}{dr}\frac{dr}{dp}\frac{dp}{d\psi} = t\frac{ds}{dr}\frac{dr}{dp},$$

this gives
$$\rho = r\frac{dr}{dp}, \qquad \ldots\ldots\ldots\ldots\ldots\ldots(7)$$

a form which is very convenient of application when the tangential-polar equation (Art. 129) is given.

Ex. 1. In the catenary

$$y = a\cosh(x/a) \qquad \ldots\ldots\ldots\ldots\ldots\ldots(8)$$

we have
$$\frac{dy}{dx} = \sinh\frac{x}{a}, \qquad \frac{d^2y}{dx^2} = \frac{1}{a}\cosh\frac{x}{a}, \qquad 1 + \left(\frac{dy}{dx}\right)^2 = \cosh^2\frac{x}{a},$$

whence
$$\rho = a\cosh^2(x/a) = y^2/a. \qquad \ldots\ldots\ldots\ldots\ldots\ldots(9)$$

Since $y = a\sec\psi$, this agrees with Art. 134, Ex. 1.

Ex. 2. In the parabola

$$r = p^2/a \qquad \ldots\ldots\ldots\ldots\ldots\ldots(10)$$

we have
$$\rho = r\frac{dr}{dp} = \frac{2p^3}{a^2} = \frac{2r^{\frac{3}{2}}}{a^{\frac{1}{2}}}. \qquad \ldots\ldots\ldots\ldots\ldots\ldots(11)$$

Ex. 3. In the central conics we have (Art. 129, Ex. 4)

$$\frac{a^2b^2}{p^2} = b^2 \pm a^2 \mp r^2, \qquad \ldots\ldots\ldots\ldots\ldots\ldots(12)$$

and therefore
$$\rho = \pm\frac{a^2b^2}{p^3}. \qquad \ldots\ldots\ldots\ldots\ldots\ldots(13)$$

Cf. Art. 134, Ex. 5.

136. Newton's Method.

In another method of treating curvature, employed by Newton [*],
a circle is described touching the given curve at P, and passing
through a neighbouring point Q on it, and we investigate the
limiting value of the radius of this circle when Q is taken infinitely
near to P.

---

[*] *Principia*, lib. i., prop. vi., cor. 3.

We can easily shew that in the limit the circle becomes identical with the 'circle of curvature' at P, as defined in Art. 133. For if C be the centre, then, since $CP = CQ$, there will be some point (P') on the curve, between P and Q, such that its distance from C is a maximum or minimum, and therefore* such that CP' is normal to the curve. In the limit P' approaches P indefinitely, and C, being the intersection of consecutive normals, will coincide with the 'centre of curvature.'

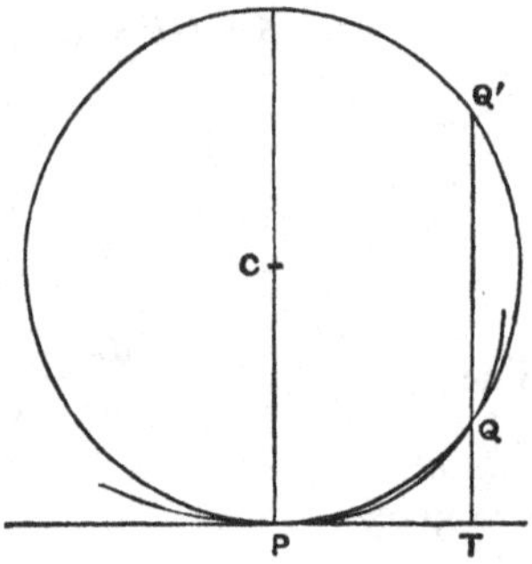

Newton's method leads to a very simple formula for the radius of curvature. Let $Q'QT$ be drawn perpendicular to the tangent at P, meeting the circle again in Q', and the tangent in T. Since

$$TP^2 = TQ \cdot TQ',$$

we have
$$2\rho = \lim TQ' = \lim \frac{TP^2}{TQ}. \qquad \ldots\ldots\ldots\ldots\ldots(1)$$

Fig. 109.

If $Q'QT$ be drawn at a definite inclination to the normal at P, instead of parallel to this normal, the limiting value of the same fraction gives the chord of curvature in the corresponding direction. It occasionally happens that the chord of curvature in some particular direction can be found with special facility; the radius of curvature can then be inferred by the formula (3) of Art. 133.

For instance, we can deduce the formula for the radius of curvature in Cartesian coordinates. Thus, referring to Fig. 42, p. 153, and denoting by q the chord of curvature parallel to the axis of y, we have

$$\frac{1}{q} = \lim \frac{QV}{PV^2} = \lim \frac{QV}{AH^2} \cdot \cos^2 \psi = \tfrac{1}{2}\phi''(a) \cos^2 \psi, \ldots(2)$$

where ψ is the inclination of the tangent at P to the axis of x. Since
$$q = 2\rho \cos \psi, \quad \tan \psi = \phi'(a),$$
it follows that

$$\frac{1}{\rho} = \phi''(a) \cos^3 \psi = \frac{\phi''(a)}{[1 + \{\phi'(a)\}^2]^{\frac{3}{2}}}. \qquad \ldots\ldots\ldots\ldots(3)$$

This is identical, except as to notation, with the formula (2) of Art. 135.

* See Art. 63, Ex. 2.

Ex. 1. In the parabola, let QR be a chord drawn parallel to the tangent at P, to meet the diameter through P in V; see Fig. 110. We have, then, from the geometry of the curve,

$$QV^2 = 4SP \cdot PV,$$

where S is the focus. Hence, for the chord of curvature (q) parallel to the axis,

$$q = \lim \frac{QV^2}{PV} = 4SP. \quad \dots\dots(4)$$

If θ be the angle which the normal at P makes with the axis, we have

$$\cos \theta = SZ/SP,$$

where SZ is the perpendicular from the focus on the tangent at P. Hence

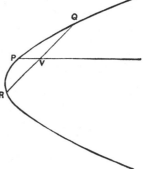

Fig. 110.

$$\rho = \tfrac{1}{2}q \sec \theta = 2\,\frac{SP^2}{SZ} = 2\,\frac{SP^{\frac{3}{2}}}{SA^{\frac{1}{2}}}, \quad \dots\dots\dots\dots(5)$$

since $SZ^2 = SA \cdot SP$, A being the vertex.

Ex. 2. In the ellipse (or hyperbola), if QR, drawn parallel to the tangent at either extremity of the diameter PCP', meet this diameter in V, we have

$$QV^2 : PV \cdot VP' = CD^2 : CP^2,$$

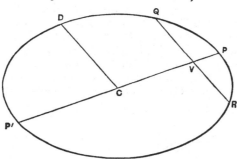

Fig. 111.

where CD is the semi-diameter conjugate to CP. Hence, for the chord of curvature (q) through the centre,

$$q = \lim \frac{QV^2}{PV} = \lim \frac{CD^2}{CP^2} \cdot VP' = 2\,\frac{CD^2}{CP}. \quad \dots\dots\dots(6)$$

If CZ be the perpendicular from the centre on the tangent at P, and θ the angle which CP makes with the normal, we have $\cos \theta = CZ/CP$, and therefore

$$\rho = \tfrac{1}{2}q \sec \theta = \frac{CD^2}{CZ}, \quad \dots\dots\dots\dots\dots\dots\dots\dots(7)$$

in agreement with Art. 134 (17).

Again, if θ' be the inclination of either focal distance to the normal at P, it is known that $\cos \theta' = CZ/CA$, where A is an extremity of the major axis. The chord of curvature (q') through either focus is therefore given by

$$q' = 2\rho \cos \theta' = 2\frac{CD^2}{CA}. \qquad\qquad\dots\dots\dots\dots\dots(8)$$

Ex. 3. To find the radius of curvature (ρ_0) at the vertex of the cycloid

$$x = a\,(\theta + \sin\theta), \quad y = a\,(1 - \cos\theta). \quad\dots\dots\dots\dots(9)$$

We have

$$\frac{x^2}{2y} = a\,(\theta + \sin\theta)^2 \div 4\sin^2 \tfrac{1}{2}\theta = a\left(1 + \frac{\sin\theta}{\theta}\right)^2 \div \left(\frac{\sin\tfrac{1}{2}\theta}{\tfrac{1}{2}\theta}\right)^2,$$

whence

$$\rho_0 = \lim_{\theta \to 0} \frac{x^2}{2y} = 4a. \qquad\qquad\dots\dots\dots\dots\dots(10)$$

137. Osculating Circle.

A slightly different way of treating the matter is based on the notion of the 'osculating circle.' If Q and R be two neighbouring points of the curve, one on each side of P^*, we consider the limiting value of the radius of the circle PQR, when Q and R are taken infinitely close to P.

We can shew that if the curvature of the given curve be continuous at P, this circle coincides in the limit with the 'circle of curvature.' For if C be the centre of the circle PQR, there will be a point P', between P and Q, such that CP' is normal to the given curve, and a point P'', between P and R, such that CP'' is normal to the curve. Let $P'C$ and $P''C$ meet the normal at P in the points C' and C'', respectively. Under the condition stated, C' and C'' will ultimately coincide with the centre of curvature at

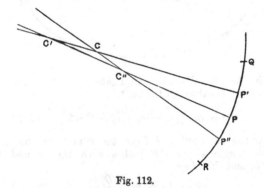

Fig. 112.

* This condition is not essential, but it simplifies the proof, and meets all ordinary requirements.

P, and, since $CC' < C'C''$, C will *à fortiori* ultimately coincide with the same point.

Since, before the limit, the circle PQR crosses the given curve three times in the neighbourhood of P, it appears that the osculating circle will in general cross the curve at the point of contact. See Fig. 116, p. 350.

If in Fig. 41, p. 151, QV meet the circle through P, Q, P' again in W, we have

$$VW = PV^2/QV,$$

and therefore, for the chord of curvature of the curve $y = \phi(x)$, parallel to the axis of y,

$$\frac{1}{q} = \lim \frac{QV}{PV^2} = \lim \frac{QV}{AH^2} \cos^2 \psi = \tfrac{1}{2}\phi''(a) \cos^2 \psi,$$

as in Art. 136 (2).

Ex. If in Fig. 110, p. 341, the circle PQR meet PV in W, we have

$$QV \cdot VR = PV \cdot VW, \text{ and therefore } VW = 4SP.$$

Hence the chord of curvature parallel to the axis of the parabola is $4SP$.

A similar argument may be used to find the chord of curvature through the centre, in the case of the ellipse (Fig. 111, p. 341).

138. Envelopes.

Suppose that we have a singly-infinite system, or family, of curves differing from one another only in the value assigned to some constant which enters into their specification. Two distinct curves of the system will in general intersect; and we consider here, more particularly, the limiting positions of the intersections when the change in the constant (or 'parameter,' of the system, as it is sometimes called), as we pass from one curve to the other, is infinitely small. On each curve we have then, in general, one or more points of 'ultimate intersection' with the consecutive curve of the system. The locus of these points of ultimate intersection is called the 'envelope' of the system.

Ex. 1. A system of circles of given radius, having their centres on a given straight line. The parameter here is the coordinate of the centre.

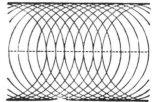

If C, C' be the centres of two circles of the system, the line joining their intersections bisects CC' at right angles. Hence the points of ultimate intersection of any circle with the consecutive circle are the extremities of the diameter which is perpendicular to the line of centres.

Fig. 113.

The envelope therefore consists of two straight lines parallel to the line of centres, at a distance equal to the given radius.

Ex. 2. A straight line including, with the coordinate axes, a triangle of constant area (k^2).

If AB, $A'B'$ be two positions of the line, intersecting in P, the triangles APA', BPB' will be equal, whence

$$PA \cdot PA' = PB \cdot PB'.$$

Hence, ultimately, when AA' is infinitely small, P will be the middle point of AB. If x, y be the coordinates of P, and ω the inclination of the axes, we have, then, $OA = 2x$, $OB = 2y$, and therefore

$$2xy \sin \omega = k^2.$$

The envelope is therefore a hyperbola having the coordinate axes as asymptotes. Fig. 114 illustrates the case of $\omega = \frac{1}{2}\pi$.

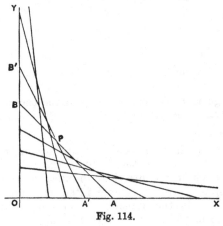

Fig. 114.

139. General Method of finding Envelopes.

The equation of any curve of the system being

$$\phi(x, y, \alpha) = 0, \qquad \dots\dots\dots\dots\dots\dots\dots(1)$$

where α is the parameter, then at the intersection with another curve

$$\phi(x, y, \alpha') = 0, \qquad \dots\dots\dots\dots\dots\dots(2)$$

we have, evidently,

$$\frac{\phi(x, y, \alpha') - \phi(x, y, \alpha)}{\alpha' - \alpha} = 0. \qquad \dots\dots\dots\dots(3)$$

When the variation $\alpha' - \alpha$ of the parameter is infinitely small, this last equation takes the form

$$\frac{\partial}{\partial \alpha} \phi(x, y, \alpha) = 0, \dots\dots\dots\dots\dots\dots\dots(4)$$

where $\partial/\partial\alpha$ is the symbol of partial differentiation with respect to α. See Art. 34.

The coordinates of the point, or points, of ultimate intersection are determined by (1) and (4) as simultaneous equations, and the locus of the ultimate intersections is to be found by elimination of α between these equations.

Ex. 1. The circles considered in Art. 138, Ex. 1 may be represented by

$$(x-a)^2 + y^2 = a^2. \quad\text{...........................(5)}$$

Differentiating with respect to a, we find

$$x - a = 0. \quad\text{...............................(6)}$$

Eliminating a between (5) and (6) we get

$$y = \pm\, a, \quad\text{................................(7)}$$

the envelope required.

Ex. 2. If a particle be projected from the origin at an elevation θ, with the velocity 'due to' a height h, the equation of the parabolic path is

$$y = x\tan\theta - \tfrac14\frac{x^2}{h}\sec^2\theta, \quad\text{....................(8)}$$

where the axes of x, y are respectively horizontal and vertical. Writing a for $\tan\theta$, we get

$$y = ax - \tfrac14\frac{x^2}{h}(1+a^2). \quad\text{.....................(9)}$$

To find the envelope of the paths for different elevations, and therefore for different values of a, we differentiate (9) with respect to a, and find

$$x - \tfrac12\frac{ax^2}{h} = 0. \quad\text{..........................(10)}$$

This is satisfied either by $x = 0$, or by $ax = 2h$. The former makes $y = 0$, and shews that the origin is part of the locus, as is otherwise obvious. The alternative result leads, on elimination of a, to

$$x^2 = 4h\,(h-y), \quad\text{..........................(11)}$$

a parabola having its axis vertical, its focus at the origin, and its vertex at an altitude h*.

140. Algebraical Method.

If in the equation　　　$\phi\,(x,\, y,\, \alpha) = 0,$　　　..........................(1)

ϕ be a rational integral function of α, the rule of the preceding Art. may be investigated otherwise as follows.

---

* This problem is interesting historically as being the first instance in which the envelope of a family of *curved* lines was obtained (Bernoulli). The general method of finding envelopes appears to be due to Leibnitz.

If we assign any particular values to x, y, the equation determines α, that is, it determines what curves of the system pass through the given point (x, y). If the equation be of the nth degree in α, the number of these curves (real or imaginary) will be n, and these n curves will in general be distinct. But if the point in question be at the intersection of two consecutive curves, two of the values of α will be coincident. Now it was shewn in Art. 50 that the condition for a double root of the equation in α is

$$\frac{\partial}{\partial \alpha} \phi (x, y, \alpha) = 0. \quad\quad\quad\quad\quad (2)$$

The ultimate intersections are therefore determined as before by (1) and (2) as simultaneous equations, and the envelope by elimination of α between them.

If the equation (1) be of the *first* degree in α, only one curve of the system passes through any assigned point, and there is of course no envelope. Examples of this are furnished by the parallel lines

$$lx + my = \alpha, \quad\quad\quad\quad\quad (3)$$

and by the concentric circles

$$x^2 + y^2 = \alpha. \quad\quad\quad\quad\quad (4)$$

If (1) be a *quadratic* in α, say

$$P\alpha^2 + 2Q\alpha + R = 0, \quad\quad\quad\quad (5)$$

where P, Q, R are given functions of x and y, the condition for equal roots is

$$PR = Q^2. \quad\quad\quad\quad\quad (6)$$

This is therefore the equation of the envelope.

Ex. 1. If the straight line

$$\frac{x}{a} + \frac{y}{\beta} = 1 \quad\quad\quad\quad\quad (7)$$

include with the coordinate axes a triangle of constant area k^2, we have

$$a\beta \sin \omega = 2k^2, \quad\quad\quad\quad\quad (8)$$

where ω is the inclination of the axes. Hence, eliminating β, the equation of the variable line is found to be

$$a^2 y \sin \omega - 2ak^2 + 2k^2 x = 0. \quad\quad\quad\quad (9)$$

Expressing that this quadratic in α has equal roots, we find for the envelope

$$2xy \sin \omega = k^2, \quad\quad\quad\quad\quad (10)$$

as in Art. 138, Ex. 2.

Ex. 2.　One leg of a right angle passes through a fixed point, and the vertex describes a fixed straight line; to find the envelope of the other leg.

If the fixed straight line be the axis of y, and the fixed point be at $(a, 0)$, the equation of the second leg is easily seen to be

$$y = mx + \frac{a}{m}, \quad \dots\dots\dots\dots\dots\dots\dots(11)$$

where m is the tangent of the inclination to the axis of x. Writing the equation in the form

$$m^2 x - my + a = 0, \quad \dots\dots\dots\dots\dots\dots(12)$$

we see that the envelope is the parabola

$$y^2 = 4ax. \quad \dots\dots\dots\dots\dots\dots\dots\dots(13)$$

141. Contact-Property of Envelopes.

The examples already given will have prepared the student for the following theorem:

The envelope of a system of curves touches (in general) at each of its points the corresponding curve of the system.

The equations
$$\phi(x, y, \alpha) = 0, \dots\dots\dots\dots\dots\dots\dots(1)$$

and
$$\frac{\partial}{\partial \alpha} \phi(x, y, \alpha) = 0, \dots\dots\dots\dots\dots\dots(2)$$

determine x, y as functions of α, say

$$x = F(\alpha), \quad y = f(\alpha), \quad \dots\dots\dots\dots\dots(3)$$

and the latter pair of equations define the envelope. If we substitute from (3) on the left-hand side of (1) we obtain a function of α which must vanish identically, and the result of differentiating this function with respect to α must also be zero. Hence, by the rule of Art. 59, $1°$, we must have

$$\frac{\partial \phi}{\partial x} \frac{dx}{d\alpha} + \frac{\partial \phi}{\partial y} \frac{dy}{d\alpha} + \frac{\partial \phi}{\partial \alpha} \frac{d\alpha}{d\alpha} = 0, \quad \dots\dots\dots\dots(4)$$

which reduces, in virtue of (2), to

$$\frac{\partial \phi}{\partial x} \frac{dx}{d\alpha} + \frac{\partial \phi}{\partial y} \frac{dy}{d\alpha} = 0, \quad \dots\dots\dots\dots(5)$$

or
$$\frac{\dfrac{dy}{d\alpha}}{\dfrac{dx}{d\alpha}} = -\frac{\dfrac{\partial \phi}{\partial x}}{\dfrac{\partial \phi}{\partial y}}. \quad \dots\dots\dots\dots\dots\dots(6)$$

Now, by Art. 61, the left-hand side of this equality is the value of dy/dx for the envelope; and the right-hand side is, by Art. 59 (10), the value of dy/dx for the curve (1). Hence at the point of ultimate intersection the curve (1) and the envelope have a common tangent line.

The geometrical basis of the theorem may be indicated as follows:

Let the figure represent portions of two curves of the system, corresponding to values a_0, a_1 of the parameter a, and intersecting

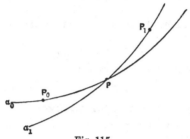

Fig. 115.

in P. Let P_0 and P_1 be the corresponding points on the envelope; viz. P_0 is the limiting position of P when, a_0 being fixed, a_1 is taken infinitely nearly equal to a_0; and P_1 is the limiting position of P when, a_1 being fixed, a_0 is taken infinitely nearly equal to a_1. Since these variations of a are in opposite senses, and since the coordinates of P are as a rule symmetric functions of a_0, a_1, the corresponding displacements of P, viz. PP_0 and PP_1, will in general, when $|a_1 - a_0|$ is very small, be in nearly opposite directions, and P_0PP_1 will be a very obtuse-angled triangle. Hence, ultimately, when $|a_1 - a_0|$ is infinitely small, the chords P_0P_1 and P_0P will coincide in direction; i.e. the tangent to the envelope is identical with the tangent to the variable curve.

The foregoing investigations break down in certain cases. As regards the analytical proof, it is plain that no inference can be drawn from (5) whenever at the point in question we have

$$\frac{\partial \phi}{\partial x} = 0, \qquad \frac{\partial \phi}{\partial y} = 0, \qquad \dots\dots\dots\dots\dots\dots\dots(7)$$

simultaneously; i.e. when the value of dy/dx for the curve (1) is not uniquely determinate. This peculiarity occurs at a 'singular point,' whether it be of the nature of a node, a cusp, or an isolated point (see Art. 119). It appears that the locus of the singular points of the given family, when such a locus exists, is included in the result of eliminating a between (1) and (2), but this locus does not in general 'touch' the given curves, in any proper sense of the word. The full investigation of this matter is beyond our limits*, but a simple example may be given. Consider the family

$$a(y - a)^2 = x^2(x + b). \qquad \dots\dots\dots\dots\dots\dots(8)$$

---

* It is given in books on Differential Equations, under the head of 'Singular Solutions.'

It appears from Art. 119 that there is a node, a cusp, or an isolated point at $(0, a)$, according as b is positive, zero, or negative. The process for finding the envelope gives $y - a = 0$ and therefore

$$x^2 (x + b) = 0. \quad\dots\dots\dots\dots\dots\dots\dots(9)$$

The line $x = 0$ gives the locus of the singular points, and does not touch the original curves; the line $x = -b$ on the other hand does so (unless $b = 0$).

In the geometrical view of the matter it was assumed that there is no other intersection of the curves a_0, a_1 in the immediate neighbourhood of P. In the case of a node we have usually *two* adjacent intersections, whose x-coordinates (for instance) are of the forms $f(a_0, a_1)$ and $f(a_1, a_0)$, respectively; but $f(a_0, a_1)$ is not a symmetric function of a_0, a_1. The argument does not therefore apply to the node locus. Again, in the case of a cusp the displacement of the point P in Fig. 115, due to an infinitesimal variation of a_0 or a_1, is found not to be of the first order; and the points P_0, P_1 are as a rule on the *same* side of P. In the neighbourhood of an isolated point there is no real intersection of consecutive curves.

142. Evolutes.

The 'evolute' of a curve is the locus of its centre of curvature. Since the centre of curvature is (Art. 133) the intersection of two consecutive normals, the evolute is also the envelope of the normals to the given curve. Hence the normals to the original curve are tangents to the evolute*.

Ex. 1. In the parabola

$$y^2 = 4ax, \quad\dots\dots\dots\dots\dots\dots\dots\dots(1)$$

we have $\qquad x = a \cot^2 \psi, \quad y = 2a \cot \psi, \quad\dots\dots\dots\dots(2)$

and (by Art. 134, Ex. 4)

$$\rho = -2a/\sin^3 \psi. \quad\dots\dots\dots\dots\dots\dots(3)$$

The coordinates of the centre of curvature are therefore

$$\left. \begin{aligned} \xi &= x - \rho \sin \psi = 3x + 2a, \\ \eta &= y + \rho \cos \psi = -y^3/4a^2 \end{aligned} \right\} . \quad\dots\dots\dots\dots(4)$$

Hence $\qquad \eta^2 = y^6/16a^4 = 4x^3/a = \tfrac{4}{27} (\xi - 2a)^3/a.$

The evolute is therefore the semi-cubical parabola

$$ay^2 = \tfrac{4}{27} (x - 2a)^3. \quad\dots\dots\dots\dots\dots\dots(5)$$

Otherwise: it is shewn in books on Conics that the equation of the normal is of the form

$$y = m (x - 2a) - am^3. \quad\dots\dots\dots\dots\dots\dots(6)$$

---

* It being evident that the exceptional cases noted at the end of Art. 141 cannot present themselves in the envelope of a *straight line*.

To find the envelope of this we differentiate partially with respect to m, and obtain

$$x - 2a = 3am^2, \quad y = 2am^3. \quad \dots\dots\dots\dots(7)$$

The elimination of m leads again to the result (5).

The curve is shewn in Fig. 116.

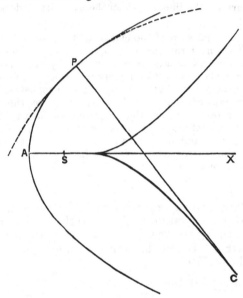

Fig. 116.

Ex. 2.   The normal at any point of the ellipse

$$x = a \cos \phi, \quad y = b \sin \phi \quad \dots\dots\dots\dots(8)$$

is

$$\frac{ax}{\cos \phi} - \frac{by}{\sin \phi} = a^2 - b^2. \quad \dots\dots\dots\dots(9)$$

Differentiating with respect to ϕ we find

$$\frac{ax}{\cos^3 \phi} = -\frac{by}{\sin^3 \phi}, = \lambda, \text{ say.} \quad \dots\dots\dots(10)$$

Substituting in (9), we have

$$\lambda = a^2 - b^2. \quad \dots\dots\dots\dots\dots(11)$$

Hence the coordinates of the centre of curvature are

$$x = \frac{a^2 - b^2}{a} \cos^3 \phi, \quad y = -\frac{a^2 - b^2}{b} \sin^3 \phi; \quad \dots\dots(12)$$

and the evolute is

$$(ax)^{\frac{2}{3}} + (by)^{\frac{2}{3}} = (a^2 - b^2)^{\frac{2}{3}}. \qquad \ldots\ldots\ldots\ldots\ldots(13)$$

This curve, which may be obtained by orthogonal projection from the astroid, is shewn in Fig. 117.

The centres of curvature at the points A, B, A', B' are E, F, E', F', respectively.

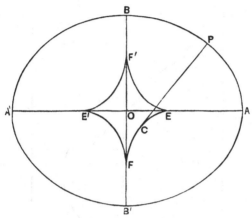

Fig. 117.

Ex. 3.　To find the evolute of a cycloid.

At any point P on the cycloid APD (Fig. 118), we have, by Art. 134, Ex. 2,

$$\rho = 2PI. \qquad \ldots\ldots\ldots\ldots\ldots\ldots\ldots\ldots\ldots(14)$$

Let the axis AB be produced to D', so that $BD' = AB$; and produce TI to meet a parallel to BI, drawn through D', in I'. If a circle be described on II' as diameter, and PI be produced to meet its circumference in P', we have $P'I = PI$, so that P' is the centre of curvature of the cycloid at P. And since the arc $P'I'$ is equal to the arc TP, and therefore to BI or $D'I'$, the locus of P' is evidently the cycloid generated by the circle $IP'I'$, supposed to roll on the under side of $D'I'$, the tracing point starting from D'. That is, the evolute is a cycloid equal to the original cycloid, and having a cusp at D'.

It appears, further, from Art. 122 (4), that the cycloidal arc $P'D$ is equal to $2IP'$, or $P'P$. Hence

$$\text{arc } D'P' + P'P = \text{const.} \qquad \ldots\ldots\ldots\ldots\ldots(15)$$

The lower cycloid in Fig. 118 is therefore an 'involute' (Art. 144) of the upper one*.

* This example is interesting historically in connection with the theory of the cycloidal pendulum. The results are due to Huyghens (1673).

Whenever a curve is defined by a relation between p and ψ, say

$$p = f(\psi), \quad\ldots\ldots\ldots\ldots\ldots\ldots\ldots\ldots\ldots(16)$$

the evolute is given by $p = f'(\psi), \quad\ldots\ldots\ldots\ldots\ldots\ldots\ldots(17)$

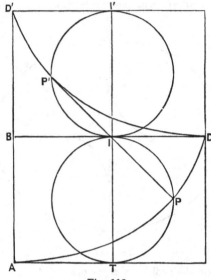

Fig. 118.

provided that in (17) the origin of ψ be supposed moved forwards through a right angle. This is seen at once on reference to Fig. 108, p. 338, since OU, the perpendicular from the origin on the tangent to the evolute, is equal to PZ, or $dp/d\psi$, when the symbols refer to the original curve.

Ex. 4.   To find the evolute of an epi- or hypo-cycloid.

If in Fig. 81, p. 297, a perpendicular p be drawn from O to TP, the tangent to the epicycloid at P, we have

$$p = OT \cos PIC = (a + 2b) \cos \tfrac{1}{2}\phi,$$

or

$$p = (a + 2b) \cos \frac{a}{a + 2b} \psi. \quad\ldots\ldots\ldots\ldots\ldots(18)$$

If the origin of ψ correspond to a cusp instead of to a vertex, the cosine of the angle must be replaced by the sine.

Hence, for the evolute, we have

$$p = -a \sin \frac{a}{a + 2b} \psi, \quad\ldots\ldots\ldots\ldots\ldots(19)$$

which can be brought to the same form as (18) by an adjustment of the origin of ψ. The evolute is therefore a similar epicycloid in which the dimensions are reduced in the ratio $a/(a + 2b)$.

For a hypocycloid we have merely to change the sign of b*.

143. Arc of an Evolute.

The difference of the radii of curvature at any two points of a curve is equal to the arc between the corresponding points of the evolute.

To prove this, let the normals at two neighbouring points P_0, P_1 of the curve meet in C; and let C_0, C_1 be the corresponding centres of curvature. By Art. 141, $C_1 C C_0$ is in general an obtuse-angled triangle; and when P_0, P_1 are taken infinitely close to one another, $C_1 C + C C_0$ is ultimately in a ratio of equality to $C_1 C_0$.

Also since the distance from C of a variable point on the curve is stationary at P_0, the difference between $C P_1$ and $C P_0$ is ultimately of the *second* order of small quantities, and may therefore be neglected. Hence

$$C_1 P_1 - C_0 P_0 = C_1 C + C C_0 = C_1 C_0.$$

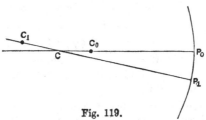

Fig. 119.

It follows that if ρ be the radius of curvature of the original curve, and σ the arc of the evolute, we have $\delta\rho = \delta\sigma$, ultimately, or

$$\frac{d\rho}{d\sigma} = 1. \qquad \dots\dots\dots\dots\dots\dots (1)$$

Hence, integrating, $\rho = \sigma + C,$ $\dots\dots\dots\dots\dots\dots\dots(2)$

where C is an arbitrary constant depending on the origin of measurement of σ.

Otherwise: by differentiation of the equations

$$\xi = x - \rho \sin\psi, \quad \eta = y + \rho \cos\psi \quad \dots\dots\dots\dots\dots(3)$$

* It appears on examination that the equation

$$p = c \cos m\psi, \quad \text{or} \quad p = c \sin m\psi,$$

represents an epi- or a hypo-cycloid according as $m \lessgtr 1$, provided we include the pericycloids among the epicycloids, in accordance with the definition of Art. 123.

The pedal of an epi- or a hypo-cycloid with respect to its centre is therefore an epicyclic of the special type referred to in Art. 125, Ex. 2. Thus Fig. 92 represents the pedal of a four-cusped epicycloid, and Fig. 94 that of a four-cusped hypocycloid.

of Art. 133, we find, since

$$dx/ds = \cos\psi, \quad dy/ds = \sin\psi, \quad d\psi/ds = 1/\rho,$$

$$\frac{d\xi}{ds} = -\frac{d\rho}{ds}\sin\psi, \quad \frac{d\eta}{ds} = \frac{d\rho}{ds}\cos\psi. \quad\text{.................(4)}$$

Hence

$$\frac{d\eta}{d\xi} = -\cot\psi, \quad\text{............................(5)}$$

which shews that the tangent to the evolute is normal to the original curve, and

$$\frac{d\sigma}{ds} = \sqrt{\left\{\left(\frac{d\xi}{ds}\right)^2 + \left(\frac{d\eta}{ds}\right)^2\right\}} = \frac{d\rho}{ds}, \quad\text{...............(6)}$$

which gives on integration the result (2).

For the case of the cycloid, this property has been already obtained in Art. 142 (15).

A curious consequence of the above theorem is that the circles of curvature of adjacent points on a curve do not in general intersect. For the distance between the centres is a chord of the evolute, and is therefore in general less than the corresponding arc, *i.e.* less than the difference of the radii.

Again, if the intrinsic equation of a curve be

$$s = f(\psi), \quad\text{............................(7)}$$

we have

$$\sigma = \rho + C = f'(\psi) + C. \quad\text{...................(8)}$$

If we alter the origin of ψ by a right angle, this is the intrinsic equation of the evolute. The additive constant may be omitted if we adjust the origin of σ.

Ex. 1. The radii of curvature of an ellipse of semi-axes a, b, at the extremities of these axes, are b^2/a and a^2/b, respectively. Hence the length of any one of the four portions into which the evolute is divided (see Fig. 117) by its cusps is

$$a^2/b - b^2/a \quad\text{or}\quad (a^3 - b^3)/ab.$$

Ex. 2. The intrinsic equation of the cycloid being

$$s = k\sin\psi, \quad\text{.............................(9)}$$

that of the evolute is

$$\sigma = k\cos\psi. \quad\text{..........................(10)}$$

The evolute is therefore an equal cycloid, as already proved.

144. Involutes, and Parallel Curves.

If a curve A be the evolute of a curve B, then B is said to be an 'involute' of A.

We say *an* involute because any given curve has an infinity of involutes. To obtain an involute we take any fixed point O on the curve, and along the tangent at a variable point P measure off a length PQ in the direction *from* O, so that

$$\text{arc } OP + PQ = \text{const.} \quad\text{.....................(1)}$$

It is easily shewn, by an inversion of the argument of Art. 143, that the tangents to the given curve are normals to the locus of Q, so that this locus fulfils the above definition of an involute. And, by varying the 'constant,' we obtain a series of involutes of the same curve.

As a concrete example we may imagine a string to be wound on a material arc of the given shape, being attached to a fixed point on it. The curve traced out by *any* point on the free portion of the string will be an involute. This is in fact the origin of the term.

Ex. 1.   The tractrix is an involute of the catenary; see Art. 120.

Ex. 2.   In an involute of a circle of radius a we have, evidently,

$$\frac{ds}{d\psi} = \rho = a\psi, \quad\dotfill(2)$$

if the origin of ψ be properly chosen. Hence, integrating,

$$s = \tfrac{1}{2}a\psi^2, \quad\dotfill(3)$$

no additive constant being required, if s be measured from the cusp ($\psi = 0$).

In this particular case (of the circle) it is evident that all the involutes are identically equal. It is therefore customary to speak of *the* involute of a circle. The curve is shown in Fig. 120

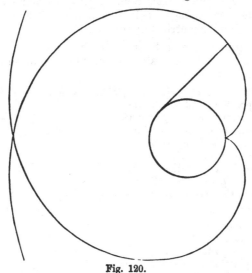

Fig. 120.

If a constant length be measured along the normal to a given curve, from the curve, the locus of the point thus determined is called a 'parallel' to the given curve.

If CP, CP' be two consecutive normals to the given curve, and

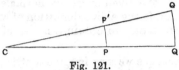

Fig. 121.

Q, Q' the corresponding points of a parallel curve, we have

$$PQ = P'Q'.$$

Since the difference between CP and CP' is of the second order of small quantities, it follows that the same holds of the difference between CQ and CQ', and thence that the angles at Q and Q' in the triangle CQQ' are ultimately right angles. Hence CQ, CQ' are normals to the parallel curve.

Hence two parallel curves have the same normals, and therefore the same evolute; in other words, parallel curves are involutes of the same curve.

Conversely, it is evident that the various involutes of any curve are a system of parallel curves.

145. Instantaneous Centre of a Moving Figure.

The theory of the displacements, in its own plane, of a figure of invariable form, though belonging properly to Kinematics, has some interesting geometrical applications.

The first proposition of the theory is that any such displacement is equivalent to a rotation about some finite or infinitely distant point.

The following is a proof. If A, B be any two points of the figure in its first position, and A', B' the same points in the second position, the new position P' of any third point originally at P is found by constructing the triangle $A'P'B'$ congruent with APB. Hence the positions of *two* points are sufficient to determine the position of the moveable figure.

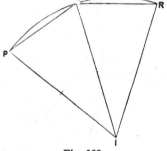

Now, considering any point whatever of the figure; let P be its initial and Q its final position; and let R be the final position of that point of the figure which was originally at Q.

Fig. 122.

Since PQ and QR are two positions of the same line, they are equal. Hence if I be the centre of the circle PQR, the triangles PIR, QIR are congruent; that is, I represents the same point in the two positions.

The displacement is therefore equivalent to a rotation about I. This point is called the 'centre of rotation.'

It may happen that PQ, QR are in a straight line. The displacement is then equivalent to a *translation* of the figure, without rotation; or, we may say, the centre of rotation is at infinity.

Next, considering any continuous motion of a plane figure in its own plane, let us fix our attention on two consecutive positions. The figure may be brought from the first of these to the second by a rotation about the proper centre. The limiting position of this point, when the two positions are taken infinitely close to one another, is called the 'instantaneous centre.'

If P, P' be consecutive positions of any the same point of the figure, and $\delta\theta$ the corresponding angle of rotation, the centre of rotation (I) is on the line bisecting PP' at right angles, and the angle PIP' is equal to $\delta\theta$. Hence, ultimately, the infinitesimal displacement of any point P at a finite distance from I is at right angles to IP and equal to $IP \cdot \delta\theta$.

If we introduce the consideration of time, and denote by δt the interval that elapses between the two positions, the limiting value of $\delta\theta/\delta t$, viz. $d\theta/dt$, is called the 'angular velocity' of the figure. The velocity of that point of the figure which coincides with the instantaneous centre I is zero, that of any other point P is at right angles to IP, and equal to $IP \cdot d\theta/dt$.

The fact that in any motion of a plane figure (of invariable form) the normals to the paths of the various points all pass through the instantaneous centre is often useful in geometrical questions. If we know the directions of displacement of two points of the figure, the instantaneous centre is determined as the intersection of the normals at these points to the respective directions. We can thence assign the directions, and relative magnitudes, of the displacements of all other points.

Again, considering any line (straight or curved) in the moving figure, it is evident that the point or points of ultimate intersection of this line with a consecutive position are the feet of the normals drawn to it from the instantaneous centre. For any other point of the line is moving in a direction making a finite angle with it.

Ex. 1. The extremities of a straight line AB of constant length describe two straight lines OX, OY at right angles to one another.

It is known that any point P of the line describes an ellipse whose principal axes are along OX, OY. The above theorem now gives us a construction for the normal to this ellipse at P; viz. if we draw AI, BI perpendicular to OX, OY, respectively, I is the instantaneous centre, and IP the required normal. See Fig. 123.

Ex. 2. In the preceding Example, the point of ultimate inter-section of the moving line AB with a consecutive position is at the

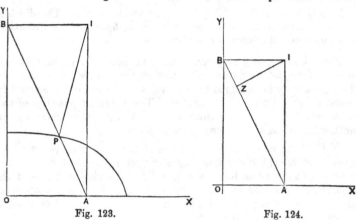

Fig. 123.   Fig. 124.

foot Z (Fig. 124) of the perpendicular from the instantaneous centre I. Now if

$$AB = k, \quad \angle OAB = \phi,$$

the coordinates of Z are given by

$$\left. \begin{aligned} x &= BZ \cos \phi = BI \cos^2 \phi = k \cos^3 \phi, \\ y &= AZ \sin \phi = AI \sin^2 \phi = k \sin^3 \phi, \end{aligned} \right\} \quad \dots\dots\dots\dots(1)$$

and the envelope of AB is therefore the astroid

$$x^{\frac{2}{3}} + y^{\frac{2}{3}} = k^{\frac{2}{3}}. \quad \dots\dots\dots\dots\dots\dots\dots(2)$$

Cf. Art. 124, Ex. 4.

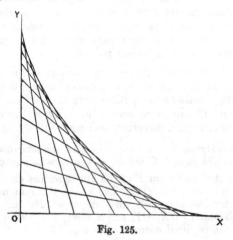

Fig. 125.

Ex. 3.   An arm *OQ* revolves about one extremity *O* with angular velocity ω; a bar is hinged to it at *Q* and is constrained to pass always through a fixed point *C*; it is required to find the velocity of this **bar**

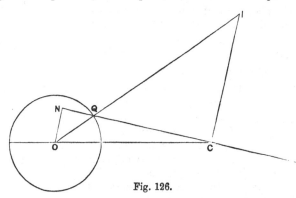

Fig. 126.

in the direction of its length.   (The arrangement is that of the crank and piston rod of a steam-engine with *oscillating* cylinder, the point *C* being on the pivot-line of the cylinder.)

The instantaneous centre is at the intersection of *OQ* produced with the perpendicular to the piston rod at *C*.   Hence, if *ON* be the perpendicular from *O* on *CQ*, produced if necessary, the velocity of the point of the rod which coincides with *C* is

$$\omega \,.\, OQ \times \frac{IC}{IQ} = \omega \,.\, OQ \times \frac{ON}{OQ} = \omega \,.\, ON. \quad \ldots\ldots\ldots\ldots(3)$$

146.   Application to Rolling Curves.

Suppose that we have two plane figures, each of invariable form, and that a curve fixed in one rolls, without sliding, on a curve fixed in the other.   Any point of either figure will then describe a curve relatively to the other; a curve so described is called a 'roulette.'

The cases where the rolling curves are circles have been considered in Arts. 122—124.

The general theory of roulettes is of some importance in Geometry and in Kinematics, owing to the fact that any continuous motion whatever of a figure in its own plane may be regarded as consisting in the rolling of a certain curve fixed relatively to the figure on a certain curve fixed in the plane.   See Art. 149.

When one plane curve rolls upon another, which is regarded as fixed, the instantaneous centre is at the point of contact.

We will suppose, in the figure, that it is the *lower* curve
which is fixed. Let A be the point of contact,
and let equal infinitely small arcs AP, AP'
($=\delta s$, say) be measured off along the two curves.
Let the normals at P and P' meet the common
normal at A in the points O and O'. Then
ultimately we have

$$OA = R, \quad O'A = R',$$

where R, R' are the radii of curvature of the
two curves at A. After an infinitely small
displacement, $P'O'$ will come into the same
straight line with OP, the two curves being
then in contact at P. Hence the angle ($\delta\theta$)
through which the rolling curve will turn,
being equal to the acute angle between OP
and $P'O'$, is equal to the sum of the angles at
O and O', so that

$$\delta\theta = \frac{\delta s}{R} + \frac{\delta s}{R'}, \quad \ldots\ldots\ldots\ldots(1)$$

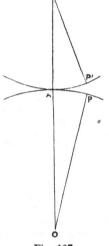

Fig. 127.

ultimately. Again, the chords AP, AP' are
ultimately equal, and they include an infinitely small angle at A.
Hence the distance PP' is ultimately of the *second order* in δs. It
follows that when δs is indefinitely diminished the limiting position
of the centre of rotation (I) coincides with A, for if it were at a
finite distance from this point, the displacement of P', being equal
to $IP' . \delta\theta$, by Art. 145, would be of the *first* order in δs.

It follows that when a curve rolls upon a fixed curve, the normals
to the paths of all points connected with the moving curve pass
through the point of contact. We have already had instances of
this result in the cycloidal and trochoidal curves discussed in
Arts. 122, 123. Again if a straight
line roll on a curve, it is normal
to the path traced out by any of
its points (Art. 144).

Further, if we consider any
line (straight or curved) which is
carried with the rolling curve, the
points of ultimate intersection
of the carried curve with its consecutive position are the feet of
the normals drawn to it from the
point of contact. And the en-

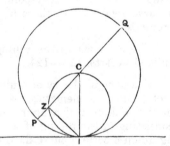

Fig. 128.

velope of the carried line is the locus of these feet.

Ex. 1. If a circle roll on a fixed straight line, any diameter envelopes a cycloid.

Let C be the centre of the rolling circle, I the point of contact, IZ the perpendicular on the diameter PQ. Since Z is on the circle whose diameter is CI, it is easily seen that if this circle be supposed to roll always with twice the angular velocity of the large circle, it will always have the same point of contact with the fixed line, and the point Z will move as if it were carried by the small circle. Its locus is therefore a cycloid.

Ex. 2. Similarly if a circle (A) roll on a fixed circle (B), the envelope of any diameter of A is an epi- or hypo-cycloid which would be generated by the rolling of a circle of half the size of A on the circumference of B.

147.　Curvature of a Point-Roulette.

To investigate the curvature of the path of any point P fixed relatively to the rolling curve, let I be the point of contact, and let I' be a consecutive point of contact, P' the corresponding position of P. Since the displacement of the point of the rolling curve which comes to I' is of the second order of small quantities, the angle through which the figure has turned is

$$\delta\theta = \angle PI'P', \quad\quad\quad\quad\dots\dots\dots\dots\dots(1)$$

ultimately. Let the normals to the path of P, viz. PI and $P'I'$, be produced to meet in C. If $\delta\psi$ be the inclination of these normals, we have

$$\delta\psi = \angle ICI' = \frac{\delta s \cos \phi}{CI}, \quad\quad\dots\dots\dots\dots(2)$$

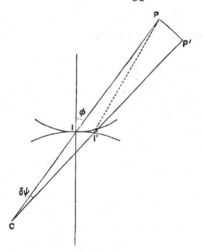

Fig. 129.

if ϕ be the angle which IP makes with the normal at I. Also, from the figure

$$\delta\psi = \angle PI'P' - \angle IPI'$$

$$= \delta\theta - \frac{\delta s \cos \phi}{PI}$$

$$= \delta s \left(\frac{1}{R} + \frac{1}{R'} - \frac{\cos\phi}{PI}\right), \quad\ldots\ldots\ldots\ldots(3)$$

by Art. 146 (1), if R and R' be the radii of curvature of the fixed and rolling curves. Equating (2) and (3), we find

$$\cos\phi \left(\frac{1}{CI} + \frac{1}{IP}\right) = \frac{1}{R} + \frac{1}{R'}. \quad\ldots\ldots\ldots\ldots(4)$$

This gives the limiting position of C, *i.e.* the centre of curvature of the path of P. The radius of curvature (ρ) is then found from

$$\rho = CP = CI + IP. \quad\ldots\ldots\ldots\ldots\ldots\ldots(5)$$

The result contained in (4) and (5) may be put in a simple geometrical form as follows. On the normal at I mark off a length IH such that

$$\frac{1}{IH} = \frac{1}{R} + \frac{1}{R'}, \quad\ldots\ldots\ldots(6)$$

and describe a circle on IH as diameter. Let IP meet this circle in Q. We have then

$$\frac{1}{IQ} = \frac{1}{IH \cos\phi} = \left(\frac{1}{R} + \frac{1}{R'}\right)\sec\phi,$$

and the relation (4) takes the form

$$\frac{1}{CI} + \frac{1}{IP} = \frac{1}{IQ}. \quad\ldots\ldots\ldots(7)$$

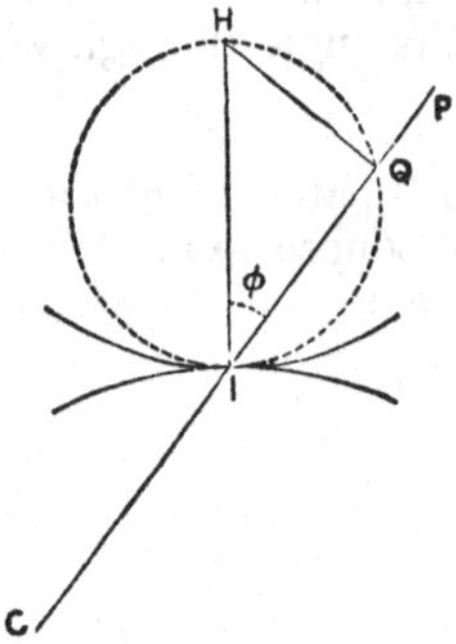

Fig. 130.

This shews that if P coincide with Q, CI is infinite; *i.e.* any point of the moving figure which lies on the circle just defined is at a point of inflexion of its path. For this reason, the circle in question is called the 'circle of inflexions.'

From (7) and (5) we find

$$CI = \frac{IP \cdot IQ}{QP}, \quad \rho = \frac{IP^2}{QP}. \quad\ldots\ldots\ldots\ldots\ldots(8)$$

The latter result shews that ρ changes sign with QP; that is, the paths of the various points of the moving figure are concave or convex to I, according to the side of the circle of inflexions on which they lie. In the standard case represented in the figures, the paths are concave or convex according as P is outside or inside the circle.

An example is furnished by the trochoidal curves figured on p. 296. The circle of inflexions has in this case half the size of the rolling circle.

We have taken as our standard case that in which the two curves are convex to one another, as in Figs. 127, 129. Any other case may be included by giving proper signs to R and R'.

The preceding theory has an application to the problem of 'rocking stones' in Statics. When one rough body rests on another, with a single point of contact, its centre of gravity must be vertically above this point. And for stability of equilibrium it is necessary that the path of the centre of gravity, in any possible rolling displacement, should be concave upwards.

Ex. 1. In the cycloid, if a be the radius of the generating circle, we have

$$R = \infty, \quad R' = a, \quad IP = 2a \cos \phi. \quad \dots\dots\dots\dots\dots (9)$$

Substituting in (4), we find

$$CI = 2a \cos \phi = IP, \quad \dots\dots\dots\dots\dots\dots (10)$$

and therefore

$$\rho = 2IP. \quad \dots\dots\dots\dots\dots\dots\dots (11)$$

Ex. 2. In the epicycloid (Art. 123) we have

$$R = a, \quad R' = b, \quad IP = 2b \cos \phi, \quad \dots\dots\dots\dots\dots (12)$$

whence

$$CI = \frac{2ab}{a + 2b} \cos \phi = \frac{a}{a + 2b} . IP, \quad \dots\dots\dots\dots (13)$$

$$\rho = \frac{2(a + b)}{a + 2b} . IP. \quad \dots\dots\dots\dots\dots (14)$$

We note that if $b = -\frac{1}{2}a$, we have $\rho = \infty$; cf. Art. 124, Ex 2.

148. Curvature of a Line-Roulette.

The curvature of a line-roulette, *i.e.* of the envelope of a straight line carried by the rolling curve, can be found still more simply. The perpendiculars IZ, $I'Z'$ let fall on two consecutive positions of the line, from the corresponding positions (in space) of the instantaneous centre, are normals to the envelope, and the angle which they make with one another at their intersection (C) is equal to the angle of rotation $\delta\theta$. Hence if ϕ be the angle which IZ makes with the normal to the rolling curve at I, and $II' = \delta s$, we have ultimately

$$\delta s \cos \phi = CI . \delta\theta. \quad \dots\dots (1)$$

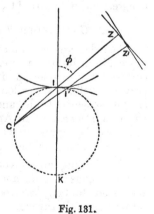

Fig. 131.

Hence, substituting the value of $\delta\theta$ from Art. 146 (1), we have

$$\frac{\cos\phi}{CI} = \frac{1}{R} + \frac{1}{R'}. \quad\quad\quad\quad\dots\dots\dots\dots\dots(2)$$

The radius of curvature of the envelope is then given by

$$\rho = CI + IZ. \quad\quad\quad\quad\dots\dots\dots\dots\dots(3)$$

If, along the normal to the rolling curve at I, but in the direction opposite to that chosen in the preceding Art., we measure off a length IK such that

$$\frac{1}{IK} = \frac{1}{R} + \frac{1}{R'}, \quad\quad\quad\quad\dots\dots\dots\dots\dots(4)$$

and describe a circle on this line as diameter, it appears from (2) that C lies on this circle; in other words, the locus of the centres of curvature of all line-roulettes, in any given position of the rolling curve, is a circle. Also, when the carried line passes through K, Z coincides with C, and C is a 'stationary point' (Art. 133) on the envelope. The aforesaid circle is therefore called the 'circle of cusps.'

Ex. 1. Regarding a cycloid as the envelope of the diameter of a circle which rolls on a fixed straight line (Art. 146, Ex. 1), we infer that the radius of curvature is double the normal.

Ex. 2. If an epicycloid be generated as the envelope of the diameter of a circle rolling on a fixed circle, then, to conform with the notation of Art. 123, we must write $R = a$, $R' = 2b$, and therefore, from (2),

$$CI = \frac{2ab}{a + 2b}\cos\phi = \frac{a}{a + 2b}. IZ,$$

in agreement with Art. 147, Ex. 2.

149. Continuous Motion of a Figure in its own Plane.

Consider any continuous series of positions of a plane figure moveable in its own plane. The instantaneous centre will have a certain locus in space, and also a certain locus in the figure. The curves so defined are called 'centrodes'; the former is distinguished as the 'space-centrode,' and the latter as the 'body-centrode.' The theorem referred to in Art. 146 is that the given motion of the figure can be represented as due to the rolling of the body-centrode, without slipping, on the space-centrode.

Considering any given position of the figure, let I be the instantaneous centre, and let I', J' be adjacent corresponding points on the body-centrode and space-centrode, respectively. Let $\delta\theta$ be the angle through which the body turns as the instantaneous centre is transferred from I to J'. We have then, ultimately, by Art. 145,

$$II' = IJ', \text{ and } I'J' = II' . \, \delta\theta.$$

The angle $I'IJ'$ therefore ultimately vanishes. The tangent lines to the two loci at I therefore coincide, and corresponding elementary arcs of the two curves are in a ratio of equality.

Ex. A straight line AB of constant length moves with its extremities on two fixed straight lines OX, OY.

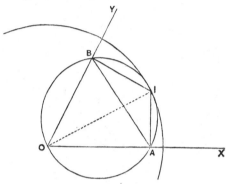

Fig. 132.

The instantaneous centre I is at the intersection of perpendiculars to OX, OY at the points A, B respectively. The points A, B lie on the circle described on OI as diameter; and since in this circle the chord AB, of given length, subtends a constant angle AOB at the circumference, the diameter is determinate. Hence the space-locus of I is a circle with centre O. Again, since the angle AIB is constant, the locus of I relative to AB is a circle whose diameter is equal to the constant value of OI. Hence the motion is equivalent to the rolling of a circle on the inside of a fixed circle of twice its size. This kind of motion has been considered in Art. 124, Ex. 2, and it has been shewn that any point P fixed relatively to AB will describe an ellipse, which in certain cases, viz. when P is on the circumference of the rolling circle, degenerates into a straight line.

150. Double Generation of Epicyclics as Roulettes.

As a further example we return to the mechanical method of compounding uniform circular motions, by means of a jointed parallelogram $OQPQ'$, referred to in Art. 125.

We will suppose for definiteness that the angular velocities n, n', of the bars OQ, OQ', have the same sign.

The instantaneous centre (I) of the bar QP will be a point in QO such that

$$n' . QI = n . OQ. \quad\quad\quad\quad\dots\dots\dots\dots\dots(1)$$

For the velocity of any point rigidly attached to QP will be made up of a translation $n \cdot OQ$ at right angles to OQ, and a rotation with angular velocity n' relatively to Q. Hence under the above condition the velocity of the point attached to QP which at the instant under consideration is at I will be zero. The two centrodes for the motion of the bar QP are therefore the circles described, with O and Q as centres, to pass through I.

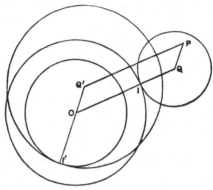

Fig. 133.

For a similar reason, the instantaneous centre (I') of the bar $Q'P$ will be a point in $Q'O$, such that

$$n \cdot Q'I' = n' \cdot OQ'. \quad\quad\dots\dots\dots\dots\dots(2)*$$

Hence, for the motion of the bar $Q'P$, the two centrodes are the circles described, with O and Q' as centres, to pass through I'.

Since P is a point on each of the bars QP, $Q'P$, we see that any direct epicyclic can be described in two ways as an epitrochoid.

In the particular case where $QP = QI$, it follows from (1) and (2) that

$$QP : OQ = QI : OQ = n : n' = OQ' : Q'I' = QP : Q'I',$$

whence $\quad\quad\quad\quad\quad\quad Q'P = OQ = Q'I'.$

The path of P is in this case an epicycloid, and we learn that any epicycloid can be generated in two ways, viz. by the rolling of either of two determinate circles on the outside of the same fixed circle†. See Fig. 134.

---

* The figure corresponds to the case of $n' > n$. If $n' < n$, I will lie in QO produced, and I' between Q' and O.

† This proposition is due to Euler (1781)

As an instance, we have the double generation of the cardioid explained in Art. 124, Exs. 1, 3.

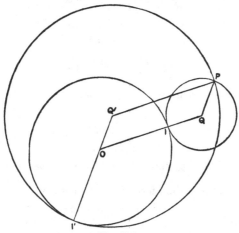

Fig. 134.

The case where the angular velocities n, n' have opposite signs may be left to the reader to examine. It will appear that any retrograde epicyclic can be generated in two distinct ways as a hypotrochoid. And, in particular, any hypocycloid can be generated in two ways by the rolling of either of two determinate circles on the inside of the same fixed circle.

EXAMPLES. XLVI.

(Curvature.)

1. Prove that the circle is the only curve whose curvature is constant.

2. Prove that the coordinates of the centre of curvature at any point (x, y) of a curve can be expressed in the forms

$$x - \frac{dy}{d\psi}, \quad y + \frac{dx}{d\psi}.$$

3. Prove that the intrinsic equation of an equiangular spiral is of the form

$$s = ae^{\psi \cot a}.$$

4. Prove that the intrinsic equation of the tractrix may be written

$$s = a \log \operatorname{cosec} \psi.$$

Prove that the curvature varies as the normal.

5. By differentiation of the formulæ

$$\frac{dx}{ds} = \cos \psi, \quad \frac{dy}{ds} = \sin \psi,$$

prove that

$$\frac{1}{\rho} = -\frac{d^2x}{ds^2} \Big/ \frac{dy}{ds} = \frac{d^2y}{ds^2} \Big/ \frac{dx}{ds},$$

and

$$\frac{1}{\rho^2} = \left(\frac{d^2x}{ds^2}\right)^2 + \left(\frac{d^2y}{ds^2}\right)^2.$$

6. If a curve be defined by the equations

$$x = F(t), \quad y = f(t),$$

prove that

$$\frac{1}{\rho} = \frac{x'y'' - y'x''}{(x'^2 + y'^2)^{\frac{3}{2}}},$$

where the accents denote differentiations with respect to t.

7. Apply the preceding formula to the cases of the ellipse

$$x = a \cos \phi, \quad y = b \sin \phi,$$

and the hyperbola

$$x = a \cosh u, \quad y = b \sinh u.$$

8. Shew how to express the coordinates x, y of a point on a curve, whose Cartesian equation is given, in terms of the inclination (ψ) of the tangent, and prove that

$$\rho = \sqrt{\left\{\left(\frac{dx}{d\psi}\right)^2 + \left(\frac{dy}{d\psi}\right)^2\right\}}.$$

9. Prove that the curve whose intrinsic equation is

$$s = k \sin \psi$$

is a cycloid. (Use the method of Art. 120 (3).)

10. Given that in the 'catenary of equal strength'

$$\rho = k \sec \psi,$$

where ψ is the inclination to the horizontal, prove that if the origin be at the lowest point

$$x = k\psi, \quad y = k \log \sec \psi,$$

the axes of x and y being horizontal and vertical.

11. Given that the intrinsic equation of a curve is

$$s = k \sin^2 \psi,$$

deduce the Cartesian equation

$$x^{\frac{2}{3}} + y^{\frac{2}{3}} = (\tfrac{2}{3}k)^{\frac{2}{3}}.$$

12. If $\rho = a^2/y$, prove that

$$y^2 = C - 2a^2 \cos \psi.$$

13. Find the curve whose intrinsic equation is

$$s = a \sec^3 \psi. \qquad\qquad [ay^2 = \tfrac{4}{9}x^3.]$$

14. If the coordinates x, y of a point on a curve be given functions of t, prove that

$$\frac{d^2x}{dt^2} = \frac{d^2s}{dt^2}\cos\psi - \frac{1}{\rho}\left(\frac{ds}{dt}\right)^2 \sin\psi,$$

$$\frac{d^2y}{dt^2} = \frac{d^2s}{dt^2}\sin\psi + \frac{1}{\rho}\left(\frac{ds}{dt}\right)^2 \cos\psi,$$

and give the kinematical interpretation of these results.

Hence shew that

$$\frac{1}{\rho^2} = \left\{\left(\frac{d^2x}{dt^2}\right)^2 + \left(\frac{d^2y}{dt^2}\right)^2 - \left(\frac{d^2s}{dt^2}\right)^2\right\} \div \left(\frac{ds}{dt}\right)^4.$$

15. Prove that, in the astroid

$$x = a\cos^3\theta, \quad y = a\sin^3\theta,$$

$$\psi = \pi - \theta,$$

and thence shew that

$$\rho = 3a\sin\theta\cos\theta.$$

16. If $\qquad\qquad x = at^2, \quad y = 2at,$

the coordinates of the centre of curvature are

$$a(2 + 3t^2), \quad -2at^3.$$

17. Prove from the Cartesian formula of Art. 135 (2) that in the rectangular hyperbola $xy = k^2$

$$\rho = \frac{(x^2 + y^2)^{\frac{3}{2}}}{2k^2}.$$

18. Also that, in the ellipse

$$\frac{x^2}{a^2} + \frac{y^2}{b^2} = 1,$$

$$\rho = \frac{(a^2 - e^2x^2)^{\frac{3}{2}}}{ab}.$$

19. Also that, in the hyperbola

$$\frac{x^2}{a^2} - \frac{y^2}{b^2} = 1,$$

$$\rho = \frac{(e^2x^2 - a^2)^{\frac{3}{2}}}{ab}.$$

20. Also that, in the parabola $y^2 = 4ax$,

$$\rho = \frac{2(a + x)^{\frac{3}{2}}}{a^{\frac{1}{2}}}.$$

21. Also that, in the semi-cubical parabola $ay^2 = x^3$,

$$\rho = \frac{(4a + 9x)^{\frac{3}{2}} x^{\frac{1}{2}}}{6a}.$$

22. Also that, in the cubical parabola $a^2y = x^3$,

$$\rho = \frac{a^2}{6x} \left(1 + 9 \frac{x^4}{a^4} \right)^{\frac{3}{2}}.$$

23. Also that, in the astroid

$$x^{\frac{2}{3}} + y^{\frac{2}{3}} = a^{\frac{2}{3}},$$

$$\rho = -3 (axy)^{\frac{1}{3}}.$$

24. Shew by differentiating the expression

$$(x - \xi)^2 + (y - \eta)^2$$

for the square of the distance of a variable point (x, y) of a curve from a fixed point (ξ, η) that when this distance is stationary the point (x, y) must be at the foot of a normal from (ξ, η) to the curve.

Also that the distance is then a minimum or maximum according as the point (ξ, η) is nearer to or further from the curve than the centre of curvature.

25. If a curve be transformed by the substitution

$$x' = ax, \quad y' = \beta y,$$

the curvature at any point is altered in the ratio

$$\frac{a\beta}{(a^2 \cos^2 \psi + \beta^2 \sin^2 \psi)^{\frac{3}{2}}},$$

where ψ is the inclination of the original curve to the axis of x.

26. Prove that

$$\frac{d\rho}{ds} = \frac{3pq^2 - (1 + p^2) r}{q^2},$$

where

$$p = dy/dx, \quad q = d^2y/dx^2, \quad r = d^3y/dx^3.$$

27. The curvature at any point of an ellipse is

$$\frac{a \cos \frac{1}{2}\phi}{rr'},$$

where r, r' are the focal distances, and ϕ is the angle between them.

28. In the rectangular hyperbola $r^2 \cos 2\theta = a^2$,

$$\rho = r^3/a^2.$$

29. In the lemniscate $r^2 = a^2 \cos 2\theta$,

$$\rho = a^2/3r.$$

30. In the curve $r^m = a^m \cos m\theta$,

$$\rho = \frac{r^2}{(m+1)\,p} = \frac{a^m}{(m+1)\,r^{m-1}}.$$

31. Apply the formula $\rho = r\,dr/dp$ to find the radius of curvature at any point of an epicycloid. (See Ex. 26, p. 330.)

Examine the case of the involute of a circle.

32. If the equation of a curve be given in the form $r = f(p)$, the chord of curvature through the pole is

$$2p\,\frac{dr}{dp}.$$

Prove that the chord of curvature through the pole of a cardioid is $1\frac{1}{3}$ times the radius vector.

33. Prove that the chord of curvature, through the pole, at any point of the curve $r^m = a^m \cos m\theta$ is $2r/(m+1)$.

34. Prove that the curvature of the pedal of a curve $r = f(p)$ with respect to the origin is

$$\frac{2}{r} - \frac{p}{r^3}\,\rho,$$

where r, p, ρ refer to the original curve.

35. Prove that the curvature at any point of the pedal of an ellipse of semi-axes a, b with respect to the centre is equal to

$$\frac{3}{r} - \frac{a^2 + b^2}{r^3},$$

where r is the radius vector of the corresponding point of the ellipse.

36. Prove the formula

$$\frac{1}{\rho} = \left\{\frac{1}{r} - \frac{1}{r}\left(\frac{dr}{ds}\right)^2 - \frac{d^2r}{ds^2}\right\} \div \left\{1 - \left(\frac{dr}{ds}\right)^2\right\}^{\frac{1}{2}};$$

and apply it to deduce the conclusions of Ex. 24.

37. Prove that in polar coordinates the condition for a stationary tangent is

$$\frac{d^2u}{d\theta^2} + u = 0,$$

where $u = 1/r$.

38. From the formula

$$\psi = \theta + \phi = \theta + \cot^{-1}\frac{1}{r}\frac{dr}{d\theta}$$

deduce the formula for curvature in polar coordinates:

$$\frac{1}{\rho} = \left\{ r^2 - r\frac{d^2r}{d\theta^2} + 2\left(\frac{dr}{d\theta}\right)^2 \right\} \div \left\{ r^2 + \left(\frac{dr}{d\theta}\right)^2 \right\}^{\frac{3}{2}}$$

$$= \left(\frac{d^2u}{d\theta^2} + u\right) \div \left\{ 1 + \left(\frac{1}{u}\frac{du}{d\theta}\right)^2 \right\}^{\frac{3}{2}},$$

where $u = 1/r$.

39. With the same notation, prove that the chord of curvature through the origin is

$$2\left\{ 1 + \left(\frac{1}{u}\frac{du}{d\theta}\right)^2 \right\} \div \left(\frac{d^2u}{d\theta^2} + u\right).$$

EXAMPLES. XLVII.

(Newton's Method.)

1. The radius of curvature of the curve

$$ay^2 = (x - a)(x - \beta)^2$$

at the point $(a, 0)$ is $(a - \beta)^2/2a$.

2. Prove by Newton's method that the radius of curvature at the vertex of the catenary

$$y = a \cosh x/a$$

is equal to a.

3. The radius of curvature of the curve

$$y^2 = a^2(a + x)/x$$

at the point $(-a, 0)$ is $\frac{1}{2}a$.

4. The radius of curvature of the 'witch'

$$y^2 = a^2(a - x)/x$$

at its vertex is $\frac{1}{2}a$.

5. Find the radius of curvature of the curve

$$a^2y^2 = x^3(a - x)$$

at the point $(a, 0)$.

6. Find the radius of curvature of the parabola

$$(x - y)^2 - 2a(x + y) + a^2 = 0$$

at the points where it touches the coordinate axes.

7. Find the radius of curvature of the curve

$$y = 4\sin x - \sin 2x$$

at the point $x = \frac{1}{2}\pi$. $\lceil 2 \cdot 795 \ldots \rceil$

8. The length of the chord of curvature, parallel to the axis of y, at the origin, in the parabola

$$y = mx + \frac{x^2}{a}$$

is $(1 + m^2)\, a$, and the equation of the circle of curvature is

$$x^2 + y^2 = (1 + m^2)\, a\, (y - mx).$$

9. Find the curvature of the curve

$$y = mx + n\, (x - a)^2\, (x - b)^2$$

at the points $(a, 0)$, $(b, 0)$.

$$\left[\frac{(1 + m^2)^{\frac{3}{2}}}{4n\, (a - b)^2} . \right]$$

10. Find the equation of the circle of curvature, at the origin, of the conic

$$y = Ax^2 + 2Hxy + By^2\, ;$$

and prove that it meets the curve again on the line

$$(A - B)\, y = 2Hx.$$

11. If the polar equation of a curve be $r = \phi\, (\theta)$, where $\phi\, (\theta)$ is an even function of θ, the curvature at the point $\theta = 0$ is

$$\frac{\phi\, (0) - \phi''\, (0)}{\{\phi\, (0)\}^2} .$$

12. Prove that in the meridian-curve $(r^2 = a^2 \cos \theta)$ of the 'solid of greatest attraction' (see Ex. 19, p. 329) the radii of curvature at the extremities of the axis are ∞ and $\frac{2}{3}a$, respectively.

13. Prove that the radius of curvature at either vertex of the lemniscate $r^2 = a^2 \cos 2\theta$ is $\frac{1}{3}a$.

14. The radii of curvature of the trochoid

$$x = a\theta + k \sin \theta, \quad y = a - k \cos \theta$$

at the points where it is nearest to and furthest from the base are

$$(a \pm k)^2/k.$$

15. Apply Newton's method to shew that the radii of curvature of the epicyclic

$$x = a_1 \cos n_1 t + a_2 \cos n_2 t, \quad y = a_1 \sin n_1 t + a_2 \sin n_2 t,$$

at the points nearest to and furthest from the centre, are

$$\frac{(n_1 a_1 \pm n_2 a_2)^2}{n_1^2 a_1 \pm n_2^2 a_2} .$$

Infer the condition that an epicyclic, at the points of nearest approach to the centre, should be concave to the centre (as in the case of the orbit of the Moon relative to the Sun).

16. Find the radius of curvature at the point $t = 0$ on the Lissajous curve

$$x = a \cos nt, \quad y = b \sin 2nt.$$

Sketch the curve. $[4b^2/a.]$

17. If a curve be referred to polar coordinates r, θ, and if the pole be on the curve, and the initial line be the tangent at the pole, prove that the diameter of curvature at the pole $= \lim r/\theta$.

Find the radius of curvature at the pole of the curve

$$r = a \cos m\theta.$$

18. If P be a point of a curve where the curvature, but not the direction of the tangent, is discontinuous, and if Q, R be neighbouring points on opposite sides of P, prove that the curvature of the circle PQR is ultimately equal to

$$\frac{m_1}{\rho_1} + \frac{m_2}{\rho_2},$$

where ρ_1, ρ_2 are the radii of curvature of the given curve on the two sides of P, and m_1, m_2 are the limiting values of the ratios PQ/QR and PR/QR, respectively.

19. The acute angle which a chord PQ of a curve makes with the tangent at P, when Q is taken infinitely close to P, is ultimately equal to $\frac{1}{2}\delta s/\rho$, where δs is the arc PQ and ρ is the radius of curvature at P.

20. Prove that if the tangents at the extremities of an infinitely small arc PQ meet in T, then TP and TQ are ultimately in a ratio of equality.

Why does it not follow that the line joining T to the middle point of PQ will be ultimately perpendicular to PQ?

21. Assuming that the radius of the circumcircle of a triangle ABC is equal to $\frac{1}{2}a/\sin A$, shew that it follows from Ex. 19 that the osculating circle coincides with the circle of curvature.

22. Prove that when the resultant force on a particle is in the direction of motion the tangent to the path is 'stationary.'

EXAMPLES. XLVIII.

(Envelopes. Evolutes.)

1. The envelope of the parabolas

$$y^2 = 4a\,(x - a),$$

where a is the parameter, is a pair of straight lines.

2. From any point P on the parabola $y^2 = 4ax$ perpendiculars PM, PN are drawn to the coordinate axes. Find the envelope of the line MN. $[y^2 = -16ax.]$

3. Find the envelope of the line
$$x \cos a + y \sin a = a \sec a ;$$
and give the geometrical interpretation of the result.

4. The envelope of the parabolas
$$ay^2 = a^2 (x - a),$$
where a is the parameter, is the curve
$$ay^2 = \tfrac{4}{27} x^3.$$

5. Circles are described on the radii vectores of a curve as diameters; prove geometrically that their envelope is the pedal of the given curve with respect to the origin.

6. Find the envelope of the circles described on the focal radii of a conic as diameters.

7. Chords of a circle are drawn through a fixed point on the circumference; prove that the envelope of the circles described on these chords as diameters is a cardioid.

8. The envelope of the circles described on the central radii of a rectangular hyperbola as diameters is a lemniscate of Bernoulli.

9. Prove that the envelope of the curves
$$P \cos a + Q \sin a = R,$$
where P, Q, R are given functions of x, y, and a is a variable parameter, is
$$P^2 + Q^2 = R^2.$$

10. Find the envelope of the circles
$$x^2 + y^2 - 2ax \cos a - 2ay \sin a = c^2,$$
and interpret the result.

11. Find the relation between p and a in order that the straight line
$$x \cos a + y \sin a = p$$
may cut the circles
$$(x - a)^2 + y^2 = b^2, \quad (x + a)^2 + y^2 = c^2$$
in chords of equal length. Prove that the envelope of the line, under this condition, is a parabola.

12. A system of ellipses of constant area have the same centre and their axes coincident in direction; prove that the envelope consists of two conjugate rectangular hyperbolas.

13. A straight line moves so that the product of the perpendiculars on it from the fixed points $(\pm c, 0)$ is constant $(= b^2)$; prove that the envelope is the ellipse
$$\frac{x^2}{b^2 + c^2} + \frac{y^2}{b^2} = 1,$$

or the hyperbola

$$\frac{x^2}{c^2 - b^2} - \frac{y^2}{b^2} = 1,$$

according as the two perpendiculars are on the same or on opposite sides of the variable line.

14. Circles are described on the double ordinates of the parabola $y^2 = 4ax$ as diameters; prove that the envelope is the parabola

$$y^2 = 4a(x + a).$$

15. Circles are described on the double ordinates of the ellipse

$$\frac{x^2}{a^2} + \frac{y^2}{b^2} = 1$$

as diameters: prove that the envelope is the ellipse

$$\frac{x^2}{a^2 + b^2} + \frac{y^2}{b^2} = 1.$$

16. A straight line moves so that the sum of the squares of the perpendiculars on it from the fixed points $(\pm c, 0)$ is constant $(= 2k^2)$; prove that the envelope is the conic

$$\frac{x^2}{k^2 - c^2} + \frac{y^2}{k^2} = 1,$$

and examine the various cases.

17. A straight line moves so that the difference of the squares of the perpendiculars on it from two fixed points is constant; prove that the envelope is a parabola.

18. Find the envelope of the ellipses

$$x = a \sin(\theta - a), \quad y = b \cos \theta,$$

where a is the parameter.

19. The envelope of the catenaries

$$y = c \cosh(x/c),$$

where c is the variable parameter, consists of two straight lines.

20. The envelope of the ellipses

$$\frac{x^2}{a^2} + \frac{y^2}{\beta^2} = 1,$$

where $$a + \beta = k,$$

is the 'astroid' $$x^{\frac{2}{3}} + y^{\frac{2}{3}} = k^{\frac{2}{3}}.$$

21. The envelope of the straight line which makes on the coordinate axes intercepts whose sum is k is the parabola

$$\sqrt{x} + \sqrt{y} = \sqrt{k}.$$

22. Two points move along the coordinate axes with different constant velocities; prove that the line joining them envelopes a parabola.

23. From any point on the ellipse

$$\frac{x^2}{a^2} + \frac{y^2}{b^2} = 1$$

perpendiculars are drawn to the coordinate axes; prove that the envelope of the straight line joining the feet of these perpendiculars is the curve

$$\left(\frac{x}{a}\right)^{\frac{2}{3}} + \left(\frac{y}{b}\right)^{\frac{2}{3}} = 1.$$

24. Find the locus of ultimate intersections of the curves

$$ay^2 = x\,(x + a)^2,$$

where a is the parameter; and examine the result.

25. If a circle of constant radius has its centre on a given curve, the envelope of the circle consists of two parallel curves.

26. If a circle of given radius touch a given curve, its envelope consists of two parallel curves.

27. If the equation of a curve be given in the form

$$r^2 = f(p),$$

that of any parallel curve is of the form

$$r^2 = f(p - c) + 2cp - c^2.$$

28. Prove that the problem of negative pedals (Art. 131) is equivalent to finding the envelope of the straight line

$$x \cos \psi + y \sin \psi = p,$$

where p is a given function of the parameter ψ.

Verify that this leads to the formulæ (4) of Art. 131.

29. Shew that the negative pedal of the parabola

$$y^2 = 4ax$$

with respect to the vertex is the curve

$$ay^2 = \tfrac{1}{27}(x - 4a)^3.$$

30. Prove by the method of envelopes that the negative pedal of a circle is an ellipse or hyperbola according as the pole is inside or outside the circle.

31. Prove geometrically that the radius of curvature at any point of an equiangular spiral subtends a right angle at the pole.

32. The evolute of an equiangular spiral is an equiangular spiral of the same angle.

33. The area enclosed by the evolute of the ellipse

$$\frac{x^2}{a^2} + \frac{y^2}{b^2} = 1$$

is $3\pi (a^2 - b^2)/8ab$.

34. The coordinates of the centre of curvature at any point of the curve

$$ay^2 = x^3$$

are $\xi = -x - \frac{9}{2}\frac{x^3}{a}, \quad \eta = 4y + \frac{4}{3}\frac{ay}{x}.$

Shew that near the origin the evolute has the form of the parabola

$$y^2 = -\frac{16}{9} ax.$$

35. Shew that if a curve has a point of inflexion the evolute has an asymptote.

Shew that the part of the evolute of the curve

$$a^2 y = x^3$$

which corresponds to the part of the curve near the origin may be represented approximately by the hyperbola

$$xy = \frac{1}{12}a^2.$$

36. The evolute of the hyperbola

$$x = a \cosh u, \quad y = b \sinh u$$

is $(ax)^{\frac{2}{3}} - (by)^{\frac{2}{3}} = (a^2 + b^2)^{\frac{2}{3}}.$

37. If rays emanating from a point O be reflected at any given curve, the reflected rays are all normal to a curve which is similar to the pedal of the given curve with respect to O, but of double the dimensions.

38. Hence shew that the caustic by reflection at a circle will be the evolute of a limaçon; and that in the particular case where the luminous point is on the circumference of the given circle the caustic is a cardioid.

39. Prove that the caustic by reflection at any curve is the evolute of the envelope of a system of circles described with the various points of the curve as centres, and all passing through the luminous point.

What is the corresponding theorem for the case of refraction?

EXAMPLES. XLIX.

(Roulettes, &c.)

1. A lamina moves in any manner in its own plane; prove that parallel straight lines in the lamina envelope parallel curves.

2. A straight line moves so as always to pass through a fixed point O, whilst a point Q on it describes a circle passing through O. Prove that the instantaneous centre is at the other extremity of the diameter through Q, and determine the two centrodes.

Deduce a construction for the normal to a limaçon; and infer that in a cardioid the normals at the extremity of any chord through the cusp meet at right angles on the perpendicular to the chord at this point.

3. A plane figure moves so that two straight lines in it touch two fixed circles; determine the two centrodes.

4. If a circle roll on a fixed circle of half the size, which it surrounds, every straight line carried by the rolling circle will envelope a circle.

5. Prove that if a plane figure move so that a straight line in it rolls on a fixed circle, the envelope of any other straight line in the figure is an involute of a circle.

6. The radius of curvature of the envelope of the straight line

$$ax + \beta y = 1,$$

where a, β are given functions of a parameter t, is

$$\pm \frac{(a^2 + \beta^2)^{\frac{3}{2}} (a''\beta' - a'\beta'')}{(a\beta' - a'\beta)^3},$$

the accents denoting differentiations with respect to t.

7. If the curve whose tangential-polar equation is $r = f(p)$ roll on a fixed straight line, the curvature of the path of the pole is

$$-\frac{d}{dp} \left(\frac{p}{r} \right),$$

where r is the radius vector to the point of contact.

8. Prove that if a parabola roll on a fixed straight line the path of the focus is a catenary.

9. Prove that if a conic roll on a fixed straight line the path of either focus is a curve such that

$$\frac{1}{\rho} + \frac{1}{n} = \frac{1}{c},$$

where ρ is the radius of curvature, n is the normal, and c is a constant.

10. If an equiangular spiral roll on a fixed straight line, the path of the pole is a straight line.

11. If the reciprocal spiral $r = a/\theta$ roll on a straight line, the path of the pole is a tractrix.

12. If any one of the Cotes' spirals

$$\frac{1}{p^2} = \frac{A}{r^2} + B$$

rolls on a straight line, the pole traces out a curve such that the curvature varies as the normal.

13. A curve rolls on a fixed straight line; prove that the arc of the roulette traced by any carried point O is equal to the corresponding arc of the pedal of the given curve with respect to O.   (Steiner.)

14. A closed oval curve rolls on a fixed straight line; prove that in a complete revolution the area swept over by the variable line which joins the point of contact to any internal carried point O is double the area of the pedal of the given curve with respect to O.
(Steiner.)

15. Prove from the theory of the instantaneous centre that when the area enclosed by a plane quadrilateral of jointed rods is stationary the quadrilateral is cyclic.

CHAPTER XI

DIFFERENTIAL EQUATIONS OF THE FIRST ORDER

151. Formation of Differential Equations.

Any relation between an independent variable x, a dependent variable y, and one or more of the derived functions

$$\frac{dy}{dx}, \quad \frac{d^2y}{dx^2}, \quad \frac{d^3y}{dx^3}, \cdots$$

is called a 'differential equation*.'

The 'order' of the equation is fixed by that of the highest differential coefficient which occurs in it. Thus a differential equation of the 'first order' is a relation between x, y, and dy/dx.

Before proceeding to methods of solution, it is instructive to consider one manner in which differential equations may arise.

If we are given a relation between the variables x, y, and an arbitrary constant C, then by differentiation we obtain an equation involving x, y, dy/dx, and C. By elimination of C between this and the original equation we obtain a differential equation of the first order.

More generally, given a relation between the variables x, y, and n arbitrary constants C_1, C_2, ... C_n, then if we differentiate n times in succession with respect to x, we have altogether $n+1$ equations between which the n arbitrary constants can be eliminated. The result is a differential equation of the nth order.

From this point of view the original equation is called the 'primitive.'

Ex. 1. If the primitive be

$$y = mx + C, \quad \dots\dots\dots\dots\dots\dots\dots(1)$$

where C is arbitrary, the differential equation is

$$\frac{dy}{dx} = m. \quad \dots\dots\dots\dots\dots\dots\dots(2)$$

---

* More particularly, it is called an 'ordinary' as distinguished from a 'partial' differential equation, *i.e.* one which involves partial derivatives of a function of two or more independent variables.

Ex. 2.   From the primitive

$$y^2 = 4ax + C \dots\dots\dots\dots\dots\dots(3)$$

we deduce

$$y \frac{dy}{dx} = 2a. \quad\dots\dots\dots\dots\dots\dots(4)$$

Ex. 3.   If the primitive be

$$x \cos a + y \sin a = a, \quad\dots\dots\dots\dots\dots(5)$$

where a is arbitrary, we deduce

$$\cos a + \frac{dy}{dx} \sin a = 0.$$

These give

$$\left(y - x \frac{dy}{dx}\right) \sin a = a, \quad \left(y - x \frac{dy}{dx}\right) \cos a = - a \frac{dy}{dx},$$

whence, squaring and adding,

$$\left(y - x \frac{dy}{dx}\right)^2 = a^2 \left\{ 1 + \left(\frac{dy}{dx}\right)^2 \right\}. \quad\dots\dots\dots\dots(6)$$

Ex. 4.   If the primitive be

$$y = Ax + B, \quad\dots\dots\dots\dots\dots\dots(7)$$

where both constants A, B are to be eliminated, we find

$$\frac{d^2 y}{dx^2} = 0. \quad\dots\dots\dots\dots\dots\dots(8)$$

Ex. 5.   From the primitive

$$(x - a)^2 + (y - \beta)^2 = a^2, \quad\dots\dots\dots\dots\dots(9)$$

where a, β are to be eliminated, we obtain

$$a^2 \left(\frac{d^2 y}{dx^2}\right)^2 = \left\{ 1 + \left(\frac{dy}{dx}\right)^2 \right\}^3. \quad\dots\dots\dots\dots(10)$$

The details of the work are given in Art. 189.

The above processes admit of a geometrical interpretation. The equations obtained by varying the arbitrary constants in the primitive represent a certain system or family of curves; the differential equation (in which these constants do not appear) expresses some property common to all these curves.

Thus in Ex. 2, above, the primitive represents a system of equal parabolas having their axes coincident with the axis of x, but their vertices at different points of it. The differential equation (4) expresses a property common to all these curves, viz. that the subnormal has a given constant value $2a$.

Again in Ex. 5, if we vary a, β in the primitive we get a doubly-infinite system of circles of given radius a, having their centres anywhere in the plane xy. The differential equation expresses that the radius of curvature has everywhere the constant value a. See Art. 135.

Other illustrations may be taken from Dynamics.

Ex. 6.   If, in the primitive

$$x = \tfrac{1}{2}gt^2 + At + B, \quad\ldots\ldots\ldots\ldots\ldots\ldots\ldots(11)$$

we vary A and B, we get a certain group or class of rectilinear motions. The differential equation

$$\frac{d^2x}{dt^2} = g \quad\ldots\ldots\ldots\ldots\ldots\ldots\ldots\ldots(12)$$

expresses a property common to the group, viz. that the acceleration has the constant value g.

Ex. 7.   Again, if the primitive be

$$x = A \cos nt + B \sin nt, \quad\ldots\ldots\ldots\ldots\ldots(13)$$

we find

$$\frac{d^2x}{dt^2} = - n^2x. \quad\ldots\ldots\ldots\ldots\ldots\ldots(14)$$

This asserts that in the whole group of motions represented by the primitive the acceleration is towards the origin of x, and varies (in a given ratio n^2) as the distance from this origin.

The preceding examples will suffice to illustrate the derivation of a differential equation from a primitive relation between x and y involving one or more arbitrary constants. In practice we are more usually confronted with the inverse problem, viz. to ascertain the most general form of relation between the variables which satisfies a given differential equation. Thus in Geometry, or in Dynamics, some general property may be propounded, whose expression takes the form of a differential equation, and it is required to determine the whole system of curves, or group of motions, which possess the property.

The process of passing from a given differential equation to the general relation between the variables which it implies is called 'solving,' or 'integrating' the equation; and the result is called the 'general solution,' or the 'complete primitive,' although the latter name is hardly appropriate from this point of view. A 'particular solution' is any relation between the variables which happens to satisfy it.

152. Equations of the First Order and First Degree.

The general type of a differential equation of the first order may be written

$$\phi\left(x, y, \frac{dy}{dx}\right) = 0. \dots\dots\dots\dots\dots(1)$$

The equation implies that y is to be a differentiable function of x, and that dy/dx is to be continuous.

The mode of derivation of a differential equation of the first order from a primitive involving an arbitrary constant, explained in Art. 151, may suggest that the general solution of (1) will in all cases consist of a relation between x and y involving an arbitrary constant. With some qualification, due to the occurrence of 'singular' solutions (Art. 161), this is in fact the case. The rigorous proof, however, is difficult, and may be passed over here without inconvenience, since in almost all cases for which practical methods of integration have been discovered the process itself contains the demonstration that the solution is of the kind indicated.

In such problems as ordinarily arise, either the left-hand side of (1) is a rational integral algebraic function of dy/dx, or the equation can be transformed so that this shall be the case. The 'degree' of the equation is then fixed by that of the highest power of dy/dx which occurs in it.

The general equation of the first degree may be written

$$M + N\frac{dy}{dx} = 0, \dots\dots\dots\dots\dots(2)$$

or

$$Mdx + Ndy = 0, \dots\dots\dots\dots\dots(3)$$

where M, N are given functions of x and y. The form (2) is also equivalent to

$$\frac{dy}{dx} = -\frac{M}{N} = \phi(x, y). \dots\dots\dots\dots(4)$$

If $\phi(x, y)$ be real and single-valued for all values of x and y, then corresponding to any point in the plane xy we have a definite direction, assigned by the equation (4). If we imagine a point, starting from any position in the plane, to move always in the direction thus indicated, it will trace out a curve, which constitutes a particular solution, or primitive, of the proposed equation. And the whole assemblage of such curves will form a singly-infinite system, each curve being determined by the point where it crosses an arbitrary line. It appears, moreover, that in the present case no two curves of the system will intersect.

We have thus a sort of intuitive proof that the complete solution of (4) will involve a single arbitrary constant*.

We proceed to give an account of the methods which have been devised for the solution of the equation (4) in various cases.

153. Methods of Solution.   One Variable absent.

1°.   The form

$$\frac{dy}{dx} = f(x), \quad \dots\dots\dots\dots\dots\dots(1)$$

where y does not appear explicitly, requires merely an ordinary integration.   Thus

$$y = \int f(x)\, dx + C, \dots\dots\dots\dots\dots(2)$$

where C is an arbitrary constant.

2°.   The equation

$$\frac{dy}{dx} = f(y), \quad \dots\dots\dots\dots\dots\dots(3)$$

in which x does not appear explicitly, may be written

$$\frac{dy}{f(y)} = dx,$$

whence

$$\int \frac{dy}{f(y)} = x + C. \quad \dots\dots\dots\dots\dots(4)$$

Ex.   To find the curves whose subtangent has a given constant value a.   We have (Art. 60)

$$y \div \frac{dy}{dx} = a,$$

or

$$\frac{dy}{y} = \frac{dx}{a}. \quad \dots\dots\dots\dots\dots\dots(5)$$

Hence

$$\log y = \frac{x}{a} + C,$$

or

$$y = be^{x/a}, \quad \dots\dots\dots\dots\dots\dots(6)$$

where $b, = e^C$, is arbitrary.

154.   Variables Separable.

A more general form is

$$F(x) + f(y)\frac{dy}{dx} = 0, \quad \dots\dots\dots\dots(1)$$

or

$$F(x)\, dx + f(y)\, dy = 0. \quad \dots\dots\dots(2)$$

---

* A rigorous proof of this was given by Cauchy.

If an equation can be brought to this form the variables are said to be 'separable.' The solution obviously is

$$\int F(x)\,dx + \int f(y)\,dy = C. \quad\dots\dots\dots\dots\dots(3)$$

Ex. 1.   To find the curves such that the normals all pass through one point.

If we adopt rectangular axes through this point as origin, the condition gives

$$\frac{dy}{dx} = -\frac{x}{y},$$

or

$$x\,dx + y\,dy = 0, \quad\dots\dots\dots\dots\dots\dots(4)$$

whence

$$x^2 + y^2 = C. \quad\dots\dots\dots\dots\dots\dots(5)$$

The required curves are therefore circles described with the origin as centre.

Ex. 2.   To find a curve such that the tangents drawn to it from any point are equal.

If we take a fixed tangent as initial line, and its point of contact as origin, then if the two tangents drawn from any point on the initial line be equal, we must have, in the notation of Art. 63,

$$\phi = \theta,$$

and therefore

$$r\frac{d\theta}{dr} = \tan\theta. \quad\dots\dots\dots\dots\dots(6)$$

Hence

$$\frac{dr}{r} = \cot\theta\,d\theta, \quad\dots\dots\dots\dots\dots(7)$$

and

$$\log r = \log\sin\theta + C,$$

or

$$r = a\sin\theta, \quad\dots\dots\dots\dots\dots\dots(8)$$

where a is arbitrary.   The circle is therefore the only curve possessing the stated property.

Ex. 3.   The equation of rectilinear motion of a particle under an attractive force varying inversely as the square of the distance from a fixed point is

$$v\frac{dv}{dx} = -\frac{\mu}{x^2}. \quad\dots\dots\dots\dots\dots(9)$$

Integrating with respect to x,

$$\tfrac{1}{2}v^2 = \frac{\mu}{x} + C. \quad\dots\dots\dots\dots\dots(10)$$

If v vanish for $x = \infty$, we have $C = 0$.   In this case the velocity with which the particle arrives at a distance a from the centre of force is $\sqrt{(2\mu/a)}$, or $\sqrt{(2ga)}$, if $g = \mu/a^2$.

This gives the velocity with which an unresisted particle, falling from rest at a great distance, would reach the Earth, provided a denote the Earth's radius, and g the value of gravity at the surface.

Ex. 4. In a suspension bridge with uniform horizontal load the form of the chain is determined by the condition that any two tangents to the curve intersect on the vertical bisecting the chord of contact.

If the lowest point be taken as origin of rectangular coordinates, and the corresponding tangent as axis of x, the subtangent of any other point must be equal to one-half the abscissa. Hence

$$y \div \frac{dy}{dx} = \tfrac{1}{2}x,$$

or
$$\frac{dy}{y} = 2\frac{dx}{x}, \qquad \dots\dots\dots\dots\dots\dots(11)$$

the integral of which is $\log y = 2 \log x + \text{const.}$,

or
$$y = x^2/a, \dots\dots\dots\dots\dots\dots\dots(12)$$

where a is arbitrary. That is, the curve formed by the chain must be a parabola with its axis vertical.

155. Exact Equations.

The case of the preceding Art. comes under the head of 'exact equations.' An equation

$$M dx + N dy = 0 \dots\dots\dots\dots\dots\dots(1)$$

is said to be 'exact' when M and N are of the forms $\partial u/\partial x$ and $\partial u/\partial y$ respectively. The form

$$\frac{\partial u}{\partial x}\,dx + \frac{\partial u}{\partial y}\,dy = 0 \dots\dots\dots\dots\dots\dots(2)$$

is equivalent to
$$du = 0, \dots\dots\dots\dots\dots\dots\dots(3)$$

and its integral is
$$u = C, \dots\dots\dots\dots\dots\dots\dots(4)$$

where C is the arbitrary constant*.

It may be shewn that every equation of the type (1) is either exact, or can be rendered exact by a suitable 'integrating factor.' The number of such factors is unlimited; for if we suppose the equation (1) to have been brought to the form (3), it will still be exact when multiplied by $f'(u)$, where $f(u)$ may be *any* function of u. The integral of

$$f'(u)\,du = 0 \dots\dots\dots\dots\dots\dots(5)$$

is
$$f(u) = C, \dots\dots\dots\dots\dots\dots(6)$$

which is obviously equivalent to (4).

---

* The rule for ascertaining whether a proposed equation of the first degree is exact is given in Art. 193.

Ex. 1. $(ax + hy + g)\, dx + (hx + by + f)\, dy = 0.$ ....(7)

This is equivalent to

$$d\left(ax^2 + 2hxy + by^2 + 2gx + 2fy\right) = 0, \qquad\ldots\ldots\ldots\ldots(8)$$

whence $ax^2 + 2hxy + by^2 + 2gx + 2fy = C.$ ....(9)

Ex. 2. $xdx + ydy = k\,(xdy - ydx).$ ....(10)

This may be written

$$d\left(x^2 + y^2\right) = 2kx^2 d\left(\frac{y}{x}\right), \qquad\ldots\ldots\ldots\ldots\ldots(11)$$

and so becomes exact on division by $x^2 + y^2$, thus

$$\frac{d\left(x^2 + y^2\right)}{x^2 + y^2} = \frac{2kd\left(\dfrac{y}{x}\right)}{1 + \dfrac{y^2}{x^2}}. \qquad\ldots\ldots\ldots\ldots(12)$$

Hence, integrating,

$$\log\left(x^2 + y^2\right) = 2k \tan^{-1}\frac{y}{x} + C. \qquad\ldots\ldots\ldots\ldots(13)$$

The equation (10) may also be solved as follows. Its form suggests the substitutions

$$x = r\cos\theta, \quad y = r\sin\theta, \qquad\ldots\ldots\ldots\ldots(14)$$

which give $xdx + ydy = rdr, \quad xdy - ydx = r^2 d\theta.$ ....(15)

The equation therefore reduces to

$$\frac{dr}{r} = kd\theta, \qquad\ldots\ldots\ldots\ldots\ldots\ldots\ldots(16)$$

whence $\log r = k\theta + C.$ ....(17)

This is obviously equivalent to (13).

Ex. 3. To find the form of a solid of revolution such that the mean centre of the volume cut off by any right section shall be at a distance from this section equal to $1/n$th of the length of the axis.

If the axis of x be that of symmetry, and y be the ordinate of the generating curve, we must have, by Art. 116 (11),

$$\frac{\displaystyle\int_0^\xi xy^2 dx}{\displaystyle\int_0^\xi y^2 dx} = \left(1 - \frac{1}{n}\right)\xi,$$

or $$\int_0^\xi xy^2 dx = \frac{n-1}{n}\,\xi \int_0^\xi y^2 dx, \qquad\ldots\ldots\ldots\ldots(18)$$

where ξ is the abscissa of the bounding section.   Hence, if η denote the radius of this section, we find, on differentiating with respect to ξ according to the rule of Art. 92,

$$\xi\eta^2 = \frac{n-1}{n}\,\xi\eta^2 + \frac{n-1}{n}\int_0^\xi y^2 dx,$$

or

$$\xi\eta^2 = (n-1)\int_0^\xi y^2 dx. \dots \dots \dots (19)$$

A second differentiation gives

$$\frac{d}{d\xi}(\xi\eta^2) = (n-1)\,\eta^2, \dots \dots (20)$$

whence

$$\frac{d\,(\xi\eta^2)}{\xi\eta^2} = (n-1)\frac{d\xi}{\xi}. \dots \dots (21)$$

Integrating, we find $\xi\eta^2 = A\xi^{n-1}.$

The generating curve must therefore be of the type

$$y^2 = Ax^{n-2}. \dots \dots (22)$$

Since we have differentiated twice with respect to ξ, the differential equation actually solved is somewhat more general than the original problem.   In fact the same differential equation would have been obtained if, instead of zero, we had had other (and distinct) constants as the lower limits of the two definite integrals in (18).   It is therefore necessary to examine *à posteriori* whether the solution finally obtained satisfies the original equation with the actual lower limits.   This is easily verified to be the case if $n > 2$.

We note that if $n = 3$, the solid is a paraboloid of revolution, and that if $n = 4$ it is a cone.

156.   Homogeneous Equation.

Let us suppose that, in the equation

$$M + N\frac{dy}{dx} = 0,$$

M, N are homogeneous functions of x and y, of the same degree.

In this case the fraction M/N is a function of y/x only, and we may write

$$\frac{dy}{dx} = f\left(\frac{y}{x}\right). \dots \dots (1)$$

If we put $y = xv$ this becomes

$$x\frac{dv}{dx} + v = f(v). \dots \dots (2)$$

The variables x, v are now separable, viz. we have

$$\frac{dx}{x} = \frac{dv}{f(v) - v}, \quad\quad\quad\quad\text{......................(3)}$$

whence

$$\log x = \int \frac{dv}{f(v) - v} + C. \quad\quad\quad\text{.....................(4)}$$

After the integration has been effected we must write $v = y/x$.

Ex.

$$(x^2 - y^2)\frac{dy}{dx} - 2xy = 0. \quad\quad\quad\text{.......................(5)}$$

Here

$$\frac{dy}{dx} = \frac{2\,\dfrac{y}{x}}{1 - \dfrac{y^2}{x^2}}, \quad\quad\quad\quad\text{............................(6)}$$

whence

$$x\frac{dv}{dx} + v = \frac{2v}{1 - v^2}, \quad\text{or}\quad x\frac{dv}{dx} = \frac{v(1 + v^2)}{1 - v^2}.$$

Hence

$$\frac{dx}{x} = \frac{1 - v^2}{v(1 + v^2)}\, dv = \left(\frac{1}{v} - \frac{2v}{1 + v^2}\right) dv. \quad\text{...............(7)}$$

Integrating, we have

$$\log x = \log v - \log(1 + v^2) + \text{const.},$$

which is equivalent to $\quad\quad x(1 + v^2) = Cv,$

or

$$x^2 + y^2 = Cy. \quad\quad\quad\quad\text{........................(8)}$$

In the geometrical interpretation, the general solution of a homogeneous differential equation must represent a system of similar and similarly situated curves, the origin being a centre of similitude. For the equation (1) shews that where the curves cross any arbitrary straight line $(y/x = m)$ through the origin, dy/dx has the same value for each, that is, the tangents are parallel.

Thus, in the above Ex. the solution represents a system of circles touching the axis of x at the origin.

If in (4) we put $\quad\quad\quad C = \log c,$

y/x or v is determined as a function of x/c. In other words, the primitive is homogeneous in respect to x, y, and c, and is therefore of the type

$$\phi\left(\frac{x}{c}, \frac{y}{c}\right) = 0. \quad\quad\quad\text{........................(9)}$$

This is in accordance with the geometrical property above stated, since if x, y, and c be altered in any the same ratio, the equation (9) is unaltered. In other words, a change in the value of c merely alters the *scale* of the curve.

157. Linear Equation of the First Order, with Constant Coefficients.

A 'linear' equation is one which involves y and its derivatives only in the first degree. Thus the linear equation of the first order is of the type

$$\frac{dy}{dx} + Py = Q, \quad \dots\dots\dots\dots\dots\dots(1)$$

where P, Q are given functions of x.

We take first the case where P is a constant, the equation being

$$\frac{dy}{dx} - ay = Q, \quad \dots\dots\dots\dots\dots\dots(2)$$

as this will be of special use to us later.

If $Q = 0$, the solution is

$$y = Ce^{ax}, \quad \dots\dots\dots\dots\dots\dots(3)$$

by Art. 38.

It appears that the factor e^{-ax} renders the left-hand side of (2) an exact differential coefficient. This gives the key to the solution in the general case where $Q \neq 0$. Thus (2) is equivalent to

$$\frac{d}{dx}(e^{-ax}y) = Qe^{-ax}, \quad \dots\dots\dots\dots\dots(4)$$

whence $\qquad e^{-ax}y = \int Qe^{-ax}dx + C,$

or $\qquad y = e^{ax}\int Qe^{-ax}dx + Ce^{ax}. \quad \dots\dots\dots\dots(5)$

In accordance with a general usage (see Art. 166), the first term on the right-hand of (5) may be called the 'particular integral,' and the second the 'complementary function.'

The following cases are important:

1°. If $\qquad\qquad Q = He^{\lambda x}, \quad \dots\dots\dots\dots\dots\dots(6)$

we have $\qquad \int Qe^{-ax}dx = H\int e^{(\lambda-a)x}dx = \dfrac{H}{\lambda - a}e^{(\lambda-a)x},$

and $\qquad\qquad y = \dfrac{H}{\lambda - a}e^{\lambda x} + Ce^{ax}. \quad \dots\dots\dots\dots(7)$

That the first term on the right-hand is a particular integral of the proposed equation is verified at once by inspection.

2°. The result (7) needs correction when $\lambda = a$, or

$$Q = He^{ax}. \quad \dots\dots\dots\dots\dots\dots(8)$$

In this case we have

$$\int Qe^{-ax}\,dx = H\int dx = Hx,$$

and
$$y = Hxe^{ax} + Ce^{ax}. \quad\ldots\ldots\ldots\ldots\ldots\ldots(9)$$

3°. If
$$Q = Hx^n e^{ax}, \quad\ldots\ldots\ldots\ldots\ldots\ldots(10)$$

we have
$$\int Qe^{-ax}\,dx = H\int x^n\,dx = \frac{Hx^{n+1}}{n+1},$$

and
$$y = \frac{Hx^{n+1}}{n+1}\,e^{ax} + Ce^{ax}. \quad\ldots\ldots\ldots\ldots(11)$$

Ex. 1. If a particle be subject to a resistance varying as the velocity, and to some other force which is a given function of the time, its equation of motion is of the type

$$\frac{dv}{dt} + kv = f(t). \quad\ldots\ldots\ldots\ldots\ldots\ldots(12)$$

The integral of this is
$$v = Ce^{-kt} + e^{-kt}\int e^{kt}f(t)\,dt. \quad\ldots\ldots\ldots\ldots\ldots(13)$$

For example, if
$$f(t) = g,$$

a constant, we have
$$v = Ce^{-kt} + \frac{g}{k}. \quad\ldots\ldots\ldots\ldots\ldots\ldots(14)$$

This might have been obtained more simply by writing the differential equation in the form

$$\frac{d}{dt}\left(v - \frac{g}{k}\right) + k\left(v - \frac{g}{k}\right) = 0, \quad\ldots\ldots\ldots\ldots(15)$$

whence
$$v - \frac{g}{k} = Ce^{-kt}. \quad\ldots\ldots\ldots\ldots\ldots\ldots(16)$$

As t increases, v tends asymptotically to the 'terminal' value g/k.

Ex. 2. If an electric current of strength x be flowing in a circuit of self-induction L and resistance R, and if E be the extraneous electro-motive force in the circuit, we have the equation

$$L\frac{dx}{dt} + Rx = E. \quad\ldots\ldots\ldots\ldots\ldots\ldots(17)$$

If E be a constant, the solution of this is

$$x = \frac{E}{R} + Ce^{-\frac{R}{L}t}, \quad\ldots\ldots\ldots\ldots\ldots\ldots(18)$$

where C is arbitrary. The current therefore tends to the constant value E/R.

If, for example, we suppose that the circuit is completed at time $t = 0$, we have to determine C so that $x = 0$ for $t = 0$; this gives

$$x = \frac{E}{R} - \frac{E}{R}e^{-\frac{R}{L}t}. \quad\ldots\ldots\ldots\ldots\ldots\ldots(19)$$

The second term represents the 'extra current at make.'

Again, if $\qquad E = E_0 \cos(pt + \epsilon),$ ....(20)

we have $\qquad \dfrac{d}{dt}\left(xe^{\frac{R}{L}t}\right) = \dfrac{E_0}{L}e^{\frac{R}{L}t}\cos(pt + \epsilon),$

whence, integrating, and dividing by $e^{\frac{R}{L}t}$, we find

$$x = Ce^{-\frac{R}{L}t} + \frac{E_0}{L}e^{-\frac{R}{L}t}\int e^{\frac{R}{L}t}\cos(pt + \epsilon)\,dt$$

$$= Ce^{-\frac{R}{L}t} + \frac{E_0}{R^2 + p^2L^2}\{R\cos(pt + \epsilon) + pL\sin(pt + \epsilon)\};\ \ldots\ldots(21)$$

see Art. 80 (14). Hence as t increases, the current settles down into the steady oscillation

$$x = \frac{E_0}{\sqrt{(R^2 + p^2L^2)}}\cos(pt + \epsilon - \epsilon_1),\ \ldots\ldots\ldots(22)$$

where $\qquad \epsilon_1 = \tan^{-1}\dfrac{pL}{R}.$ ....(23)

The effect of the self-induction (L) is therefore to diminish the amplitude of the current in the ratio

$$R/\sqrt{(R^2 + p^2L^2)},$$

and to retard its phase by ϵ_1.

158. General Linear Equation of the First Order.

We return to the general linear equation of the first order,

$$\frac{dy}{dx} + Py = Q.\ \ldots\ldots\ldots\ldots\ldots(1)$$

If $Q = 0$, we have

$$\frac{1}{y}\frac{dy}{dx} + P = 0,\ \ldots\ldots\ldots\ldots(2)$$

whence $\qquad \log y + \int P dx = A,$

or $\qquad ye^{\int P dx} = C.$ ....(3)

This shews that $e^{\int P dx}$ is an integrating factor of (1), since

$$e^{\int P dx}\left(\frac{dy}{dx} + Py\right) = \frac{d}{dx}\left(ye^{\int P dx}\right).$$

Hence (1) may be written

$$\frac{d}{dx}\left(ye^{\int P dx}\right) = Qe^{\int P dx}.\ \ldots\ldots\ldots\ldots(4)$$

Integrating, we find

$$ye^{\int P dx} = \int Qe^{\int P dx}\,dx + C.\ \ldots\ldots\ldots\ldots(5)$$

The integrating factor will often suggest itself on inspection of the equation, without recourse to the above rule.

Ex. 1. $$\frac{dy}{dx} + y \cot x = 2 \cos x. \quad \dotfill (6)$$

Here $\qquad P = \cot x, \quad \int P dx = \log \sin x, \quad e^{\int P dx} = \sin x.$

Hence, multiplying by $\sin x$,

$$\frac{d}{dx} (y \sin x) = 2 \sin x \cos x, \quad \dotfill (7)$$

$$y \sin x = \sin^2 x + C,$$

$$y = \sin x + \frac{C}{\sin x}. \quad \dotfill (8)$$

Ex. 2. $$(1 - x^2) \frac{dy}{dx} - xy = 1. \quad \dotfill (9)$$

Dividing by $1 - x^2$, we have

$$\frac{dy}{dx} - \frac{x}{1 - x^2} y = \frac{1}{1 - x^2}. \quad \dotfill (10)$$

Here $\qquad P = -\frac{x}{1 - x^2}, \quad \int P dx = \tfrac{1}{2} \log (1 - x^2), \quad e^{\int P dx} = \sqrt{(1 - x^2)}.$

Multiplying (10) by the integrating factor, we get

$$\sqrt{(1 - x^2)} \frac{dy}{dx} - \frac{x}{\sqrt{(1 - x^2)}} y = \frac{1}{\sqrt{(1 - x^2)}},$$

or $$\frac{d}{dx} \{\sqrt{(1 - x^2)} \, y\} = \frac{1}{\sqrt{(1 - x^2)}}. \quad \dotfill (11)$$

Hence, integrating,

$$\sqrt{(1 - x^2)} \, y = \sin^{-1} x + C,$$

or $$y = \frac{\sin^{-1} x}{\sqrt{(1 - x^2)}} + \frac{C}{\sqrt{(1 - x^2)}}. \quad \dotfill (12)$$

Ex. 3. $$\frac{dy}{dx} + n \frac{y}{x} = x^m. \quad \dotfill (13)$$

The integrating factor here is obvious. The steps are

$$x^n \frac{dy}{dx} + n x^{n-1} y = x^{m+n},$$

$$x^n y = \frac{x^{m+n+1}}{m + n + 1} + C,$$

$$y = \frac{x^{m+1}}{m + n + 1} + C x^{-n}. \quad \dotfill (14)$$

159. Orthogonal Trajectories.

Suppose that we have a singly-infinite family of curves

$$\phi(x, y, C) = 0, \quad \dots\dots\dots\dots\dots\dots(1)$$

where C is a variable parameter, and that it is required to determine the curves which cut these everywhere at right angles.

We first form the differential equation of the family, by differentiation of (1) with respect to x, and elimination of C. See Art. 151.

If two curves cut at right angles, and if ψ, ψ' be the angles which the tangents at the intersection make with the axis of x, we have $\psi - \psi' = \pm \tfrac{1}{2}\pi$, and therefore

$$\tan \psi = - \cot \psi'.$$

Hence the differential equation of one family is obtained from that of the other by writing

$$-1 \Big/ \frac{dy}{dx} \quad \text{for} \quad \frac{dy}{dx}.$$

Otherwise: if dx, dy be the projections of an element of one of the curves (1) we have

$$\frac{\partial \phi}{\partial x} dx + \frac{\partial \phi}{\partial y} dy = 0. \quad \dots\dots\dots\dots\dots(2)$$

Hence, if dx, dy be the projections of an element of the orthogonal curve through the point (x, y), we have

$$\frac{dx}{\partial \phi / \partial x} = \frac{dy}{\partial \phi / \partial y}. \quad \dots\dots\dots\dots\dots(3)$$

The differential equation of the trajectories is then obtained by elimination of C between (1) and (3).

If the equation of the given family of curves be in polar coordinates, thus

$$f(r, \theta, C) = 0, \quad \dots\dots\dots\dots\dots(4)$$

and if ϕ, ϕ' denote the angles which the tangents to the original curve and to the trajectory make with the radius vector, we have in like manner

$$\tan \phi = - \cot \phi'.$$

Hence the differential equation of one system is obtained from that of the other by writing

$$-\frac{1}{r}\frac{dr}{d\theta} \quad \text{for} \quad \frac{r d\theta}{dr}.$$

Or, differentiating (4), we have

$$\frac{\partial f}{\partial r}\,dr + \frac{1}{r}\frac{\partial f}{\partial \theta}\,r d\theta = 0, \quad\dots\dots\dots\dots\dots(5)$$

and therefore, for the trajectory,

$$\frac{\partial f}{\partial r}\,r d\theta - \frac{1}{r}\frac{\partial f}{\partial \theta}\,dr = 0. \quad\dots\dots\dots\dots\dots(6)$$

The elimination of C between (4) and (6) leads to the differential equation of the required system.

Ex. 1.   To find the orthogonal trajectories of the rectangular hyperbolas

$$xy = C. \quad\dots\dots\dots\dots\dots\dots\dots\dots(7)$$

Differentiating, we find $\qquad x\,dy + y\,dx = 0, \quad\dots\dots\dots\dots(8)$

and therefore, for the trajectories,

$$x\,dx - y\,dy = 0, \quad\dots\dots\dots\dots\dots(9)$$

whence $\qquad\qquad\qquad x^2 - y^2 = C'. \quad\dots\dots\dots\dots\dots(10)$

This represents a system of rectangular hyperbolas whose axes coincide in direction with the asymptotes of the former system.

Ex. 2.   To find the curves orthogonal to the circles

$$x^2 + y^2 + 2\mu y - k^2 = 0, \quad\dots\dots\dots\dots\dots(11)$$

where μ is the variable parameter.

Differentiating, we have

$$x\,dx + (y + \mu)\,dy = 0,$$

and therefore, for the trajectory,

$$x\,dy - (y + \mu)\,dx = 0.$$

Eliminating μ between this and (11), we find

$$2xy\,\frac{dy}{dx} + (x^2 - y^2 - k^2) = 0, \quad\dots\dots\dots(12)$$

or $\qquad\qquad\qquad x\,\frac{d\,(y^2)}{dx} - y^2 = -x^2 + k^2. \quad\dots\dots\dots(13)$

This is linear, with y^2 as the independent variable. The integrating factor, as found by the rule of Art. 158, or by inspection, is $1/x^2$. Introducing this we have

$$\frac{d}{dx}\left(\frac{y^2}{x}\right) = -1 + \frac{k^2}{x^2},$$

whence $\qquad\qquad\qquad \frac{y^2}{x} = -x - \frac{k^2}{x} + 2\lambda,$

or $\qquad\qquad\qquad x^2 + y^2 - 2\lambda x + k^2 = 0, \quad\dots\dots\dots\dots(14)$

λ being arbitrary.

The original equation represents a system of coaxial circles, cutting the axis of x in the points $(\pm k, 0)$. The trajectories (14) consist of a second system of coaxial circles having these as 'limiting points'; viz. if we put $\lambda = \pm k$ we get the point-circles

$$(x \mp k)^2 + y^2 = 0 ; \quad \dots\dots\dots\dots\dots(15)$$

see Fig. 135.

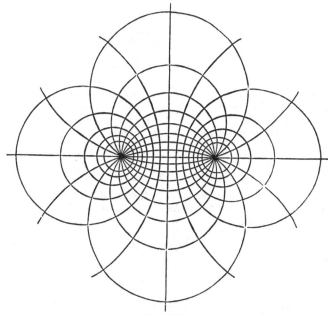

Fig. 135.

Ex. 3. In the circles

$$r = c \cos \theta, \quad \dots\dots\dots\dots\dots(16)$$

which pass through the origin, and have their centres on the initial line, we have

$$\frac{dr}{r} = - \tan \theta \, d\theta, \quad \dots\dots\dots\dots\dots(17)$$

and therefore, for the trajectory,

$$r d\theta = \tan \theta \, dr, \quad \text{or} \quad \frac{dr}{r} = \cot \theta \, d\theta. \quad \dots\dots\dots(18)$$

Integrating, we find $\log r = \log \sin \theta + \text{const.}$,

or $r = c' \sin \theta, \quad \dots\dots\dots\dots\dots\dots\dots(19)$

which represents another system of circles, passing through the origin, and touching the initial line.

160.   Equations of Degree higher than the First.

The general type of an equation of the first order and nth degree is

$$p^n + P_1 p^{n-1} + P_2 p^{n-2} + \ldots + P_{n-1} p + P_n = 0, \quad \ldots\ldots(1)$$

where
$$p = \frac{dy}{dx}, \quad \ldots\ldots\ldots\ldots\ldots\ldots\ldots\ldots(2)$$

and $P_1, P_2, \ldots, P_n$ are given functions of x and y. It is usually implied that these functions are algebraic, and rational.

The equation (1), being of the nth degree in p, indicates that n branches of the primitive curves go through any assigned point in the plane xy. Some of these branches may of course be imaginary, and for some ranges of x and y all may be imaginary. There may also be a real locus of points at which two of the values of p coincide; this locus is of special importance in the higher development of the subject.

For example, in the equation of the *second* degree,

$$p^2 + Pp + Q = 0, \quad \ldots\ldots\ldots\ldots\ldots\ldots\ldots\ldots(3)$$

the values of p will be real and distinct, coincident, or imaginary, according as $P^2 \gtreqless 4Q$. And the locus of points at which the two values of p coincide is the curve $P^2 = 4Q$.

If the left-hand side of (1), considered as a function of p, can be resolved into linear factors, thus

$$(p - p_1)(p - p_2) \ldots (p - p_n) = 0, \quad \ldots\ldots\ldots\ldots(4)$$

where $p_1, p_2, \ldots p_n$ are known functions of x and y, the complete solution will consist of the aggregate of the solutions of the several equations

$$\frac{dy}{dx} = p_1, \quad \frac{dy}{dx} = p_2, \ldots, \frac{dy}{dx} = p_n. \quad \ldots\ldots\ldots\ldots(5)$$

Ex. $$xy p^2 - (x^2 - y^2) p - xy = 0. \quad \ldots\ldots\ldots\ldots\ldots(6)$$

This is equivalent to
$$(xp + y)(yp - x) = 0 ; \quad \ldots\ldots\ldots\ldots\ldots\ldots(7)$$

and the solutions of
$$xp + y = 0, \quad yp - x = 0,$$

are, respectively, $$xy = C, \quad x^2 - y^2 = C. \quad \ldots\ldots\ldots\ldots\ldots(8)$$

The product of the two values of p given by (6) is -1. This shews *à priori* that the two branches of the primitive curves which pass through any point (x, y) will be at right angles to one another. Cf. Art. 159, Ex. 1.

161. Clairaut's form.

When the equation (1) of Art. 160 cannot be conveniently resolved into its linear factors, we may in certain cases have recourse to other methods. These are for the most part of somewhat limited utility, and are accordingly passed over here; but an exception may be made in favour of Clairaut's form, which is very simple in theory, and moreover often presents itself in questions where a curve is defined by some property of the tangent.

If we write p for dy/dx, the form in question is

$$y = xp + f(p). \quad \ldots\ldots\ldots\ldots\ldots(1)$$

It was proved in Art. 60 that the intercepts (α, β) made by the tangent to a curve on the axes of x and y are given by

$$\alpha = (xp - y)/p, \quad \beta = y - xp, \quad \ldots\ldots\ldots\ldots(2)$$

respectively. Hence any equation of the form (1) expresses a relation between either intercept and the direction of the tangent, or (again) between the two intercepts*. Now it is evident that this relation is satisfied by any *straight line* whose intercepts have the given relation. Along any such straight line we have

$$p = C, \quad \ldots\ldots\ldots\ldots\ldots\ldots(3)$$

and we thus get the solution

$$y = Cx + f(C), \quad \ldots\ldots\ldots\ldots\ldots(4)$$

involving an arbitrary constant C.

But the equation will also be satisfied by the curve which has the family (4) of straight lines as its tangents; in other words, by the *envelope* of this family. This envelope is found by expressing that (4), considered as an equation in C, has a pair of equal roots, *i.e.* by eliminating C between (4) and

$$x + f'(C) = 0; \quad \ldots\ldots\ldots\ldots\ldots(5)$$

see Art. 139.

The more usual method of deducing the above solutions is to differentiate (1) with respect to x; thus

$$p = \frac{dy}{dx} = p + \{x + f'(p)\}\frac{dp}{dx},$$

whence

$$\{x + f'(p)\}\frac{dp}{dx} = 0. \quad \ldots\ldots\ldots\ldots(6)$$

---

* The equation is equivalent to

$$\beta = f(-\beta/a), \quad \text{or} \quad \phi(a, \beta) = 0.$$

This requires, either that

$$\frac{dp}{dx} = 0, \quad\dots\dots\dots\dots\dots\dots\dots(7)$$

or that $\qquad\qquad x + f'(p) = 0. \quad\dots\dots\dots\dots\dots(8)$

The former result makes $p = C$, and

$$y = Cx + f(C). \quad\dots\dots\dots\dots\dots(9)$$

The alternative result (8), combined with (1), leads, on elimination of p, to a particular relation between x and y. Since the result of eliminating p between (1) and (8) must be the same as that of eliminating C between (4) and (5), we identify this second solution with the envelope aforesaid.

The solution (9), involving an arbitrary constant C, is called the 'complete primitive.' The second, or envelope-solution, is not included in the complete primitive, *i.e.* it cannot be derived from it by giving a particular value to C. It is therefore called a 'singular solution*.'

Ex. To find the curve whose pedal with respect to the point $(a, 0)$ as pole is the straight line $x = 0$.

The expression of this property is

$$a = p\beta,$$

where β is the intercept on the axis of y, whence

$$y = xp + \frac{a}{p}. \quad\dots\dots\dots\dots\dots(10)$$

This is satisfied by any one of the family of straight lines

$$y = Cx + \frac{a}{C}, \quad\dots\dots\dots\dots\dots(11)$$

and also by their envelope $\qquad y^2 = 4ax; \quad\dots\dots\dots\dots\dots(12)$

see Art. 140, Ex. 2.

EXAMPLES. L.

(Formation of Differential Equations.)

1. If $\qquad\qquad y = Ax^2 + B,$

prove that $\qquad\qquad x\frac{d^2y}{dx^2} - \frac{dy}{dx} = 0.$

---

* The general theory of singular solutions of equations of degree higher than the first must be sought for in books specially devoted to the subject of Differential Equations. It is closely related to, but not altogether co-extensive with, the theory of envelopes.

2. If
$$y = Ax^2 + Bx,$$

prove that
$$\frac{d^2y}{dx^2} - \frac{2}{x}\frac{dy}{dx} + \frac{2y}{x^2} = 0.$$

3. If
$$y = Ae^{kx} + Be^{-kx},$$

prove that
$$\frac{d^2y}{dx^2} - k^2y = 0.$$

4. If
$$y = Ae^{ax} + Be^{\beta x},$$

prove that
$$\frac{d^2y}{dx^2} - (a + \beta)\frac{dy}{dx} + a\beta y = 0.$$

5. If
$$y = (A + Bx)\,e^{kx},$$

prove that
$$\frac{d^2y}{dx^2} - 2k\frac{dy}{dx} + k^2y = 0.$$

6. If
$$x = e^{-\frac{1}{2}kt}\,(A\cos nt + B\sin nt),$$

prove that
$$\frac{d^2x}{dt^2} + k\frac{dx}{dt} + (n^2 + \tfrac{1}{4}k^2)\,x = 0.$$

7. If
$$V = \frac{A}{r} + B,$$

prove that
$$\frac{d^2V}{dr^2} + \frac{2}{r}\frac{dV}{dr} = 0.$$

8. If
$$V = A\log r + B,$$

prove that
$$\frac{d^2V}{dr^2} + \frac{1}{r}\frac{dV}{dr} = 0.$$

9. If
$$\phi = \frac{Ae^{kr} + Be^{-kr}}{r},$$

prove that
$$\frac{d^2\phi}{dr^2} + \frac{2}{r}\frac{d\phi}{dr} - k^2\phi = 0.$$

10. If
$$\phi = \frac{A\cos kr + B\sin kr}{r},$$

prove that
$$\frac{d^2\phi}{dr^2} + \frac{2}{r}\frac{d\phi}{dr} + k^2\phi = 0.$$

11. If $y = (A + Bx)\cos kx + (C + Dx)\sin kx,$

prove that
$$\frac{d^4y}{dx^4} + 2k^2\frac{d^2y}{dx^2} + k^4y = 0.$$

12. If $y = A\cosh kx + B\sinh kx + C\cos kx + D\sin kx,$

prove that
$$\frac{d^4y}{dx^4} = k^4y.$$

13. If

$$y = \left(A \cosh \frac{kx}{\sqrt{2}} + B \sinh \frac{kx}{\sqrt{2}} \right) \cos \frac{kx}{\sqrt{2}}$$
$$+ \left(C \cosh \frac{kx}{\sqrt{2}} + D \sinh \frac{kx}{\sqrt{2}} \right) \sin \frac{kx}{\sqrt{2}},$$

prove that

$$\frac{d^4y}{dx^4} + k^4y = 0.$$

14. If

$$y = A \sin^{-1} x + B,$$

prove that

$$(1 - x^2) \frac{d^2y}{dx^2} - x \frac{dy}{dx} = 0.$$

15. If

$$y = (\sin^{-1} x)^2 + A \sin^{-1} x + B,$$

prove that

$$(1 - x^2) \frac{d^2y}{dx^2} - x \frac{dy}{dx} = 2.$$

16. If

$$y = A \cos (\log x) + B \sin (\log x),$$

prove that

$$x^2 \frac{d^2y}{dx^2} + x \frac{dy}{dx} + y = 0.$$

17. If

$$y = A \{ x + \sqrt{(x^2 - 1)} \}^n + B \{ x - \sqrt{(x^2 - 1)} \}^n,$$

prove that

$$(x^2 - 1) \frac{d^2y}{dx^2} + x \frac{dy}{dx} - n^2y = 0.$$

18. Shew that the primitive

$$y = mx + \frac{a}{m},$$

where m is arbitrary, leads to

$$x \left(\frac{dy}{dx} \right)^2 - y \frac{dy}{dx} + a = 0.$$

19. If

$$2cy + c^2 = x^2,$$

where c is arbitrary, prove that

$$x \left(\frac{dy}{dx} \right)^2 - 2y \frac{dy}{dx} - x = 0.$$

20. The differential equation of all parabolas having their axes parallel to the axis of y is

$$\frac{d^3y}{dx^3} = 0.$$

21. The differential equation of all parabolas having their axis of symmetry coincident with the axis of x is

$$y \frac{d^2y}{dx^2} + \left(\frac{dy}{dx} \right)^2 = 0.$$

22. The differential equation of all conics having their principal axes coincident with the coordinate axes is

$$xy \frac{d^2y}{dx^2} + x \left(\frac{dy}{dx}\right)^2 - y \frac{dy}{dx} = 0.$$

23. Prove that the differential equation of all circles touching the axis of x at the origin is

$$\frac{dy}{dx} = \frac{2xy}{x^2 - y^2}.$$

24. Prove that the differential equation of all conics touching the axis of y at the origin, and having their centres on the axis of x, is

$$x^2 y \frac{d^2y}{dx^2} + \left(x \frac{dy}{dx} - y\right)^2 = 0.$$

25. If
$$y = \frac{x^2 + a^2}{x^2 + b^2},$$

prove that
$$x(y-1)\frac{d^2y}{dx^2} = 2x\left(\frac{dy}{dx}\right)^2 + (y-1)\frac{dy}{dx}.$$

26. Prove that the differential equation of all hyperbolas which pass through the origin, and have their asymptotes parallel to the coordinate axes, is

$$xy \frac{d^2y}{dx^2} - 2x \left(\frac{dy}{dx}\right)^2 + 2y \frac{dy}{dx} = 0.$$

27. Prove that the equation

$$\frac{d^2y}{dt^2} + n^2y = f(t)$$

is satisfied by

$$y = \frac{1}{n} \sin nt \int f(t) \cos nt \, dt - \frac{1}{n} \cos nt \int f(t) \sin nt \, dt,$$

and that this is the complete solution.

EXAMPLES. LI.

(Equations of the First Order.)

1. Integrate $\dfrac{dy}{dx} = \dfrac{y}{x}.$ $[y = Cx.]$

2. Integrate $\dfrac{dy}{dx} = \dfrac{y}{x^2 - 1}.$ $\left[y^2 = C\dfrac{x-1}{x+1}.\right]$

3. Integrate $\dfrac{dy}{dx} = \cot x \cot y.$ $[\sin x \cos y = C.]$

4. Integrate $x^2 \dfrac{dy}{dx} + y = 1.$ $[y = 1 + Ce^{1/x}.]$

5. Solve $m(y + b) dx + n(x + a) dy = 0.$

$$[(x + a)^m (y + b)^n = C.]$$

6. Solve $\dfrac{dy}{dx} = \dfrac{1 + y^2}{1 + x^2}.$ $\left[y = \dfrac{x + C}{1 - Cx}.\right]$

7. Solve $(1 + y^2) dx - xy (1 + x^2) dy = 0.$

$$[(1 + x^2)(1 + y^2) = Cx^2.]$$

8. Solve $\dfrac{dy}{dx} = \dfrac{y(1 - y^2)}{x(1 - x^2)}.$

$$[y^2 (1 - x^2) = Cx^2 (1 - y^2).]$$

9. Solve $\dfrac{dy}{dx} = (x + y)^2.$ $[x + y = \tan(x + a).]$

10. Solve $(x + y)^2 \left(x \dfrac{dy}{dx} + y\right) = xy \left(1 + \dfrac{dy}{dx}\right).$

11. Find the curves in which the angle between the tangent and the radius is one-half the vectorial angle (θ).

[The cardioids $r = a(1 - \cos \theta)$.]

12. Find the curves in which the perpendicular from the origin on the tangent is equal to the abscissa of the point of contact.

[The circles $r = 2a \cos \theta.$]

13. Find the curves such that the portion of the tangent included between the coordinate axes is bisected at the point of contact.

[The hyperbolas $xy = C.$]

14. Find the curves in which the subtangent varies as the abscissa. $[y = Cx^m.]$

15. Prove that if the subnormal bears a constant ratio to the abscissa the curve is a conic.

16. Find the curves in which the perpendicular from the foot of the ordinate to the tangent has a constant length a.

[The catenaries $y = a \cosh (x - a)/a.$]

17. Find the curve in which the polar subtangent is constant $(= a)$. $[r = a/(\theta - a).]$

18. Find the curve in which the polar subnormal is constant $(= a)$. $[r = a(\theta - a).]$

19. Find the curves such that the area included between any two ordinates is proportional to the intercepted arc.

[The catenaries $y = a \cosh (x - a)/a.$]

20. Find the curves such that the area included between any ordinate, the axis of x, and the curve is $1/n$th of the rectangle contained by the ordinate and the corresponding abscissa.

$$[y = Cx^{n-1}.]$$

21. Find the form of a solid of revolution in order that the volume cut off by any right section may be $1/n$th of the product of the area of this section into the length of the axis.

[The equation of the generating curve must be $y^2 = Ax^{n-1}$.]

22. In a suspension-rod of uniform strength the area of the cross-section (S) varies as the total stress across it; prove that if x be measured vertically downwards the relation between S and x must be of the form

$$S = A - B \int_0^x S\,dx.$$

Hence shew that the form of the rod must be that generated by the revolution of a curve of the type

$$y = be^{-x/a}$$

about the axis of x.

23. Find the form of a curve, symmetrical with respect to the axis of x, such that the mean centre of the area cut off by any double ordinate shall be at a distance from this ordinate equal to $1/n$th of the length of the axis. $[y = Cx^{n-2}.]$

24. Solve $\quad (x^3 + 3xy^2)\,dx + (y^3 + 3x^2y)\,dy = 0.$

25. Solve $\quad x\,dx + y\,dy = a^2 \dfrac{x\,dy - y\,dx}{x^2 + y^2}.$

$$\left[x^2 + y^2 = 2a^2 \tan^{-1}\frac{y}{x} + C. \right]$$

26. Solve $\quad x\dfrac{dy}{dx} - y = \dfrac{x}{a}\,\sqrt{(x^2 + y^2)}. \qquad \left[y = x\sinh\dfrac{x-a}{a}. \right]$

27. Solve $\quad x\dfrac{dy}{dx} - y = \sqrt{(x^2 + y^2)}.$

Give the geometrical interpretations of the differential equation and of its primitive. $[x^2 = 2Cy + C^2.]$

28. Solve $\quad \dfrac{dx}{x^2 - 2xy} = \dfrac{dy}{y^2 - 2xy}. \qquad [xy\,(x - y) = C.]$

29. Solve $\quad x\dfrac{dy}{dx} + y = xy. \qquad\qquad [xy = Ce^x.]$

30. Solve $\quad x\dfrac{dy}{dx} - y = xy. \qquad\qquad [y = Cxe^x.]$

31. Solve $\quad \dfrac{dy}{dx} = \dfrac{y\,(x + y)}{x\,(y - x)}. \qquad\qquad [xy = Ce^{y/x}.]$

32. Solve $\quad (x^2 - 3y^2)\,x\,dx = (y^2 - 3x^2)\,y\,dy.$

$$[(x^2 + y^2)^2 = C\,(x^2 - y^2).]$$

33. Shew that the equation

$$\frac{dy}{dx} = \frac{ax + by + c}{a'x + b'y + c'}$$

is rendered homogeneous by the substitutions

$$ax + by + c = \xi, \quad a'x + b'y + c' = \eta.$$

34. Shew that an equation of the type

$$\frac{dy}{dx} = f(ax + by)$$

may be solved by the substitution

$$ax + by = z.$$

35. Shew how to solve any equation of the type

$$\frac{dy}{dx} = f\left(\frac{ax + by + c}{a'x + b'y + c'}\right).$$

EXAMPLES. LII.

(Linear Equation.)

1. Solve $\dfrac{dy}{dx} + y \tan x = \sec x.$ $\qquad [y = \sin x + C \cos x.]$

2. Solve $(1 - x^2)\dfrac{dy}{dx} + xy = ax.$ $\qquad [y = a + C \sqrt{(1 - x^2)}.]$

3. Solve $x \dfrac{dy}{dx} + x + y = 0.$ $\qquad [x^2 + 2xy = C.]$

4. Solve $\dfrac{dy}{dx} + \dfrac{y}{x} = 1.$ $\qquad [x^2 - 2xy = C.]$

5. Solve $\dfrac{dy}{dx} + 2xy = 1 + 2x^2.$ $\qquad [y = x + Ce^{-x^2}.]$

6. Solve $x \dfrac{dy}{dx} + \dfrac{1 - x^2}{1 + x^2} y = 1.$

7. Solve $a \dfrac{du}{d\theta} + au \tan \theta = \tan \theta.$ $\qquad [au = 1 + C \cos \theta.]$

8. Solve $\dfrac{dy}{dx} = y \tan x - 2 \sin x.$ $\qquad [y = \cos x + C \sec x.]$

9. Solve $x(1 - x^2)\dfrac{dy}{dx} + (2x^2 - 1) y = x^3.$

$$[y = x + Cx \sqrt{(1 - x^2)}.]$$

10. Shew that the equation

$$\frac{dy}{dx} + Py = Qy^n$$

is made linear by the substitution

$$y^{-(n-1)} = z.$$

(Bernoulli's equation.)

11. Solve $\quad x\dfrac{dy}{dx} + y = y^2 \log x.$ $\qquad \left[\dfrac{1}{y} = 1 + \log x + Cx.\right]$

12. Solve $\quad \cos x\dfrac{dy}{dx} - y \sin x + y^2 = 0.$ $\quad \left[\dfrac{1}{y} = \sin x + C \cos x.\right]$

13. If the two plates of a condenser of capacity C are connected by a wire of resistance R (and zero self-induction), the equation connecting the charge (q) with the electromotive force (E) is

$$E = R\frac{dq}{dt} + \frac{q}{C}.$$

Integrate this in the cases $E = 0,\ E = \text{const.},\ E = E_0 \cos (pt + \epsilon).$

EXAMPLES. LIII.

(Orthogonal Trajectories.)

1. Find the orthogonal trajectories of the straight lines
$$y = Cx. \qquad [\text{The circles } x^2 + y^2 = C.]$$

2. Find the orthogonal trajectories of the curves
$$a^{n-1} y = x^n. \qquad [\text{The conics } x^2 + ny^2 = C.]$$

3. Find the orthogonal trajectories of the circles
$$x^2 + y^2 = 2cy. \qquad [\text{The circles } x^2 + y^2 = 2c'x.]$$

4. Find the curves for which
$$\frac{dy}{dx} = \frac{y^3 + 3x^2y}{x^3 + 3xy^2},$$

and determine their orthogonal trajectories.
$$[(x^2 - y^2)^2 = Cxy\ ;\ x^4 + 6x^2y^2 + y^4 = C.]$$

5. Prove that the differential equation of the confocal parabolas
$$y^2 = 4a\,(x + a),$$

is $\qquad yp^2 + 2xp - y = 0,$

where $p = dy/dx$.

Shew that this coincides with the differential equation of the orthogonal curves; and interpret the result.

6.   Prove that the differential equation of the confocal conics

$$\frac{x^2}{a^2 + \lambda} + \frac{y^2}{b^2 + \lambda} = 1,$$

is $$xy\,p^2 + (x^2 - y^2 - a^2 + b^2)\,p - xy = 0.$$

Shew that this coincides with the differential equation of the orthogonal curves, and interpret the result.

7.   A system of rectangular hyperbolas pass through the fixed points $(\pm\,a,\ 0)$ and have the origin as centre; prove that their orthogonal trajectories are the Cassini's ovals

$$(x^2 + y^2)^2 = 2a^2\,(x^2 - y^2) + C.$$

8.   Prove that the differential equation of the involutes of the parabola $y^2 = 4ax$ is

$$x + y\,\frac{dy}{dx} + a\left(\frac{dy}{dx}\right)^2 = 0.$$

9.   Prove that the differential equation of the involutes of the circle $x^2 + y^2 = a^2$ is

$$x^2 - a^2 + 2xy\,\frac{dy}{dx} + (y^2 - a^2)\left(\frac{dy}{dx}\right)^2 = 0.$$

10.   Find the orthogonal trajectories of the cardioids

$$r = a\,(1 - \cos\theta).$$

$$[\text{The cardioids } r = b\,(1 + \cos\theta).]$$

11.   Prove that the orthogonal trajectories of the curves

$$r^m = a^m \cos m\theta$$

are the curves $$r^m = b^m \sin m\theta.$$

Interpret the cases of $m = 1,\ -1,\ 2,\ -2,\ \frac{1}{2},\ -\frac{1}{2}$, respectively.

12.   Prove that the orthogonal trajectories of the curves

$$r^2 = A \cos\theta$$

are the curves

$$r = B \sin^2\theta.$$

13.   If in bipolar coordinates (Art. 132) the equation of a family of curves be

$$f(r,\ r') = C,$$

the differential equation of the orthogonal trajectories is

$$r\,\frac{\partial f}{\partial r}\,d\theta = r'\,\frac{\partial f}{\partial r'}\,d\theta'.$$

Hence shew that the orthogonal trajectories of the circles

$$r/r' = C,$$

are the circles $$\theta + \theta' = C.$$

14. Also that the orthogonal trajectories of the Cassini's ovals

$$rr' = C,$$

are the rectangular hyperbolas

$$\theta - \theta' = C.$$

15. Also that the orthogonal trajectories of the equipotential curves

$$\frac{1}{r} - \frac{1}{r'} = C,$$

are the magnetic curves

$$\cos \theta + \cos \theta' = C.$$

EXAMPLES. LIV.

(Equations of Higher Degree.)

1. Solve $\left(\dfrac{dy}{dx}\right)^2 - (a + \beta)\dfrac{dy}{dx} + a\beta = 0.$

$$[y = ax + C, \quad y = \beta x + C.]$$

2. Solve $\left(\dfrac{dy}{dx}\right)^2 = \sin^2 x.$ $[y = C \pm \cos x.]$

3. Solve $\left(\dfrac{dy}{dx}\right)^2 = m^2 y^2.$ $[y = Ce^{\pm mx}].$

4. Solve $y^2 \left(\dfrac{dy}{dx}\right)^2 = 4a^2.$ $[y^2 = C \pm 4ax.]$

5. Solve $x \left(\dfrac{dy}{dx}\right)^2 = a.$ $[y = C \pm 2\sqrt{(ax)}.]$

6. Solve $(1 - x^2) \left(\dfrac{dy}{dx}\right)^2 = 1.$ $[y = C \pm \sin^{-1} x.]$

7. Solve $\dfrac{dy}{dx}\left(\dfrac{dy}{dx} + y\right) = x\,(x + y).$

$$[y = \tfrac{1}{2}x^2 + C, \quad y = 1 - x + Ce^{-x}.]$$

8. Solve $\dfrac{dy}{dx}\left(\dfrac{dy}{dx} + x\right) = (x + y)\,y.$

$$[y = Ce^x, \quad y = 1 - x + Ce^{-x}.]$$

9. Solve $x \left(\dfrac{dy}{dx}\right)^2 - 2y \dfrac{dy}{dx} - x = 0.$ $[x^2 = 2Cy + C^2.]$

10. Solve $y \left(\dfrac{dy}{dx}\right)^2 + 2x \dfrac{dy}{dx} - y = 0.$

$$[\sqrt{(x^2 + y^2)} = C \pm x.]$$

11.   Find the curve such that the product of the intercepts made by the tangent on the coordinate axes is constant $(= k^2)$.

[The hyperbola $4xy = k^2$.]

12.   Find the curve such that the perpendicular from the origin on any tangent is equal to a.   [The circle $x^2 + y^2 = a^2$.]

13.   Solve $y = xp + \sqrt{(b^2 + a^2 p^2)}$.

$$\left[\text{Singular solution}: \frac{x^2}{a^2} + \frac{y^2}{b^2} = 1. \right]$$

14.   Find the curve such that the product of the perpendiculars from the points $(\pm c, 0)$ on any tangent is equal to b^2.

$$\left[\text{The conics } \frac{x^2}{b^2 + c^2} + \frac{y^2}{b^2} = 1, \quad \frac{x^2}{c^2 - b^2} - \frac{y^2}{b^2} = 1. \right]$$

15.   Find the curve such that the tangent intercepts on the perpendiculars to the axis of x at the points $(\pm a, 0)$ lengths whose product is b^2.

$$\left[\text{The conics } \frac{x^2}{a^2} \pm \frac{y^2}{b^2} = 1. \right]$$

16.   Solve $y = xp + ap(1 - p)$.

[Singular solution : $(x + a)^2 = 4ay$.]

17.   Solve $(x - a)p^2 + (x - y)p - y = 0$.

[Singular solution: $(x + y)^2 = 4ay$.]

18.   Find the curve such that the sum of the intercepts made by the tangent on the coordinate axes is equal to a.

[The parabola $(x - y)^2 - 2a(x + y) + a^2 = 0$.]

19.   Shew that any differential equation of the type

$$x + y \frac{dy}{dx} = f\left(\frac{dy}{dx}\right)$$

represents a system of parallel curves.

20.   Shew that any differential equation of the type

$$f\left(x, y, p - \frac{1}{p}\right) = 0$$

represents two systems of orthogonal curves.

CHAPTER XII

DIFFERENTIAL EQUATIONS OF THE SECOND ORDER

162. Equations of the Type $d^2y/dx^2 = f(x)$.

This chapter is devoted principally to differential equations of the second order, and especially to such types as are of most frequent occurrence in the geometrical and physical applications of the Calculus. Occasionally, the methods will admit of extension to equations of higher order.

We begin by the consideration of a few special types, and afterwards proceed to the study of the linear equation. The linear equation with *constant* coefficients is treated in the next chapter.

We take, first, the type

$$\frac{d^2y}{dx^2} = f(x). \qquad \qquad \text{.........................(1)}$$

This requires merely two ordinary integrations with respect to x; thus

$$\frac{dy}{dx} = \int f(x)\,dx + A,$$

$$y = \int \{\textstyle\int f(x)\,dx\}\,dx + Ax + B, \quad \text{..............(2)}$$

where the constants A, B are arbitrary.

Ex. 1. The dynamical equation

$$\frac{d^2x}{dt^2} = f(t), \qquad \text{...........................(3)}$$

which determines the motion of a particle in a straight line under a force which is a given function of the time, is of the above type, with merely a difference of notation.

In the case of a particle subject to a constant acceleration g we have

$$\frac{d^2x}{dt^2} = g, \qquad \text{...............................(4)}$$

whence
$$\frac{dx}{dt} = gt + A,$$

$$x = \tfrac{1}{2}gt^2 + At + B. \quad \dots\dots\dots\dots\dots(5)$$

Again, if
$$\frac{d^2x}{dt^2} = f \sin nt, \quad \dots\dots\dots\dots\dots\dots(6)$$

the force varying as a simple-harmonic function of the time, we have

$$\frac{dx}{dt} = -\frac{f}{n} \cos nt + A,$$

$$x = -\frac{f}{n^2} \sin nt + At + B. \quad \dots\dots\dots\dots\dots(7)$$

The constants A, B which occur in these problems may be adjusted so that at any chosen instant the particle shall be in a given position and have a given velocity.

Ex. 2. To solve the equation

$$\mathfrak{B} \frac{d^2y}{dx^2} - W(l - x) = 0, \quad \dots\dots\dots\dots\dots(8)$$

subject to the conditions that $y = 0$ and $dy/dx = 0$ for $x = 0$. This is the problem of determining the flexure of a bar which is clamped in a horizontal position at one end ($x = 0$) and supports a given weight (W) at the other end ($x = l$).

Two successive integrations of (8) give

$$\mathfrak{B} \frac{dy}{dx} = W(lx - \tfrac{1}{2}x^2) + A,$$

$$\mathfrak{B}y = W(\tfrac{1}{2}lx^2 - \tfrac{1}{6}x^3) + Ax + B, \quad \dots\dots\dots(9)$$

where A, B are arbitrary. The terminal conditions require that $A = 0$, $B = 0$, whence

$$y = \tfrac{1}{2}\frac{W}{\mathfrak{B}} x^2 (l - \tfrac{1}{3}x). \quad \dots\dots\dots\dots\dots(10)$$

163. Equations of the Type $d^2y/dx^2 = f(y)$.

If the equation be of the type

$$\frac{d^2y}{dx^2} = f(y), \quad \dots\dots\dots\dots\dots\dots(1)$$

a first integral may be obtained in two ways.

In one of these we multiply both sides by dy/dx, and then integrate with respect to x; thus

$$\frac{dy}{dx}\frac{d^2y}{dx^2} = f(y)\frac{dy}{dx},$$

$$\tfrac{1}{2}\left(\frac{dy}{dx}\right)^2 = \int f(y)\frac{dy}{dx}\,dx + A = \int f(y)\,dy + A. \quad \dots\dots(2)$$

The second method is to introduce a special symbol (p) for dy/dx. Since this makes

$$\frac{d^2y}{dx^2} = \frac{dp}{dx} = \frac{dp}{dy}\frac{dy}{dx} = p\frac{dp}{dy}, \quad \dots\dots\dots\dots(3)$$

we have, in place of (1),

$$p\frac{dp}{dy} = f(y), \quad \dots\dots\dots\dots\dots\dots(4)$$

which may be regarded as an equation of the first order, with p as dependent, and y as independent, variable. Integrating (3) with respect to y, we have

$$\tfrac{1}{2}p^2 = \int f(y)\,dy + A, \quad \dots\dots\dots\dots\dots(5)$$

which is equivalent to (2).

To complete the solution, we write (2) in the form

$$\frac{dy}{\sqrt{\{2\int f(y)\,dy + 2A\}}} = \pm\,dx. \quad \dots\dots\dots\dots(6)$$

The variables are here separated (Art. 154); but on account mainly of the occurrence of the radical the further integration is often impracticable, even with comparatively simple forms of the function $f(y)$.

A very important case is where $f(y)$ is a linear function of y, so that the equation takes the shape

$$\frac{d^2y}{dx^2} + ay = b. \quad \dots\dots\dots\dots\dots\dots(7)$$

By a change of dependent variable, writing $y_1 + b/a$ for y, and afterwards omitting the suffix, this is reduced to the somewhat simpler form

$$\frac{d^2y}{dx^2} + ay = 0. \quad \dots\dots\dots\dots\dots(8)$$

The first integral of this is

$$\left(\frac{dy}{dx}\right)^2 + ay^2 = C. \quad \dots\dots\dots\dots\dots(9)$$

If a be positive, we may write

$$m = \sqrt{a}, \quad C = m^2\alpha^2, \quad \dots\dots\dots\dots\dots(10)$$

it being evident that, if we are concerned solely with real quantities, C must be positive. Thus

$$\frac{dy}{\sqrt{(\alpha^2 - y^2)}} = \pm\,mdx, \quad \dots\dots\dots\dots(11)$$

whence
$$\cos^{-1}\frac{y}{\alpha} = \pm\,(mx + \epsilon),$$

or
$$y = \alpha\cos\,(mx + \epsilon). \quad\text{......................(12)}$$

This is the complete solution of (8), and involves the two arbitrary constants α, ϵ. If we put

$$A = \alpha\cos\epsilon,\quad B = -\,\alpha\sin\epsilon, \quad\text{...............(13)}$$

we obtain the equivalent form

$$y = A\cos mx + B\sin mx. \quad\text{..................(14)}$$

These results are exceedingly important, and should be remembered.

The case where a is negative, $= -\,m^2$, say, can be treated in a similar manner, and we should find, as the complete solution

$$y = A\cosh mx + B\sinh mx, \quad\text{...............(15)}$$

where $m = \sqrt{(-\,a)}$. A simpler method of treating this case will however be given later.

The type (1) is of very great importance in Dynamics. Thus, the equation of rectilinear motion of a particle subject to a force which is a given function of its position only is of the form

$$\frac{d^2x}{dt^2} = f(x), \quad\text{.............................(16)}$$

which is identical with (1), if regard be had to the difference of notation.

The first method of integration consists in multiplying both sides by dx/dt, thus

$$\frac{dx}{dt}\frac{d^2x}{dt^2} = f(x)\frac{dx}{dt},$$

and integrating both sides with respect to t. In this way we obtain

$$\tfrac{1}{2}\left(\frac{dx}{dt}\right)^2 = \int f(x)\frac{dx}{dt}\,dt + C = \int f(x)\,dx + C, \quad\text{.........(17)}$$

which is the 'equation of energy.'

The second method consists in writing v for dx/dt, and therefore $v\,dv/dx$ for d^2x/dt^2; cf. Art. 32. Thus

$$v\frac{dv}{dx} = f(x).$$

Hence, integrating with respect to x, we have

$$\tfrac{1}{2}v^2 = \int f(x)\,dx + C, \quad\text{........................(18)}$$

in agreement with (17).

Ex. 1. If a particle be attracted to the origin with a force varying as the distance, the equation of motion is

$$\frac{d^2x}{dt^2} = -\mu x. \qquad\qquad (19)$$

This is of the special type (8), and the solution is

$$x = a \cos\left(\sqrt{\mu}\, t + \epsilon\right). \qquad\qquad (20)$$

This represents a 'simple-harmonic' motion. The values of x and dx/dt both recur whenever $\sqrt{\mu}\, t$ increases by 2π; the period of oscillation is therefore $2\pi/\sqrt{\mu}$. The arbitrary constants a and ϵ are in this problem known as the 'amplitude' and the 'epoch,' respectively.

The equation of motion of *any* 'conservative' dynamical system having one degree of freedom, when *slightly* disturbed from a position of stable equilibrium, is also of the type (19). For example, the accurate equation of motion of a pendulum is

$$l\frac{d^2\theta}{dt^2} = -g \sin\theta, \qquad\qquad (21)$$

where g is the acceleration of gravity, and l is a certain length depending on the structure of the pendulum. In the case of a 'simple' pendulum l is the length of the string. If the extreme angular deviation from the equilibrium position be small, we may write θ for $\sin\theta$, thus

$$\frac{d^2\theta}{dt^2} = -\frac{g}{l}\,\theta. \qquad\qquad (22)$$

The solution of this equation is

$$\theta = a \cos\left(\sqrt{\frac{g}{l}}\cdot t + \epsilon\right), \qquad\qquad (23)$$

and the period is therefore $2\pi\sqrt{(l/g)}$.

The accurate equation (21) can be integrated once by the method above explained; we thus find

$$\tfrac{1}{2}l\left(\frac{d\theta}{dt}\right)^2 = g\cos\theta + C, \qquad\qquad (24)$$

but the second integration cannot be effected (except in the particular case of $C = g$) without the introduction of elliptic functions.

Ex. 2. If a particle move in a straight line under an attraction varying as the inverse square of the distance from the origin, we have

$$\frac{d^2x}{dt^2} = -\frac{\mu}{x^2}, \qquad\qquad (25)$$

whence, as in Art. 154, Ex. 3,

$$\tfrac{1}{2}\left(\frac{dx}{dt}\right)^2 = \frac{\mu}{x} + C. \qquad\qquad (26)$$

If the particle start from rest at the distance a, we have $C = -\mu/a$, and

$$\frac{dx}{dt} = -\left(\frac{2\mu}{a}\right)^{\frac{1}{2}} \cdot \left(\frac{a-x}{x}\right)^{\frac{1}{2}}, \quad \ldots\ldots\ldots\ldots\ldots(27)$$

the *minus* sign being taken since the velocity is towards the origin. The second integration is facilitated by the substitution

$$x = a \cos^2 \theta. \quad \ldots\ldots\ldots\ldots\ldots\ldots\ldots(28)$$

Separating the variables, we find

$$(1 + \cos 2\theta)\, d\theta = \left(\frac{2\mu}{a^3}\right)^{\frac{1}{2}} dt, \quad \ldots\ldots\ldots\ldots(29)$$

$$\theta + \tfrac{1}{2} \sin 2\theta = \left(\frac{2\mu}{a^3}\right)^{\frac{1}{2}} t + C. \quad \ldots\ldots\ldots\ldots(30)$$

As x diminishes from a to 0, θ increases from 0 to $\frac{1}{2}\pi$. Hence the time (t_1) of falling from rest at the distance a into the centre of force is given by

$$t_1 = \frac{\pi}{2\sqrt{2}} \frac{a^{\frac{3}{2}}}{\sqrt{\mu}}. \quad \ldots\ldots\ldots\ldots\ldots\ldots(31)$$

164. Equations involving only the First and Second Derivatives.

If the equation be of the type

$$\phi\left(\frac{d^2y}{dx^2}, \frac{dy}{dx}\right) = 0, \quad \ldots\ldots\ldots\ldots\ldots(1)$$

i.e. the variables x, y do not appear (explicitly), then, writing p for dy/dx, we have

$$\phi\left(\frac{dp}{dx}, p\right) = 0, \quad \ldots\ldots\ldots\ldots\ldots(2)$$

which is an equation of the first order with p as dependent variable.

The equation (1) may also be reduced to an equation of the first order, with y as independent variable, by writing as in Art. 163,

$$p \frac{dp}{dy} \text{ for } \frac{d^2y}{dx^2};$$

thus

$$\phi\left(p \frac{dp}{dy}, p\right) = 0. \quad \ldots\ldots\ldots\ldots(3)$$

Ex. 1. To find the curves whose radius of curvature is constant ($= a$, say).

By Art. 135 we have

$$\frac{\dfrac{d^2y}{dx^2}}{\left\{1 + \left(\dfrac{dy}{dx}\right)^2\right\}^{\frac{3}{2}}} = \pm\frac{1}{a}, \quad \ldots\ldots\ldots\ldots(4)$$

or
$$\frac{dp}{(1 + p^2)^{\frac{3}{2}}} = \pm \frac{dx}{a}. \qquad \text{......................(5)}$$

Integrating this we have (Art. 77 (13))
$$\frac{p}{(1 + p^2)^{\frac{1}{2}}} = \pm \frac{x - a}{a}, \qquad \text{....................(6)}$$

where a is an arbitrary constant. This gives
$$\frac{dy}{dx} = p = \pm \frac{x - a}{\sqrt{\{a^2 - (x - a)^2\}}}, \qquad \text{.................(7)}$$

whence
$$y - \beta = \pm \sqrt{\{a^2 - (x - a)^2\}}, \qquad \text{....................(8)}$$

if β be the arbitrary constant introduced by this last integration. The result may be written
$$(x - a)^2 + (y - \beta)^2 = a^2, \qquad \text{......................(9)}$$

and so represents a family of circles of radius a.

This investigation is given merely as an example of the general method; the problem itself can be solved more easily in other ways.

Ex. 2. To determine the rectilinear motion of a particle subject to a force which is a given function of the velocity.

The equation of motion is of the form
$$\frac{d^2x}{dt^2} = f\left(\frac{dx}{dt}\right), \qquad \text{...........................(10)}$$

which evidently comes under the type (1). Writing v for dx/dt, we have
$$\frac{dv}{dt} = f(v), \qquad \frac{dv}{f(v)} = dt, \qquad \int \frac{dv}{f(v)} = t + C. \qquad \text{...........(11)}$$

For example, if the particle be subject solely to a resistance varying as the velocity, we have
$$\frac{dv}{dt} = -kv, \qquad \text{..........................(12)}$$

whence
$$\frac{dx}{dt} = v = Ae^{-kt}, \qquad x = -\frac{1}{k}Ae^{-kt} + B. \qquad \text{...............(13)}$$

Hence, whatever the circumstances of projection, x will approach asymptotically, as t increases, to a limiting value B.

Again, if the resistance vary as the square of the velocity, we have
$$\frac{dv}{dt} = -kv^2, \qquad -\frac{dv}{v^2} = kdt, \qquad \frac{1}{v} = kt + A. \qquad \text{...........(14)}$$

Hence
$$\frac{dx}{dt} = v = \frac{1}{kt + A}, \qquad \text{and} \qquad x = \frac{1}{k}\log(kt + A) + B. \qquad \text{......(15)}$$

We see that, although v tends asymptotically to zero, there is now no limit to the space described.

If we follow the alternative method, the equation (10) is replaced by

$$v \frac{dv}{dx} = f(v). \qquad \ldots\ldots\ldots\ldots\ldots\ldots(16)$$

Thus, in the case of resistance varying as the velocity, we get

$$\frac{dv}{dx} = -k, \quad v = -kx + C. \qquad \ldots\ldots\ldots\ldots\ldots(17)$$

Hence
$$\frac{dx}{dt} + kx = C, \qquad \ldots\ldots\ldots\ldots\ldots\ldots(18)$$

and therefore, by Art. 157,

$$x = \frac{C}{k} + De^{-kt}, \qquad \ldots\ldots\ldots\ldots\ldots\ldots(19)$$

where C, D are arbitrary constants. This agrees with (13).

Again, if the resistance vary as the square of the velocity, we have

$$\frac{dv}{dx} = -kv, \quad v = Ce^{-kx}. \qquad \ldots\ldots\ldots\ldots\ldots(20)$$

Hence
$$e^{kx} \frac{dx}{dt} = C, \quad \frac{1}{k} e^{kx} = Ct + D \qquad \ldots\ldots\ldots\ldots(21)$$

or
$$kx = \log (kCt + kD), \qquad \ldots\ldots\ldots\ldots\ldots(22)$$

a form not really distinct from (15), as may be verified by putting

$$A = kD/C, \quad kB = \log C.$$

165. Equations with one Variable absent.

1°. If the *dependent* variable do not appear explicitly, the equation being of the type

$$\phi \left(\frac{d^2y}{dx^2}, \frac{dy}{dx}, x \right) = 0, \qquad \ldots\ldots\ldots\ldots(1)$$

then, writing p for dy/dx, we have an equation of the first order in p, viz.

$$\phi \left(p, \frac{dp}{dx}, x \right) = 0. \qquad \ldots\ldots\ldots\ldots(2)$$

If the solution of this be put in the form

$$p = f(x, A), \qquad \ldots\ldots\ldots\ldots\ldots(3)$$

where A is the arbitrary constant, a second integration gives

$$y = \int f(x, A)\, dx + B. \qquad \ldots\ldots\ldots\ldots(4)$$

That one of the arbitrary constants would occur as an addition to y might have been anticipated à *priori*, since the equation (1) is unaltered when we write $y + C$ for y.

2°. If the *independent* variable do not appear explicitly, the type being

$$\phi\left(\frac{d^2y}{dx^2}, \ \frac{dy}{dx}, \ y\right) = 0, \quad \dots\dots\dots\dots(5)$$

we write as in Art. 163 (3)

$$\frac{dy}{dx} = p, \ \frac{d^2y}{dx^2} = p\frac{dp}{dy}, \quad \dots\dots\dots\dots(6)$$

and obtain

$$\phi\left(p\frac{dp}{dy}, \ p, \ y\right) = 0, \quad \dots\dots\dots\dots(7)$$

an equation of the first order between p and y.

If the solution of this can be put in the form

$$p = f(y, A), \dots\dots\dots\dots\dots\dots(8)$$

the next integration gives

$$\int \frac{dy}{f(y, A)} = x + B. \quad \dots\dots\dots\dots(9)$$

Here, again, it might have been anticipated from the form of the given equation (5) that one of the arbitrary constants would consist in an addition to x.

Ex. 1.
$$(1 - x^2)\frac{d^2y}{dx^2} - x\frac{dy}{dx} = 0. \quad \dots\dots\dots\dots(10)$$

Writing this in the form

$$\frac{1}{p}\frac{dp}{dx} = \frac{x}{1 - x^2},$$

we find
$$\log p = -\tfrac{1}{2}\log(1 - x^2) + \text{const.},$$

or
$$\frac{dy}{dx} = p = \frac{A}{\sqrt{(1 - x^2)}}. \quad \dots\dots\dots\dots(11)$$

Hence
$$y = A\sin^{-1}x + B. \quad \dots\dots\dots\dots(12)$$

Ex. 2. In the theory of Attractions we meet with the equation

$$\frac{d^2V}{dr^2} + \frac{2}{r}\frac{dV}{dr} = 0. \quad \dots\dots\dots\dots(13)$$

Regarding dV/dr as the dependent variable, we have

$$\frac{\dfrac{d^2V}{dr^2}}{\dfrac{dV}{dr}} + \frac{2}{r} = 0, \quad \dots\dots\dots\dots(14)$$

whence
$$\log\frac{dV}{dr} + 2\log r = \text{const.,}$$

27—2

or
$$\frac{dV}{dr} = \frac{A}{r^2}. \quad \dots\dots\dots\dots\dots\dots(15)$$

Integrating again, we find

$$V = -\frac{A}{r} + B. \quad \dots\dots\dots\dots\dots(16)$$

Ex. 3.   To find the curves in which the radius of curvature is equal to the normal, but lies on the opposite side of the curve.

Referring to Arts. 60, 135, we see that the expression of the above condition is

$$\frac{\left\{1 + \left(\frac{dy}{dx}\right)^2\right\}^{\frac{3}{2}}}{\frac{d^2y}{dx^2}} = y\left\{1 + \left(\frac{dy}{dx}\right)^2\right\}^{\frac{1}{2}}. \quad \dots\dots\dots(17)$$

Simplifying, and making the substitutions (6), we find

$$\frac{p}{1 + p^2}\frac{dp}{dy} = \frac{1}{y}. \quad \dots\dots\dots\dots(18)$$

Hence $\frac{1}{2}\log(1 + p^2) = \log y + \text{const.}, \quad 1 + p^2 = \frac{y^2}{c^2}, \quad \dots\dots(19)$

where c^2 is written for the arbitrary constant, which must evidently be positive.   This gives

$$\frac{dy}{dx} = p = \pm\frac{\sqrt{(y^2 - c^2)}}{c}. \quad \dots\dots\dots\dots(20)$$

Separating the variables, we have

$$\frac{dy}{\sqrt{(y^2 - c^2)}} = \pm\frac{dx}{c}, \quad \cosh^{-1}\frac{y}{c} = \pm\frac{(x - a)}{c}, \quad \dots\dots(21)$$

where a is the second arbitrary constant.   Hence, finally,

$$y = c\cosh\frac{x - a}{c}, \quad \dots\dots\dots\dots(22)$$

a family of catenaries.   Cf. Art. 134, Ex. 1.

166.   Linear Equation of the Second Order.

A linear equation of the nth order is one which involves the dependent variable and its first n derivatives in the first degree only, without products.   Thus, the general linear equation of the second order may be written

$$\frac{d^2y}{dx^2} + P\frac{dy}{dx} + Qy = V, \quad \dots\dots\dots\dots(1)$$

where P, Q, V are given functions of x.

There are several important properties common to all linear equations.   We give the proofs for the equation of the second order, but the generalization will be evident.

1°. The complete solution of (1) may be written

$$y = u + w, \qquad\dots\dots\dots\dots\dots\dots(2)$$

where w is any function whatever which satisfies (1) as it stands and u is the general solution of the equation

$$\frac{d^2y}{dx^2} + P\frac{dy}{dx} + Qy = 0, \qquad\dots\dots\dots\dots\dots(3)$$

which differs from (1) by the absence of the right-hand member.

For, assuming that $y = u + w$, where w satisfies (1), and u is to be determined, we find, on substitution in (1),

$$\frac{d^2u}{dx^2} + P\frac{du}{dx} + Qu + \frac{d^2w}{dx^2} + P\frac{dw}{dx} + Qw = V,$$

or, since

$$\frac{d^2w}{dx^2} + P\frac{dw}{dx} + Qw = V, \qquad\dots\dots\dots\dots(4)$$

by hypothesis,

$$\frac{d^2u}{dx^2} + P\frac{du}{dx} + Qu = 0; \qquad\dots\dots\dots\dots(5)$$

i.e. the function u must satisfy (3).

The two parts which make up the general solution of (1), viz. w and u, are called the 'particular integral,' and the 'complementary function,' respectively. It is to be observed that the particular integral may be any solution whatever of the original equation; the simpler it is, the better. The complementary function must be the most general solution of (3), and will involve two arbitrary constants.

2°. If u_1, u_2 be any two solutions of (3), the equation will also be satisfied by

$$y = C_1u_1 + C_2u_2, \qquad\dots\dots\dots\dots\dots(6)$$

where C_1, C_2 are arbitrary constants. This is easily verified by substitution.

Hence if the functions u_1, u_2 are 'independent,' i.e. one is not merely a constant multiple of the other, the formula (6) gives a solution of (3) involving two arbitrary constants.

3°. If a particular integral (v) of the equation (3) be known, the complete solution of (1) is reduced by the substitution

$$y = zv, \qquad\dots\dots\dots\dots\dots\dots(7)$$

to the integration of an equation of the first order in dz/dx. For (1) becomes

$$v\frac{d^2z}{dx^2} + \left(2\frac{dv}{dx} + Pv\right)\frac{dz}{dx} + \left(\frac{d^2v}{dx^2} + P\frac{dv}{dx} + Qv\right)z = V,$$

which reduces, in virtue of the hypothesis, to

$$v \frac{d^2z}{dx^2} + \left(2\frac{dv}{dx} + Pv\right)\frac{dz}{dx} = V. \quad \dots\dots\dots\dots(8)$$

This is linear, of the first order, with dz/dx as the dependent variable.

In particular, if $V = 0$, we have

$$\frac{\dfrac{d^2z}{dx^2}}{\dfrac{dz}{dx}} + \frac{2}{v}\frac{dv}{dx} + P = 0, \quad \dots\dots\dots\dots\dots(9)$$

whence $\qquad \log \dfrac{dz}{dx} + 2\log v + \int P dx = \text{const.},$

or $\qquad\qquad\qquad \dfrac{dz}{dx} = \dfrac{A}{v^2} e^{-\int P dx}. \quad \dots\dots\dots\dots(10)$

Hence $\qquad\qquad\qquad z = A\int \dfrac{e^{-\int P dx}}{v^2}\,dx + B. \quad \dots\dots\dots(11)$

The complete solution of (3) is therefore

$$y = A v \int \frac{e^{-\int P dx}}{v^2}\,dx + Bv. \quad \dots\dots\dots\dots(12)$$

We add a few examples of the integration of linear equations, by various artifices. The method of integration by series will be noticed in Chapter XIV.

Ex. 1. In the theory of Sound, and in other branches of Mathematical Physics, we meet with the equation

$$\frac{d^2\phi}{dr^2} + \frac{2}{r}\frac{d\phi}{dr} + k^2\phi = 0. \quad \dots\dots\dots\dots(13)$$

If we multiply by r, this is seen to be equivalent to

$$\frac{d^2(r\phi)}{dr^2} + k^2(r\phi) = 0. \quad \dots\dots\dots\dots(14)$$

Hence, by Art. 163, $\quad r\phi = A\cos kr + B\sin kr,$

or $\qquad\qquad\qquad \phi = \dfrac{A\cos kr + B\sin kr}{r}. \quad \dots\dots\dots(15)$

Ex. 2. $\qquad\qquad (1-x^2)\dfrac{d^2y}{dx^2} - x\dfrac{dy}{dx} + y = 0. \quad \dots\dots\dots(16)$

A particular solution is obviously $y = x$. We therefore put

$$y = xz, \quad \dots\dots\dots\dots\dots\dots(17)$$

which leads to
$$x(1-x^2)\frac{d^2z}{dx^2} + (2 - 3x^2)\frac{dz}{dx} = 0. \quad \dots\dots\dots\dots(18)$$

Separating the variables, we have

$$\frac{\dfrac{d^2z}{dx^2}}{\dfrac{dz}{dx}} + \frac{2}{x} - \frac{x}{1-x^2} = 0, \quad \dots\dots\dots\dots\dots(19)$$

whence
$$\frac{dz}{dx} = \frac{A}{x^2\sqrt{(1-x^2)}}, \quad \dots\dots\dots\dots\dots(20)$$

$$z = -A\frac{\sqrt{(1-x^2)}}{x} + B. \quad \dots\dots\dots\dots\dots(21)$$

The complete solution of (16) has therefore the form

$$y = A\sqrt{(1-x^2)} + Bx. \quad \dots\dots\dots\dots\dots(22)$$

Ex. 3.
$$(1 + x^2)\frac{d^2y}{dx^2} + 3x\frac{dy}{dx} + y = 0. \quad \dots\dots\dots\dots(23)$$

This happens to be an 'exact equation,' *i.e.* the left-hand side is the exact differential coefficient of a function of x, y, and dy/dx, for it may be written

$$\left\{(1 + x^2)\frac{d^2y}{dx^2} + 2x\frac{dy}{dx}\right\} + \left\{x\frac{dy}{dx} + y\right\} = 0.$$

The integral
$$(1 + x^2)\frac{dy}{dx} + xy = A. \quad \dots\dots\dots\dots\dots(24)$$

This is linear, of the first order, and the integrating factor is seen to be $1/\sqrt{(1 + x^2)}$. We thus find

$$\frac{d}{dx}\{\sqrt{(1 + x^2)} \cdot y\} = \frac{A}{\sqrt{(1 + x^2)}},$$

$$\sqrt{(1 + x^2)} \cdot y = A\sinh^{-1}x + B. \quad \dots\dots\dots\dots(25)$$

EXAMPLES. LV.

1.
$$x\frac{d^2y}{dx^2} = 1. \qquad [y = x\log x + Ax + B.]$$

2.
$$\frac{d^2y}{dx^2} = xe^x. \qquad [y = (x - 2)e^x + Ax + B.]$$

3.
$$x^2\frac{d^2y}{dx^2} = a. \qquad \left[y = a\log\frac{a}{x} + Ax + B.\right]$$

4. The differential equation for the deflection of a horizontal beam subject only to its weight and to the pressures of its supports is

$$\mathfrak{B}\,\frac{d^4y}{dx^4} = w,$$

where w is the weight per unit length. Integrate this, on the supposition that w is constant, and determine the constants so that $y = 0$, $d^2y/dx^2 = 0$ both for $x = 0$ and for $x = l$. (This is the case of a uniform beam of length l supported at its ends.) $[\mathfrak{B}y = \frac{1}{24}wx\,(l-x)\,(l^2 + lx - x^2).]$

5. Solve the same equation subject to the conditions that $y = 0$, $dy/dx = 0$ for $x = 0$ and $x = l$. (This is the case of a beam clamped at both ends.) $[\mathfrak{B}y = \frac{1}{24}wx^2\,(l-x)^2.]$

6. Solve the equation of Ex. 4, subject to the conditions that $y = 0$, $dy/dx = 0$ for $x = 0$, and $d^2y/dx^2 = 0$, $d^3y/dx^3 = 0$ for $x = l$. (This is the case of a beam clamped at one end and free at the other.)
$$[\mathfrak{B}y = \tfrac{1}{24}wx^2\,(6l^2 - 4lx + x^2).]$$

7. Solve the equation

$$\frac{d^2x}{dt^2} = -\mu x + f,$$

and interpret the result. $[x = f/\mu + a\cos\,(\sqrt{\mu}\,t + \epsilon).]$

8. Shew that the solution of the equation of motion of a particle moving in a straight line under a force of repulsion varying as the distance, viz.

$$\frac{d^2x}{dt^2} = \mu x,$$

is of one or other of the types :

$$x = a\cosh\,(\sqrt{\mu}\,t + \epsilon), \quad x = a\sinh\,(\sqrt{\mu}\,t + \epsilon), \quad x = ae^{\pm\sqrt{\mu}\,t + \epsilon} ;$$

and interpret these results.

9. A particle moves from rest at a distance a towards a centre of force whose accelerative effect is $\mu \times (\text{dist.})^{-3}$. Prove that the time of falling to the centre is $a^2/\sqrt{\mu}$.

10. Obtain a first integral of the general differential equation of central orbits, viz.

$$\frac{d^2u}{d\theta^2} + u = \frac{P}{h^2u^2},$$

where P is a given function of u. $\left[\left(\dfrac{du}{d\theta}\right)^2 + u^2 = 2\displaystyle\int\frac{P}{h^2u^2}\,du + C.\right]$

11. Solve the equation $\dfrac{d^2r}{dt^2} = \dfrac{\mu}{r^3}$,

and shew that the solution is equivalent to

$$r^2 = A + 2Bt + Ct^2,$$

where A, B, C are connected by the relation

$$AC - B^2 = \mu.$$

12.
$$a \frac{d^2y}{dx^2} = \frac{dy}{dx}.$$
$$[y = A + Be^{x/a}.]$$

13.
$$2a \frac{dy}{dx} \frac{d^2y}{dx^2} = 1.$$
$$[a(y-\beta)^2 = \tfrac{4}{9}(x-a)^3.]$$

14.
$$a \frac{d^2y}{dx^2} + \left(\frac{dy}{dx}\right)^2 = 0.$$
$$\left[y = a \log \frac{x-a}{\beta}.\right]$$

15.
$$a \frac{d^2y}{dx^2} = \left\{1 + \left(\frac{dy}{dx}\right)^2\right\}^{\frac{1}{2}}$$
$$\left[y - \beta = a \cosh \frac{x-a}{a}.\right]$$

16.
$$\frac{d^2y}{dx^2} + \left(\frac{dy}{dx}\right)^2 + 1 = 0.$$
$$[y = \beta + \log \cos (x-a).]$$

17.
$$\frac{d^4y}{dx^4} = \frac{d^2y}{dx^2}.$$
$$[y = A + Bx + Ce^x + De^{-x}.]$$

18.
$$\frac{d^4y}{dx^4} + \frac{d^2y}{dx^2} = 0.$$
$$[y = A + Bx + C \cos x + D \sin x.]$$

19.
$$\frac{d^2V}{dr^2} + \frac{1}{r} \frac{dV}{dr} = 0.$$
$$[V = A \log r + B.]$$

20.
$$x \frac{d^2y}{dx^2} = \frac{dy}{dx}.$$
$$[y = Ax^2 + B.]$$

21.
$$\frac{d^2y}{dx^2} + y \frac{dy}{dx} = 0.$$
$$[y = 2\beta \tanh \beta (x-a).]$$

22.
$$(1 + x^2) \frac{d^2y}{dx^2} + 1 + \left(\frac{dy}{dx}\right)^2 = 0.$$
$$\left[y = \beta + \left(1 + \frac{1}{a^2}\right) \log (1 + ax) - \frac{x}{a}.\right]$$

23.
$$(1 + x^2) \frac{d^2y}{dx^2} + x \frac{dy}{dx} = 0.$$
$$[y = A + B \sinh^{-1} x.]$$

24.
$$(x^2 - 1) \frac{d^2y}{dx^2} + x \frac{dy}{dx} = 0.$$
$$[y = A + B \cosh^{-1} x.]$$

25.
$$(1 + x^2) \frac{d^2y}{dx^2} + 2x \frac{dy}{dx} = 0.$$
$$[y = A \tan^{-1} x + B.]$$

26.
$$(1 - y) \frac{d^2y}{dx^2} + 2 \left(\frac{dy}{dx}\right) = 0.$$
$$\left[y = \frac{x+A}{x+B}.\right]$$

27.
$$y \frac{d^2y}{dx^2} = 2 \left(\frac{dy}{dx}\right)^2.$$
$$\left[y = \frac{A}{x+B}.\right]$$

28.
$$y \frac{d^2y}{dx^2} = 1 - \left(\frac{dy}{dx}\right)^2.$$
$$[y^2 = x^2 + Ax + B.]$$

29. $(1 - x^2)\dfrac{d^2y}{dx^2} - x\dfrac{dy}{dx} = 2.$ $[y = (\sin^{-1} x)^2 + A \sin^{-1} x + B.]$

30. $\dfrac{d}{d\mu}\left\{(1 - \mu^2)\dfrac{du}{d\mu}\right\} = 0.$ $[u = A + B \tanh^{-1} \mu.]$

31. $\dfrac{d}{d\mu}\left\{(1 - \mu^2)\dfrac{du}{d\mu}\right\} + 2u = 0.$ $[u = A\mu + B (1 - \mu \tanh^{-1} \mu).]$

32. Find the curves in which the radius of curvature is equal to the normal, on the same side of the curve.

[The circles $(x - a)^2 + y^2 = \beta^2.$]

33. Find the curves in which the radius of curvature is double the normal, on the opposite side of the curve.

[The parabolas $(x - a)^2 = 4\beta (y - \beta).$]

34. Find the curves in which the projection of the radius of curvature on the axis of y has a given constant value (a).

$$\left[y = \beta \log \sec \frac{x - a}{a}.\right]$$

35. Find the curves in which the radius of curvature varies as the cube of the normal.

[Conics having the axis of x as an axis of symmetry.]

36. $(1 + x)\dfrac{d^2y}{dx^2} + x\left(\dfrac{dy}{dx}\right)^2 = \dfrac{dy}{dx}.$

$$\left[y = \beta + \log (x^2 - a^2) - \frac{1}{a}\log \frac{x + a}{x - a}.\right]$$

37. $(1 - x^2)\dfrac{d^2y}{dx^2} - \dfrac{1}{x}\dfrac{dy}{dx} + x^2 = 0.$

$[y = A + B \sqrt{(1 - x^2)} + \tfrac{1}{2}x^2.]$

38. $\dfrac{d^2y}{dx^2} + P\dfrac{dy}{dx} + \dfrac{dP}{dx}y = 0.$ $[y = e^{-\int P dx} (A \int e^{\int P dx} dx + B).]$

39. $\dfrac{d^2y}{dx^2} + \dfrac{2}{x}\dfrac{dy}{dx} - y = 0.$ $\left[y = \dfrac{Ae^x + Be^{-x}}{x}.\right]$

40. $(1 + x^2)\dfrac{d^2y}{dx^2} - 2x\dfrac{dy}{dx} + 2y = 0.$ $[y = Ax + B (1 - x^2).]$

41. $(1 + x^2)\dfrac{d^2y}{dx^2} + x\dfrac{dy}{dx} = y.$ $[y = Ax + B \sqrt{(1 + x^2)}.]$

42. Solve the equation

$$(x^2 - 1)\frac{d^2y}{dx^2} - 2x\frac{dy}{dx} + 2y = 0,$$

one solution of which is $y = x.$ $[y = Ax + B (x + 1)^2.]$

43. Solve the equation

$$x \frac{d^2y}{dx^2} - (n - x) \frac{dy}{dx} - ny = 0,$$

one solution of which is $y = e^{-x}$. $\qquad \left[y = Ae^{-x} + Be^{-x} \int x^n e^x dx. \right]$

44. Change the independent variable in the equation

$$(x^2 - 1) \frac{d^2y}{dx^2} + x \frac{dy}{dx} - y = 0$$

to z, where $z = \cosh^{-1} x$; and solve the equation.

$$\left[y = A \cos z + B \sin z. \right]$$

45. $\qquad \dfrac{d^2y}{dx^2} + 2n \cot nx \dfrac{dy}{dx} + (m^2 - n^2) y = 0.$

$$\left[y = \frac{A \cos mx + B \sin mx}{\sin nx} . \right]$$

46. $\qquad (1 - x^2) \dfrac{d^2y}{dx^2} - 3x \dfrac{dy}{dx} - y = 0. \qquad \left[y = \dfrac{A \sin^{-1} x + B}{\sqrt{(1 - x^2)}} . \right]$

47. $\qquad x^2 y \dfrac{d^2y}{dx^2} + \left(y - x \dfrac{dy}{dx} \right)^2 = 0. \qquad \left[y^2 = Ax + Bx^2. \right]$

48. $\qquad xy \dfrac{d^2y}{dx^2} + x \left(\dfrac{dy}{dx} \right)^2 - y \dfrac{dy}{dx} = 0. \qquad \left[y^2 = Ax^2 + B. \right]$

49. $\qquad 3 \dfrac{d^2y}{dx^2} \dfrac{d^4y}{dx^4} = 5 \left(\dfrac{d^3y}{dx^3} \right)^2. \qquad \left[(y - Ax - B)^2 = Cx + D. \right]$

50. $\qquad 2 \dfrac{dy}{dx} \dfrac{d^3y}{dx^3} = 3 \left(\dfrac{d^2y}{dx^2} \right)^2. \qquad \left[y = A + \dfrac{B}{x + C} . \right]$

CHAPTER XIII

LINEAR EQUATIONS WITH CONSTANT COEFFICIENTS

167. Equation of the Second Order. Complementary Function.

Linear equations with constant coefficients occur so frequently in Mathematical Physics as to call for a somewhat detailed treatment. This is much facilitated by the use of a few simple properties of the operator D, or d/dx, where x is the independent variable.

It has been shewn in Art. 29 that the operation indicated by D is 'distributive,' viz. u, v being any functions of x, we have

$$D(u+v) = Du + Dv. \quad \dots\dots\dots\dots\dots(1)$$

Again, if a be a constant, we have

$$(D+a)u = \frac{du}{dx} + au = au + \frac{du}{dx} = (a+D)u, \quad \dots\dots(2)$$

and

$$D(au) = a\frac{du}{dx} = aDu; \quad \dots\dots\dots\dots(3)$$

so that the operator D, in conjunction with constant multipliers, obeys the 'commutative' laws.

Further, the symbol D is subject to the 'index-law,' viz.

$$D^m D^n u = D^{m+n} u. \quad \dots\dots\dots\dots\dots(4)$$

Hence the operator D, both by itself, and in conjunction with constant multipliers, is subject to the fundamental laws of ordinary Algebra. We can therefore assume at once, so far as they have a meaning in the present application, all the results which in Algebra follow from these laws.

For example, if λ_1, λ_2 be any constants, we have

$$
\begin{aligned}
(D - \lambda_1)(D - \lambda_2)u &= (D - \lambda_1)\left(\frac{du}{dx} - \lambda_2 u\right) \\
&= \frac{d}{dx}\left(\frac{du}{dx} - \lambda_2 u\right) - \lambda_1\left(\frac{du}{dx} - \lambda_2 u\right) \\
&= \frac{d^2 u}{dx^2} - (\lambda_1 + \lambda_2)\frac{du}{dx} + \lambda_1\lambda_2 u \\
&= \{D^2 - (\lambda_1 + \lambda_2)D + \lambda_1\lambda_2\}u. \quad \dots\dots\dots\dots(5)
\end{aligned}
$$

We proceed to the equation of the second order, which occurs very frequently in dynamical problems. To find the complementary function, we have to solve an equation of the form

$$\frac{d^2y}{dx^2} + a\frac{dy}{dx} + by = 0, \quad \dots\dots\dots\dots(6)$$

or

$$(D^2 + aD + b)\,y = 0. \quad \dots\dots\dots\dots(7)$$

If $\frac{1}{4}a^2 > b$, this is equivalent to

$$(D - \lambda_1)(D - \lambda_2)\,y = 0, \quad \dots\dots\dots(8)$$

where λ_1, λ_2 are the roots of

$$\lambda^2 + a\lambda + b = 0, \quad \dots\dots\dots\dots(9)$$

viz.

$$\left.\begin{array}{c}\lambda_1\\ \lambda_2\end{array}\right\} = -\tfrac{1}{2}a \pm \sqrt{(\tfrac{1}{4}a^2 - b)}. \quad \dots\dots\dots(10)$$

If we write

$$(D - \lambda_2)\,y = z, \quad \dots\dots\dots\dots(11)$$

the equation (8) becomes

$$(D - \lambda_1)\,z = 0, \quad \dots\dots\dots\dots(12)$$

a linear equation of the first order. The solution of this is, by Art. 157,

$$z = Ae^{\lambda_1 x}. \quad \dots\dots\dots\dots(13)$$

Substituting in (11), we have

$$(D - \lambda_2)\,y = Ae^{\lambda_1 x}, \quad \dots\dots\dots(14)$$

whence, by Art. 157, 1°,

$$y = C_1 e^{\lambda_1 x} + C_2 e^{\lambda_2 x}, \quad \dots\dots\dots(15)$$

if $C_1 = A/(\lambda_1 - \lambda_2)$. Since A is arbitrary, the constants C_1, C_2 are both arbitrary; and the process shews that (15) is the most general solution of the proposed equation (6).

If $\frac{1}{4}a^2 = b$, the roots of the equation (9) in λ are coincident, and the equation (14) takes the form

$$(D - \lambda)\,y = Ae^{\lambda x}. \quad \dots\dots\dots(16)$$

The general solution of this, as found in Art. 157, 2°, is

$$y = (Ax + B)\,e^{\lambda x}. \quad \dots\dots\dots(17)$$

If $\frac{1}{4}a^2 < b$, the values of λ_1, λ_2 which satisfy (9) are imaginary, but we can still obtain, by the foregoing process, a symbolical solution of (6) involving $\sqrt{(-1)}$. Into the question as to what meaning can be attributed to such a result it is not necessary to enter here, as the difficulty can be evaded in the following manner.

If we write $\qquad\qquad y = e^{mx}z,$ ....(18)

where m is as yet undetermined, we have

$$Dy = e^{mx}(D+m)z, \quad D^2y = e^{mx}(D^2 + 2mD + m^2)z, \; ...(19)$$

so that the equation becomes

$$\{D^2 + (2m+a)D + m^2 + am + b\}z = 0. \qquad ....(20)$$

If we now put $m = -\tfrac{1}{2}a$ this reduces to the form of Art. 163 (8), thus

$$\frac{d^2z}{dx^2} + (b - \tfrac{1}{4}a^2)z = 0. \qquad ....(21)$$

The solution of this, when $b > \tfrac{1}{4}a^2$, has been shewn in Art. 163 to be

$$z = A\cos\beta x + B\sin\beta x, \qquad ....(22)$$

where $\qquad\qquad \beta = \surd(b - \tfrac{1}{4}a^2).$ ....(23)

Hence the solution of (6), in the present case, is

$$y = e^{-\frac{1}{2}ax}(A\cos\beta x + B\sin\beta x). \qquad ....(24)$$

This is also equivalent to

$$y = Ce^{-\frac{1}{2}ax}\cos(\beta x + \epsilon), \qquad ....(25)$$

where the constants C, ϵ are arbitrary.

To summarize our results:

    (a)   If $\tfrac{1}{4}a^2 > b$, the solution of (6) is

$$y = C_1e^{\lambda_1 x} + C_2e^{\lambda_2 x},$$

where λ_1, λ_2 are the roots of

$$\lambda^2 + a\lambda + b = 0;$$

    (b)   If $\tfrac{1}{4}a^2 = b$, the solution is

$$y = (Ax + B)e^{-\frac{1}{2}ax};$$

    (c)   If $\tfrac{1}{4}a^2 < b$, the solution is

$$y = e^{-\frac{1}{2}ax}(A\cos\beta x + B\sin\beta x),$$

if $\qquad\qquad \beta^2 = b - \tfrac{1}{4}a^2.$

Ex. 1. $\qquad\qquad \dfrac{d^2y}{dx^2} + \dfrac{dy}{dx} - 6y = 0.$ ....(26)

The equation in λ is

$$\lambda^2 + \lambda - 6 = 0, \quad \text{whence} \quad \lambda = 2, \; \text{or} \; -3.$$

Hence $\qquad\qquad y = Ae^{2x} + Be^{-3x}.$ ....(27)

Ex. 2. $\qquad\qquad \dfrac{d^2y}{dx^2} + 4\dfrac{dy}{dx} + 4y = 0.$

The equation in λ, viz. $\qquad (\lambda + 2)^2 = 0,$

has the double root -2. Hence

$$y = (Ax + B)e^{-2x}. \qquad ....(28)$$

Ex. 3. The free oscillations of a pendulum in a medium whose resistance varies as the velocity are determined by an equation of the form

$$\frac{d^2x}{dt^2} + k\frac{dx}{dt} + \mu x = 0, \quad\dots\dots\dots\dots\dots\dots(29)$$

where k is a coefficient of friction. The same equation also serves to represent the motion of a galvanometer-needle as affected by the viscosity of the air, and by the electro-magnetic action of the currents induced by its motion in neighbouring masses of metal.

When regard is had to the difference of notation, the solution of (29), when the friction falls below a certain limit, is

$$x = Ce^{-\frac{1}{2}kt}\cos(n_1 t + \epsilon), \quad\dots\dots\dots\dots\dots(30)$$

where

$$n_1 = \sqrt{(\mu - \tfrac{1}{4}k^2)}. \quad\dots\dots\dots\dots\dots(31)$$

The motion represented by (30) may be described as a simple-harmonic vibration of period $2\pi/n_1$, whose amplitude diminishes asymptotically to zero according to the law $e^{-\frac{1}{2}kt}$.

The solution (30) assumes that $k^2 < 4\mu$. When $k^2 > 4\mu$, the proper form is

$$x = Ae^{\lambda_1 t} + Be^{\lambda_2 t}, \quad\dots\dots\dots\dots\dots(32)$$

where λ_1, λ_2 are the roots of

$$\lambda^2 + k\lambda + \mu = 0. \quad\dots\dots\dots\dots\dots(33)$$

By hypothesis, these roots are real. Since their product (μ) is positive they have the same sign; and since their sum ($-k$) is negative, the sign is *minus*. Hence the displacement x sinks asymptotically to zero after passing once (at most) through this value. This case is realized in a 'dead-beat' galvanometer, or in a pendulum swinging in a very viscous fluid.

In the critical case of $k^2 = 4\mu$, we have

$$x = (A + Bt)e^{-\frac{1}{2}kt}. \quad\dots\dots\dots\dots\dots(34)$$

The first factor increases (in absolute value) indefinitely with t, whilst the second diminishes. The decrease of the second factor prevails however over the increase of the first, and the limiting value of the product, for $t \to \infty$, is zero. Cf. Art. 43 (2).

168. Determination of Particular Integrals.

We have next to consider the problem of finding a particular integral of the linear equation of the second order with constant coefficients, when this equation has a right-hand member, thus:

$$(D^2 + aD + b)y = V, \quad\dots\dots\dots\dots\dots(1)$$

where V is a given function of x. As already remarked, *any* particular integral, however obtained, will serve the purpose. Thus, we may omit from the particular integral any terms which occur in the complementary function, since these will contribute nothing to the left-hand side of (1). Conversely, if for any purpose it is convenient to do so, we may add to the particular integral any groups of terms taken from the complementary function.

Again, if V be composed of a series of terms, the problem consists in finding values of y which, when substituted on the left-hand side of (1), will reproduce the several terms, and adding the results.

It will be sufficient here to notice the most useful cases.

1°. If V contains a term

$$He^{ax}, \dots\dots\dots\dots\dots\dots(2)$$

the corresponding term of the particular integral is

$$y = \frac{H}{a^2 + a\alpha + b}\, e^{ax}. \dots\dots\dots\dots(3)$$

For if we perform the operation $D^2 + aD + b$ on the right-hand side of this, we reproduce (2).

This rule fails if $a^2 + a\alpha + b = 0$, $i.e.$ if e^{ax} be one of the terms which occur in the complementary function. Using the notation of the preceding Art., we will suppose that $\alpha = \lambda_1$, so that the equation to be solved is

$$(D - \lambda_1)(D - \lambda_2)\, y = He^{\lambda_1 x}. \dots\dots\dots\dots(4)$$

If we write, for a moment,

$$(D - \lambda_2)\, y = z, \dots\dots\dots\dots\dots(5)$$

this takes the form $\qquad (D - \lambda_1)\, z = He^{\lambda_1 x}. \dots\dots\dots\dots(6)$

It was found in Art. 157, 2°, that a particular solution of (6) is

$$z = Hxe^{\lambda_1 x}. \dots\dots\dots\dots\dots(7)$$

It remains only to solve

$$(D - \lambda_2)\, y = Hxe^{\lambda_1 x}. \dots\dots\dots\dots(8)$$

The integrating factor is $e^{-\lambda_2 x}$; thus

$$D(ye^{-\lambda_2 x}) = Hxe^{(\lambda_1 - \lambda_2) x}. \dots\dots\dots\dots(9)$$

Integrating the right-hand member by parts, and omitting a term already included in the complementary function, we find

$$ye^{-\lambda_2 x} = \frac{H}{\lambda_1 - \lambda_2}\, xe^{(\lambda_1 - \lambda_2) x},$$

or $\qquad\qquad y = \frac{H}{\lambda_1 - \lambda_2}\, xe^{\lambda_1 x}. \dots\dots\dots\dots(10)$

A further modification is necessary if α be a *double* root of the equation $D^2 + aD + b = 0$. The equation to be solved has now the form

$$(D - \lambda)^2 y = He^{\lambda x}. \quad\ldots\ldots\ldots\ldots\ldots(11)$$

The first step is as before, but in place of (8) we have

$$(D - \lambda) y = Hxe^{\lambda x}. \quad\ldots\ldots\ldots\ldots\ldots(12)$$

We found in Art. 157, 3°, that a particular integral of this is

$$y = \tfrac{1}{2}Hx^2 e^{\lambda x}. \quad\ldots\ldots\ldots\ldots\ldots(13)$$

The *forms* of these results being once established, the student will probably find it the easiest and safest course to assume

$$y = Ce^{ax}, \quad y = Cxe^{ax}, \quad \text{or} \quad y = Cx^2 e^{ax}, \quad\ldots\ldots(14)$$

as the case may be, and to determine the value of C by actual substitution in the equation

$$(D^2 + aD + b)\, y = He^{ax}. \quad\ldots\ldots\ldots\ldots(15)$$

The work is facilitated by formulæ to be given in Art. 169.

2°. If V contains terms of the form

$$H \cos \alpha x + K \sin \alpha x, \quad\ldots\ldots\ldots\ldots\ldots(16)$$

we may assume $\quad y = A \cos \alpha x + B \sin \alpha x. \quad\ldots\ldots\ldots\ldots(17)$

Substituting in (1), we obtain, on the left-hand side,

$$(- \alpha^2 A + a\alpha B + bA) \cos \alpha x + (- \alpha^2 B - a\alpha A + bB) \sin \alpha x.$$

Hence the terms (16) are reproduced, provided

$$(- \alpha^2 + b)\, A + a\alpha B = H, \quad - a\alpha A + (- \alpha^2 + b)\, B = K. \ldots(18)$$

Except in the particular case where $a = 0$, $\alpha^2 = b$, which will be considered presently, these equations determine A and B; thus

$$A = \frac{(- \alpha^2 + b)\, H - a\alpha K}{(\alpha^2 - b)^2 + a^2 \alpha^2}, \quad B = \frac{a\alpha H + (- \alpha^2 + b)\, K}{(\alpha^2 - b)^2 + a^2 \alpha^2}. \ (19)$$

The foregoing results simplify when the coefficient a in the differential equation is zero; a particular integral of

$$\frac{d^2 y}{dx^2} + by = H \cos \alpha x + K \sin \alpha x \quad\ldots\ldots\ldots(20)$$

being obviously

$$y = \frac{H}{b - \alpha^2} \cos \alpha x + \frac{K}{b - \alpha^2} \sin \alpha x. \quad\ldots\ldots\ldots(21)$$

A singularity arises, however, when $\alpha^2 = b$. To find the proper form in this case we may assume

$$y = u \cos \alpha x + v \sin \alpha x. \quad\ldots\ldots\ldots\ldots(22)$$

This makes

$$\frac{d^2y}{dx^2} + \alpha^2 y = (2\alpha Dv + D^2 u)\cos \alpha x + (-2\alpha Du + D^2 v)\sin \alpha x. \quad (23)$$

Hence (20) will be satisfied in this case, provided

$$Du = -K/2\alpha, \quad Dv = H/2\alpha,$$

or $\qquad\qquad u = -\dfrac{Kx}{2\alpha}, \quad v = \dfrac{Hx}{2\alpha}. \qquad\qquad$ ....(24)

A particular integral is, therefore,

$$y = \frac{H}{2\alpha} x \sin \alpha x - \frac{K}{2\alpha} x \cos \alpha x. \qquad (25)$$

Ex. 1. $\qquad\qquad \dfrac{d^2y}{dx^2} + \dfrac{dy}{dx} - 6y = e^{3x} + e^{-3x}. \qquad$ ....(26)

As in Art. 167, Ex. 1, the complementary function is

$$y = Ae^{2x} + Be^{-3x}.$$

If we assume $y = Ce^{3x}$, we find on substitution that the first term on the right-hand side of (26) is reproduced provided $C = \frac{1}{6}$. The second term comes under one of the exceptional cases above discussed, since -3 is a root of $\lambda^2 + \lambda - 6 = 0$. If we assume $y = Cxe^{-3x}$, we find that the term in question is reproduced, provided $C = -\frac{1}{5}$.

The complete solution of (26) is therefore

$$y = Ae^{2x} + Be^{-3x} + \tfrac{1}{6}e^{3x} - \tfrac{1}{5}xe^{-3x}. \qquad (27)$$

Ex. 2. $\qquad\qquad \dfrac{d^2y}{dx^2} + 4\dfrac{dy}{dx} + 4y = e^{2x} + e^{-2x}. \qquad$ ....(28)

The complementary function has been found in Art. 167, Ex. 2, to be

$$y = (Ax + B)\, e^{-2x}.$$

To reproduce the first term on the right hand, we assume $y = Ce^{2x}$, and find $C = \frac{1}{16}$. The second term corresponds to a double root in the equation for λ; assuming, therefore, $y = Cx^2 e^{-2x}$, we find $C = \frac{1}{2}$.

Hence the complete solution of (28) is

$$y = (Ax + B)\, e^{-2x} + \tfrac{1}{16}e^{2x} + \tfrac{1}{2}x^2 e^{-2x}. \qquad (29)$$

Ex. 3.   To find a particular integral of

$$\frac{d^2x}{dt^2} + k\frac{dx}{dt} + \mu x = f\cos(pt + \epsilon). \qquad (30)$$

This is the equation of motion of a pendulum subject to a resistance varying as the velocity, and acted on by a force which is a simple-harmonic function of the time.

We assume
$$x = A \cos(pt + \epsilon) + B \sin(pt + \epsilon), \quad \ldots\ldots\ldots\ldots(31)$$
and find, on substitution,
$$\left.\begin{array}{l} -p^2 A + kpB + \mu A = f, \\ -p^2 B - kpA + \mu B = 0, \end{array}\right\} \quad \ldots\ldots\ldots\ldots\ldots(32)$$

whence
$$A = \frac{\mu - p^2}{(\mu - p^2)^2 + k^2 p^2} f, \quad B = \frac{kp}{(\mu - p^2)^2 + k^2 p^2} f. \quad \ldots\ldots\ldots(33)$$

If we put
$$A = R \cos \epsilon_1, \quad B = R \sin \epsilon_1, \quad \ldots\ldots\ldots\ldots(34)$$
the solution (31) takes the form
$$x = R \cos(pt + \epsilon - \epsilon_1), \quad \ldots\ldots\ldots\ldots\ldots(35)$$

where
$$R = \frac{f}{\sqrt{\{(\mu - p^2)^2 + k^2 p^2\}}}, \quad \epsilon_1 = \tan^{-1} \frac{kp}{\mu - p^2}. \quad \ldots\ldots\ldots(36)$$

We have thus determined the 'forced oscillations' due to the given periodic force. The 'free oscillations,' which are in general superposed on these, are given by the complementary function (Art. 167, Ex. 3). Unless $k = 0$ they gradually die out as t increases.

Ex. 4. For the forced oscillations of an unresisted pendulum we have
$$\frac{d^2 x}{dt^2} + n^2 x = f \cos(pt + \epsilon). \quad \ldots\ldots\ldots\ldots\ldots(37)$$

A particular integral is
$$x = \frac{f}{n^2 - p^2} \cos(pt + \epsilon). \quad \ldots\ldots\ldots\ldots\ldots(38)$$

This fails when $p = n$. Assuming that in this case
$$x = Ct \sin(nt + \epsilon), \quad \ldots\ldots\ldots\ldots\ldots(39)$$
we find on substitution that (37) is satisfied provided
$$2nC = f, \quad \text{or} \quad C = f/2n. \quad \ldots\ldots\ldots\ldots\ldots(40)$$

The interpretation of (39) is that, if an unresisted pendulum be acted on by a periodic force whose period coincides with that natural to the pendulum, the amplitude of the forced oscillations will at first* increase proportionally to the time.

169. Properties of the Operator D.

The methods of Arts. 167, 168 admit of extension to the general linear equation with constant coefficients
$$\frac{d^n y}{dx^n} + A_1 \frac{d^{n-1} y}{dx^{n-1}} + A_2 \frac{d^{n-2} y}{dx^{n-2}} + \ldots + A_{n-1} \frac{dy}{dx} + A_n y = V, \ldots(1)$$

---

* Usually, in the physical applications, the equation (37) is approximate only, being obtained by the neglect of powers of x higher than the first. Hence when the amplitude increases beyond a certain limit, the equation ceases to apply, even approximately, to the subsequent motion.

or, as we may write it for shortness,

$$f(D) y = V, \quad \dots\dots\dots\dots\dots\dots\dots(2)$$

$f(D)$ denoting a rational integral function of D. We shall however content ourselves with indicating how a solution of (1), involving n distinct arbitrary constants, can be obtained when $V = 0$, and how a particular integral can be found for certain forms of V. The proof that the solution thus arrived at is the most general which the equation admits of is omitted; but from the point of view of practical applications it is sufficient to have at our disposal the proper number of arbitrary constants required to satisfy the remaining conditions of the problem.

The following properties of the operator D will be useful.

1°.   If $\psi(D)$ be any rational integral function of D, say

$$\psi(D) \equiv A_0 D^n + A_1 D^{n-1} + A_2 D^{n-2} + \dots + A_{n-1} D + A_n, \dots(3)$$

then $$\psi(D) e^{\lambda x} = \psi(\lambda) e^{\lambda x}. \quad \dots\dots\dots\dots\dots(4)$$

For $$D^r e^{\lambda x} = \lambda^r e^{\lambda x},$$

and thus the several terms of $\psi(D)$ give rise to the several terms of $\psi(\lambda)$ as factors of $e^{\lambda x}$.

2°.   With the same meaning of $\psi(D)$, if u be any function of x, then

$$\psi(D) . e^{\lambda x} u = e^{\lambda x} . \psi(D + \lambda) u. \quad \dots\dots\dots\dots(5)$$

For we find in succession

$$D . e^{\lambda x} u = e^{\lambda x} (D + \lambda) u,$$
$$D^2 . e^{\lambda x} u = D \{ e^{\lambda x} (D + \lambda) u \} = e^{\lambda x} (D + \lambda)(D + \lambda) u$$
$$= e^{\lambda x} (D + \lambda)^2 u,$$

and so on; the general result being

$$D^r . e^{\lambda x} u = e^{\lambda x} (D + \lambda)^r u.$$

Hence the several terms of the operator $\psi(D)$ give rise to the corresponding terms of the result given in (5).

3°.   If $\psi(D)$ contain only even powers of D, it may be denoted by $\phi(D^2)$. It appears from Art. 64 that if

$$u = A \cos \alpha x + B \sin \alpha x, \quad \dots\dots\dots\dots(6)$$

then $$D^2 u = - \alpha^2 u,$$

and therefore $$\phi(D^2) u = \phi(-\alpha^2) u. \quad \dots\dots\dots\dots(7)$$

170. General Linear Equation with Constant Coefficients. Complementary Function.

To obtain solutions of the equation

$$f(D)\, y = 0, \qquad \dots\dots\dots\dots\dots\dots\dots(1)$$

we remark that if $f(D)$ be resolved into any two rational integral factors, thus

$$f(D) \equiv \phi(D)\,.\,\psi(D), \qquad \dots\dots\dots\dots\dots\dots(2)$$

the equation (1) is obviously satisfied by any solution of

$$\psi(D)\, y = 0. \qquad \dots\dots\dots\dots\dots\dots\dots(3)$$

And since the factors are commutative (Art. 167), it is also satisfied by any solution of

$$\phi(D)\, y = 0. \qquad \dots\dots\dots\dots\dots\dots\dots(4)$$

Hence (1) is satisfied by the sum of any solutions of (3) and (4). Continuing the resolution we see that if

$$f(D) \equiv \phi_1(D)\,.\,\phi_2(D)\,.\,\phi_3(D), \qquad \dots\dots\dots\dots(5)$$

the equation (1) will be satisfied by the sum of the several solutions of

$$\phi_1(D)\, y = 0, \quad \phi_2(D)\, y = 0, \quad \phi_3(D)\, y = 0, \dots\dots(6)$$

By a theorem of Algebra, already referred to in Art. 85, the function $f(D)$ can be resolved (if its coefficients be real) into real factors of the first and second degrees, the sum of the degrees of the several factors being equal to the degree (n, say) of the function. Moreover the factors of the first degree are of the forms

$$D - \lambda_1, \quad D - \lambda_2, \quad D - \lambda_3, \dots$$

where $\lambda_1, \lambda_2, \lambda_3, \dots$ are the real roots of

$$f(\lambda) = 0. \qquad \dots\dots\dots\dots\dots\dots\dots(7)$$

If λ be a simple root of (7), one of the equations (6) is of the form

$$(D - \lambda)\, y = 0, \qquad \dots\dots\dots\dots\dots\dots(8)$$

the integral of which is known to be

$$y = C e^{\lambda x}. \qquad \dots\dots\dots\dots\dots\dots\dots(9)$$

And if the roots of (7) be *all* real, say they are $\lambda_1, \lambda_2, \dots \lambda_n$, a solution of (1) involving n arbitrary constants is

$$y = C_1 e^{\lambda_1 x} + C_2 e^{\lambda_2 x} + \dots + C_n e^{\lambda_n x}; \qquad \dots\dots\dots(10)$$

cf. Art. 167 (15).

If the equation (7) has a *multiple* root, two or more of the terms on the right hand of (10) coalesce, and the number of distinct solutions thus obtained is less than n. To supply the deficiency, we remark that, if λ be an r-fold root of (7), $f(D)$ contains a factor $(D - \lambda)^r$. To solve

$$(D - \lambda)^r y = 0, \quad \text{......................(11).}$$

we assume $$y = e^{\lambda x} z, \quad \text{..........................(12)}$$

which makes

$$(D - \lambda)^r y = (D - \lambda)^r . e^{\lambda x} z = e^{\lambda x} D^r z,$$

by Art. 169, 2°, and the solution of $D^r z = 0$ is obviously

$$z = B_0 + B_1 x + B_2 x^2 + \ldots + B_{r-1} x^{r-1},$$

whence $$y = (B_0 + B_1 + B_2 x^2 + \ldots + B_{r-1} x^{r-1})\, e^{\lambda x}. \quad \text{.........(13)}$$

We have here r arbitrary constants, corresponding to the r-fold factor of $f(D)$. Cf. Art. 167 (17).

If $f(D)$ contains, once only, an irreducible *quadratic* factor $D^2 + aD + b$, where $\frac{1}{4}a^2 < b$, then part of the solution of (1) consists of the solution of

$$(D^2 + aD + b)\, y = 0. \quad \text{...................(14)}$$

If we put $$y = e^{-\frac{1}{2}ax} z, \quad \beta^2 = b - \tfrac{1}{4}a^2, \quad \text{...............(15)}$$

we have, by Art. 169 (5),

$$(D^2 + aD + b)\, y = \{(D + \tfrac{1}{2}a)^2 + \beta^2\}\, e^{-\frac{1}{2}ax} z$$
$$= e^{-\frac{1}{2}ax} (D^2 + \beta^2)\, z.$$

And the solution of $$(D^2 + \beta^2)\, z = 0$$

is $$z = E \cos \beta x + F \sin \beta x.$$

We thus find $$y = e^{-\frac{1}{2}ax} (E \cos \beta x + F \sin \beta x), \quad \text{............(16)}$$

as in Art. 167 (24). Hence for every distinct quadratic factor of $f(D)$ we obtain a solution involving two arbitrary constants.

Finally, if $f(D)$ contains an irreducible quadratic factor which occurs r times, we have to solve

$$(D^2 + aD + b)^r y = 0, \quad \text{...................(17)}$$

or, making the substitutions (15), as before,

$$(D^2 + \beta^2)^r z = 0. \quad \text{.....................(18)}$$

To solve this, we assume

$$z = u \cos \beta x + v \sin \beta x. \quad \text{.................(19)}$$

Now, by actual differentiation, we find

$$(D^2 + \beta^2) . u \cos \beta x = 2\beta . Du . \cos (\beta x + \tfrac{1}{2}\pi) + \ldots,$$
$$(D^2 + \beta^2)^2 . u \cos \beta x = (2\beta)^2 . D^2 u . \cos (\beta x + \pi) + \ldots,$$

and, generally,

$$(D^2 + \beta^2)^r \cdot u \cos \beta x = (2\beta)^r \cdot D^r u \cdot \cos (\beta x + \tfrac{1}{2} r \pi) + \ldots, \quad \ldots (20)$$

where only the terms of lowest order in the derivatives of u are expressed. Similarly

$$(D^2 + \beta^2)^r \cdot v \sin \beta x = (2\beta)^r \cdot D^r v \cdot \sin (\beta x + \tfrac{1}{2} r \pi) + \ldots \quad \ldots (21)$$

Hence the equation (18) is satisfied, provided

$$D^r u = 0, \quad D^r v = 0, \ldots\ldots\ldots\ldots\ldots\ldots (22)$$

i.e.
$$\left. \begin{aligned} u &= E_0 + E_1 x + E_2 x^2 + \ldots + E_{r-1} x^{r-1}, \\ v &= F_0 + F_1 x + F_2 x^2 + \ldots + F_{r-1} x^{r-1}. \end{aligned} \right\} \quad \ldots\ldots\ldots (23)$$

The complete solution of (17) is therefore

$$y = (E_0 + E_1 x + E_2 x^2 + \ldots + E_{r-1} x^{r-1}) e^{-\frac{1}{2} a x} \cos \beta x$$
$$+ (F_0 + F_1 x + F_2 x^2 + \ldots + F_{r-1} x^{r-1}) e^{-\frac{1}{2} a x} \sin \beta x, \ldots (24)$$

involving $2r$ arbitrary constants.

Ex. 1.
$$\frac{d^3 y}{dx^3} = \frac{d^2 y}{dx^2}. \quad \ldots\ldots\ldots\ldots\ldots\ldots (25)$$

This may be written $\qquad D^2 (D - 1) y = 0,$

and the complete solution is accordingly made up of the solutions of $D^2 y = 0$, $(D - 1) y = 0$, thus

$$y = A + Bx + Ce^x. \quad \ldots\ldots\ldots\ldots\ldots (26)$$

Ex. 2.
$$\frac{d^4 y}{dx^4} = m^4 y. \quad \ldots\ldots\ldots\ldots\ldots\ldots (27)$$

This may be written

$$(D - m)(D + m)(D^2 + m^2) y = 0.$$

Adding together the solutions of

$$(D - m) y = 0, \quad (D + m) y = 0, \quad (D^2 + m^2) y = 0,$$

we obtain $\qquad y = A e^{mx} + B e^{-mx} + E \cos mx + F \sin mx. \quad \ldots\ldots (28)$

Ex. 3.
$$\frac{d^4 y}{dx^4} + \frac{d^2 y}{dx^2} + y = 0. \quad \ldots\ldots\ldots\ldots\ldots (29)$$

This is equivalent to

$$(D^2 + D + 1)(D^2 - D + 1) y = 0.$$

Hence

$$y = e^{-\frac{1}{2} x} \left(A \cos \frac{\sqrt{3}}{2} x + B \sin \frac{\sqrt{3}}{2} x \right) + e^{\frac{1}{2} x} \left(A' \cos \frac{\sqrt{3}}{2} x + B' \sin \frac{\sqrt{3}}{2} x \right).$$
$$\ldots\ldots (30)$$

Ex. 4. $$\frac{d^4y}{dx^4} + 2m^2 \frac{d^2y}{dx^2} + m^4y = 0. \quad\dots\dots\dots(31)$$

Writing this in the form
$$(D^2 + m^2)^2 \, y = 0,$$

we find $\quad y = (E_0 + E_1x) \cos mx + (F_0 + F_1x) \sin mx. \quad\dots\dots(32)$

171. Particular Integrals.

We proceed to the determination of a particular integral of the equation

$$f(D) \, y = V, \quad\dots\dots\dots\dots\dots\dots(1)$$

in the two most important cases.

1°. If V contain a term
$$He^{ax}, \quad\dots\dots\dots\dots\dots\dots(2)$$

the corresponding term in the particular integral is

$$y = \frac{H}{f(\alpha)} \, e^{ax}, \quad\dots\dots\dots\dots\dots(3)$$

for this makes $\quad f(D) \, y = \frac{H}{f(\alpha)} f(D) \, e^{ax} = He^{ax},$

by Art. 169 (4).

The rule fails when α is a root of

$$f(D) = 0. \quad\dots\dots\dots\dots\dots\dots(4)$$

If it be a simple root, we may write

$$f(D) \equiv \phi(D)(D - \alpha), \quad\dots\dots\dots\dots(5)$$

where $\phi(D)$ does not contain the factor $D - \alpha$. The equation

$$\phi(D)(D - \alpha) \, y = He^{ax} \quad\dots\dots\dots\dots(6)$$

is satisfied, provided

$$(D - \alpha) \, y = \frac{H}{\phi(\alpha)} \, e^{ax},$$

and we have seen, in Art. 157, 2°, that a particular integral of this is

$$y = \frac{H}{\phi(\alpha)} \, xe^{ax}. \quad\dots\dots\dots\dots(7)$$

If α be an r-fold root of (4), we may write

$$f(D) \equiv \phi(D)(D - \alpha)^r, \quad\dots\dots\dots(8)$$

where $\phi(D)$ does not contain $D - \alpha$ as a factor. The equation

$$\phi(D)(D - \alpha)^r \, y = He^{ax} \quad\dots\dots\dots\dots(9)$$

is satisfied, provided

$$(D - \alpha)^r y = \frac{H}{\phi(\alpha)} e^{\alpha x}.$$

Now if we put $y = e^{\alpha x} z,$

we have, by Art. 169 (5),

$$(D - \alpha)^r y = e^{\alpha x} D^r z,$$

whence $D^r z = \dfrac{H}{\phi(\alpha)}.$

This is evidently satisfied by

$$z = \frac{H}{r! \, \phi(\alpha)} x^r.$$

A particular integral of (9) is therefore

$$y = \frac{H}{r! \, \phi(\alpha)} x^r e^{\alpha x}. \quad \dots\dots\dots\dots\dots(10)$$

2°.   Let V contain terms of the type

$$H \cos \alpha x + K \sin \alpha x. \quad \dots\dots\dots\dots(11)$$

Since the operation $f(D)$, performed on

$$y = A \cos \alpha x + B \sin \alpha x, \quad \dots\dots\dots\dots(12)$$

must result in an expression of the same form with altered coefficients, a particular integral can in general be found by substituting this value of y in the equation

$$f(D) y = H \cos \alpha x + K \sin \alpha x, \quad \dots\dots\dots(13)$$

and determining A and B by equating coefficients of $\cos \alpha x$ and $\sin \alpha x$.

In one very frequent case, the values of A and B can be written down at once, viz. when the equation is of the type

$$\phi(D^2) y = H \cos \alpha x + K \sin \alpha x, \quad \dots\dots\dots(14)$$

i.e. $f(D)$ contains only even powers of D. We have, then, by Art. 169, 3°,

$$y = \frac{H}{\phi(-\alpha^2)} \cos \alpha x + \frac{K}{\phi(-\alpha^2)} \sin \alpha x. \quad \dots\dots(15)$$

This result fails if $\phi(-\alpha^2) = 0$, *i.e.* if $\phi(D^2)$ contain $D^2 + \alpha^2$ as a factor, in which case terms of the type (11) occur in the complementary function. If the factor $D^2 + \alpha^2$ occur once only, we write

$$\phi(D^2) \equiv \chi(D^2)(D^2 + \alpha^2). \quad \dots\dots\dots\dots(16)$$

Now the equation

$$\chi(D^2)(D^2+\alpha^2)y = H\cos\alpha x + K\sin\alpha x \quad \ldots\ldots(17)$$

will be satisfied if

$$(D^2+\alpha^2)y = \frac{H}{\chi(-\alpha^2)}\cos\alpha x + \frac{K}{\chi(-\alpha^2)}\sin\alpha x \ldots\ldots(18)$$

The problem is thus reduced to one already solved under Art 168, 2°, viz. we have the particular integral

$$y = \frac{H}{2\alpha\chi(-\alpha^2)}x\sin\alpha x - \frac{K}{2\alpha\chi(-\alpha^2)}x\cos\alpha x. \quad\ldots\ldots(19)$$

If the factor $D^2+\alpha^2$ occur r times in $\phi(D^2)$, we write

$$\phi(D^2) \equiv \chi(D^2)(D^2+\alpha^2)^r, \quad\ldots\ldots\ldots\ldots(20)$$

and the problem is reduced to finding a particular integral of

$$(D^2+\alpha^2)^r y = \frac{H}{\chi(-\alpha^2)}\cos\alpha x + \frac{K}{\chi(-\alpha^2)}\sin\alpha x. \ldots\ldots(21)$$

If we assume

$$y = u\cos(\alpha x - \tfrac{1}{2}r\pi) + v\sin(\alpha x - \tfrac{1}{2}r\pi), \quad\ldots\ldots\ldots(22)*$$

we find

$$(D^2+\alpha^2)^r y = (2\alpha)^r . D^r u . \cos\alpha x + (2\alpha)^r . D^r v . \sin\alpha x + \ldots$$

by Art. 170 (20), (21). Hence (21) is satisfied provided

$$D^r u = \frac{H}{(2\alpha)^r\chi(-\alpha^2)}, \quad D^r v = \frac{K}{(2\alpha)^r\chi(-\alpha^2)}, \quad\ldots(23)$$

or

$$u = \frac{Hx^r}{r!(2\alpha)^r\chi(-\alpha^2)}, \quad v = \frac{Kx^r}{r!(2\alpha)^r\chi(-\alpha^2)}. \quad\ldots\ldots(24)$$

A particular integral is therefore

$$y = \frac{x^r}{r!(2\alpha)^r\chi(-\alpha^2)}\{H\cos(\alpha x - \tfrac{1}{2}r\pi) + K\sin(\alpha x - \tfrac{1}{2}r\pi)\}. \ldots(25)$$

In the general case, where $f(D)$ contains both odd and even powers of D, the assumption (12) fails in like manner if, and only if, $f(D)$ contains the factor $D^2+\alpha^2$. Writing

$$f(D) = \chi(D)(D^2+\alpha^2)^r, \quad\ldots\ldots\ldots\ldots\ldots(26)$$

where $\chi(D)$ does not contain the factor $D^2+\alpha^2$, we first obtain a particular integral of the equation

$$\chi(D)y = H\cos\alpha x + K\sin\alpha x, \quad\ldots\ldots\ldots\ldots(27)$$

in the form

$$y = H_1\cos\alpha x + K_1\sin\alpha x. \quad\ldots\ldots\ldots\ldots(28)$$

---

* The assumption $y = u\cos\alpha x + v\sin\alpha x$ would serve equally well, but the form in the text enables us to write the final result in a more compact manner.

It then remains only to solve

$$(D^2 + a^2)^r y = H_1 \cos ax + K_1 \sin ax. \quad \dots\dots\dots\dots(29)$$

This has been treated above.

3°. Another case where a particular integral can be obtained is that of

$$f(D) y = X, \quad \dots\dots\dots\dots\dots(30)$$

where X is a rational integral function, of (say) degree r. We put $y = x^m v$, where m is the lowest index of D which occurs in $f(D)$, and v is a rational integral function of x, of degree r. The coefficients in v are then determined by substitution.

172. Homogeneous Linear Equation.

An equation of the type

$$x^n \frac{d^n y}{dx^n} + A_1 x^{n-1} \frac{d^{n-1}y}{dx^{n-1}} + \dots + A_r x^r \frac{d^r y}{dx^r} + \dots$$

$$+ A_{n-1} x \frac{dy}{dx} + A_n y = V, \quad \dots\dots\dots(1)$$

is sometimes called a 'homogeneous' linear equation. The complementary function in this case consists in general of a series of terms of the form Cx^m, where C is arbitrary, and the values of m are to be found by substitution on the left-hand side. Moreover, if V contains a term Hx^p, the corresponding term in the particular integral will in general be Bx^p, provided B be properly determined.

To see the truth of these statements we may take the homogeneous linear equation of the second order. To solve

$$x^2 \frac{d^2 y}{dx^2} + ax \frac{dy}{dx} + by = 0, \quad \dots\dots\dots(2)$$

we assume $y = Cx^m. \quad \dots\dots\dots\dots(3)$

This will satisfy the equation provided

$$\{m(m-1) + am + b\} Cx^m = 0. \quad \dots\dots\dots(4)$$

Equating to zero the expression in { }, we have a quadratic in m. If m_1, m_2 be the roots of this, the solution is

$$y = C_1 x^{m_1} + C_2 x^{m_2}. \quad \dots\dots\dots(5)$$

Again, a particular integral of the equation

$$x^2 \frac{d^2 y}{dx^2} + ax \frac{dy}{dx} + by = Hx^p \quad \dots\dots\dots(6)$$

will be $y = Bx^p, \quad \dots\dots\dots\dots(7)$

provided $\{p(p-1) + ap + b\} B = H. \quad \dots\dots\dots(8)$

Complications arise when the equation in m has imaginary roots, or when it has coincident roots; there are, further, difficulties in connection with the particular integral when V contains terms of the type x^p, where p is a root of the equation in m. To avoid special investigation, we shew how the equation (1) can always, by a change of independent variable, be transformed into a linear equation with constant coefficients.

If we put $\qquad\qquad x = e^\theta,$ ....(9)

then, u being any function whatever of x, we have

$$\frac{du}{dx} = \frac{du}{d\theta} \div \frac{dx}{d\theta} = e^{-\theta} \frac{du}{d\theta},$$

or $\qquad\qquad\qquad\qquad x \frac{du}{dx} = \frac{du}{d\theta}.$ ....(10)

We will denote the operator $d/d\theta$, which is seen to be equivalent to $x\,d/dx$, by ϑ. Then, D standing as usual for d/dx, we have

$$xD\,(x^m\,D^m\,u) = x^{m+1}\,D^{m+1}\,u + m x^m\,D^m u,$$

or

$$x^{m+1}\,D^{m+1}\,u = (xD - m)\,(x^m\,D^m\,u) = (\vartheta - m)\,(x^m\,D^m\,u). \ ...(11)$$

Putting $m = 0, 1, 2, \ldots$ in this formula we can express

$$xDu, \quad x^2 D^2 u, \quad x^3 D^3 u, \ldots$$

in succession in terms of $\vartheta u, \vartheta^2 u, \vartheta^3 u, \ldots$. Thus, since the operator $\vartheta, = d/d\theta$, is commutative,

$$xDu = \vartheta u,$$
$$x^2 D^2 u = \vartheta\,(\vartheta - 1)\,u,$$
$$x^3 D^3 u = \vartheta\,(\vartheta - 1)\,(\vartheta - 2)\,u,$$

and so on, the general formula being

$$x^r D^r u = \vartheta\,(\vartheta - 1)\,(\vartheta - 2)\,\ldots\,(\vartheta - r + 1)\,u. \ \ldots\ldots(12)$$

If we substitute for the several terms of (1) their values as given by (12), we get a linear equation with constant coefficients, of the form

$$f(\vartheta)\,y = V, \quad \text{or} \quad f\!\left(\frac{d}{d\theta}\right) y = V. \ \ldots\ldots\ldots(13)$$

Ex. 1. $\qquad\qquad\qquad \dfrac{d^2 V}{dr^2} + \dfrac{2}{r}\dfrac{dV}{dr} = 0.$ ....(14)

If we multiply by r^2 this comes under the form (1); thus

$$r^2 \frac{d^2 V}{dr^2} + 2r \frac{dV}{dr} = 0. \ \ldots\ldots\ldots\ldots\ldots(15)$$

Assuming $$V = Cr^m,$$

we find $$m(m-1) + 2m = 0, \quad \text{or} \quad m(m+1) = 0.$$

The admissible values of m are 0 and -1; and the solution is therefore

$$V = A + \frac{B}{r}. \quad \dots \dots \dots \dots \dots \dots \dots (16)$$

Cf. Art. 165, Ex. 2.

Ex. 2. $$x^2 \frac{d^2y}{dx^2} + 2x \frac{dy}{dx} - 2y = x^2. \quad \dots \dots \dots \dots \dots (17)$$

To find the complementary function, we assume $y = Cx^m$, and obtain

$$m(m-1) + 2m - 2 = 0, \quad \text{or} \quad (m-1)(m+2) = 0,$$

whence $m = 1$ or -2. Again, a particular integral is $y = Cx^2$, provided

$$(2-1)(2+2)C = 1, \quad \text{or} \quad C = \tfrac{1}{4}.$$

Hence $$y = Ax + \frac{B}{x^2} + \tfrac{1}{4}x^2. \quad \dots \dots \dots \dots \dots (18)$$

Ex. 3. $$x^2 \frac{d^2y}{dx^2} - x \frac{dy}{dx} + y = x. \quad \dots \dots \dots \dots (19)$$

This becomes $$\{\vartheta(\vartheta-1) - \vartheta + 1\} y = e^\theta,$$

or $$(\vartheta-1)^2 y = e^\theta.$$

Allowing for the difference of notation, the solution of this is, by Art. 167,

$$y = (A + B\theta) e^\theta + \tfrac{1}{2}\theta^2 e^\theta,$$

or, in terms of x,

$$y = (A + B \log x) x + \tfrac{1}{2}x (\log x)^2. \quad \dots \dots \dots \dots (20)$$

Ex. 4. $$x^2 \frac{d^2y}{dx^2} + x \frac{dy}{dx} + y = x^3. \quad \dots \dots \dots \dots (21)$$

This gives $$(\vartheta^2 + 1) y = e^{3\theta},$$

$$y = A \cos \theta + B \sin \theta + \tfrac{1}{10}e^{3\theta}$$

$$= A \cos (\log x) + B \sin (\log x) + \tfrac{1}{10}x^3. \quad \dots \dots \dots \dots (22)$$

173. Simultaneous Differential Equations.

In dynamical and other problems we often meet with systems of simultaneous differential equations, involving two or more functions of a single independent variable, and their derivatives, the number of equations being always equal to that of the dependent variables. We denote the dependent variables by the letters $x, y, \dots$, and the independent variable by t.

Without entering into questions of general theory, it will be sufficient here to give a few examples exhibiting the methods which are most generally useful.

In the first place, it may happen that each of the given equations involves only one of the dependent variables, and so can be treated separately.

Ex. 1.   In the case of a projectile under gravity, if the axes of x and y be horizontal and vertical, we have

$$\frac{d^2x}{dt^2} = 0, \quad \frac{d^2y}{dt^2} = -g. \quad\quad\quad\quad\dots\dots\dots\dots\dots(1)$$

Hence
$$x = A + A't, \quad y = B + B't - \tfrac{1}{2}gt^2. \quad\dots\dots\dots\dots(2)$$

The arbitrary constants A, A', B, B' enable us to satisfy the four initial conditions as to position and velocity.

Ex. 2.   In the case of a particle attracted to a fixed centre (the origin) with a force varying as the distance, we have

$$\frac{d^2x}{dt^2} = -\mu x, \quad \frac{d^2y}{dt^2} = -\mu y, \quad\quad\quad\dots\dots\dots\dots\dots(3)$$

whence

$$x = A_1 \cos\sqrt{\mu}t + A_2 \sin\sqrt{\mu}t, \quad y = B_1 \cos\sqrt{\mu}t + B_2 \sin\sqrt{\mu}t.$$

If we eliminate t, we find

$$(B_1 x - A_1 y)^2 + (B_2 x - A_2 y)^2 = (A_1 B_2 - A_2 B_1)^2, \quad\dots\dots\dots(4)$$

which shews that the path is an ellipse.

If the given equations, which we will suppose to be n in number, are not of this simple type, then by differentiations, and algebraical manipulation, we may eliminate all the dependent variables x, y, z, $\dots$, save one (say x). If, after integration of the resulting equation, we substitute the general value of x in the original system, we shall find that this now reduces to $n - 1$ equations involving the $n - 1$ dependent variables y, z, $\dots$. The process can be repeated until each dependent variable in turn has been determined as a function of t and of arbitrary constants.

In particular cases a more symmetrical procedure is possible. We content ourselves with a few illustrations taken from physical problems.

Ex. 3.   If x, y be the coordinates of any point in a rigid plane which is rotating with angular velocity n about the origin, we have the equations

$$\frac{dx}{dt} = -ny, \quad \frac{dy}{dt} = nx. \quad\quad\quad\quad\dots\dots\dots\dots\dots(5)$$

Eliminating y, we have

$$\frac{d^2x}{dt^2} = -n\frac{dy}{dt} = -n^2x,$$

whence
$$x = a\cos(nt + \epsilon), \quad\quad\quad\quad\dots\dots\dots\dots\dots(6)$$

the constants a and ϵ being arbitrary. Substituting in the first of the equations (5), we find

$$y = a \sin (nt + \epsilon). \dots\dots\dots\dots\dots\dots(7)$$

The results (6) and (7) shew that each point describes a circle about the origin with angular velocity n.

Ex. 4. In the theory of electro-magnetic induction we meet with the equations

$$\left.\begin{array}{l} L \dfrac{dx}{dt} + M \dfrac{dy}{dt} + Rx = E, \\[2mm] M \dfrac{dx}{dt} + N \dfrac{dy}{dt} + Sy = F. \end{array}\right\} \quad \dots\dots\dots\dots\dots(8)$$

Here x and y denote the currents in two circuits subject to mutual influence; R and S are the resistances of the circuits, L and N the coefficients of self-induction, M that of mutual induction, and E, F are the extraneous electro-motive forces.

Let us first suppose that $E = 0$, $F = 0$. The equations are then satisfied by

$$x = Ae^{\lambda t}, \quad y = Be^{\lambda t}, \quad \dots\dots\dots\dots\dots\dots(9)$$

provided

$$\left.\begin{array}{l} (L\lambda + R)\,A + M\lambda B = 0, \\[1mm] M\lambda A + (N\lambda + S)\,B = 0. \end{array}\right\} \dots\dots\dots\dots\dots(10)$$

Eliminating the ratio $A : B$, we have

$$(L\lambda + R)(N\lambda + S) - M^2\lambda^2 = 0,$$

or

$$(LN - M^2)\,\lambda^2 + (LS + NR)\,\lambda + RS = 0. \dots\dots\dots(11)$$

Since

$$(LS + NR)^2 - 4RS\,(LN - M^2)$$
$$= (LS - NR)^2 + 4M^2RS,$$

a positive quantity, it appears that the roots of the quadratic (11) are always real. Again, for physical reasons, LN is necessarily greater than M^2. Hence (11) shews that the two values of λ must have the same sign (since their product is positive), and further that this sign must be negative (since the sum is negative). Hence, denoting the roots by $-\lambda_1$, $-\lambda_2$, we have the solutions

$$\left.\begin{array}{l} x = A_1 e^{-\lambda_1 t}, \quad y = B_1 e^{-\lambda_1 t}, \\[1mm] x = A_2 e^{-\lambda_2 t}, \quad y = B_2 e^{-\lambda_2 t}, \end{array}\right\} \quad \dots\dots\dots\dots(12)$$

and

where the relation between A_1 and B_1, or A_2 and B_2, is given by either of equations (10) with $-\lambda_1$ or $-\lambda_2$ written for λ. The arbitrary constants therefore virtually reduce to two. On account of their linear character, the equations (8), with $E = F = 0$, are satisfied by the sums of the above values of x and y, respectively. The result represents the decay of free currents initially present in the circuits.

If E and F are not zero, but given *constants*, a particular integral of (8) is evidently

$$x = \frac{E}{R}, \qquad y = \frac{F}{S};$$

and the complete solution is

$$\left. \begin{aligned} x &= \frac{E}{R} + A_1 e^{-\lambda_1 t} + A_2 e^{-\lambda_2 t}, \\ y &= \frac{F}{S} + B_1 e^{-\lambda_1 t} + B_2 e^{-\lambda_2 t}, \end{aligned} \right\} \quad \dots\dots\dots\dots(13)$$

where the relations between A_1 and B_1, and between A_2 and B_2 are as above indicated.

The first terms in these values of x and y represent the steady currents due to the given electro-motive forces; the remaining terms represent the effects of induction. Since we have, virtually, two arbitrary constants, these can be determined so as to make the actual currents have any given initial values.

Another important case is where E is a simple-harmonic function of the time, and F is zero. Putting, then,

$$E = E_0 \cos pt, \quad F = 0, \quad \dots\dots\dots\dots\dots(14)$$

a particular integral of the equations (8) may be found by assuming

$$\left. \begin{aligned} x &= A \cos pt + A' \sin pt, \\ y &= B \cos pt + B' \sin pt. \end{aligned} \right\} \quad \dots\dots\dots\dots(15)$$

On substitution we find, equating separately the coefficients of $\cos pt$ and $\sin pt$,

$$\left. \begin{aligned} pLA' + pMB' + RA &= E_0, \\ -pLA - pMB + RA' &= 0, \\ pMA' + pNB' + SB &= 0, \\ -pMA - pNB + SB' &= 0. \end{aligned} \right\} \quad \dots\dots\dots\dots(16)$$

These formulæ give A, A', B, B', and so determine the electrical oscillations in the two circuits due to the given periodic electro-motive force. The free currents are given by terms of the same form as in (12). Their values depend on the initial circumstances, and in any case they die out as t increases.

Ex. 5. As a final example, we take the equations

$$\left. \begin{aligned} A \frac{d^2x}{dt^2} + H \frac{d^2y}{dt^2} + ax + hy &= X, \\ H \frac{d^2x}{dt^2} + B \frac{d^2y}{dt^2} + hx + by &= Y, \end{aligned} \right\} \quad \dots\dots\dots\dots(17)$$

which determine the motion of any 'conservative' dynamical system, having two degrees of freedom, in the neighbourhood of a position of equilibrium.

To find the free motion, we put $X = 0$, $Y = 0$, and assume

$$x = Fe^{\lambda t}, \quad y = Ge^{\lambda t}. \quad\dots\dots\dots\dots(18)$$

We thus get
$$\left.\begin{array}{l} (A\lambda^2 + a) F + (H\lambda^2 + h) G = 0, \\ (H\lambda^2 + h) F + (B\lambda^2 + b) G = 0. \end{array}\right\} \quad\dots\dots\dots(19)$$

Eliminating the ratio $F : G$, we have

$$(A\lambda^2 + a)(B\lambda^2 + b) - (H\lambda^2 + h)^2 = 0, \dots\dots\dots(20)$$

or $\quad (AB - H^2)\lambda^4 + (Ab + Ba - 2Hh)\lambda^2 + (ab - h^2) = 0. \quad\dots\dots(21)$

This is a quadratic in λ^2.

The expressions

$$\tfrac{1}{2}\left\{ A\left(\frac{dx}{dt}\right)^2 + 2H\frac{dx}{dt}\frac{dy}{dt} + B\left(\frac{dy}{dt}\right)^2 \right\}, \quad\dots\dots\dots(22)$$

and
$$\tfrac{1}{2}(ax^2 + 2hxy + by^2), \quad\dots\dots\dots\dots(23)$$

represent the kinetic and potential energies of the system, respectively. The former is essentially positive; hence A, B are positive and $AB > H^2$. It follows that the left-hand side of (20) or (21) will be positive both for $\lambda^2 = +\infty$ and for $\lambda^2 = -\infty$, whilst for $\lambda^2 = 0$ the sign is that of $ab - h^2$. Also, from (20), it appears that the left-hand member is negative for $\lambda^2 = -a/A$ and for $\lambda^2 = -b/B$.

Hence if the expression (23) be essentially negative, so that a, b are negative, whilst $ab - h^2$ is positive, the equation (21) is satisfied by two positive values of λ^2, one of which is greater, and the other less, than either of the quantities $-a/A$, $-b/B$. Denoting these roots by λ_1^2, λ_2^2, we have the solutions

$$\left.\begin{array}{l} x = F_1 e^{\lambda_1 t} + F_1' e^{-\lambda_1 t} + F_2 e^{\lambda_2 t} + F_2' e^{-\lambda_2 t}, \\ y = G_1 e^{\lambda_1 t} + G_1' e^{-\lambda_1 t} + G_2 e^{\lambda_2 t} + G_2' e^{-\lambda_2 t}. \end{array}\right\} \quad\dots\dots\dots(24)$$

Of the eight coefficients, only four are arbitrary. The ratio $F_1 : G_1$, which is the same as $F_1' : G_1'$, is determined by (19), with λ_1^2 written for λ^2. Similarly, the ratio $F_2 : G_2$ or $F_2' : G_2'$ is determined by the same equations, with λ_2^2 written for λ^2. The four arbitrary constants which remain may be utilized to give any prescribed initial values to x, y, dx/dt, dy/dt. It appears that x and y will increase indefinitely with t, unless the initial circumstances be specially adjusted to make F_1 and F_2 vanish. Hence if the potential energy in the equilibrium position be greater than in any neighbouring position, an arbitrarily started disturbance will in general increase indefinitely; so that the equilibrium position is unstable.

If, on the other hand, the expression (23) be essentially positive, so that a, b, $ab - h^2$ are positive, the roots of the quadratic in λ^2 will both be negative, viz. one will lie between 0 and the numerically smaller of the two quantities $-a/A$, $-b/B$, and the other will lie between the numerically greater of these quantities and $-\infty$. This indicates that in place of (18) the proper assumption now is

$$x = F \cos pt + F' \sin pt, \quad y = G \cos pt + G' \sin pt. \quad \ldots\ldots(25)$$

This leads to equations of the forms (19) and (21), with $-p^2$ written for λ^2. It follows that the two roots of the quadratic in p^2 will be real and positive. Denoting them by p_1^2 and p_2^2, we get the solutions

$$\left.\begin{array}{l} x = F_1 \cos p_1 t + F_1' \sin p_1 t + F_2 \cos p_2 t + F_2' \sin p_2 t, \\ y = G_1 \cos p_1 t + G_1' \sin p_1 t + G_2 \cos p_2 t + G_2' \cos p_2 t, \end{array}\right\} \ldots\ldots(26)$$

where the ratios F_1/G_1, F_1'/G_1', F_2/G_2, F_2'/G_2' are determined in the same manner as before. Since $F_1'/G_1' = F_1/G_1$, and $F_2'/G_2' = F_2/G_2$, the results may also be written

$$\left.\begin{array}{l} x = F_1 \cos (p_1 t + \epsilon_1) + F_2 \cos (p_2 t + \epsilon_2), \\ y = G_1 \cos (p_1 t + \epsilon_1) + G_2 \cos (p_2 t + \epsilon_2), \end{array}\right\} \ldots\ldots\ldots(27)$$

where F_1/G_1 and F_2/G_2 are determinate. It appears that when the potential energy in the equilibrium position is less than in any neighbouring position, a slight disturbance will merely cause the system to oscillate about the equilibrium position, which is therefore stable.

The two roots of the quadratic in λ^2 (or in p^2) have been assumed to be distinct. It may be proved that they cannot be equal unless $a/A = b/B = h/H$; and that if these conditions be fulfilled the solution is of one or other of the two types:

$$x = Fe^{\lambda t} + F' e^{-\lambda t}, \quad y = Ge^{\lambda t} + G' e^{-\lambda t}, \quad \ldots\ldots\ldots(28)$$

$$x = F \cos pt + F' \sin pt, \quad y = G \cos pt + G' \sin pt, \quad \ldots\ldots(29)$$

where the four constants are in each case independent.

Finally, we have the case where the expression (23) for the potential energy may be sometimes positive and sometimes negative. In this case $ab - h^2$ is negative, and one root of the quadratic in λ^2 is positive, the other negative. The complete solution is now of the type

$$\left.\begin{array}{l} x = Fe^{\lambda t} + F' e^{-\lambda t} + F''' \cos pt + F'''' \sin pt, \\ y = Ge^{\lambda t} + G' e^{-\lambda t} + G'' \cos pt + G''' \sin pt. \end{array}\right\} \ldots\ldots(30)$$

It is clear that an arbitrary disturbance will in general increase indefinitely, so that the equilibrium position must be reckoned as unstable.

A slightly different method of treating the question is to assume

$$y = \mu x. \quad \ldots\ldots\ldots\ldots\ldots\ldots\ldots\ldots\ldots(31)$$

The equations to be solved now take the form

$$\left.\begin{array}{l} (A + \mu H) \dfrac{d^2 x}{dt^2} + (a + \mu h) x = 0, \\[2mm] (H + \mu B) \dfrac{d^2 x}{dt^2} + (h + \mu b) x = 0. \end{array}\right\} \ldots\ldots\ldots(32)$$

These are both satisfied by

$$x = Fe^{\lambda t}, \quad \dots\dots\dots\dots\dots\dots(33)$$

provided

$$\frac{A + \mu H}{a + \mu h} = \frac{H + \mu B}{h + \mu b} = -\frac{1}{\lambda^2}. \quad \dots\dots\dots\dots(34)$$

Hence μ is determined by the quadratic

$$(Hb - Bh)\,\mu^2 + (Ab - Ba)\,\mu + (Ah - Ha) = 0. \quad \dots\dots(35)$$

If μ_1, μ_2 be the roots of this, the corresponding values of λ^2 are given by (34). In this way we obtain two solutions, which, on account of the linearity of the differential equations, we may superpose.

If we eliminate μ from (34), we get the same quadratic to determine λ^2 as before. Hence the condition for the reality of the roots of (35) must be the same as in the case of the quadratic (21). This is easily verified.

If λ^2 be negative, the solutions are of the types

$$\left.\begin{aligned}
x &= F_1 \cos\,(p_1 t + \epsilon_1), \quad y = \mu_1 F_1 \cos\,(p_1 t + \epsilon_1), \\
x &= F_2 \cos\,(p_2 t + \epsilon_2), \quad y = \mu_2 F_2 \cos\,(p_2 t + \epsilon_2),
\end{aligned}\right\} \quad \dots\dots(36)$$

where F_1, F_2, ϵ_1, ϵ_2 are arbitrary. Either of these by itself represents what is called a 'normal mode' of vibration of the system.

To find the *forced* oscillations when the extraneous forces X, Y are of the type

$$X = a \cos\,(nt + \epsilon), \quad Y = \beta \sin\,(nt + \epsilon), \quad \dots\dots\dots(37)$$

we may assume

$$x = F \cos\,(nt + \epsilon), \quad y = G \sin\,(nt + \epsilon), \quad \dots\dots\dots(38)$$

and determine the coefficients F, G by substitution in (17). A case of failure may arise, when the expression (23) is essentially positive, owing to n^2 coinciding with one of the roots of the quadratic in p^2.

EXAMPLES. LVI.

(Constant Coefficients.)

1. $\dfrac{d^2 y}{dx^2} + \dfrac{dy}{dx} = 0.$ $[y = A + Be^{-x}.]$

2. $\dfrac{d^2 y}{dx^2} - 3\dfrac{dy}{dx} - 10y = 0.$ $[y = Ae^{5x} + Be^{-2x}.]$

3. $2\dfrac{d^2 y}{dx^2} - 3\dfrac{dy}{dx} + y = 0.$ $[y = Ae^{x} + Be^{\frac{1}{2}x}.]$

4. $\dfrac{d^3 y}{dx^3} - 6\dfrac{d^2 y}{dx^2} + 11\dfrac{dy}{dx} - 6y = 0.$ $[y = Ae^{x} + Be^{2x} + Ce^{3x}.]$

5. $\dfrac{d^4y}{dx^4} = m^4 y.$ $[y = A e^{mx} + B e^{-mx} + C \cos mx + D \sin mx.]$

6. $\dfrac{d^3y}{dx^3} - \dfrac{d^2y}{dx^2} + \dfrac{dy}{dx} - y = 0.$ $[y = A e^x + B \cos x + C \sin x.]$

7. $\dfrac{d^2y}{dx^2} - 2m \dfrac{dy}{dx} + (m^2 + n^2) y = 0.$

$$[y = e^{mx} (A \cos nx + B \sin nx).]$$

8. $\dfrac{d^2y}{dx^2} - 4 \dfrac{dy}{dx} + 13y = 0.$ $[y = e^{2x} (A \cos 3x + B \sin 3x).]$

9. $\dfrac{d^2y}{dx^2} + 2 \dfrac{dy}{dx} + 5y = 0.$ $[y = e^{-x} (A \cos 2x + B \sin 2x).]$

10. $\dfrac{d^2y}{dx^2} - 6 \dfrac{dy}{dx} + 13y = 0.$ $[y = e^{3x} (A \cos 2x + B \sin 2x).]$

11. $\dfrac{d^4y}{dx^4} + m^4 y = 0.$

$$\left[y = \left(A \cosh \frac{mx}{\sqrt{2}} + B \sinh \frac{mx}{\sqrt{2}} \right) \cos \frac{mx}{\sqrt{2}} \right.$$
$$\left. + \left(A' \cosh \frac{mx}{\sqrt{2}} + B' \sinh \frac{mx}{\sqrt{2}} \right) \sin \frac{mx}{\sqrt{2}}. \right]$$

12. $\dfrac{d^3y}{dx^3} + \dfrac{d^2y}{dx^2} - \dfrac{dy}{dx} - y = 0.$ $[y = A e^x + (B + Cx) e^{-x}.]$

13. $\dfrac{d^3y}{dx^3} - \dfrac{d^2y}{dx^2} - \dfrac{dy}{dx} + y = 0.$ $[y = (A + Bx) e^x + C e^{-x}.]$

14. $\dfrac{d^3y}{dx^3} - 3 \dfrac{d^2y}{dx^2} + 4y = 0.$ $[y = (A + Bx) e^{2x} + C e^{-x}.]$

15. $\dfrac{d^3y}{dx^3} - 3 \dfrac{dy}{dx} + 2y = 0.$ $[y = (A + Bx) e^x + C e^{-2x}.]$

16. $\dfrac{d^3y}{dx^3} - 3 \dfrac{d^2y}{dx^2} + 3 \dfrac{dy}{dx} - y = 0.$ $[y = (A + Bx + Cx^2) e^x.]$

17. $\dfrac{d^4y}{dx^4} + (m^2 + n^2) \dfrac{d^2y}{dx^2} + m^2 n^2 y = 0.$

$$[y = A \cos (mx + a) + B \cos (nx + \beta).]$$

18.   Shew that the solution of the equation

$$\frac{d^2x}{dt^2} + k \frac{dx}{dt} - \mu x = 0$$

is of the form $x = A e^{-at} + B e^{\beta t},$

where a, β are both positive (if k and μ are positive) and $a > \beta$.

19.
$$\frac{d^2y}{dx^2} - 2m\frac{dy}{dx} + m^2y = \sin nx.$$

$$\left[y = \frac{(m^2 - n^2)\sin nx + 2mn\cos nx}{(m^2 + n^2)^2} + \&c. \right]$$

20. $\dfrac{d^3y}{dx^3} - \dfrac{d^2y}{dx^2} - \dfrac{dy}{dx} + y = \cosh x.$ $\quad \left[y = \frac{1}{8}x^2e^x + \frac{1}{8}xe^{-x} + \&c. \right]$

21. $\dfrac{d^2y}{dx^2} + \dfrac{dy}{dx} = 1 + \cosh x.$ $\quad \left[y = x + \frac{1}{4}e^x - \frac{1}{2}xe^{-x} + \&c. \right]$

22. $\dfrac{d^4y}{dx^4} - m^4y = \cos kx + \cosh kx.$

$$\left[y = \frac{1}{k^4 - m^4}(\cos kx + \cosh kx) + \&c. \right]$$

23. $(D^4 - m^4)\,y = \cosh mx + \cos mx.$

$$\left[y = \frac{x}{4m^3}(\sinh x - \sin mx) + \&c. \right]$$

24. $D^2(D^2 - 1)\,y = \cosh x.$ $\quad \left[y = \frac{1}{2}x\sinh x + \&c. \right]$

25. $(D^2 + m^2)^2\,y = \cosh mx + \cos mx.$

$$\left[y = \frac{1}{4m^4}\cosh mx - \frac{1}{8m^2}x^2\cos mx. \right]$$

26. Find the values of x and dx/dt from the equation

$$\frac{d^2x}{dt^2} + 2n\frac{dx}{dt} + n^2x = f\sin pt,$$

subject to the conditions $x = 0$, $dx/dt = 0$ for $t = 0$.

$$\left[x = \frac{f}{R'}\{\sin(pt - 2\epsilon) + (pt + \sin 2\epsilon)\,e^{-nt}\}, \right.$$

$$\frac{dt}{dx} = \frac{f\sin\epsilon}{R}\{\cos(pt - 2\epsilon) - (nt + \cos 2\epsilon)\,e^{-nt}\},$$

where $\qquad R = \surd(p^2 + n^2), \quad \epsilon = \tan^{-1}(p/n).$ $\Big]$

EXAMPLES. LVII.

(Homogeneous Equations.)

1. $x^3\dfrac{d^3y}{dx^3} + x\dfrac{d^2y}{dx^2} - 4\dfrac{dy}{dx} = 0.$ $\quad \left[y = A + \dfrac{B}{x} + Cx^3. \right]$

2. Solve $\qquad \dfrac{d^2V}{dr^2} + \dfrac{1}{r}\dfrac{dV}{dr} = 0$

as a homogeneous linear equation. $\qquad [\, V = A\log r + B. \,]$

3 $$x^2 \frac{d^3y}{dx^3} - 2\frac{dy}{dx} = 0. \qquad [y = A + B \log x + Cx^3.]$$

4. $$x^2 \frac{d^2y}{dx^2} - 2y = x. \qquad \left[y = Ax^2 + \frac{B}{x} - \tfrac{1}{2}x. \right]$$

5. $$x^2 \frac{d^2y}{dx^2} + 4x\frac{dy}{dx} + 2y = \frac{1}{x}. \qquad \left[y = \frac{A}{x} + \frac{B}{x^2} + \frac{\log x}{x}. \right]$$

6. $$x^2 \frac{d^2y}{dx^2} + x\frac{dy}{dx} - y = x^3. \qquad \left[y = Ax + \frac{B}{x} + \tfrac{1}{8}x^3. \right]$$

7. $$x^2 \frac{d^2y}{dx^2} - 3x\frac{dy}{dx} + 4y = x^3. \qquad [y = (A + B \log x)\, x^2 + x^3.]$$

8. $$4x^2 \frac{d^2y}{dx^2} + 4x\frac{dy}{dx} = y. \qquad [y = Ax^{\frac{1}{2}} + Bx^{-\frac{1}{2}}.]$$

9. $$\left(\frac{d^2}{dr^2} - \frac{2}{r^2} \right)^2 f(r) = 0. \qquad \left[f(r) = \frac{A}{r} + Br + Cr^2 + Dr^4. \right]$$

10. Prove that

$$f\left(x\frac{d}{dx} \right) x^m u = x^m f\left(x\frac{d}{dx} + m \right) u.$$

11. Prove that

$$f\left(x\cdot\frac{d}{dx} \right) x^m \log x = x^m \{ f(m) \log x + f'(m) \}.$$

EXAMPLES. LVIII.

(Simultaneous Equations.)

1. $$\frac{dx}{dt} + 7x - y = 0, \quad \frac{dy}{dt} + 2x + 5y = 0.$$
$$[x = \tfrac{1}{2} \{(A + B) \sin t + (A - B) \cos t\}\, e^{-6t},$$
$$y = (A \cos t + B \sin t)\, e^{-6t}.]$$

2. $$\frac{dx}{dt} = 3x - y, \quad \frac{dy}{dt} = x + y.$$
$$[x = (A + Bt)\, e^{2t}, \quad y = (A - B + Bt)\, e^{2t}.]$$

3. $$\frac{dx}{dt} + 5x + y = e^t, \quad \frac{dy}{dt} + 3y - x = e^{2t}.$$
$$[x = (A + Bt)\, e^{-4t} + \tfrac{4}{25} e^t - \tfrac{1}{36} e^{2t},$$
$$y = -(A + B + Bt)\, e^{-4t} + \tfrac{1}{25} e^t + \tfrac{7}{36} e^{2t}.]$$

4. Solve $\dfrac{dx}{dt} = x^2 + xy, \quad \dfrac{dy}{dt} = y^2 + xy.$

$$\left[x = \frac{1+B}{A-2t}, \quad y = \frac{1-B}{A-2t}. \right]$$

5. $\dfrac{d^2x}{dt^2} + 3x - 4y + 3 = 0, \quad \dfrac{d^2y}{dt^2} + x - y + 5 = 0.$

$$[x = (A + Bt)\cos t + (A' + B't)\sin t - 17,$$
$$y = \tfrac{1}{2}(A + B' + Bt)\cos t + \tfrac{1}{2}(A' - B + B't)\sin t - 12.]$$

6. Shew that the integrals of

$$\frac{dx}{dt} = ax + hy, \quad \frac{dy}{dt} = hx + by,$$

are $\qquad x = A_1 e^{\lambda_1 t} + A_2 e^{\lambda_2 t}, \quad y = \mu_1 A_1 e^{\lambda_1 t} + \mu_2 A_2 e^{\lambda_2 t},$

where μ_1, μ_2 are the roots of

$$\mu^2 + \frac{a-b}{h}\mu - 1 = 0,$$

and $\qquad \lambda_1 = a + h\mu_1, \quad \lambda_2 = a + h\mu_2.$

7. Solve $\dfrac{d^2x}{dt^2} + m^2 y = 0, \quad \dfrac{d^2y}{dt^2} - m^2 x = 0.$

$$\left[x = A e^{\frac{mt}{\sqrt{2}}} \cos\left(\frac{mt}{\sqrt{2}} + a\right) + B e^{-\frac{mt}{\sqrt{2}}} \cos\left(\frac{mt}{\sqrt{2}} + \beta\right), \right.$$
$$\left. y = A e^{\frac{mt}{\sqrt{2}}} \sin\left(\frac{mt}{\sqrt{2}} + a\right) - B e^{-\frac{mt}{\sqrt{2}}} \sin\left(\frac{mt}{\sqrt{2}} + \beta\right). \right]$$

8. Solve $\dfrac{d^2x}{dt^2} = ax + hy, \quad \dfrac{d^2y}{dt^2} = hx + by.$

9. Solve $\dfrac{d^2x}{dt^2} + m^2 x - n^2 y = 0, \quad \dfrac{d^2y}{dt^2} + m^2 y - n^2 x = 0.$

$$[x + y = A \cos \sqrt{(m^2 - n^2)}\, t + A' \sin \sqrt{(m^2 - n^2)}\, t,$$
$$x - y = B \cos \sqrt{(m^2 + n^2)}\, t + B' \sin \sqrt{(m^2 + n^2)}\, t.]$$

10. Determine the constants in the solution of the simultaneous equations

$$\frac{d^2x}{dt^2} = \mu x, \quad \frac{d^2y}{dt^2} = \mu y$$

so that, for $t = 0$,

$$x = a, \quad y = 0, \quad \frac{dx}{dt} = 0, \quad \frac{dy}{dt} = V.$$

$$[x = a \cosh \sqrt{\mu}\, t, \quad y = V/\sqrt{\mu} . \sinh \sqrt{\mu}\, t.]$$

11. The equations of motion of electricity in a circuit of self-induction L, and resistance R, which is interrupted by a condenser of capacity C, are

$$L \frac{dx}{dt} + Rx = -\frac{q}{C}, \quad \frac{dq}{dt} = x,$$

where x is the current, and q the charge of the condenser. Find the condition that the discharge should be oscillatory. $[L > \frac{1}{4}R^2C.]$

12. Solve $\quad \dfrac{d^2x}{dt^2} = 0, \quad \dfrac{d^2y}{dt^2} = -\dfrac{\mu}{y^3},$

and shew that the solution represents a conic symmetrical with respect to the axis of x.

13. Solve $\quad \dfrac{d^2\dot{x}}{dt^2} = 0, \quad \dfrac{d^2y}{dt^2} = \mu y^3,$

and prove that the curves represented by the solution include a family of hyperbolas.

14. Solve $\quad \dfrac{d^2x}{dt^2} = -n\dfrac{dy}{dt} + f, \quad \dfrac{d^2y}{dt^2} = n\dfrac{dx}{dt}.$

$$\left[x = a + a\cos\left(nt + \epsilon\right), \quad y = \beta + \frac{f}{n}t + a\sin\left(nt + \epsilon\right). \right]$$

15. Solve $\quad \dfrac{d^2x}{dt^2} - 2n\dfrac{dy}{dt} + m^2x = 0,$

$$\frac{d^2y}{dt^2} + 2n\frac{dx}{dt} + m^2y = 0.$$

$$[x = \quad A\cos\left(pt + \epsilon\right) + A'\cos\left(p't + \epsilon'\right),$$
$$y = -A\sin\left(pt + \epsilon\right) + A'\sin\left(p't + \epsilon'\right),$$

$$\text{where } \left.\begin{matrix} p \\ p' \end{matrix}\right\} = \sqrt{(m^2 + n^2)} \pm n.]$$

16. Solve $\quad \left.\begin{matrix} \dfrac{d^2x}{dt^2} + (\mu + 1)\dfrac{g}{a}x - \mu\dfrac{g}{b}y = 0, \\[2mm] \dfrac{d^2x}{dt^2} + \dfrac{d^2y}{dt^2} + \dfrac{g}{b}y = 0 \end{matrix}\right\}.$

(The equations of motion of a double pendulum, the lengths of the upper and lower strings being a and b, and μ being the ratio of the mass of the lower to that of the upper particle.)

Prove that the periods of the normal vibrations are $2\pi/p_1$, $2\pi/p_2$, if p_1^2, p_2^2 be the roots of

$$p^4 - (1 + \mu)g\left(\frac{1}{a} + \frac{1}{b}\right)p^2 + (1 + \mu)\frac{g^2}{ab} = 0.$$

Shew that the roots of this quadratic in p^2 are real, positive, and distinct.

CHAPTER XIV

DIFFERENTIATION AND INTEGRATION OF POWER-SERIES

174. Statement of the Question.

The main object of this Chapter is to justify, under proper conditions, the application of the processes of differentiation and integration to functions expressed by 'power-series,' *i.e.* by series of the type

$$A_0 + A_1 x + A_2 x^2 + \ldots + A_n x^n + \ldots, \quad \ldots\ldots\ldots\ldots(1)$$

where the coefficients are constants. Thus, if $S(x)$ denote the sum of this series, assumed to be convergent for all values of x extending over a certain range, we have to examine under what conditions it can be asserted that $S(x)$ is a continuous and differentiable function of x, and that, moreover,

$$S'(x) = A_1 + 2A_2 x + 3A_3 x^2 + \ldots + n A_n x^{n-1} + \ldots, \quad \ldots(2)$$

and

$$\int_0^x S(x)\, dx = A_0 x + \tfrac{1}{2} A_1 x^2 + \tfrac{1}{3} A_2 x^3 + \ldots + \frac{1}{n+1} A_n x^{n+1} + \ldots, \quad \ldots(3)$$

respectively.

If the number of terms in (1) were *finite*, the statements in question would need no proof, beyond what has already been indicated in the course of this work (see Arts. 29, 74), but it must be remembered that the word 'sum' as applied to an infinite series bears an artificial sense, and that we are not entitled to assume without examination that statements which are true when the word has one meaning remain true when it is used in another.

It is convenient to have a notation for the sum of the first n terms of the series (1). We write

$$S_n(x) = A_0 + A_1 x + A_2 x^2 + \ldots + A_{n-1} x^{n-1}, \quad \ldots\ldots\ldots(4)$$

a rational integral function of degree $n-1$. This is called a 'partial sum,' and its graphical representation is called an 'approximation curve.' An example of such curves is given in Fig. 136, p. 485. If, further, we put

$$S(x) = S_n(x) + R_n(x), \quad \ldots\ldots\ldots\ldots\ldots(5)$$

the quantity $R_n(x)$ is called the 'remainder after n terms.' It is of course the sum of the series

$$A_n x^n + A_{n+1} x^{n+1} + \ldots \ldots \ldots \ldots \ldots (6)$$

By hypothesis, the sequence

$$S_1(x), \quad S_2(x), \quad S_3(x), \ldots, \quad \ldots \ldots \ldots \ldots (7)$$

has, for any value of x for which the series is convergent, the limiting value $S(x)$. It follows that the sequence

$$R_1(x), \quad R_2(x), \quad R_3(x), \ldots \quad \ldots \ldots \ldots \ldots (8)$$

has the limiting value zero.

It is to be noted that in each of the questions above propounded we have to deal with what is known as a 'double limit.' Thus, to establish the continuity of $S(x)$ for $x = a$ we have to shew that

$$\lim_{n \to \infty} \lim_{x \to a} S_n(x) = \lim_{x \to a} \lim_{n \to \infty} S_n(x). \ldots \ldots (9)$$

Again, the formulæ (2) and (3) may be written

$$\frac{d}{dx} \lim_{n \to \infty} S_n(x) = \lim_{n \to \infty} \frac{d}{dx} S_n(x), \quad \ldots \ldots \ldots (10)$$

and $$\int_0^x \{\lim_{n \to \infty} S_n(x)\} \, dx = \lim_{n \to \infty} \int_0^x S_n(x) \, dx, \quad \ldots (11)$$

respectively. Since a derived function is the limit of a quotient, and a definite integral is the limit of a sum, these forms also come under the description of a double limit. It is not to be assumed, and it is not *necessarily* true, that the result is independent of the order in which the two operations of proceeding to a limit are performed[*].

175. Derivation of the Logarithmic Series.

There are one or two cases where the questions above raised can be answered without difficulty, the form of $R_n(x)$ being known; and the results are of great importance.

By actual division we have, as in the theory of the Geometric Progression,

$$\frac{1}{1+t} = 1 - t + t^2 - \ldots + (-)^{n-1} t^{n-1} + (-)^n \frac{t^n}{1+t}, \quad \ldots (1)$$

provided $t \neq -1$. We will suppose that x is positive. We have, then, from (1),

$$\log(1+x) = \int_0^x \frac{dt}{1+t} = x - \frac{x^2}{2} + \frac{x^3}{3} - \ldots$$

$$+ (-)^{n-1} \frac{x^n}{n} + (-)^n \int_0^x \frac{t^n dt}{1+t}. \quad \ldots \ldots (2)$$

[*] For an example see Art. 193.

The value of the integral in the last term is increased if we replace the denominator in the integrand by its least value, viz. unity. The integral is therefore less than

$$\int_0^x t^n dt, \quad \text{or} \quad \frac{x^{n+1}}{n+1}.$$

If x be less than unity, or even equal to unity, this tends with increasing n to the limit 0.

Hence if x be positive and $\not> 1$ we have

$$\log(1+x) = x - \frac{x^2}{2} + \frac{x^3}{3} - \ldots + (-)^{n-1}\frac{x^n}{n} + \ldots, \ldots \ldots (3)$$

the series extending to infinity.

In particular, putting $x = 1$,

$$\log 2 = 1 - \tfrac{1}{2} + \tfrac{1}{3} - \tfrac{1}{4} + \ldots \quad \ldots \ldots \ldots \ldots \ldots (4)$$

This result, though exact, is not suited for numerical calculation, on account of the slow convergence of the series. It may be shewn that about 10^n terms would be required to obtain a result accurate to n places of decimals. A more practical formula is given by (12) below.

Again, we have

$$\frac{1}{1-t} = 1 + t + t^2 + \ldots + t^{n-1} + \frac{t^n}{1-t}, \quad \ldots \ldots (5)$$

provided $t \neq 1$. Hence if x be positive and less than unity,

$$\int_0^x \frac{dt}{1-t} = x + \frac{x^2}{2} + \frac{x^3}{3} + \ldots + \frac{x^n}{n} + \int_0^x \frac{t^n dt}{1-t}. \quad \ldots \ldots (6)$$

The integral on the right-hand side is increased if we replace the denominator by the least value which it has within the range of integration, i.e. by $1 - x$. Hence

$$\int_0^x \frac{t^n dt}{1-t} < \frac{1}{1-x}\int_0^x t^n dt < \frac{x^{n+1}}{(n+1)(1-x)}. \quad \ldots \ldots (7)$$

Since x is by hypothesis less than unity, this tends with increasing n to the limit 0. Moreover, since

$$\int_0^x \frac{dt}{1-t} = \left[-\log(1-t)\right]_0^x = -\log(1-x), \quad \ldots \ldots (8)$$

we have

$$\log(1-x) = -x - \frac{x^2}{2} - \frac{x^3}{3} - \ldots - \frac{x^n}{n} - \ldots, \quad \ldots \ldots (9)$$

to infinity.

The results (3) and (9) are combined in the statement that

$$\log(1+x) = x - \frac{x^2}{2} + \frac{x^3}{3} - \ldots + (-)^{n-1}\frac{x^n}{n} + \ldots, \ldots(10)$$

for values of x ranging from -1 exclusively to $+1$ inclusively. The series on the right is known as the 'logarithmic series *.'

If x be positive and less than unity, we have from (3) and (10), by subtraction,

$$\log\frac{1+x}{1-x} = 2\left(x + \frac{x^3}{3} + \frac{x^5}{5} + \ldots\right). \ldots\ldots\ldots\ldots(11)$$

If in this formula we put $x = 1/(2m+1)$, we obtain

$$\log(m+1) - \log m = \log\frac{m+1}{m}$$

$$= 2\left\{\frac{1}{2m+1} + \frac{1}{3(2m+1)^3} + \frac{1}{5(2m+1)^5} + \ldots\right\}.\ldots\ldots(12)$$

This series is very convergent, even for $m=1$. Putting $m=1, 2, 3, \ldots,$ we obtain the values of

$$\log 2, \quad \log 3 - \log 2, \quad \log 4 - \log 3, \ldots$$

in succession, and thence the values of the logarithms (to base e) of the natural numbers 2, 3, ... When log 10 has been found, its reciprocal gives the modulus μ by which logarithms to base e must be multiplied in order to convert them into logarithms to base 10†.

Ex. 1. If $n > 1$, we have

$$\log\frac{n+1}{n} = \log\left(1 + \frac{1}{n}\right) = \frac{1}{n} - \frac{1}{2n^2} + \frac{1}{3n^3} - \ldots \quad \ldots\ldots(13)$$

Since the terms are alternately positive and negative, and tend to the limit 0, the sum is less than $1/n$, by Art. 5.

Again,

$$\log\frac{n}{n-1} = -\log\left(1 - \frac{1}{n}\right) = \frac{1}{n} + \frac{1}{2n^2} + \frac{1}{3n^3} + \ldots, \quad \ldots\ldots(14)$$

which is obviously greater than $1/n$.

Hence

$$\frac{1}{n} > \log\frac{n+1}{n} > \frac{1}{n+1}. \ldots\ldots\ldots\ldots\ldots\ldots(15)$$

---

* It was apparently first given by N. Mercator in 1668.

† The most rapid way of determining μ is by means of the identity

$$\log 10 = 3\log 2 + \log\tfrac{5}{4}.$$

The two logarithms on the right hand are found by putting $m=1$, $m=4$, in (12).

Ex. 2.   Let us write

$$u_n = 1 + \frac{1}{2} + \frac{1}{3} + \ldots + \frac{1}{n} - \log n,$$

$$v_n = 1 + \frac{1}{2} + \frac{1}{3} + \ldots + \frac{1}{n} - \log (n + 1),$$

$$\Bigg\}\qquad \ldots\ldots\ldots\ldots(16)$$

n being now of course a positive integer.   We have

$$u_n - u_{n+1} = \log \frac{n+1}{n} - \frac{1}{n+1} > 0, \quad \ldots\ldots\ldots\ldots(17)$$

by (15).   Again,

$$v_{n+1} - v_n = \frac{1}{n+1} - \log \frac{n+2}{n+1} > 0, \quad \ldots\ldots\ldots\ldots(18)$$

also by (15).   Moreover

$$u_n - v_n = \log \frac{n+1}{n}, \quad \ldots\ldots\ldots\ldots\ldots(19)$$

which lies between 0 and $1/n$.   Hence the quantities

$$u_1, \quad u_2, \quad u_3, \quad \ldots, \quad u_n, \quad \ldots \ldots\ldots\ldots\ldots\ldots(20)$$

form a descending sequence, and

$$v_1, \quad v_2, \quad v_3, \quad \ldots, \quad v_n, \quad \ldots \ldots\ldots\ldots\ldots\ldots(21)$$

an ascending sequence.   Since each member of (20) is, by (19), greater than the corresponding member of (21), the sequence (20) has a lower limit (Art. 2), and the sequence (21) an upper limit.   And since

$$\lim_{n \to \infty} (u_n - v_n) = 0, \quad \ldots\ldots\ldots\ldots\ldots(22)$$

these limits must be the same.   Hence

$$\lim_{n \to \infty} \left(1 + \frac{1}{2} + \frac{1}{3} + \ldots + \frac{1}{n} - \log n\right) = \gamma, \quad \ldots\ldots\ldots(23)$$

where γ is a certain constant (known as 'Euler's constant').   Since $v_1 = 1 - \log 2$, which is > 0, γ is positive.   Its value has been ascertained to be $\cdot 57721566\ldots$*.

176.   Gregory's Series.

Since

$$\frac{1}{1+t^2} = 1 - t^2 + t^4 - \ldots + (-)^{n-1}t^{2n-2} + (-)^n \frac{t^{2n}}{1+t^2}, \quad \ldots(1)$$

we have

$$\int_0^x \frac{dt}{1+t^2} = x - \frac{x^3}{3} + \frac{x^5}{5} - \ldots + (-)^{n-1}\frac{x^{2n-1}}{2n-1} + (-)^n \int_0^x \frac{t^{2n}}{1+t^2}\, dt.$$

$$\ldots\ldots(2)$$

---

* The method of calculation is beyond our scope.

If x is positive and $\not> 1$, the latter integral is less than

$$\int_0^1 t^{2n}dt, \quad \text{or} \quad \frac{1}{2n+1}, \quad\ldots\ldots\ldots\ldots\ldots(3)$$

and therefore tends with increasing n to the limit 0. Hence

$$\tan^{-1}x = x - \frac{x^3}{3} + \frac{x^5}{5} - \ldots + (-)^{n-1}\frac{x^{2n-1}}{2n-1} + \ldots, \quad\ldots(4)$$

that value of $\tan^{-1}x$ being understood which starts from zero with x. This is known as 'Gregory's series*.' Since both sides of (4) change sign with x, the equality holds for values of x ranging from -1 to $+1$, inclusively.

Putting $x = 1$, we have

$$\tfrac{1}{4}\pi = 1 - \tfrac{1}{3} + \tfrac{1}{5} - \tfrac{1}{7} + \ldots. \quad\ldots\ldots\ldots\ldots(5)$$

This series converges very slowly, and has been superseded for the calculation of π by others. Euler used the identity

$$\tfrac{1}{4}\pi = \tan^{-1}\tfrac{1}{2} + \tan^{-1}\tfrac{1}{3}, \quad\ldots\ldots\ldots\ldots\ldots(6)$$

which gives

$$\tfrac{1}{4}\pi = \left(\frac{1}{2} - \frac{1}{3\,.\,2^3} + \frac{1}{5\,.\,2^5} - \ldots\right) + \left(\frac{1}{3} - \frac{1}{3\,.\,3^3} + \frac{1}{5\,.\,3^5} - \ldots\right)\ldots(7)$$

Machin had previously employed the formula

$$\tfrac{1}{4}\pi = 4\tan^{-1}\tfrac{1}{5} - \tan^{-1}\tfrac{1}{239}, \quad\ldots\ldots\ldots\ldots(8)$$

which, like (6), is proved in most elementary books on Trigonometry. This leads to

$$\tfrac{1}{4}\pi = 4\left(\frac{1}{5} - \frac{1}{3\,.\,5^3} + \frac{1}{5\,.\,5^5} - \ldots\right) - \left(\frac{1}{239} - \frac{1}{3\,.\,239^3} + \frac{1}{5\,.\,239^5} - \ldots\right).$$
$$\ldots\ldots\ldots(9)$$

On account of the importance of the matter, it is worth while to give the details of the calculation of π from Machin's formula. To calculate $\tan^{-1}\tfrac{1}{5}$, we first draw up the following table:

n	$\dfrac{1}{5^n}$	$\pm\dfrac{1}{n\,.\,5^n}$
1	·200 000 000 0	+ ·200 000 000 0
3	8 000 000 0	− 2 666 666 7
5	320 000 0	+ 64 000 0
7	12 800 0	− 1 828 6
9	512 0	+ 56 9
11	20 5	− 1 9
13	8	+ 1

* After the discoverer, James Gregory (1671).

The sum of the positive terms in the last column is

$$+ \cdot 200\ 064\ 057\ 0,$$

and that of the negative terms is

$$- \cdot 002\ 668\ 497\ 2.$$

Hence $\tan^{-1}\frac{1}{5} = \cdot 197\ 395\ 559\ 8.$

Again to calculate $\tan^{-1}\frac{1}{239}$ we have the table :

n	$\dfrac{1}{239^n}$	$\pm\ \dfrac{1}{n\,.\,239^n}$
1	$\cdot 004\ 184\ 100\ 4$	$+\ \cdot 004\ 184\ 100\ 4$
3	$73\ 2$	$-\qquad\qquad 24\ 4$

This makes $\tan^{-1}\frac{1}{239} = \cdot 004\ 184\ 076\ 0.$

Hence $\frac{1}{4}\pi = 4\tan^{-1}\frac{1}{5} - \tan^{-1}\frac{1}{239}$

$$= + \cdot 789\ 582\ 239\ 2$$
$$- \cdot 004\ 184\ 076\ 0$$
$$= \cdot 785\ 398\ 163\ 2,$$
$$\pi = 3\cdot 141\ 592\ 652\ 8.$$

The last figures are of course liable to error. To estimate the possible error of the final result, we remark that in the calculation of $\tan^{-1}\frac{1}{5}$ there were five errors, each not exceeding half a unit in the last place, and that there are two such errors in the computation of $\tan^{-1}\frac{1}{239}$. The errors, therefore, in the inferred value of π, even if cumulative, cannot exceed

$$4 \times (4 \times 5 \times \tfrac{1}{2} + 2 \times \tfrac{1}{2}),\ = 44$$

times the unit of the last place. The first seven decimal places cannot therefore be affected, and we can assert that the last three must lie within the limits 484 and 572. As a matter of fact the errors are not all in the same direction, and the correct value of π to ten places is

$$\pi = 3\cdot 141\ 592\ 653\ 6.$$

177. Convergence of Power-Series.

Proceeding now to the discussion of the general questions raised in Art. 174, we have first to consider the matter of convergence.

If the terms of an infinite series are functions of a variable x, it may happen that the series is convergent for all values of x without restriction, as in the case of the exponential series (Art. 37), or it may be convergent only for values of x belonging to a certain continuous range. If this range extends from $x = a$ to $x = b$, *inclusively*, it is said to be 'closed,' and is denoted by (a, b). If both terminal points a, b are excluded from the statement, the range is said to be 'open' at both ends, and is denoted by $[a, b]$. If the first or second terminal point alone is excluded, this is indicated by the notation $[a, b)$ or $(a, b]$, as the case may be*. For example, the logarithmic series has been shewn to be convergent over the range $[-1, 1)$, whilst Gregory's series is convergent over the range $(-1, 1)$.

The most generally useful test of convergence, in the case of a power-series

$$A_0 + A_1 x + A_2 x^2 + \ldots + A_n x^n + \ldots \ldots\ldots\ldots\ldots(1)$$

is the 'ratio-test.' It is obvious that if, after some finite number of terms, the ratio of each term to the preceding is less in absolute value than some quantity k which is itself less than unity, the series is essentially convergent. For the successive terms then diminish more rapidly than those of a geometric progression whose common ratio is k.

In particular, the series (1) will be essentially convergent if

$$\lim_{n \to \infty} \left| \frac{A_{n+1}}{A_n} x \right| < 1. \quad \ldots\ldots\ldots\ldots\ldots(2)\dagger$$

For if this condition be fulfilled, and the limit in question be k', we can by taking n sufficiently great ensure that for this and for all greater values of n the ratio

$$\left| \frac{A_{n+1}}{A_n} x \right|$$

shall be less than any assigned quantity k which lies between k' and 1.

If the condition (2) is satisfied, it follows that the series

$$A_1 + 2A_2 x + 3A_3 x^2 + \ldots + nA_n x^{n-1} + \ldots, \ldots\ldots\ldots(3)$$

and $\qquad A_0 x + \dfrac{A_1 x^2}{2} + \dfrac{A_2 x^3}{3} + \ldots + \dfrac{A_n x^{n+1}}{n+1} + \ldots, \quad \ldots\ldots\ldots(4)$

the terms of which are derived from those of (1) by differentiation and integration, respectively, will also be essentially convergent. For in the case of (3) we have

$$\lim_{n \to \infty} \left| \frac{(n+1) A_{n+1}}{n A_n} x \right| = \lim_{n \to \infty} \left(1 + \frac{1}{n} \right)$$

$$\times \lim_{n \to \infty} \left| \frac{A_{n+1}}{A_n} x \right| = k'. \ldots\ldots(5)$$

The case of (4) is still more obvious.

---

* These notations are due to Prof. F. S. Carey.

† This form of test is known as d'Alembert's.

Again, leaving aside any particular test, let us suppose merely that the series (1) is known to be convergent for a particular value α of x. Since the terms must diminish indefinitely, we must have

$$\lim_{n \to \infty} | A_n \alpha^n | = 0. \quad \dots\dots\dots\dots\dots(6)$$

It follows that the series will be essentially convergent for all values of x such that $|x| < |\alpha|$. For, writing the series in the form

$$A_0 + A_1 \alpha \left(\frac{x}{\alpha}\right) + A_2 \alpha^2 \left(\frac{x}{\alpha}\right)^2 + \dots + A_n \alpha^n \left(\frac{x}{\alpha}\right)^n + \dots, \quad \dots(7)$$

and denoting by M the greatest value of $| A_n \alpha^n |$, we see that the several terms of (7) are in absolute value less than the corresponding terms of the convergent geometrical series

$$M (1 + t + t^2 + \dots + t^n + \dots), \quad \dots\dots\dots\dots(8)$$

where $$t = | x/\alpha |.$$

The series (7) is therefore itself essentially convergent.

Hence if the series (1) be convergent for any one value (α) of x, other than 0, it will be convergent over the range $[-\alpha, \alpha)$, and essentially convergent over the range $[-\alpha, \alpha]$.

It follows also from (6) that the series (3) and (4) will be essentially convergent over the range $[-\alpha, \alpha]$. For if β be any quantity less in absolute value than α, we have in the case of (3)

$$\lim_{n \to \infty} | n A_n \beta^{n-1} | = \frac{1}{|\beta|} \lim_{n \to \infty} | A_n \alpha^n | \times \lim_{n \to \infty} n \left| \frac{\beta}{\alpha} \right|^n. \quad (9)$$

The former of the two limits on the right-hand side vanishes by hypothesis, and the second in virtue of Art. 43 (3).

In the case of (4) we have

$$\lim_{n \to \infty} \left| \frac{A_n \alpha^{n+1}}{n+1} \right| = | \alpha | \lim_{n \to \infty} \left| \frac{A_n \alpha^n}{n+1} \right| = 0, \quad \dots\dots(10)$$

à fortiori.

Ex. 1.   The series

$$1 + \frac{1}{2} x + \frac{1 \cdot 3}{2 \cdot 4} x^2 + \frac{1 \cdot 3 \cdot 5}{2 \cdot 4 \cdot 6} x^3 + \dots \quad \dots \quad \dots\dots\dots(11)$$

is convergent for $x = -1$, by Art. 5. It is therefore essentially convergent if $|x| < 1$. It may be shewn to be divergent for $x = 1$. It is therefore convergent over the range $(-1, 1]$, but *essentially* convergent only over the range $[-1, 1]$.

Ex. 2. The series

$$\frac{x^2}{1 \cdot 2} + \frac{x^3}{2 \cdot 3} + \frac{x^4}{3 \cdot 4} + \dots \quad \dots\dots\dots\dots(12)$$

is convergent, in virtue of the test (2), for $|x| < 1$. It is also easily seen to be convergent for $x = \pm 1$, and is therefore convergent over the range $(-1, 1)$.

But the argument above given only entitles us to assert that the series

$$x + \frac{x^2}{2} + \frac{x^3}{3} + \dots, \quad \dots\dots\dots\dots(13)$$

which is derived from (12) by differentiation, is convergent over the range $[-1, 1]$. It is, however, obviously convergent for $x = -1$, by Art. 5. For $x = 1$ it is divergent (Art. 175).

178. Continuity of a Power-Series.

We will now assume that the series

$$S(x) = A_0 + A_1 x + A_2 x^2 + \dots + A_n x^n + \dots \quad \dots\dots\dots(1)$$

is known to be essentially convergent over a range $[-\alpha, \alpha]$. If x and x' be any two points of this range we have, by Art. 5, 1°, 3°,

$$S(x') - S(x) = (x' - x)\left\{ A_1 + 2A_2 \frac{x + x'}{2} + 3A_3 \frac{x^2 + xx' + x'^2}{3} + \dots \right.$$
$$\left. + nA_n \frac{x^{n-1} + x^{n-2}x' + \dots + x'^{n-1}}{n} + \dots \right\}. \quad \dots(2)$$

If x and x' have the same sign, the fraction

$$\frac{x^{n-1} + x^{n-2}x' + \dots + x'^{n-1}}{n}$$

lies between x^{n-1} and x'^{n-1}. The several terms in { } are therefore intermediate in absolute value to the corresponding terms in the two series

$$A_1 + 2A_2 x + 3A_3 x^2 + \dots + nA_n x^{n-1} + \dots, \quad \dots\dots(3)$$

and $\qquad A_1 + 2A_2 x' + 3A_3 x'^2 + \dots + nA_n x'^{n-1} + \dots. \quad \dots\dots(4)$

It has been shewn that on the above hypothesis these series are essentially convergent. It follows that the expression in { } in (2) is finite. Hence

$$\lim_{x' \to x} \{ S(x') - S(x) \} = 0, \quad \dots\dots\dots\dots(5)$$

i.e. $S(x)$ is continuous for all values of x belonging to the range $[-\alpha, \alpha]$[*].

We can infer also that the power-series (3) and (4) are continuous over the range $[-\alpha, \alpha]$.

* It may happen that the series (1) and (3) are known to be essentially convergent when x is a terminal point of the range of convergence of (1). In that case we can assert the continuity of $S(x)$ up to this value of x inclusively.

179. Differentiation of a Power-Series.

With the same notation as in the preceding Art., and on the same assumption, we have

$$\frac{S(x') - S(x)}{x' - x} = A_1 + 2A_2\frac{x + x'}{2} + 3A_3\frac{x^2 + xx' + x'^2}{3} + \dots$$
$$+ nA_n\frac{x^{n-1} + x^{n-2}x' + \dots + x'^{n-1}}{n} + \dots \quad \dots(1)$$

Since x' is to be made ultimately equal to x, we may suppose that it has the same sign.

Let us first suppose that all the coefficients A_n are positive, and that x is also positive. The series on the right-hand side of (1) is then intermediate in value between the series (3) and (4) of the preceding Art., and since the sum of (3) is a continuous function of x, it follows that

$$S'(x) = \lim_{x' \to x}\frac{S(x') - S(x)}{x' - x}$$
$$= A_1 + 2A_2x + 3A_3x^2 + \dots + nA_nx^{n-1} + \dots \quad \dots(2)$$

This holds for all points of the range $[-a, a]$.

The same result would obviously follow if the coefficients A_n were all negative.

Let us next suppose that x is negative, the coefficients A_n being still assumed to be positive. The preceding argument will then apply separately to the two series formed by taking alternate terms in $S(x)$, viz.

$$A_0 + A_2x^2 + A_4x^4 + \dots + A_{2n}x^{2n} + \dots, \dots\dots\dots\dots(3)$$

and $\quad A_1x + A_3x^3 + A_5x^5 + \dots + A_{2n+1}x^{2n+1} + \dots, \quad \dots\dots\dots(4)$

since the terms of (3) are all positive, and those of (4) all negative. Their derived functions are therefore equal to the sums of the series obtained by differentiating them respectively term by term. The result (2) then follows by addition, since the series are essentially convergent.

Finally, if the coefficients A_n are not all of the same sign, we can resolve $S(x)$ into the sum of two series, the coefficients in one of which are all positive, and in the other all negative. The foregoing argument applies to each of these, and therefore to the combination.

It will be observed that the assumption that the series with which we are concerned are all *essentially* convergent is vital to the whole argument.

Ex. It is known that if $|x| < 1$

$$\frac{1}{1-x} = 1 + x + x^2 + \ldots + x^n + \ldots \ldots \ldots \ldots \ldots \ldots \ldots (5)$$

Differentiating both sides, we obtain

$$\frac{1}{(1-x)^2} = 1 + 2x + 3x^2 + \ldots + nx^{n-1} + \ldots \ldots \ldots \ldots (6)$$

A second differentiation gives

$$\frac{1}{(1-x)^3} = \tfrac{1}{2}\{1\cdot2 + 2\cdot3x + \ldots + 3\cdot4x^2 + \ldots + (n-1)nx^{n-2} + \ldots\}.$$
$$\ldots\ldots(7)$$

180. Integration of a Power-Series.

Using the same notation as in Arts. 177–179, let

$$I(x) = A_0 x + A_1 \frac{x^2}{2} + A_2 \frac{x^3}{3} + \ldots + A_n \frac{x^{n+1}}{n+1} + \ldots \ldots (1)$$

On the present assumption that $S(x)$ is essentially convergent over the range $[-\alpha, \alpha]$, this will also be essentially convergent over the same range. Hence, by Art. 179 we have

$$I'(x) = A_0 + A_1 x + A_2 x^2 + \ldots + A_n x^n + \ldots = S(x). \ldots(2)$$

Hence
$$\int_0^x S(x)\,dx = \left[I(x) \right]_0^x = I(x). \ldots\ldots\ldots\ldots(3)$$

Ex. 1. If $|x| < 1$, we have by the Binomial Theorem (Art. 182)

$$\frac{1}{\sqrt{(1-x^2)}} = 1 + \frac{1}{2} x^2 + \frac{1\cdot3}{2\cdot4} x^4 + \frac{1\cdot3\cdot5}{2\cdot4\cdot6} x^6 + \ldots \ldots\ldots\ldots(4)$$

Hence, integrating term by term between the limits 0 and x,

$$\sin^{-1} x = x + \frac{1}{2}\frac{x^3}{3} + \frac{1\cdot3}{2\cdot4}\frac{x^5}{5} + \frac{1\cdot3\cdot5}{2\cdot4\cdot6}\frac{x^7}{7} + \ldots \ldots\ldots\ldots(5)$$

This series is due to Newton.

If we put $x = \frac{1}{2}$ we get

$$\pi = 6\left\{ \frac{1}{2} + \frac{1}{2\cdot3\cdot2^3} + \frac{1\cdot3}{2\cdot4\cdot5\cdot2^5} + \ldots \right\}, \ldots\ldots\ldots\ldots(6)$$

from which π can be calculated without much trouble.

Ex. 2. If $|x| < 1$,
$$\log(1+x) = x - \frac{x^2}{2} + \frac{x^3}{3} - \ldots \ldots\ldots\ldots\ldots\ldots(7)$$

Integrating this between the limits 0 and x,

$$(1 + x) \log (1 + x) - x = \frac{x^2}{1 \cdot 2} - \frac{x^3}{2 \cdot 3} + \frac{x^4}{3 \cdot 4} - \dots \quad \dots\dots(8)$$

It follows from a remark made in Art. 178 (footnote) that the function on the right-hand is continuous up to the limit $x = 1$. We infer that

$$\frac{1}{1 \cdot 2} - \frac{1}{2 \cdot 3} + \frac{1}{3 \cdot 4} - \dots = 2 \log 2 - 1 = \cdot 38629 \dots.$$

181. Integration of Differential Equations by Series.

Given a linear differential equation, with coefficients which are rational integral functions of the independent variable (x), it is often possible to obtain a solution in the form of an ascending power-series, thus

$$y = A_0 + A_1 x + A_2 x^2 + \dots + A_n x^n + \dots \quad \dots\dots(1)$$

If we assume, provisionally, that this series is convergent for a certain range of x, it can be differentiated once, twice, ... with respect to x, by the theorem of Art. 179. Substituting in the differential equation, we find that this can be satisfied if certain relations between the coefficients A_0, A_1, A_2, ... are fulfilled. In this way we obtain a series involving one or more arbitrary constants; and if this series proves to be in fact essentially convergent, we have obtained a-solution of the proposed differential equation. Whether it is the *complete* solution, or how far it may require to be supplemented, are of course distinct questions, which remain to be discussed independently.

The following is an important example.

Let the equation be

$$\frac{d^2 y}{dx^2} + y = 0. \dots\dots\dots\dots(2)$$

Assuming the form (1), and substituting, we find

$$(1 \cdot 2 A_2 + A_0) + (2 \cdot 3 A_3 + A_1) x + (3 \cdot 4 A_4 + A_2) x^2 + \dots$$
$$+ \{(n - 1) n A_n + A_{n-2}\} x^{n-2} + \dots = 0, \dots(3)$$

which is satisfied identically, provided

$$A_2 = -\frac{1}{1 \cdot 2} A_0, \qquad A_3 = -\frac{1}{2 \cdot 3} A_1,$$

$$A_4 = -\frac{1}{3 \cdot 4} A_2 = \frac{1}{4!} A_0, \quad A_5 = -\frac{1}{4 \cdot 5} A_3 = \frac{1}{5!} A_1,$$

and, generally,

$$A_{2n} = -\frac{1}{(2n-1)\,2n}\,A_{2n-2} = (-)^n\frac{1}{2n\,!}\,A_0,$$
$$A_{2n+1} = -\frac{1}{2n\,(2n+1)} = (-)^n\frac{1}{(2n+1)!}\,A_1. \qquad \Bigg\} \quad \text{.........(4)}$$

We thus obtain the solution

$$y = A_0\left(1 - \frac{x^2}{2!} + \frac{x^4}{4!} - \ldots\right) + A_1\left(x - \frac{x^3}{3!} + \frac{x^5}{5!} - \ldots\right). \quad (5)$$

The series in brackets are easily seen to be essentially convergent, and their sums therefore continuous, for all values of x.

It has been shewn in Art. 163 that the complete solution of (2) is

$$y = A\cos x + B\sin x. \quad \text{.....................(6)}$$

Hence, given A_0 and A_1, it must be possible to determine A and B so that the expressions (5) and (6) shall be identical.

For example, putting $A_0 = 1$, $A_1 = 0$, we must have

$$1 - \frac{x^2}{2!} + \frac{x^4}{4!} - \ldots = A\cos x + B\sin x,$$

and, changing the sign of x,

$$1 - \frac{x^2}{2!} + \frac{x^4}{4!} - \ldots = A\cos x - B\sin x.$$

Hence we must have $B = 0$, and putting $x = 0$, we find $A = 1$. We thus obtain the formula

$$\cos x = 1 - \frac{x^2}{2!} + \frac{x^4}{4!} - \ldots. \quad \text{.................(7)}$$

In the same way, if we put $A_0 = 0$, $A_1 = 1$, we find $A = 0$, $B = 1$, and therefore

$$\sin x = x - \frac{x^3}{3!} + \frac{x^5}{5!} - \ldots. \quad \text{.................(8)}$$

The foregoing method is, for various reasons, not always practicable. It may also lead to a solution which is *incomplete*; thus in the case of the linear equation of the second order, the method may yield only *one* series, with one arbitrary constant. This occurs not infrequently in the physical applications of the subject. The solution may, in this case, be completed, at least symbolically, by the method of Art. 166, 3°.

182. Expansions by means of Differential Equations.

The method of the preceding Art. may sometimes be utilized to obtain the expansion of a given function in a power-series, provided we can form a linear differential equation, with rational integral coefficients, which the function satisfies*.

For example, let
$$y = (1 + x)^m, \qquad\qquad\dots\dots\dots\dots(1)$$
where m may be integral or fractional, positive or negative. Taking logarithms of both sides, and then differentiating, we have
$$\frac{1}{y}\frac{dy}{dx} = \frac{m}{1+x},$$
or
$$(1 + x)\frac{dy}{dx} - my = 0. \qquad\dots\dots\dots\dots(2)$$

Assuming
$$y = A_0 + A_1 x + A_2 x^2 + \dots + A_n x^n + \dots, \qquad\dots\dots(3)$$
and substituting, we have
$$(1 + x)(A_1 + 2A_2 x + \dots + nA_n x^{n-1} + \dots)$$
$$- m(A_0 + A_1 x + A_2 x^2 + \dots + A_n x^n + \dots) = 0,$$
or $(A_1 - mA_0) + \{2A_2 - (m-1)A_1\}x + \{3A_3 - (m-2)A_2\}x^2 + \dots$
$$+ \{nA_n - (m-n+1)A_{n-1}\}x^{n-1} + \dots = 0, \qquad\dots\dots(4)$$
which is satisfied identically provided
$$A_1 = \frac{m}{1}A_0,$$
$$A_2 = \frac{m-1}{2}A_1 = \frac{m(m-1)}{1.2}A_0,$$
$$A_3 = \frac{m-2}{3}A_2 = \frac{m(m-1)(m-2)}{1.2.3}A_0,$$
and, generally,
$$A_n = \frac{m-n+1}{n}A_{n-1} = \frac{m(m-1)\dots(m-n+1)}{1.2.3\dots n}A_0. \dots(5)$$
We thus obtain
$$y = A_0\left\{1 + \frac{m}{1}x + \frac{m(m-1)}{2!}x^2 + \dots\right.$$
$$\left. + \frac{m(m-1)\dots(m-n+1)}{n!}x^n + \dots\right\}, \dots(6)$$
as a solution of (2); and it is easily verified that the series is convergent so long as $|x| < 1$.

---

* This method was first employed by Newton, to whom the series for $\cos x$ and $\sin x$ are also due. The manner of obtaining these series was, however, different.

Now if we retrace the steps by which the differential equation (2) was formed, we see that its complete solution is

$$y = C(1 + x)^m, \quad \dots\dots\dots\dots\dots\dots(7)$$

where C is arbitrary. Hence (6) must be equivalent to (7), and putting $x = 0$ in both, we see that $C = A_0$. Hence

$$(1 + x)^m = 1 + \frac{m}{1}x + \frac{m(m-1)}{2!}x^2 + \dots$$

$$+ \frac{m(m-1)\dots(m-n+1)}{n!}x^n + \dots, \quad \dots\dots(8)$$

for all values of x such that $|x| < 1$. This is the well-known 'Binomial Expansion*.'

Ex. As a further example, we take the function

$$y = \frac{\sin^{-1} x}{\sqrt{(1 - x^2)}}. \quad \dots\dots\dots\dots\dots\dots(9)$$

Multiplying up by $\sqrt{(1 - x^2)}$, and then differentiating, we find

$$\sqrt{(1 - x^2)}\frac{dy}{dx} - \frac{x}{\sqrt{(1 - x^2)}}y = \frac{1}{\sqrt{(1 - x^2)}},$$

or

$$(1 - x^2)\frac{dy}{dx} - xy = 1. \quad \dots\dots\dots\dots\dots(10)$$

Assuming

$$y = A_0 + A_1 x + A_2 x^2 + \dots + A_n x^n + \dots, \quad \dots\dots\dots(11)$$

we find

$$(1 - x^2)(A_1 + 2A_2 x + 3A_3 x^2 + \dots + nA_n x^{n-1} + \dots)$$

$$- x(A_0 + A_1 x + A_2 x^2 + \dots + A_n x^n + \dots) = 1, \dots(12)$$

or

$$(A_1 - 1) + (2A_2 - A_0)x + (3A_3 - 2A_1)x^2 + \dots$$

$$+ \{nA_n - (n-1)A_{n-2}\}x^{n-1} + \dots = 0, \dots(13)$$

which is satisfied identically, provided

$$\left.\begin{array}{ll} A_1 = 1, & A_2 = \dfrac{1}{2}A_0, \\[2mm] A_3 = \dfrac{2}{3}A_1 = \dfrac{2}{3}, & A_4 = \dfrac{3}{4}A_2 = \dfrac{1 \cdot 3}{2 \cdot 4}A_0, \\[2mm] A_5 = \dfrac{4}{5}A_3 = \dfrac{2 \cdot 4}{3 \cdot 5}, & A_6 = \dfrac{5}{6}A_4 = \dfrac{1 \cdot 3 \cdot 5}{2 \cdot 4 \cdot 6}A_0, \\[2mm] \dots\dots\dots\dots & \dots\dots\dots\dots \end{array}\right\} \quad \dots\dots(14)$$

We thus obtain the solution

$$y = x + \frac{2}{3}x^3 + \frac{2 \cdot 4}{3 \cdot 5}x^5 + \frac{2 \cdot 4 \cdot 6}{3 \cdot 5 \cdot 7}x^7 + \dots$$

$$+ A_0\left(1 + \frac{1}{2}x^2 + \frac{1 \cdot 3}{2 \cdot 4}x^4 + \frac{1 \cdot 3 \cdot 5}{2 \cdot 4 \cdot 6}x^6 + \dots\right). \quad \dots(15)$$

---

* Newton (1676). The cases of $x = \pm 1$ would require special investigation.

Now if we retrace the steps by which the linear equation (10) was formed, we see that its general solution is

$$\sqrt{(1-x^2)} \cdot y = \sin^{-1} x + A, \quad \dots\dots\dots\dots(16)^*$$

or

$$y = \frac{\sin^{-1} x}{\sqrt{(1-x^2)}} + \frac{A}{\sqrt{(1-x^2)}}, \quad \dots\dots\dots\dots(17)$$

and, by the nature of the case, (15) must be included in this form. If we put $x = 0$ in (15) and (17), we see that $A = A_0$. The identity of the two expressions for y then requires that

$$\frac{\sin^{-1} x}{\sqrt{(1-x^2)}} = x + \frac{2}{3} x^3 + \frac{2 \cdot 4}{3 \cdot 5} x^5 + \dots, \quad \dots\dots\dots\dots(18)$$

and

$$\frac{1}{\sqrt{(1-x^2)}} = 1 + \frac{1}{2} x^2 + \frac{1 \cdot 3}{2 \cdot 4} x^4 + \dots. \quad \dots\dots\dots\dots(19)$$

These series are both of them convergent for $|x| < 1$.

The result (19) is a mere reproduction of the binomial expansion of $(1-x^2)^{-\frac{1}{2}}$.

If we put $x = \sin\theta$, the former series may be written

$$\theta = \sin\theta\cos\theta\left(1 + \frac{2}{3}\sin^2\theta + \frac{2 \cdot 4}{3 \cdot 5}\sin^4\theta + \dots\right). \quad \dots\dots(20)$$

Again, if we put $\tan\theta = z$, we obtain the form

$$\tan^{-1} z = \frac{z}{1+z^2}\left\{1 + \frac{2}{3}\frac{z^2}{1+z^2} + \frac{2 \cdot 4}{3 \cdot 5}\left(\frac{z^2}{1+z^2}\right)^2 + \dots\right\}. \quad \dots(21)$$

This series has been made the basis of several ingenious methods of calculating π. It may be shewn, for example, that

$$\tfrac{1}{4}\pi = 5\tan^{-1}\tfrac{1}{7} + 2\tan^{-1}\tfrac{3}{79},$$

whence

$$\pi = \frac{28}{10}\left\{1 + \frac{2}{3}\left(\frac{2}{100}\right) + \frac{2 \cdot 4}{3 \cdot 5}\left(\frac{2}{100}\right)^2 + \dots\right\}$$
$$+ \frac{30336}{100000}\left\{1 + \frac{2}{3}\left(\frac{144}{100000}\right) + \frac{2 \cdot 4}{3 \cdot 5}\left(\frac{144}{100000}\right)^3 + \dots\right\}. \quad \dots(22)$$

These series are rapidly convergent, and are otherwise very convenient for computation, owing to the powers of 10 in the denominators†.

Another remarkable series follows by integration from (18), viz.

$$\tfrac{1}{2}(\sin^{-1}x)^2 = \frac{x^2}{2} + \frac{2}{3}\frac{x^4}{4} + \frac{2 \cdot 4}{3 \cdot 5}\frac{x^6}{6} + \dots. \quad \dots\dots\dots\dots(23)$$

---

* See also Art. 158, Ex. 2.
† For the history of these series, see Glaisher, *Mess. of Math.*, t. ii., p. 119 (1873).

EXAMPLES. LIX.

(Logarithmic Series.)

1. If m, n be positive quantities such that $2n > m > n$, prove that $\log (m/n)$ lies between $(m-n)/m$ and $(m-n)/n$.

2. Obtain the following results by calculation from the series of Art. 175:

$$\log 2 = \cdot693\ 147\ 181 \qquad \log\ 7 = 1\cdot945\ 910\ 149$$
$$\log 3 = 1\cdot098\ 612\ 289 \qquad \log\ 8 = 2\cdot079\ 441\ 542$$
$$\log 4 = 1\cdot386\ 294\ 361 \qquad \log\ 9 = 2\cdot197\ 224\ 577$$
$$\log 5 = 1\cdot609\ 437\ 912 \qquad \log 10 = 2\cdot302\ 585\ 093$$
$$\log 6 = 1\cdot791\ 759\ 469 \qquad \mu = \cdot434\ 294\ 482.$$

3. Prove that
$$\log\ 2 = 7a - 2b + 3c,$$
$$\log\ 3 = 11a - 3b + 5c,$$
$$\log\ 5 = 16a - 4b + 7c,$$
and thence
$$\log 10 = 23a - 6b + 10c,$$
where
$$a = \frac{1}{10} + \frac{1}{2\cdot10^2} + \frac{1}{3\cdot10^3} + \ldots = \cdot1053605157,$$
$$b = \frac{4}{100} + \frac{4^2}{2\cdot100^2} + \frac{4^3}{3\cdot100^3} + \ldots = \cdot0408219945,$$
$$c = \frac{1}{80} - \frac{1}{2\cdot80^2} + \frac{1}{3\cdot80^3} - \ldots = \cdot0124225200.$$

Apply this to find $\log 10$. (Adams.)

4. Prove that
$$\log\ 2 = 7P + 5Q + 3R,$$
$$\log\ 3 = 11P + 8Q + 5R,$$
$$\log\ 5 = 16P + 12Q + 7R,$$
and thence
$$\log 10 = 23P + 17Q + 10R,$$
where
$$P = 2\left(\frac{1}{31} + \frac{1}{3\cdot31^3} + \frac{1}{5\cdot31^5} + \ldots\right) = \cdot0645385211,$$
$$Q = 2\left(\frac{1}{49} + \frac{1}{3\cdot49^3} + \frac{1}{5\cdot49^5} + \ldots\right) = \cdot0408219945,$$
$$R = 2\left(\frac{1}{161} + \frac{1}{3\cdot161^3} + \frac{1}{5\cdot161^5} + \ldots\right) = \cdot0124225200.$$

Apply this to find $\log 10$. (Glaisher.)

5. If $|x| < 1$, prove that
$$\tanh^{-1} x = x + \tfrac{1}{3}x^3 + \tfrac{1}{5}x^5 + \ldots.$$

6. Prove that
$$\lim_{n \to \infty} \left(1 + \tfrac{1}{3} + \tfrac{1}{5} + \ldots + \frac{1}{2n-1} - \tfrac{1}{2}\log n \right) = \tfrac{1}{2}\gamma + \log 2.$$

7. If p, q are positive integers, prove that
$$\lim_{n \to \infty} \left(\frac{1}{pn+1} + \frac{1}{pn+2} + \ldots + \frac{1}{qn} \right) = \log\left(\frac{q}{p}\right).$$

8. Prove that if x be large, and positive,
$$\log \cosh x = x - \log 2 + e^{-2x},$$
approximately.

9. Also that
$$\log \tanh x = -2e^{-2x},$$
approximately.

EXAMPLES. LX.

(Differentiation and Integration of Series.)

1. Prove by repeated differentiation of the identity
$$\frac{1}{1-x} = 1 + x + x^2 + x^3 + \ldots,$$
where $|x| < 1$, that, if m be a positive integer
$$(1-x)^{-m} = 1 + mx + \frac{m(m+1)}{1 \cdot 2} x^2 + \frac{m(m+1)(m+2)}{1 \cdot 2 \cdot 3} x^3 + \ldots.$$

2. If $|x| < 1$, prove that
$$\frac{x^3}{1 \cdot 2} - \frac{x^4}{3 \cdot 4} + \frac{x^6}{5 \cdot 6} - \ldots = x \tan^{-1} x - \tfrac{1}{2}\log(1+x^2).$$
Hence shew that
$$1 - \tfrac{1}{2} - \tfrac{1}{3} + \tfrac{1}{4} + \tfrac{1}{5} - \ldots = \cdot 43882 \ldots.$$

3. Prove that if $|x| < 1$
$$\frac{x^3}{1 \cdot 3} - \frac{x^5}{3 \cdot 5} + \frac{x^7}{5 \cdot 7} - \ldots = \tfrac{1}{2}(1+x^2)\tan^{-1} x - \tfrac{1}{2}x.$$

4. Prove that the sum of the series
$$\frac{1}{1 \cdot 3 \cdot 5} + \frac{1}{5 \cdot 7 \cdot 9} + \frac{1}{7 \cdot 9 \cdot 11} + \ldots$$
is $\cdot 071349 \ldots.$

5. Prove that

$$\frac{1}{1.2.3} + \frac{1}{3.4.5} + \frac{1}{5.6.7} + \dots = \cdot 193147\dots,$$

$$\frac{1}{1.2.3} - \frac{1}{3.4.5} + \frac{1}{5.6.7} - \dots = \cdot 153426\dots.$$

6. Prove that

$$\int_0^1 \frac{x^{m-1}}{1+x^n} \, dx = \frac{1}{m} - \frac{1}{m+n} + \frac{1}{m+2n} - \frac{1}{m+3n} + \dots.$$

7. Prove that

$$\int_0^x \frac{\sin x}{x} \, dx = x - \frac{x^3}{3.3!} + \frac{x^5}{5.5!} - \dots,$$

$$\int_0^x \frac{\sinh x}{x} \, dx = x + \frac{x^3}{3.3!} + \frac{x^5}{5.5!} + \dots.$$

8. Prove that

$$\int_0^1 \frac{dx}{\sqrt{(1-x^4)}} = 1 + \frac{1}{2} \cdot \frac{1}{5} + \frac{1.3}{2.4} \cdot \frac{1}{9} + \frac{1.3.5}{2.4.6} \cdot \frac{1}{13} + \dots.$$

9. Prove that

$$\int_a^b \frac{e^x}{x} \, dx = \log \frac{b}{a} + (b-a) + \frac{b^2-a^2}{2.2!} + \frac{b^3-a^3}{3.3!} + \dots.$$

10. Prove that

$$\pi = 2\sqrt{3} \left(1 - \frac{1}{3.3} + \frac{1}{5.3^2} - \frac{1}{7.3^3} + \dots \right).$$

EXAMPLES. LXI.

(Integration of Differential Equations by Series.)

1. Assuming the series for $\sin x$, prove Huyghens' rule for calculating approximately the length of a circular arc, viz.: From eight times the chord of half the arc subtract the chord of the whole arc, and divide the result by three.

Prove that in an arc of 45° the proportional error is less than 1 in 20000.

2. Obtain a particular solution of the equation

$$x \frac{d^2y}{dx^2} + \frac{dy}{dx} + my = 0$$

in the form

$$y = A \left(1 - \frac{mx}{1^2} + \frac{m^2x^2}{1^2.2^2} - \frac{m^3x^3}{1^2.2^2.3^2} + \dots \right).$$

3. Obtain a particular solution of the equation

$$\frac{d^2\phi}{dr^2} + \frac{1}{r}\cdot\frac{d\phi}{dr} + k^2\phi = 0$$

in the form $\qquad \phi = A\left(1 - \frac{k^2 r^2}{2^2} + \frac{k^4 r^4}{2^2 \cdot 4^2} - \ldots\right).$

4. Integrate $\qquad (1 - x^2)\dfrac{d^2y}{dx^2} - x\dfrac{dy}{dx} = 0,$

by series, and deduce the expansion of $\sin^{-1}x$ (Art. 180 (5)).

5. Prove that $\qquad y = \sinh^{-1}x$

satisfies the differential equation

$$(1 + x^2)\frac{d^2y}{dx^2} + x\frac{dy}{dx} = 0.$$

Hence shew that, for $|x| < 1$,

$$\log\{x + \sqrt{(1 + x^2)}\} = x - \frac{1}{2}\frac{x^3}{3} + \frac{1 \cdot 3}{2 \cdot 4}\frac{x^5}{5} - \ldots\ .$$

6. Obtain a solution of the equation

$$x\frac{d^2y}{dx^2} + (a - x)\frac{dy}{dx} - y = 0$$

in the form $y = Cu$, where

$$u = 1 + \frac{x}{a} + \frac{x^2}{a\,(a+1)} + \frac{x^3}{a\,(a+1)\,(a+2)} + \ldots\ .$$

Prove that the equation

$$x\frac{d^2y}{dx^2} + (a + x)\frac{dy}{dx} - (1 - a)\,y = 0$$

is satisfied by $y = Ce^{-x}u$.

7. Obtain a solution of the equation

$$\frac{d}{d\mu}\left\{(1 - \mu^2)\frac{du}{d\mu}\right\} + n\,(n+1)\,u = 0$$

in the form

$$u = A\left(1 - \frac{n\,(n+1)}{2\,!}\mu^2 + \frac{(n-2)\,n\,(n+1)\,(n+3)}{4\,!}\mu^4 - \ldots\right)$$

$$+ B\left(\mu - \frac{(n-1)\,(n+2)}{3\,!}\mu^3 + \frac{(n-3)\,(n-1)\,(n+2)\,(n+4)}{5\,!}\mu^5 - \ldots\right).$$

8. Obtain a solution, by a series, of

$$(1 - x^2)\frac{d^2y}{dx^2} + n(n-1)y = 0,$$

and give the symbolical expression of the complete solution.

9. Obtain a solution of the equation

$$x(1-x)\frac{d^2y}{dx^2} + \{\gamma - (a+\beta+1)x\}\frac{dy}{dx} - a\beta y = 0$$

in the form

$$y = A\left(1 + \frac{a\beta}{1\cdot\gamma}x + \frac{a(a+1)\beta(\beta+1)}{1\cdot2\cdot\gamma(\gamma+1)}x^2\right.$$
$$\left. + \frac{a(a+1)(a+2)\beta(\beta+1)(\beta+2)}{1\cdot2\cdot3\cdot\gamma(\gamma+1)(\gamma+2)}x^3 + \dots\right).$$

10. Obtain a solution of the equation

$$\frac{d^2\phi}{dr^2} + \frac{1}{r}\frac{d\phi}{dr} + \left(k^2 - \frac{n^2}{r^2}\right)\phi = 0$$

in the form

$$\phi = Ar^n\left(1 - \frac{k^2r^2}{2(2n+2)} + \frac{k^4r^4}{2\cdot4(2n+2)(2n+4)} - \dots\right).$$

11. Obtain the solution of

$$\frac{d^2R}{dr^2} + \frac{2(n+1)}{r}\frac{dR}{dr} + k^2R = 0$$

in the form

$$R = A\left(1 - \frac{k^2r^2}{2(2n+3)} + \frac{k^4r^4}{2\cdot4(2n+3)(2n+5)} - \dots\right)$$
$$+ Br^{-2n-1}\left(1 - \frac{k^2r^2}{2(1-2n)} + \frac{k^4r^4}{2\cdot4(1-2n)(3-2n)} - \dots\right).$$

12. If

$$y = \sin(m\sin^{-1}x),$$

prove that

$$(1-x^2)\frac{d^2y}{dx^2} - x\frac{dy}{dx} + m^2y = 0.$$

Hence shew that

$$\frac{\sin m\theta}{m\sin\theta} = 1 - \frac{m^2-1}{3!}\sin^2\theta + \frac{(m^2-1)(m^2-3^2)}{5!}\sin^4\theta - \dots,$$

$$\cos m\theta = 1 - \frac{m^2}{2!}\sin^2\theta + \frac{m^2(m^2-2^2)}{4!}\sin^4\theta - \dots.$$

13.   If $\log y = a \sin^{-1} x$, prove that

$$(1 - x^2) \frac{d^2y}{dx^2} = x \frac{dy}{dx} + a^2 y \; ;$$

and expand y in a series of ascending powers of x.

$$\left[y = 1 + ax + \frac{a^2 x^2}{2\,!} + \frac{a\,(a^2 + 1^2)}{3\,!}\,x^3 + \frac{a^2\,(a^2 + 2^2)}{4\,!}\,x^4 + \dots \right]$$

14.   If
$$y = \frac{\log\,(1 + x)}{1 + x},$$

prove that
$$(1 + x)^2 \frac{dy}{dx} + (1 + x)\,y = 1.$$

Hence shew that

$$y = x - \left(1 + \tfrac{1}{2}\right) x^2 + \left(1 + \tfrac{1}{2} + \tfrac{1}{3}\right) x^3 - \dots.$$

Shew that the series is convergent if $|x| < 1$.

15.   Prove that if $|x| < 1$,

$$(\tan^{-1} x)^2 = x^2 - \tfrac{1}{2}\left(1 + \tfrac{1}{3}\right) x^4 + \tfrac{1}{3}\left(1 + \tfrac{1}{3} + \tfrac{1}{5}\right) x^6 - \dots.$$

CHAPTER XV

TAYLOR'S THEOREM

183. Form of the Expansion.

Let $f(x)$ be any function of x which admits of expansion in a convergent power-series for all values of x within certain limits $\pm \alpha$. It has been proved, in Art. 179, that the derived function $f'(x)$ will be given by a similar series, obtained by differentiating the original series term by term, for all values of x between $\pm \alpha$. By a second application of the theorem cited, the value of $f''(x)$ will be obtained, for values of x between the above limits, by differentiating the series for $f'(x)$ term by term. And so on. Hence, writing

$$f(x) = A_0 + A_1 x + A_2 x^2 + \ldots + A_n x^n + \ldots, \quad \ldots\ldots\ldots\ldots\ldots(1)$$

we have

$$\left.\begin{aligned}
f'(x) &= A_1 + 2A_2 x + \ldots + nA_n x^{n-1} + \ldots, \\
f''(x) &= 2 . 1 A_2 + \ldots + n(n-1) A_n x^{n-2} + \ldots, \\
&\ldots\ldots\ldots\ldots\ldots\ldots\ldots\ldots\ldots\ldots\ldots\ldots\ldots\ldots\ldots\ldots\ldots\ldots\ldots\ldots\ldots\ldots\ldots\ldots \\
f^{(n)}(x) &= n(n-1) \ldots 2 . 1 A_n + \ldots,
\end{aligned}\right\} \ldots(2)$$

Putting $x = 0$ in these equations, we find

$$A_0 = f(0),\, A_1 = f'(0),\, A_2 = \frac{1}{2!} f''(0), \ldots,\, A_n = \frac{1}{n!} f^{(n)}(0), \ldots, \ldots(3)$$

where the symbols $f(0), f'(0), f''(0), \ldots$ are used to express that x is put $= 0$ *after* the differentiations have been performed.

The original expansion may now be written

$$f(x) = f(0) + x f'(0) + \frac{x^2}{2!} f''(0) + \ldots + \frac{x^n}{n!} f^{(n)}(0) + \ldots. \quad \ldots(4)$$

This investigation was given by Maclaurin*.

---

* *Treatise on Fluxions* (1742). The theorem had been previously noticed by Stirling.

It will be noticed that the proof depends entirely on the initial assumption that the function $f(x)$ admits of being expanded in a convergent power-series. The question as to when, and under what limitations, such an expansion is possible will be discussed presently (Arts. 185—187).

If we write $$\phi(a + x) = f(x), \dots\dots\dots\dots\dots\dots(5)$$
we can deduce the form of the expansion of $\phi(a + x)$ in a power-series, when such expansion is possible. For if we write, for a moment,
$$u = a + x,$$
we have $\qquad f(x) = \phi(u),$

$$f'(x) = \frac{d}{dx}\phi(u) = \frac{d}{du}\phi(u) \cdot \frac{du}{dx} = \phi'(u),$$

$$f''(x) = \frac{d}{dx}\phi'(u) = \frac{d}{du}\phi'(u) \cdot \frac{du}{dx} = \phi''(u),$$

and so on (Art. 32, 1°). Hence, putting $x = 0$, $u = a$, we find
$$f(0) = \phi(a), f'(0) = \phi'(a), f''(0) = \phi''(a), \dots, f^{(n)}(0) = \phi^{(n)}(a);$$
$$\dots\dots\dots\dots(6)$$
so that (4) takes the form

$$\phi(a + x) = \phi(a) + x\phi'(a) + \frac{x^2}{1 \cdot 2}\phi''(a) + \dots + \frac{x^n}{n!}\phi^{(n)}(a) + \dots. \ (7)$$

This is known as Taylor's Theorem[*]. We have deduced it from Maclaurin's Theorem, but the two theorems are only slightly different expressions of the same result. Thus assuming (7), we deduce Maclaurin's expansion if we put $a = 0$[†].

184. Particular Cases.

Before proceeding to a more fundamental treatment of the problem suggested in the preceding Art., the student will do well to make himself familiar with the mode of formation of the series. In the following examples the possibility of the expansion is assumed to begin with; and the results obtained are therefore not to be considered as established, at all events by this method.

1°. If $$\phi(a) = a^m, \dots\dots\dots\dots\dots\dots(1)$$
we have
$$\phi'(a) = ma^{m-1}, \ \phi''(a) = m(m-1)a^{m-2}, \dots,$$
$$\phi^{(n)}(a) = m(m-1)\dots(m-n+1)a^{m-n}, \dots\dots(2)$$

---

[*] Given (under a slightly different form) as a corollary from a theorem in Finite Differences, *Methodus Incrementorum* (1716).

[†] The virtual identity of (4) with Taylor's Theorem was clearly recognized by Maclaurin.

Taylor's formula then gives

$$(a+x)^m = a^m + ma^{m-1}x + \frac{m(m-1)}{1 \cdot 2} a^{m-2}x^2 + \dots$$

$$+ \frac{m(m-1)\dots(m-n+1)}{1 \cdot 2 \dots m} a^{m-n}x^n + \dots, \dots (3)$$

which is the well-known Binomial Expansion.

That Taylor's Theorem cannot hold in all cases, without quali-fication, is shewn by the fact that the series on the right-hand is divergent if $|x| > |a|$. For $|x| < |a|$ the series is convergent, but it is not legitimate to affirm on the basis of the investigation of Art. 183 that its sum is then equal to $(a+x)^m$*. A valid proof of the equality has been given in Art. 182.

2°. The exponential function $E(x)$ was defined in Art. 36 as that solution of the equation

$$f'(x) = f(x) \dots (4)$$

which is equal to unity for $x = 0$. From this we have at once

$$f^{(n)}(x) = f^{(x)}. \dots (5)$$

Hence $\qquad f(0) = 1, \quad f^{(n)}(0) = 1. \dots (6)$

Maclaurin's expansion is therefore

$$E(x) = 1 + x + \frac{x^2}{1 \cdot 2} + \frac{x^3}{1 \cdot 2 \cdot 3} + \dots + \frac{x^n}{n!} + \dots (7)$$

3°. Let $\qquad f(x) = \cos x. \dots (8)$

It was shewn in Art. 64 that this makes

$$f^{(n)}(x) = \cos(x + \tfrac{1}{2}n\pi),$$

so that $\qquad f(0) = 1, \quad f^{(n)}(0) = \cos \tfrac{1}{2}n\pi. \dots (9)$

Hence $f^{(n)}(0)$ vanishes when n is odd, and is equal to ± 1 when n is even, according as $\tfrac{1}{2}n$ is even or odd. Substituting in Maclaurin's formula, we get

$$\cos x = 1 - \frac{x^2}{2!} + \frac{x^4}{4!} - \dots + (-)^n \frac{x^{2n}}{2n!} + \dots (10)$$

4°. Let $\qquad f(x) = \sin x. \dots (11)$

This makes $\qquad f^{(n)}(x) = \sin(x + \tfrac{1}{2}n\pi),$

so that

$$f(0) = 0, \quad f^{(n)}(0) = \sin \tfrac{1}{2}n\pi = \sin\{\tfrac{1}{2}(n-1)\pi + \tfrac{1}{2}\pi\}. \dots (12)$$

---

* There are in fact cases where Taylor's expansion is convergent, whilst the sum is *not* equal to $\phi(a+x)$.

Hence $f^{(n)}(0)$ vanishes when n is even, and is equal to ± 1 when n is odd, according as $\frac{1}{2}(n-1)$ is even or odd. Maclaurin's formula then gives

$$\sin x = x - \frac{x^3}{3!} + \frac{x^5}{5!} - \dots + (-)^n \frac{x^{2n+1}}{(2n+1)!} + \dots \dots (13)$$

The results (10) and (13) have been established rigorously in Art. 181.

5°.   Let $\qquad\qquad f(x) = \log(1+x). \quad\dots\dots\dots\dots\dots\dots(14)$

This makes

$$f'(x) = \frac{1}{1+x}, \text{ and, for } n > 1, f^{(n)}(x) = \frac{(-)^{n-1}(n-1)!}{(1+x)^n}.$$

Hence $f(0) = 0, f'(0) = 1$, and, for $n > 1$,

$$f^{(n)}(0) = (-)^{n-1}(n-1)! \quad\dots\dots\dots\dots\dots(15)$$

Substituting in Maclaurin's formula, we get

$$\log(1+x) = x - \frac{x^2}{2} + \frac{x^3}{3} - \dots + (-)^{n-1}\frac{x^n}{n} + \dots \dots (16)$$

Cf. Art. 175.

When a general formula for the nth derivative of the given function is not known, the only plan is to calculate the derivatives in succession as far as may be considered necessary. The later stages of the work may sometimes be contracted by omitting terms which will contribute nothing to the final result, so far as it is proposed to carry it.

Ex.   To expand $\tan x$ as far as x^7.

Putting $\qquad\qquad f(x) = \tan x,$

we find in succession

$$f'(x) = \sec^2 x = 1 + \tan^2 x,$$
$$f''(x) = 2 \tan x \sec^2 x = 2 \tan x + 2 \tan^3 x,$$
$$f'''(x) = (2 + 6 \tan^2 x) \sec^2 x = 2 + 8 \tan^2 x + 6 \tan^4 x,$$
$$f^{iv}(x) = (16 \tan x + 24 \tan^3 x) \sec^2 x$$
$$= 16 \tan x + 40 \tan^3 x + 24 \tan^5 x,$$
$$f^{v}(x) = (16 + 120 \tan^2 x + 120 \tan^4 x) \sec^2 x$$
$$= 16 + 136 \tan^2 x + 240 \tan^4 x + 120 \tan^6 x,$$
$$f^{vi}(x) = 272 \tan x \sec^2 x + \&c.,$$
$$f^{vii}(x) = 272 \sec^4 x + \&c.,$$

where, in the last two lines, terms have been omitted which will contribute nothing to the value of $f^{\text{vii}}(0)$. Hence

$$f \quad (0) = 0, \qquad f' \ (0) = 1,$$
$$f'' (0) = 0, \qquad f''' (0) = 2,$$
$$f^{\text{iv}} (0) = 0, \qquad f^{\text{v}} \ (0) = 16,$$
$$f^{\text{vi}} (0) = 0, \qquad f^{\text{vii}} (0) = 272,$$

and the expansion is

$$\tan x = x + \frac{2x^3}{3!} + \frac{16x^5}{5!} + \frac{272x^7}{7!} + \dots$$

$$= x + \tfrac{1}{3} x^3 + \tfrac{2}{15} x^5 + \tfrac{17}{315} x^7 + \dots \, . \quad \dots\dots\dots(17)$$

That *odd* powers, only, of x would appear in this expansion might have been anticipated from the fact that $\tan x$ changes sign with x.

185. Proof of Maclaurin's and Taylor's Theorems. Remainder after n Terms.

Let $f(x)$ be a function of x which, together with its first $n-1$ derivatives, is continuous for values of x ranging from 0 to h, inclusively; and let us write

$$f(x) = \Phi_n(x) + R_n(x), \quad \dots\dots\dots\dots\dots(1)$$
where

$$\Phi_n(x) = f(0) + xf'(0) + \frac{x^2}{2!} f''(0) + \dots + \frac{x^{n-1}}{(n-1)!} f^{(n-1)}(0); \dots(2)$$

i.e. $\Phi_n(x)$ is the sum of the first n terms of Maclaurin's expansion, and $R_n(x)$ is at present merely a symbol for the difference, whatever it is, between $f(x)$ and $\Phi_n(x)$. The object aimed at, in any rigorous investigation of Maclaurin's Theorem, is to find (if possible) limits to the value of $R_n(x)$; in other words, to find limits to the error committed when $f(x)$ is replaced by the sum of the first n terms of Maclaurin's formula. If we can, in any given case, shew that, by taking n great enough, a point can always be reached after which the values of $R_n(x)$ will all be less than any assigned magnitude, however small, then Maclaurin's series is necessarily convergent, and its sum to infinity is $f(x)$. It is evident that the argument cannot be pushed to this conclusion if $f(x)$ or any of its derivatives be discontinuous for any value of x belonging to the range considered.

The notion of representing a function $f(x)$ approximately by a rational integral function of assigned degree, say

$$A_0 + A_1 x + A_2 x^2 + \dots + A_{n-1} x^{n-1}, \quad \dots\dots\dots\dots(3)$$

has already been utilized in Art. 114. The plan there adopted was to determine the n coefficients $A_0, A_1, A_2, \dots A_{n-1}$ so that the function (3) should be equal to $f(x)$ for n assigned values of x, which were distributed at equal intervals over a certain range. In the present case, the n values of x are taken to be ultimately coincident with 0; in other words, the coefficients are chosen so as to make the function (3) and its first $n-1$ derivatives coincide respectively with $f(x)$ and its first $n-1$ derivatives for the particular value $x=0$. The result of this determination is, by Art. 183, the function $\Phi_n(x)$.

In the graphical representation, the parabolic curve $y=\Phi_n(x)$ is determined so as to have contact of the $(n-1)$th order (see Art. 189) with a given curve $y=f(x)$ at the point $x=0$; and the problem is, to find limits to the possible deviation of one curve from the other, as measured by the difference of the ordinates, for values of x lying within a certain range. This is illustrated by Fig. 136, which shews the curve

$$y = \log(1+x), \quad \dots\dots\dots\dots\dots\dots(4)$$

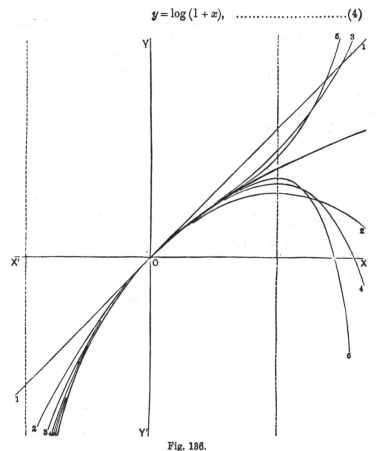

Fig. 136.

and (by thinner lines) the 'approximation curves'

$$y = x, \quad y = x - \tfrac{1}{2}x^2, \quad y = x - \tfrac{1}{2}x^2 + \tfrac{1}{3}x^3, \ldots, \quad \ldots\ldots\ldots(5)$$

obtained by taking 1, 2, 3, ... terms of the 'logarithmic series' (Art. 184 (16)). The dotted lines correspond to $x = \pm 1$, and so mark out the range of convergence of the latter series.

It appears, from the conditions which $\Phi_n(x)$ has been made to satisfy, that $R_n(x)$ and its first $n-1$ derivatives will be continuous from $x = 0$ to $x = h$, and will all vanish for $x = 0$. Now we can shew that any function which satisfies *these* conditions, and has a finite nth derivative, must lie between

$$A \frac{x^n}{n!} \quad \text{and} \quad B \frac{x^n}{n!},$$

where A and B are the lower and upper limits to the values which the nth derivative assumes in the interval from 0 to h.

For, let $F(x)$ be such a function. By hypothesis, we have

$$F(0) = 0, \quad F'(0) = 0, \quad F''(0) = 0, \quad \ldots, \quad F^{(n-1)}(0) = 0, \ldots(6)$$

and

$$A < F^{(n)}(x) < B, \ldots\ldots\ldots\ldots\ldots\ldots(7)$$

the latter condition holding from $x = 0$ to $x = h$. It follows from (7), by Art. 91, 4°, that

$$\int_0^x A \, dx < \int_0^x F^{(n)}(x) \, dx < \int_0^x B \, dx^*,$$

or, since $F^{(n-1)}(0) = 0$,

$$Ax < F^{(n-1)}(x) < Bx. \ldots\ldots\ldots\ldots\ldots(8)$$

By a second application of the theorem referred to, we have

$$\int_0^x Ax \, dx < \int_0^x F^{(n-1)}(x) \, dx < \int_0^x Bx \, dx,$$

or, since $F^{(n-2)}(0) = 0$,

$$A \frac{x^2}{2!} < F^{(n-2)}(x) < B \frac{x^2}{2!}. \ldots\ldots\ldots\ldots(9)$$

A similar argument applies to shew that

$$A \frac{x^3}{3!} < F^{(n-3)}(x) < B \frac{x^3}{3!}, \ldots\ldots\ldots\ldots(10)$$

and so on, until we arrive at the result

$$A \frac{x^n}{n!} < F(x) < B \frac{x^n}{n!}. \ldots\ldots\ldots\ldots(11)$$

---

* Provided x be positive. If x be negative the inequalities must be reversed, but the final conclusion is unaffected.

Hence we may write

$$F(x) = C\frac{x^n}{n!}, \qquad \ldots\ldots\ldots\ldots\ldots(12)$$

where C is some quantity between A and B.

In the present application, since $\Phi_n(x)$ is a rational integral function of degree $n-1$, its nth derivative is zero (Art. 64); and the nth derivative of $R_n(x)$ is therefore, by (1), equal to $f^{(n)}(x)$, if this latter derivative exists. We infer, then, that

$$R_n(x) = C\frac{x^n}{n!}, \qquad \ldots\ldots\ldots\ldots\ldots(13)$$

where C is some quantity intermediate to the greatest and least values which $f^{(n)}(x)$ assumes in the interval from 0 to h. And if, as we will suppose, this latter derivative is continuous from $x = 0$ to $x = h$, there will be some value of x, between 0 and h, for which $f^{(n)}(x)$ is equal to C. Denoting this value by θh, we have

$$R_n(x) = \frac{x^n}{n!}f^{(n)}(\theta h), \qquad \ldots\ldots\ldots\ldots(14)$$

where all we know as to the value of θ is that it lies between 0 and 1.

The formula (14) holds from $x = 0$ to $x = h$, inclusively. Putting $x = h$, and substituting in (1), we obtain

$$f(h) = f(0) + hf'(0) + \frac{h^2}{2!}f''(0) + \ldots + \frac{h^{n-1}}{(n-1)!}f^{(n-1)}(0)$$
$$+ \frac{h^n}{n!}f^{(n)}(\theta h). \qquad \ldots\ldots\ldots(15)$$

In this form Maclaurin's Theorem is exact, subject to the hypothesis that $f(x)$ and its derivatives up to the order n, inclusively, are continuous over the range from 0 to h. The conditions, however, that $f^{(n)}(x)$ is to exist, and to be continuous over the above range, include the rest.

If we write

$$f(x) = \phi(a + x), \qquad \ldots\ldots\ldots\ldots(16)$$

we deduce

$$\phi(a+h) = \phi(a) + h\phi'(a) + \frac{h^2}{2!}\phi''(a) + \ldots + \frac{h^{n-1}}{(n-1)!}\phi^{(n-1)}(a)$$
$$+ \frac{h^n}{n!}\phi^{(n)}(a + \theta h), \qquad \ldots\ldots\ldots(17)$$

where as before, $1 > \theta > 0$.

This is an accurate form of Taylor's Theorem. It holds on the assumption that $\phi^{(n)}(x)$ exists and is continuous from $x = a$ to $x = a + h$, inclusively.

The last terms in (15) and (17) are known as Lagrange's forms of the 'remainder' in the respective theorems.

The formula (17) is a generalization of some results obtained in the course of this treatise. For example, putting $n = 1$, we get

$$\phi(a + h) = \phi(a) + h\phi'(a + \theta h) ; \quad \ldots\ldots\ldots(18)$$

and, putting $n = 2$,

$$\phi(a + h) = \phi(a) + h\phi'(a) + \tfrac{1}{2}h^2\phi''(a + \theta h). \quad \ldots\ldots(19)$$

These agree with Art. 56 (9) and Art. 70 (23), respectively.

186. Another Proof.

The proof of Taylor's (or Maclaurin's) theorem which is most frequently given follows the lines of Art. 70, 2°.

Considering any given curve

$$y = f(x), \quad \ldots\ldots\ldots\ldots\ldots\ldots(1)$$

we compare with it the curve

$$y = A_0 + A_1 x + A_2 x^2 + \ldots + A_{n-1}x^{n-1} + A_n x^n, \quad \ldots\ldots(2)$$

in which the $n + 1$ coefficients are assumed to be determined so as to make the two curves intersect at $x = 0$ and $x = h$, and, further, so as to make the values of

$$\frac{dy}{dx}, \quad \frac{d^2 y}{dx^2}, \quad \ldots \quad \frac{d^{n-1} y}{dx^{n-1}}$$

respectively the same in the two curves at the point $x = 0$. These conditions give

$$A_0 = f(0), \quad A_1 = f'(0), \quad A_2 = \frac{1}{2!}f''(0), \ldots,$$

$$A_{n-1} = \frac{1}{(n-1)!}f^{(n-1)}(0), \ldots\ldots\ldots(3)$$

as before, and

$$f(h) = A_0 + A_1 h + A_2 h^2 + \ldots + A_{n-1}h^{n-1} + A_n h^n, \quad \ldots(4)$$

this latter equation determining A_n.

Denoting by $F(x)$ the difference of the ordinates of the two curves, it appears that

$$F(0) = 0, \quad F'(0) = 0, \quad F''(0) = 0, \ldots F^{(n-1)}(0) = 0, \ldots(5)$$

and

$$F(h) = 0. \quad \ldots\ldots\ldots\ldots\ldots\ldots(6)$$

Since $F(x)$ vanishes for $x = 0$ and $x = h$, it follows, under the usual conditions, that $F'(x)$ vanishes for some value of x between 0 and h, say for $x = \theta_1 h$, where $1 > \theta_1 > 0$. Again, since $F'(x)$ vanishes for $x = 0$ and $x = \theta_1 h$, $F''(x)$ will vanish for some value of x between 0 and $\theta_1 h$, say for $x = \theta_2 h$, where $\theta_1 > \theta_2 > 0$. Proceeding in this way we find that $F^{(n-1)}(x)$ vanishes for $x = 0$ and $x = \theta_{n-1}h$, where $1 > \theta_{n-1} > 0$, and hence that

$$F^{(n)}(\theta h) = 0, \quad \dots\dots\dots\dots\dots\dots(7)$$

where $1 > \theta > 0$. Now, on reference to (1) and (2), we see that

$$F^{(n)}(x) = f^{(n)}(x) - n! \, A_n. \quad \dots\dots\dots\dots(8)$$

It follows from (7) that

$$A_n = \frac{1}{n!} f^{(n)}(\theta h). \quad \dots\dots\dots\dots\dots(9)$$

Hence, substituting from (3) and (9) in (4) we obtain

$$f(h) = f(0) + hf'(0) + \frac{h^2}{2!}f''(0) + \dots + \frac{h^{n-1}}{(n-1)!}f^{(n-1)}(0)$$
$$+ \frac{h^n}{n!}f^{(n)}(\theta h), \quad \dots\dots\dots\dots(10)$$

as before. The conditions of validity are as stated in Art. 185, after equation (15)*.

187. Cauchy's Form of Remainder.

Another form for the remainder after n terms may be obtained as follows.

If $F(x)$ be subject to the conditions

$$F(0) = 0, \; F'(0) = 0, \; F''(0) = 0, \; \dots, \; F^{(n-1)}(0) = 0, \; \dots(1)$$

we have, by integration by parts,

$$\int_0^h \left(1 - \frac{x}{h}\right)^{n-1} F^{(n)}(x)\,dx = \left[\left(1 - \frac{x}{h}\right)^{n-1} F^{(n-1)}(x)\right]_0^h$$
$$+ \frac{n-1}{h}\int_0^h \left(1 - \frac{x}{h}\right)^{n-2} F^{(n-1)}(x)\,dx = \frac{n-1}{h}\int_0^h \left(1 - \frac{x}{h}\right)^{n-2} F^{(n-1)}(x)\,dx,$$
$$\dots\dots\dots(2)$$

since the integrated term vanishes at both limits. Performing this process $n - 1$ times, we obtain

$$\int_0^h \left(1 - \frac{x}{h}\right)^{n-1} F^{(n)}(x)\,dx = \frac{(n-1)!}{h^{n-1}}\int_0^h F'(x)\,dx = \frac{(n-1)!}{h^{n-1}} F(h),$$

or

$$F(h) = \frac{h^{n-1}}{(n-1)!}\int_0^h \left(1 - \frac{x}{h}\right)^{n-1} F^{(n)}(x)\,dx. \quad \dots\dots\dots(3)$$

* The foregoing proof is substantially that given by Homersham Cox, *Camb. and Dub. Math. Journ.*, 1851.

Since the function under the $\int$ sign is continuous, we infer, by Art. 91, 3°, that

$$F(h) = \frac{h^n}{(n-1)!}(1-\theta)^{n-1} F^{(n)}(\theta h), \quad \ldots\ldots\ldots\ldots(4)$$

where $1 > \theta > 0$.

It appears, then, that the last term in Art. 185 (15) may be replaced by

$$\frac{h^n}{(n-1)!}(1-\theta)^{n-1} f^{(n)}(\theta h), \quad \ldots\ldots\ldots\ldots\ldots(5)$$

and the last term in (17) by

$$\frac{h^n}{(n-1)!}(1-\theta)^{n-1} \phi^{(n)}(a+\theta h). \quad \ldots\ldots\ldots\ldots(6)$$

These forms of remainder are due to Cauchy.

188. Derivation of Certain Expansions.

We proceed to consider the value of the remainder for various forms of $f(x)$, or $\phi(x)$; and in particular to examine under what circumstances it tends, with increasing n, to the limit 0. In this way we are enabled to demonstrate several very important expansions; but it is right to warn the student that the method has a somewhat restricted application, since the general form of the nth derivative of a given function can be ascertained in only a few cases. Moreover, even when the method is successful, it is often far from being the most instructive way of arriving at the final result.

1°. If
$$f(x) = \cos x, \quad \ldots\ldots\ldots\ldots\ldots\ldots(1)$$

we have
$$\frac{x^n}{n!} f^{(n)}(\theta x) = \frac{x^n}{n!} \cos(\theta x + \tfrac{1}{2}n\pi). \quad \ldots\ldots\ldots\ldots(2)$$

The limiting value of the fraction $x^n/n!$ is zero, and the cosine lies always between ± 1. Hence the expansion (10) of Art. 184 holds for all values of x.

The same reasoning applies in the case of $\sin x$.

2°. If
$$f(x) = (1+x)^m, \quad \ldots\ldots\ldots\ldots\ldots(3)$$

we find
$$\frac{x^n}{n!} f^{(n)}(\theta x) = \frac{m(m-1)\ldots(m-n+1)}{1.2\ldots n} \cdot \frac{x^n}{(1+\theta x)^{n-m}} \ldots(4)$$

This may be regarded as the product of $(1+\theta x)^m$ into n factors of the type

$$\frac{m-r+1}{r} \cdot \frac{x}{1+\theta x} \quad \text{or} \quad \left(-1+\frac{m+1}{r}\right)\frac{x}{1+\theta x}. \quad \ldots\ldots(5)$$

If $1 > x > 0$, the fraction $x/(1 + \theta x)$ lies between 0 and x, and since the first factor in (5) tends with increasing r to the limiting value -1, it appears that by taking n great enough the value of the expression (4) can be made less than any assignable magnitude. Hence, for $1 > x > 0$, we may write

$$(1+x)^m = 1 + mx + \frac{m(m-1)}{1.2} x^2 + \ldots\ldots\ldots\ldots(6)$$

ad infinitum.

We cannot make the same inference when x is negative, even if $|x| < 1$. For if we put $x = -x_1$, the fraction $x_1/(1 - \theta x_1)$ is less than 1 only if $\theta < (1 - x_1)/x_1$. And if $x_1 > \frac{1}{2}$, we have no warrant for assuming that θ lies below this value.

Cauchy's form of remainder (Art. 187 (5)) is now of service. We have, in place of (4),

$$\frac{m(m-1)\ldots(m-n+1)}{1.2\ldots(n-1)} \frac{(1-\theta)^{n-1}x^n}{(1+\theta x)^{n-m}} \cdot \ldots\ldots\ldots(7)$$

This is equal to $mx(1 + \theta x)^{m-1}$ multiplied into $n-1$ factors of the type

$$\left(-1 + \frac{m}{r}\right) \frac{x - \theta x}{1 + \theta x} \cdot \ldots\ldots\ldots\ldots\ldots(8)$$

If x be positive this expression tends to a limit between 0 and $-x$. Hence if $x < 1$, the remainder tends to the limit 0, as before.

If $x = -x_1$, where $1 > x_1 > 0$, the expression (8) becomes

$$\left(1 - \frac{m}{r}\right) \frac{1-\theta}{1 - \theta x_1} x_1, \ldots\ldots\ldots\ldots\ldots(9)$$

which tends to a limit between 0 and x_1. We conclude that the remainder (7) tends to the limit 0 for all values of x between -1 and 1.

3°. If $\qquad f(x) = \log(1 + x), \ldots\ldots\ldots\ldots\ldots(10)$

we find $\qquad \frac{x^n}{n!} f^{(n)}(\theta x) = \frac{(-1)^{n-1}}{n} \left(\frac{x}{1 + \theta x}\right)^n. \ldots\ldots\ldots(11)$

The limiting value of the first factor is 0, and, if x be positive and $\not> 1$, $x/(1 + \theta x) \not> 1$. Hence the limiting value of (11), for $n \to \infty$, is zero, and the expansion (16) of Art. 184 is valid from $x = 0$ to $x = 1$, inclusively. Cf. Fig. 136.

The above form of remainder does not enable us to determine the case of x negative, even when $|x| < 1$. In Cauchy's form, we have, in place of (11)

$$(-1)^{n-1} \frac{x}{1 + \theta x} \left(\frac{x - \theta x}{1 + \theta x}\right)^{n-1} \ldots\ldots\ldots\ldots(12)$$

If $x = -x_1$, where $1 > x_1 > 0$, this becomes

$$-\frac{x_1}{1 - \theta x_1} \cdot \left(\frac{x_1 - \theta x_1}{1 - \theta x_1}\right)^{n-1} . \quad \text{...............(13)}$$

Since $(x_1 - \theta x_1)/(1 - \theta x_1) < x_1$, this tends with increasing n to the limit 0.

189. Applications of Taylor's Theorem. Order of Contact of Curves.

If two curves intersect in *two* points, and if, by continuous modification of one curve, these two points be made to coalesce into a single point P, then the two curves are said ultimately to have contact 'of the first order' at P. An instance is the contact of a curve with its tangent line. And whenever two curves have contact of the first order, they have a common tangent line.

Again, if two curves intersect in *three* points, and if by continuous modification of one curve these three points are made to coalesce into a single point P, then the two curves are said ultimately to have contact 'of the second order' at P. An instance is the contact of a curve with its osculating circle (Art. 137).

Let us suppose that the two curves

$$y = \phi(x), \quad y = \psi(x) \quad \text{....................(1)}$$

intersect at the points for which $x = x_0, x_1, x_2$, respectively. The function

$$F(x) \equiv \phi(x) - \psi(x), \quad \text{....................(2)}$$

which represents the difference of the ordinates of the two curves, will vanish for $x = x_0, x_1, x_2$. Hence, on the usual assumptions as to the continuity of $F(x)$ and $F'(x)$, the derived function $F'(x)$ will, by the theorem of Art. 49, vanish for some value of x between x_0 and x_1, say for $x = x_0'$, and again for some value of x between x_1 and x_2, say for $x = x_1'$. Hence, by another application of the theorem referred to, if $F''(x)$ be continuous in the interval from x_0' to x_1', it will vanish for some value of x between x_0' and x_1', say for $x = x_0''$. Hence if, by continuous modification of one of the curves (1), the three points $x = x_0, x_1, x_2$ be made to coalesce into the one point $x = x_0$, the values of $F(x_0)$, $F'(x_0)$, $F''(x_0)$ will all be zero; *i.e.* we shall have

$$\phi(x) = \psi(x), \quad \phi'(x) = \psi'(x), \quad \phi''(x) = \psi''(x), \quad \text{......(3)}$$

simultaneously, for $x = x_0$.

In other words, if two curves have contact of the second order at any point, the values of

$$y, \frac{dy}{dx}, \frac{d^2y}{dx^2}$$

will at that point be respectively identical for the two curves.

Ex. To determine the circle having contact of the second order with the curve

$$y = \phi(x) \quad \dots\dots\dots\dots\dots\dots\dots(4)$$

at a given point.

The equation of a circle with centre (ξ, η) and radius ρ is

$$(x - \xi)^2 + (y - \eta)^2 = \rho^2. \quad \dots\dots\dots\dots\dots(5)$$

If we differentiate this twice with respect to x, we find

$$x - \xi + (y - \eta)\frac{dy}{dx} = 0, \quad \dots\dots\dots\dots\dots(6)$$

$$1 + \left(\frac{dy}{dx}\right)^2 + (y - \eta)\frac{d^2y}{dx^2} = 0. \quad \dots\dots\dots\dots(7)$$

In these results, y is regarded as a function of x determined by the equation (5). But if the circle have contact of the second order with the curve (4) at the point (x, y), the values of y, dy/dx, and d^2y/dx^2 will be the same for the circle as for the curve. We may therefore suppose that in (5), (6), (7) the values of x, y, dy/dx, d^2y/dx^2 refer to the curve (4). These equations then determine the circle uniquely, viz. we find that the coordinates of the centre are

$$\xi = x - \frac{\left\{1 + \left(\frac{dy}{dx}\right)^2\right\}\frac{dy}{dx}}{\frac{d^2y}{dx^2}}, \quad \eta = y + \frac{1 + \left(\frac{dy}{dx}\right)^2}{\frac{d^2y}{dx^2}}, \quad \dots\dots(8)$$

and that the radius is

$$\rho = \frac{\left\{1 + \left(\frac{dy}{dx}\right)^2\right\}^{\frac{3}{2}}}{\frac{d^2y}{dx^2}}; \quad \dots\dots\dots\dots\dots(9)$$

cf. Art. 135.

The above considerations may be extended, and we may say that if two curves intersect in $n + 1$ consecutive points, or have contact 'of the nth order,' the values of

$$y, \frac{dy}{dx}, \frac{d^2y}{dx^2}, \dots, \frac{d^ny}{dx^n}$$

must be respectively identical for the two curves at the point in question.

The investigations of Arts. 185, 186 give a measure of the degree of closeness of two curves in the neighbourhood of a contact of the nth order. By hypothesis we have at the point $x = a$ (say)

$$\phi(a) = \psi(a), \quad \phi'(a) = \psi'(a), \quad \dots, \quad \phi^{(n)}(a) = \psi^{(n)}(a), \dots(10)$$

and therefore, with $F(x)$ defined by (2),

$$F(a) = 0, \quad F'(a) = 0, \quad F''(a) = 0, \quad \dots, \quad F^{(n)}(a) = 0. \dots(11)$$

It follows that, under the usual conditions,

$$F(a+h) = \frac{h^{n+1}}{(n+1)!} F^{(n+1)}(a+\theta h), \quad \ldots\ldots\ldots(12)$$

where $1 > \theta > 0$. Hence, if h be infinitely small, the difference of the ordinates is in general a small quantity of the order $n+1$. Moreover, it will or will not change sign with h, according as n is even or odd.

For example, the deviation of a curve from a tangent line, in the neighbourhood of the point of contact, is in general a small quantity of the second order, and the curve does not in general cross the tangent at the point. Again, the deviation of a curve from the osculating circle is a small quantity of the third order, and the curve in general crosses the circle. See Fig. 116, p. 350. But if the contact with the circle be of the fourth order, as at the vertex of a conic, the curve does not cross the circle. The same thing is further illustrated in Fig. 136, p. 485, where the curves numbered 1, 3, 5 do not cross the curve $y = \log(1+x)$ at the origin, whilst the curves numbered 2, 4, 6 do cross it.

190. Maxima and Minima.

If $\phi(x)$ be a function of x which with its first and second derivatives is finite and continuous for all values of the variable considered, we have

$$\phi(a+h) - \phi(a) = h\phi'(a) + \frac{h^2}{2!}\phi''(a+\theta h), \quad \ldots\ldots(1)$$

where $1 > \theta > 0$. By taking h sufficiently small, the second term on the right-hand can in general be made smaller in absolute value than the first, and $\phi(a+h) - \phi(a)$ will then have the same sign as $h\phi'(a)$, and will therefore change sign with h.

Now if $\phi(a)$ is a maximum or a minimum value of $\phi(x)$, the difference

$$\phi(a+h) - \phi(a)$$

must have the same sign for sufficiently small values of h, whether h be positive or negative. Hence we cannot, under the present conditions, have a maximum or a minimum unless $\phi'(a) = 0$.

Let us now suppose that $\phi'(a) = 0$, so that (1) reduces to

$$\phi(a+h) - \phi(a) = \frac{h^2}{2!}\phi''(a+\theta h). \quad \ldots\ldots\ldots(2)$$

When h is sufficiently small, the sign of the right-hand will be that of $\phi''(a)$. Hence if this be *positive* we shall have $\phi(a+h) > \phi(a)$, whether h be positive or negative; *i.e.* $\phi(a)$ is a *minimum*. Similarly, if $\phi''(a)$ be *negative* we have a *maximum*.

If $\phi''(a)$ vanish simultaneously with $\phi'(a)$, it is necessary to continue the expansion in (1) further. To take at once the general case, if we have

$$\phi'(a) = 0, \quad \phi''(a) = 0, \quad \ldots, \quad \phi^{(n-1)}(a) = 0, \ldots\ldots\ldots(3)$$

simultaneously, but $\qquad \phi^{(n)}(a) \neq 0, \ldots\ldots\ldots\ldots\ldots\ldots\ldots(4)$

then $\qquad \phi(a+h) - \phi(a) = \dfrac{h^n}{n!} \phi^{(n)}(a+\theta h). \ldots\ldots\ldots(5)$

If h be small enough, the sign is that of $h^n \phi^{(n)}(a)$. If n be odd, this changes sign with h, and we have neither a maximum nor a minimum. But if n be even, we have a maximum or minimum, according as $\phi^{(n)}(a)$ is negative or positive.

In words, $\phi(x)$ is either a maximum or a minimum for a given value of x if the first derivative which does not vanish for this value of x be of even order, but not otherwise. And the function is a maximum or a minimum according as this derivative is negative or positive.

Ex. 1. $\qquad\qquad \phi(x) = \cosh x + \cos x. \ldots\ldots\ldots\ldots\ldots\ldots(6)$
This makes

$$\phi'(x) = \sinh x - \sin x, \qquad \phi''(x) = \cosh x - \cos x,$$
$$\phi'''(x) = \sinh x + \sin x, \qquad \phi^{iv}(x) = \cosh x + \cos x.$$

The first derivative which does not vanish for $x = 0$ is $\phi^{iv}(x)$. And since $\phi^{iv}(0)$ is positive, we infer that $\phi(0)$ is a minimum value of $\phi(x)$.

This is also obvious from the expansion

$$\phi(x) = 2\left(1 + \frac{x^4}{4!} + \frac{x^8}{8!} + \ldots\right). \ldots\ldots\ldots\ldots(7)$$

Ex. 2. Let $\qquad V = b \cos \theta - c \cos 2\theta. \ldots\ldots\ldots\ldots\ldots(8)$
This makes

$$\frac{dV}{d\theta} = -b \sin \theta + 2c \sin 2\theta = \sin \theta \, (4c \cos \theta - b),$$

$$\frac{d^2 V}{d\theta^2} = -b \cos \theta + 4c \cos 2\theta = \cos \theta \, (4c \cos \theta - b) - 4c \sin^2 \theta,$$

$$\frac{d^3 V}{d\theta^3} = \quad b \sin \theta - 8c \sin 2\theta,$$

$$\frac{d^4 V}{d\theta^4} = \quad b \cos \theta - 16c \cos 2\theta.$$

For brevity, consider only angles in the first quadrant. If $b > 4c$, the only stationary value of V is when $\theta = 0$, and since this makes $d^2 V/d\theta^2 < 0$, V is then a maximum.

If $b < 4c$, V is a minimum when $\theta \doteq 0$, and there is a maximum for $\theta = \cos^{-1}(b/4c)$.

If $b = 4c$, $dV/d\theta$, $d^2V/d\theta^2$, $d^3V/d\theta^3$ all vanish for $\theta = 0$, whilst $d^4V/d\theta^4$ is negative. Hence V is then a maximum.

This example occurs in the discussion of the possible positions of equilibrium of a square plate resting in a vertical plane between two smooth pegs at the same level. If b be the length of the diagonal of the square, c the distance between the pegs, V is proportional to the potential energy when the diagonal which crosses the line of the pegs makes an angle θ with the vertical. For equilibrium V must be stationary, and for stability it must be a minimum.

191. Infinitesimal Geometry of Plane Curves.

Let the tangent and normal at any point O of a plane curve be taken as axes of coordinates; it is required to express the coordinates of a neighbouring point P of the curve in terms of the arc OP, $= s$, say.

If, for brevity, we use accents to denote differentiations with respect to s, we have, as in Art. 111,

$$x' = \cos\psi, \quad y' = \sin\psi, \quad \ldots\ldots\ldots\ldots(1)$$

and thence

$$\left.\begin{array}{ll} x'' = -\sin\psi \cdot \psi', & x''' = -\cos\psi \cdot \psi'^2 - \sin\psi \cdot \psi'', \ldots \\ y'' = \cos\psi \cdot \psi', & y''' = -\sin\psi \cdot \psi'^2 + \cos\psi \cdot \psi'', \ldots \end{array}\right\} \ldots(2)$$

and so on.

Now, by Maclaurin's Theorem,

$$\left.\begin{array}{l} x = x_0 + \dfrac{s}{1}x_0' + \dfrac{s^2}{1.2}x_0'' + \ldots, \\[2mm] y = y_0 + \dfrac{s}{1}y_0' + \dfrac{s^2}{1.2}y_0'' + \ldots, \end{array}\right\} \ldots\ldots\ldots(3)$$

where the suffix is used to mark the values which the respective quantities assume for $s = 0$. But, putting $\psi = 0$ in (1) and (2), we have

$$\left.\begin{array}{lll} x_0' = 1, & x_0'' = 0, & x_0''' = -\dfrac{1}{\rho^2}, \ldots \\[2mm] y_0' = 0, & y_0'' = \dfrac{1}{\rho}, & y_0''' = -\dfrac{1}{\rho^2}\dfrac{d\rho}{ds}, \ldots \end{array}\right\} \ldots\ldots\ldots(4)$$

where $1/\rho$ has been written for $d\psi/ds$. Hence

$$x = s - \frac{s^3}{6\rho^2} + \ldots, \quad y = \frac{s^2}{2\rho} - \frac{s^3}{6\rho^2}\frac{d\rho}{ds} + \ldots, \quad \ldots\ldots(5)$$

where ρ and $d\rho/ds$ refer to the origin.

These formulæ are useful in various questions of 'infinitesimal geometry.'

Ex. 1. Thus the second formula in (5) shews that the deviation of the curve from the osculating circle at O is ultimately

$$\tfrac{1}{6}\frac{s^3}{\rho^2}\frac{d\rho}{ds}, \quad\dots\dots\dots\dots\dots\dots\dots\dots(6)$$

since $d\rho/ds = 0$ for the circle.

And, generally, for all purposes where s^2 can be neglected, the curve may be replaced by its osculating circle.

Ex. 2. Again, the normal at P meets the normal at O in a point whose distance from O is $y + x\cot\psi$. If we neglect terms of the second order in s, this

$$= \frac{s}{\dfrac{s}{\rho} - \tfrac{1}{2}\dfrac{s^2}{\rho^3}\dfrac{d\rho}{ds}} = \rho\left(1 + \tfrac{1}{2}\frac{s}{\rho}\frac{d\rho}{ds}\right).$$

Hence the distance of the intersection from the centre of curvature at s is ultimately

$$\tfrac{1}{2}s\frac{d\rho}{ds}. \quad\dots\dots\dots\dots\dots\dots\dots\dots(7)$$

When ρ is a maximum or minimum we have (in general) $d\rho/ds = 0$, and the distance is of a higher order, the evolute having then a cusp at the point corresponding to O.

EXAMPLES. LXII.

(Expansions.)

1.
$$\cosh x \cos x = 1 - \frac{2^2 x^4}{4!} + \frac{2^4 x^8}{8!} - \dots,$$

$$\sinh x \sin x = x^2 - \frac{2^3 x^6}{6!} + \frac{2^5 x^{10}}{10!} - \dots.$$

2.
$$\cosh x \sin x = x + \frac{2x^3}{3!} - \frac{2^2 x^5}{5!} - \dots,$$

$$\sinh x \cos x = x - \frac{2x^3}{3!} - \frac{2^2 x^5}{5!} + \dots.$$

3.
$$e^x \cos x = 1 + x - \frac{2x^3}{3!} - \frac{2^2 x^4}{4!} - \frac{2^2 x^5}{5!} + \frac{2^3 x^7}{7!} + \dots,$$

$$e^x \sin x = x + x^2 + \frac{2x^3}{3!} - \frac{2^2 x^5}{5!} - \frac{2^3 x^6}{6!} - \frac{2^3 x^7}{7!} + \dots,$$

4.
$$\sec x = 1 + \frac{x^2}{2!} + \frac{5x^4}{4!} + \frac{61x^6}{6!} + \dots.$$

5.
$$\log \sec x = \frac{x^2}{2} + \frac{x^4}{12} + \frac{x^6}{45} + \dots.$$

6.
$$\log \cosh x = \tfrac{1}{2}x^2 - \tfrac{1}{12}x^4 + \tfrac{1}{45}x^6 - \dots.$$

7. $\tanh x = x - \frac{1}{3}x^3 + \frac{2}{15}x^5 - \frac{17}{315}x^7 + \ldots.$

8. $\cos^2 x = 1 - x^2 + \dfrac{2^3 x^4}{4!} - \dfrac{2^5 x^6}{6!} + \ldots.$

9. $\cos^n x = 1 - \dfrac{n}{2!}x^2 + \dfrac{n(3n-2)}{4!}x^4 - \ldots.$

10. $\left(\dfrac{\sin x}{x}\right)^n = 1 - \dfrac{n}{3!}x^2 + \dfrac{n(5n-2)}{3.5!}x^4 - \ldots.$

11. $\dfrac{x}{\sin x} = 1 + \dfrac{x^2}{3!} + \dfrac{14x^4}{6!} + \ldots,$

          $\dfrac{x}{\sinh x} = 1 - \dfrac{x^2}{3!} + \dfrac{14x^4}{6!} - \ldots.$

12. $\dfrac{x}{e^x - 1} = 1 - \frac{1}{2}x + \frac{1}{6}\dfrac{x^2}{2!} - \frac{1}{30}\dfrac{x^4}{4!} + \ldots.$

13. $\tan\left(\frac{1}{4}\pi + x\right) = 1 + 2x + 2x^2 + \frac{8}{3}x^3 + \frac{10}{3}x^4 + \ldots.$

14. $\log \tan\left(\frac{1}{4}\pi + x\right) = 2x + \frac{4}{3}x^3 + \frac{4}{3}x^5 + \ldots.$

15. $\log(1 + \sin x) = x - \frac{1}{2}x^2 + \frac{1}{6}x^3 - \frac{1}{12}x^4 + \frac{1}{24}x^5 - \ldots.$

16. $\log(1 + e^x) = \log 2 + \frac{1}{2}x + \frac{1}{8}x^2 - \frac{1}{192}x^4 + \ldots.$

17. $\dfrac{3\sin x}{2 + \cos x} = x - \dfrac{x^5}{180} + \ldots.$

18. $\log \sec^2 \frac{1}{2}\theta = \dfrac{1}{2}\dfrac{\sin^2\theta}{2} + \dfrac{1.3}{2.4}\dfrac{\sin^4\theta}{4} + \dfrac{1.3.5}{2.4.6}\dfrac{\sin^6\theta}{6} + \ldots.$

19.   If $D = d/dx$, prove that

          $D^n e^{x\cos a}\cos(x\sin a) = e^{x\cos a}\cos(x\sin a + na).$

Hence shew that

          $e^{x\cos a}\cos(x\sin a) = 1 + x\cos a + \dfrac{x^2}{2!}\cos 2a + \dfrac{x^3}{3!}\cos 3a + \ldots.$

20.   Draw graphs of the functions

          $x, \quad x - \dfrac{x^3}{3!}, \quad x - \dfrac{x^3}{3!} + \dfrac{x^5}{5!},$

respectively, and compare them with the graph of $\sin x$.

21.   Draw graphs of the functions

          $1 - \dfrac{x^2}{2!}, \quad 1 - \dfrac{x^2}{2!} + \dfrac{x^4}{4!},$

respectively, and compare them with the graph of $\cos x$.

22.   Prove that, in the formula (17) of Art. 185, the limiting value of θ, when h is indefinitely diminished, is in general $1/(n+1)$.

23. Prove that when h is sufficiently small the error in Simpson's formula (Art. 114 (8)) of approximate integration is

$$\frac{1}{90}h^5 \frac{d^4y}{dx^4},$$

nearly.

24. Prove that the mean value of a function $\phi(x)$ over the range extending from $x = a - h$ to $x = a + h$ is

$$\phi(a) + \frac{h^2}{3!}\phi''(a) + \frac{h^4}{5!}\phi^{iv}(a) + \dots .$$

Shew that this falls short of the arithmetic mean of the values at the extremities of the range by

$$\frac{h^2}{1 . 3}\phi''(a) + \frac{h^4}{3!5}\phi^{iv}(a) + \dots .$$

25. Shew that if, from a given curve, another curve be constructed whose ordinate, for any value a of x, is the mean of the ordinates of the first curve over the range bounded by $x = a \pm h$, where h is a given small constant, the ordinate of the second curve exceeds that of the first by one-third the sagitta of the arc whose extremities are $x = a \pm h$.

EXAMPLES. LXIII.

(Geometrical Applications.)

1. Prove that if the expansions (4) of Art. 191 be carried to the order s^4, the results are

$$x = s - \frac{1}{6}\frac{s^3}{\rho^2} + \frac{1}{8}\frac{s^4}{\rho^3}\frac{d\rho}{ds} + \dots ,$$

$$y = \frac{1}{2}\frac{s^2}{\rho} - \frac{1}{6}\frac{s^3}{\rho^2}\frac{d\rho}{ds} - \frac{1}{24}\frac{s^4}{\rho^3}\left\{1 - 2\left(\frac{d\rho}{ds}\right)^2 + \rho\frac{d^2\rho}{ds^2}\right\} + \dots .$$

2. If the values of x, y, referred to the tangent and normal at a point O of a curve, be developed in terms of ψ, the inclination of the tangent to the axis of x, the results are

$$x = \rho\psi + \frac{1}{2}\frac{d\rho}{d\psi}\psi^2 - \frac{1}{6}\left(\rho - \frac{d^2\rho}{d\psi^2}\right)\psi^3 + \dots ,$$

$$y = \frac{1}{2}\rho\psi^2 + \frac{1}{3}\frac{d\rho}{d\psi}\psi^3 + \dots .$$

3. Prove that, with the same axes, the coordinates of the centre of curvature are

$$\xi = -\frac{1}{2}\frac{d\rho}{d\psi}\psi^2 - \frac{1}{3}\frac{d^2\rho}{d\psi^2}\psi^3 + \dots ,$$

$$\eta = \rho + \frac{d\rho}{d\psi}\psi + \frac{1}{2}\frac{d^2\rho}{d\psi^2}\psi^3 + \dots .$$

4. If O, P be two adjacent points on a curve, and PQ be drawn perpendicular to the chord OP, to meet the normal at O in Q, then ultimately $OQ = 2\rho$.

5. If equal small lengths OP, OT be measured along the *arc* of a curve, and along the tangent at the point O, and if R be the limiting position of the intersection of PT produced with the normal at O, then $OR = 3\rho$.

6. If P, Q be adjacent points on a curve, and a point T be taken on the tangent at P such that PT is equal to the *chord* PQ, and if R be the intersection of TQ with the normal at P, prove that the limiting value of PR is 4ρ.

7. The perpendicular to a chord OP at its middle point meets the normal at O at a distance from the centre of curvature ultimately equal to $\frac{1}{3}s d\rho/ds$, where $s = OP$.

8. The tangents at two adjacent points P, Q of a curve meet in T, and V is the middle point of the chord PQ. Prove that TV makes with the normal to the curve an angle $\tan^{-1}\left(\frac{1}{3}d\rho/ds\right)$.

9. Prove that if PQ be a small arc (s) of a curve, the arc exceeds the chord by $\frac{1}{24}s^3/\rho^2$, and the sum of the tangents at P and Q exceeds the arc by $\frac{1}{12}s^3/\rho^2$.

10. Prove that the form of a curve near a cusp is given by
$$ay^2 = x^3,$$
approximately, where $a = \frac{2}{9}d\rho/d\psi$.

11. Prove that the form of a curve near a point of inflexion is given by
$$y = \frac{1}{6}\frac{dc}{ds}x^3,$$
approximately, where c is the curvature.

12. If P be any point on a curve, the form of the evolute of the part near P, referred to the centre of curvature at P as origin, is in general given by
$$ay = x^2,$$
where $a = 2d\rho/d\psi$.

13. Prove that if P be a point of maximum or minimum curvature, the form of the evolute is
$$ay^2 = x^3,$$
approximately, where $a = \frac{2}{9}d^2\rho/d\psi^2$.

CHAPTER XVI

FUNCTIONS OF SEVERAL INDEPENDENT VARIABLES

192. Partial Derivatives of Various Orders.

If u be a function of two or more independent variables $x, y, \ldots$, the partial derivatives

$$\frac{\partial u}{\partial x}, \quad \frac{\partial u}{\partial y}, \ldots \quad \ldots\ldots\ldots\ldots\ldots\ldots\ldots(1)$$

will themselves in general be functions of $x, y, \ldots$, and be susceptible of differentiation with respect to these several variables.

Thus, if
$$u = \phi(x, y), \ldots\ldots\ldots\ldots\ldots\ldots\ldots\ldots(2)$$
we can form the second derivatives

$$\frac{\partial}{\partial x}\left(\frac{\partial u}{\partial x}\right), \quad \frac{\partial}{\partial y}\left(\frac{\partial u}{\partial x}\right), \quad \frac{\partial}{\partial x}\left(\frac{\partial u}{\partial y}\right), \quad \frac{\partial}{\partial y}\left(\frac{\partial u}{\partial y}\right),$$

or, as they are usually written,

$$\frac{\partial^2 u}{\partial x^2}, \quad \frac{\partial^2 u}{\partial y\partial x}, \quad \frac{\partial^2 u}{\partial x\partial y}, \quad \frac{\partial^2 u}{\partial y^2}. \quad \ldots\ldots\ldots\ldots(3)$$

It will be noticed that there is (primarily) a distinction of meaning between the second and third of these symbols, the operations indicated being performed in inverse orders in the two cases. It will be shewn, however, in Art. 193 that under certain conditions, which are generally satisfied in practice, the results are identical.

The first derivatives of $\phi(x, y)$ are sometimes denoted by

$$\phi_x(x, y), \quad \phi_y(x, y), \quad \ldots\ldots\ldots\ldots\ldots(4)$$

and the second derivatives (3) by

$$\phi_{xx}(x, y), \quad \phi_{yx}(x, y), \quad \phi_{xy}(x, y), \quad \phi_{yy}(x, y). \quad \ldots\ldots(5)$$

These are often abbreviated into

$$\phi_x, \quad \phi_y, \quad \ldots\ldots\ldots\ldots\ldots\ldots\ldots(6)$$

and
$$\phi_{xx}, \quad \phi_{yx}, \quad \phi_{xy}, \quad \phi_{yy}, \quad \ldots\ldots\ldots\ldots\ldots(7)$$
respectively.

Ex. 1. If
$$u = Ax^m y^n, \quad \dots\dots\dots\dots(8)$$

we have
$$\frac{\partial u}{\partial x} = mAx^{m-1}y^n, \quad \frac{\partial u}{\partial y} = nAx^m y^{n-1}, \quad \dots\dots\dots(9)$$

$$\frac{\partial^2 u}{\partial x^2} = m(m-1)Ax^{m-2}y^n, \quad \frac{\partial^2 u}{\partial y^2} = n(n-1)Ax^m y^{n-2},$$

$$\frac{\partial^2 u}{\partial y \partial x} = mnAx^{m-1}y^{n-1} = \frac{\partial^2 u}{\partial x \partial y}. \quad \dots \quad \dots\dots\dots(10)$$

Ex. 2. If
$$z = a \tan^{-1} \frac{y}{x}, \quad \dots\dots\dots\dots(11)$$

we find
$$\frac{\partial z}{\partial x} = -\frac{ay}{x^2 + y^2}, \quad \frac{\partial z}{\partial y} = \frac{ax}{x^2 + y^2}, \quad \dots\dots\dots(12)$$

$$\frac{\partial^2 z}{\partial x^2} = \frac{2axy}{(x^2+y^2)^2}, \quad \frac{\partial^2 z}{\partial y \partial x} = \frac{a(y^2 - x^2)}{(x^2+y^2)^2} = \frac{\partial^2 z}{\partial x \partial y}, \quad \frac{\partial^2 z}{\partial y^2} = \frac{-2axy}{(x^2+y^2)^2}. \quad \dots(13)$$

193. Proof of the Commutative Property.

Let
$$u = \phi(x, y), \quad \dots\dots\dots\dots\dots(1)$$

and let us suppose that the functions

$$u, \quad \frac{\partial u}{\partial x}, \quad \frac{\partial u}{\partial y}, \quad \frac{\partial^2 u}{\partial y \partial x}, \quad \frac{\partial^2 u}{\partial x \partial y} \quad \dots\dots\dots(2)$$

are continuous (Art. 34) over a finite range of the variables, including the values considered. We proceed to shew that, under these conditions,

$$\frac{\partial^2 u}{\partial y \partial x} = \frac{\partial^2 u}{\partial x \partial y}. \quad \dots\dots\dots\dots(3)$$

To this end, we consider the fraction

$$\chi(h, k) = \frac{\phi(x+h, y+k) - \phi(x+h, y) - \phi(x, y+k) + \phi(x, y)}{hk},$$
$$\dots\dots(4)$$

in which x, y are regarded as fixed, whilst h, k will (finally) be made infinitely small.

Let us write, for a moment,
$$F(x) = \phi(x, y+k) - \phi(x, y). \quad \dots\dots\dots(5)$$

By the mean-value theorem of Art. 56 (9), we have
$$F(x+h) - F(x) = hF'(x + \theta_1 h), \quad \dots\dots\dots(6)$$

or, in full,
$$\{\phi(x+h, y+k) - \phi(x+h, y)\} - \{\phi(x, y+k) - \phi(x, y)\}$$
$$= h\{\phi_x(x+\theta_1 h, y+k) - \phi_x(x+\theta_1 h, y)\}, \quad \dots\dots(7)$$

where $1 > \theta_1 > 0$, the value of y not being varied in this process. Hence

$$\chi(h,\ k) = \frac{\phi_x(x + \theta_1 h,\ y + k) - \phi_x(x + \theta_1 h,\ y)}{k}. \qquad \ldots\ldots(8)$$

If we now write

$$f(y) = \phi_x(x + \theta_1 h,\ y), \ldots\ldots\ldots\ldots\ldots\ldots(9)$$

we have, by a second application of the theorem referred to,

$$f(y + k) - f(y) = kf'(y + \theta_2 k), \quad \ldots\ldots\ldots\ldots(10)$$

or

$$\phi_x(x + \theta_1 h,\ y + k) - \phi_x(x + \theta_1 h,\ y) = k\phi_{yx}(x + \theta_1 h,\ y + \theta_2 k).$$
$$\ldots\ldots(11)$$

Hence $$\chi(h,\ k) = \phi_{yx}(x + \theta_1 h,\ y + \theta_2 k), \quad \ldots\ldots\ldots\ldots(12)$$

where θ_1, θ_2 lie between 0 and 1.

By a similar process we could shew that

$$\chi(h,\ k) = \phi_{xy}(x + \theta_1' h,\ y + \theta_2' k), \quad \ldots\ldots\ldots\ldots(13)$$

where θ_1', θ_2' also lie between 0 and 1.

These results are exact, provided $x + h$, $y + k$ lie within the range of the variables for which the conditions above postulated hold. If we now diminish h and k indefinitely, it follows from the comparison of (12) and (13), and from the continuity of the derivatives, that

$$\phi_{yx}(x,\ y) = \phi_{xy}(x,\ y), \quad \ldots\ldots\ldots\ldots(14)$$

as was to be proved*.

It follows from the above theorem that in the case of a function of any number of independent variables $x, y, z, \ldots$ the operations

$$\frac{\partial}{\partial x},\ \frac{\partial}{\partial y},\ \frac{\partial}{\partial z},\ \cdots$$

or, as we may denote them for shortness,

$$D_x,\ D_y,\ D_z, \ldots$$

are in general commutative, *i.e.* the result of any number of them is independent of the order in which they are performed.

For example,

$$D_x D_y D_z u = D_x(D_y D_z u) = D_x(D_z D_y)\, u = D_x D_z(D_y u) = D_z D_x D_y u = \text{etc.}$$

---

* This proof appears to be due to Ossian Bonnet. An alternative proof is indicated in Art. 194.

It appears from (4) that

$$\lim_{h \to 0} \chi(h, k) = \frac{1}{k} \lim_{h \to 0} \frac{\phi(x+h, y+k) - \phi(x, y+k)}{h}$$

$$- \frac{1}{k} \lim_{h \to 0} \frac{\phi(x+h, y) - \phi(x, y)}{h}$$

$$= \frac{\phi_x(x, y+k) - \phi_x(x, y)}{k}. \qquad \ldots\ldots\ldots\ldots\ldots\ldots(15)$$

Hence

$$\lim_{k \to 0} \lim_{h \to 0} \chi(h, k) = \phi_{yx}(x, y). \qquad \ldots\ldots\ldots\ldots(16)$$

Similarly, we find

$$\lim_{h \to 0} \lim_{k \to 0} \chi(h, k) = \phi_{xy}(x, y). \qquad \ldots\ldots\ldots\ldots(17)$$

If, then, we could assume that the limiting value of the fraction (4), when h and k are indefinitely diminished, is unique, and independent of the *order* in which these quantities are made to vanish, the theorem (3) would follow at once. A simple example shews, however, that the assumption is not legitimate without further examination. If

$$f(h, k) = \frac{h^2 - k^2}{h^2 + k^2},$$

we have

$$\lim_{k \to 0} \lim_{h \to 0} f(h, k) = -1, \qquad \lim_{h \to 0} \lim_{k \to 0} f(h, k) = +1.$$

Ex. A necessary condition that

$$M dx + N dy \qquad \ldots\ldots\ldots\ldots\ldots\ldots\ldots(18)$$

should be an exact differential (Art. 155) is

$$\frac{\partial M}{\partial y} = \frac{\partial N}{\partial x}. \qquad \ldots\ldots\ldots\ldots\ldots\ldots\ldots\ldots(19)$$

For if the expression (18) be equal to du, we have

$$M = \frac{\partial u}{\partial x}, \qquad N = \frac{\partial u}{\partial y}, \qquad \ldots\ldots\ldots\ldots\ldots(20)$$

and therefore each of the partial derivatives in (19) is equal to $\partial^2 u / \partial y \partial x$ or $\partial^2 u / \partial x \partial y$.

Conversely, we can shew that, if the condition (19) hold, (18) will be an exact differential. Let v denote the function $\int M dx$, obtained by integrating as if y were constant. We have, then,

$$\frac{\partial v}{\partial x} = M, \qquad \ldots\ldots\ldots\ldots\ldots\ldots(21)$$

and therefore

$$\frac{\partial N}{\partial x} = \frac{\partial M}{\partial y} = \frac{\partial^2 v}{\partial x \partial y},$$

or

$$\frac{\partial}{\partial x} \left(N - \frac{\partial v}{\partial y} \right) = 0. \qquad \ldots\ldots\ldots\ldots\ldots(22)$$

This shews (Art. 56) that the function $N - \partial v/\partial y$ is constant so far as x is concerned and is therefore a function of y only. Denoting its value by $f'(y)$, we have

$$N = \frac{\partial v}{\partial y} + f'(y). \qquad \ldots\ldots\ldots\ldots\ldots\ldots(23)$$

Hence, if we write
$$u = v + f(y), \qquad \ldots\ldots\ldots\ldots\ldots\ldots(24)$$
we have, by (21) and (23),

$$\frac{\partial u}{\partial x} = M, \quad \frac{\partial u}{\partial y} = N, \qquad \ldots\ldots\ldots\ldots\ldots(25)$$

and therefore
$$M\,dx + N\,dy = du. \qquad \ldots\ldots\ldots\ldots\ldots(26)$$

194. Extension of Taylor's Theorem.

Let $\phi(x, y)$ be a function of x and y which, with its derivatives up to a certain order, is continuous for all values of the variables considered. It may be required to find the expansion of

$$\phi(a + h,\ b + k) \qquad \ldots\ldots\ldots\ldots\ldots(1)$$

in ascending powers of h and k. We shall in the first place give a direct investigation of the expansion as far as the terms of the *second* degree in h and k.

First expanding in powers of h, we have, by Taylor's Theorem,

$$\phi(a + h,\ b + k) = \phi(a,\ b + k) + h\phi_x(a,\ b + k)$$
$$+ \tfrac{1}{2}h^2\phi_{xx}(a,\ b + k) + \ldots. \quad \ldots\ldots(2)$$

Again, by the same theorem,

$$\left.\begin{aligned}
\phi(a,\ b + k) &= \phi(a,\ b) + k\phi_y(a,\ b) + \tfrac{1}{2}k^2\phi_{yy}(a,\ b) + \ldots \\
\phi_x(a,\ b + k) &= \phi_x(a,\ b) + k\phi_{yx}(a,\ b) + \ldots \\
\phi_{xx}(a,\ b + k) &= \phi_{xx}(a,\ b) + \ldots.
\end{aligned}\right\} \ldots(3)$$

Substituting in (2), we find

$$\phi(a + h,\ b + k) = \phi(a,\ b) + \{h\phi_x(a,\ b) + k\phi_y(a,\ b)\}$$
$$+ \tfrac{1}{2}\{h^2\phi_{xx}(a,\ b) + 2hk\phi_{yx}(a,\ b) + k^2\phi_{yy}(a,\ b)\} + \ldots. \quad \ldots\ldots(4)$$

If we regard the forms of the several 'remainders' (Art. 185) in the preliminary expansions, it appears that the remainder in (4) will be of the form

$$\frac{1}{3!}\{Rh^3 + 3Sh^2k + 3Thk^2 + Uk^3\}, \qquad \ldots\ldots\ldots\ldots(5)$$

where R, S, T, U are functions of a, b, h, k which remain finite when h, k are indefinitely diminished. The remainder is therefore of the *third* order in h, k.

The conditions for the validity of the foregoing result are that $\phi(x, y)$ and its derivatives up to the third order should be continuous for all values of the variable considered.

With a slight change of notation we may write (4) in the form

$$\phi(x + h, y + k) = \phi(x, y) + \left(h\frac{\partial\phi}{\partial x} + k\frac{\partial\phi}{\partial y}\right)$$
$$+ \tfrac{1}{2}\left(h^2\frac{\partial^2\phi}{\partial x^2} + 2hk\frac{\partial^2\phi}{\partial x\partial y} + k^2\frac{\partial^2\phi}{\partial y^2}\right) + \ldots, \quad \ldots(6)$$

where, on the right hand, ϕ stands for $\phi(x, y)$. A still more compact form is

$$\phi(x + h, y + k) = \phi(x, y) + (h\phi_x + k\phi_y)$$
$$+ \tfrac{1}{2}(h^2\phi_{xx} + 2hk\phi_{xy} + k^2\phi_{yy}) + \ldots \ldots \ldots(7)$$

Again, if u be any function of the independent variables x, y, and if, as in Art. 57, δu denote the increment of u due to given increments δx, δy of these variables, the formula is equivalent to

$$\delta u = \frac{\partial u}{\partial x}\,\delta x + \frac{\partial u}{\partial y}\,\delta y + \tfrac{1}{2}\left\{\frac{\partial^2 u}{\partial x^2}(\delta x)^2 + 2\frac{\partial^2 u}{\partial x\partial y}\,\delta x\delta y + \frac{\partial^2 u}{\partial y^2}(\delta y)^2\right\} + \ldots.$$
$$\ldots\ldots(8)^*$$

It may be remarked that in the proof of (4) it was not necessary to assume that

$$\phi_{yx}(a, b) = \phi_{xy}(a, b). \quad \ldots\ldots\ldots\ldots\ldots\ldots(9)$$

If we had begun by expanding (1) in powers of k (instead of h) we should have arrived at a result similar to (4), but with $\phi_{xy}(a, b)$ in place of $\phi_{yx}(a, b)$. From a comparison of the two forms we can obtain an independent proof of the theorem of Art. 193.

195. General Term of the Expansion.

An independent investigation, giving the general term of the expansion (7), is as follows. We write $h = \alpha t$, $k = \beta t$, and

$$F(t) = \phi(x + h, y + k) = \phi(x + \alpha t, y + \beta t). \quad \ldots\ldots(1)$$

Regarded as a function of t, this can be expanded by Maclaurin's theorem, and the general term is

$$\frac{t^n}{n!}F'^{(n)}(0). \quad \ldots\ldots\ldots\ldots\ldots\ldots(2)$$

---

* The extension of the investigations of this Art. to cases where there are three or more independent variables will be obvious.

Now if we put for a moment $x + \alpha t = u$, $y + \beta t = v$, we have

$$\frac{\partial \phi}{\partial x} = \frac{\partial \phi}{\partial u}\frac{\partial u}{\partial x} = \frac{\partial \phi}{\partial u}, \quad \frac{\partial \phi}{\partial y} = \frac{\partial \phi}{\partial v}\frac{\partial v}{\partial y} = \frac{\partial \phi}{\partial v}, \quad \dots\dots\dots(3)$$

where ϕ is written for $\phi(u, v)$. Hence

$$F'(t) = \frac{\partial \phi}{\partial u}\frac{\partial u}{\partial t} + \frac{\partial \phi}{\partial v}\frac{\partial v}{\partial t} = \alpha\frac{\partial \phi}{\partial x} + \beta\frac{\partial \phi}{\partial y}$$

$$= \left(\alpha\frac{\partial}{\partial x} + \beta\frac{\partial}{\partial y}\right)\phi(u, v). \quad\dots\dots\dots\dots(4)$$

The result is evidently a function of u and v, hence, by a repetition of the argument,

$$F''(t) = \left(\alpha\frac{\partial}{\partial x} + \beta\frac{\partial}{\partial y}\right)\left(\alpha\frac{\partial}{\partial x} + \beta\frac{\partial}{\partial y}\right)\phi(u, v)$$

$$= \left(\alpha\frac{\partial}{\partial x} + \beta\frac{\partial}{\partial y}\right)^2\phi(u, v), \quad\dots\dots\dots\dots(5)$$

and, generally,

$$F^{(n)}(t) = \left(\alpha\frac{\partial}{\partial x} + \beta\frac{\partial}{\partial y}\right)^n\phi(u, v), \quad\dots\dots\dots(6)$$

where the operator admits of expansion by the Binomial Theorem, in virtue of the commutative property of the operators $\partial/\partial x$ and $\partial/\partial y$. Since t only occurs in the combinations $x + \alpha t$, $y + \beta t$, it is evidently immaterial in (6) whether we put $t = 0$ before or after the differentiations indicated on the right-hand side. The general term of our expansion is therefore

$$\frac{t^n}{n!}F^{(n)}(0) = \frac{t^n}{n!}\left(\alpha\frac{\partial}{\partial x} + \beta\frac{\partial}{\partial y}\right)^n\phi(x, y) = \frac{1}{n!}\left(h\frac{\partial}{\partial x} + k\frac{\partial}{\partial y}\right)^n\phi(x, y)$$

$$= \frac{1}{n!}\left(h^n\frac{\partial^n\phi}{\partial x^n} + nh^{n-1}k\frac{\partial^n\phi}{\partial x^{n-1}\partial y} + \frac{n(n-1)}{1\cdot 2}h^{n-2}k^2\frac{\partial^n\phi}{\partial x^{n-2}\partial y^2} + \dots\right),$$

$$\dots\dots(7)$$

where ϕ is now written for $\phi(x, y)$.

Ex. To prove that if $\phi(x, y)$ be a homogeneous function of x, y, of degree m, we have

$$x\phi_x + y\phi_y = m\phi, \quad\dots\dots\dots\dots\dots\dots(8)$$

$$x^2\phi_{xx} + 2xy\phi_{xy} + y^2\phi_{yy} = m(m-1)\phi. \quad\dots\dots\dots(9)$$

The general definition of a homogeneous function of degree m is that if x and y be altered in any ratio μ, the function is altered in the ratio μ^m, or

$$\phi(\mu x, \mu y) = \mu^m\phi(x, y). \quad\dots\dots\dots\dots\dots(10)$$

In this equality, let us put $\mu = 1 + t$. Since

$$\phi(x + xt,\, y + yt) = \phi(x, y) + t\,(x\phi_x + y\phi_y)$$
$$+ \tfrac{1}{2}t^2\,(x^2\phi_{xx} + 2xy\phi_{xy} + y^2\phi_{yy}) + \ldots$$

by (8), and

$$(1 + t)^m\,\phi(x, y) = \left(1 + mt + \frac{m\,(m-1)}{1 \cdot 2}\,t^2 + \ldots\right)\phi,$$

by the Binomial Theorem, the results (9) and (10) will follow, on equating coefficients of t and t^2. More generally, equating coefficients of t^n, and making use of (7), we find

$$x^n\frac{\partial^n\phi}{\partial x^n} + nx^{n-1}y\,\frac{\partial^n\phi}{\partial x^{n-1}\partial y} + \frac{n\,(n-1)}{1 \cdot 2}\,x^{n-2}y^2\,\frac{\partial^n\phi}{\partial x^{n-2}\partial y^2} + \ldots$$
$$= m\,(m-1)\,(m-2)\ldots(m-n+1)\,\phi. \quad\ldots\ldots(11)$$

This is 'Euler's Theorem of Homogeneous Functions,' for the case of two independent variables. The extension to three or more independent variables will be obvious.

196. Maxima and Minima of a Function of Two Variables. Geometrical Interpretation.

We may utilize the generalized form of Taylor's Theorem to carry a step further the discussion (see Art. 53) of the maxima and minima of a function (u) of two independent variables (x, y).

It appears from Art. 194 (8) that when δx, δy are continually diminished in absolute value, preserving any given ratio to one another, the sign of δu is ultimately that of

$$\frac{\partial u}{\partial x}\,\delta x + \frac{\partial u}{\partial y}\,\delta y. \quad\ldots\ldots\ldots\ldots\ldots\ldots(1)$$

Unless $\partial u/\partial x$ and $\partial u/\partial y$ both vanish, the sign of (1) is reversed by reversing the signs of δx and δy. Hence for some variations δu will be positive, and for others negative. In other words, u cannot be a maximum or minimum unless we have

$$\frac{\partial u}{\partial x} = 0, \quad \frac{\partial u}{\partial y} = 0, \quad\ldots\ldots\ldots\ldots\ldots\ldots(2)*$$

simultaneously.

Let us now suppose the conditions (2) to be fulfilled. We have, then,

$$\delta u = \tfrac{1}{2}\left\{\frac{\partial^2 u}{\partial x^2}\,(\delta x)^2 + 2\,\frac{\partial^2 u}{\partial x\partial y}\,\delta x\delta y + \frac{\partial^2 u}{\partial y^2}\,(\delta y)^2\right\} + \ldots\ldots\ldots(3)$$

---

* It is assumed in the investigation of Art. 194 that these derivatives are continuous and therefore finite. That is, we exclude *ab initio* the two-dimensional analogues of the cases considered in Art. 51.

When δx and δy are sufficiently small, the sign of δu will be that of the terms written. Now it is known from Algebra that the sign of a homogeneous quadratic function

$$A\xi^2 + 2H\xi\eta + B\eta^2 \quad\ldots\ldots\ldots\ldots\ldots\ldots(4)$$

is invariable, if (and only if)

$$AB > H^2, \quad\ldots\ldots\ldots\ldots\ldots\ldots\ldots(5)$$

and that the sign is then that of A (or B). We infer that when the conditions (2) are satisfied δu will have the same sign for all values of $|\delta x|$ and $|\delta y|$ not exceeding certain limits, provided

$$\frac{\partial^2 u}{\partial x^2}\frac{\partial^2 u}{\partial y^2} > \left(\frac{\partial^2 u}{\partial x \partial y}\right)^2, \quad\ldots\ldots\ldots\ldots\ldots(6)$$

and that the sign will then be that of $\partial^2 u/\partial x^2$ and $\partial^2 u/\partial y^2$. And u will be a maximum or minimum according as this sign is negative or positive.

If

$$\frac{\partial^2 u}{\partial x^2}\frac{\partial^2 u}{\partial y^2} < \left(\frac{\partial^2 u}{\partial x \partial y}\right)^2, \quad\ldots\ldots\ldots\ldots\ldots(7)$$

then for some values of the ratio $\delta y/\delta x$ the increment of u will be positive, for others negative, and the value of u, though 'stationary' (cf. Art. 51) is neither a maximum nor a minimum.

If

$$\frac{\partial^2 u}{\partial x^2}\frac{\partial^2 u}{\partial y^2} = \left(\frac{\partial^2 u}{\partial x \partial y}\right)^2, \quad\ldots\ldots\ldots\ldots\ldots(8)$$

the terms which appear on the right-hand side of (3) are equal to $\pm$ the square of a linear function of δx and δy, and therefore vanish for a particular value of the ratio $\delta y/\delta x$. Since δu is then of the third order it appears that there is in general neither a maximum nor a minimum, but the question cannot be absolutely decided without continuing the expansion further. The same remark applies when the second derivatives $\partial^2 u/\partial x^2$, $\partial^2 u/\partial x \partial y$, $\partial^2 u/\partial y^2$ all vanish.

The preceding investigation has an interesting geometrical interpretation. If, as in Art. 34, z be the vertical ordinate of a surface, and x, y rectangular coordinates in a horizontal plane, the first condition for a point of maximum or minimum altitude is that

$$\frac{\partial z}{\partial x} = 0, \quad \frac{\partial z}{\partial y} = 0 \quad\ldots\ldots\ldots\ldots\ldots\ldots(9)$$

simultaneously. Since these equations ensure that δz shall be of the second order in δx, δy, it follows that at the point (P, say,) in question the tangent line to every vertical section through P will be horizontal; in other words, we have a horizontal *tangent plane*.

We have next to examine whether the surface cuts the tangent plane at P. Along the line of intersection (if any), we shall have

$\delta z = 0$, and therefore from (3), if we put $\delta y = m\delta x$, and finally make δx vanish, the directions of the tangent lines at P to the curve of intersection are determined by

$$\frac{\partial^2 z}{\partial x^2} + 2\frac{\partial^2 z}{\partial x \partial y} m + \frac{\partial^2 z}{\partial y^2} m^2 = 0. \quad \ldots\ldots\ldots\ldots(10)$$

This quadratic in m will have imaginary roots if

$$\frac{\partial^2 z}{\partial x^2}\frac{\partial^2 z}{\partial y^2} > \left(\frac{\partial^2 z}{\partial x \partial y}\right)^2; \quad \ldots\ldots\ldots\ldots\ldots(11)$$

the surface then, in the immediate neighbourhood of P, will lie wholly on one side of the tangent plane, and the contour-line at P reduces to a point. Hence P will be a point of maximum or minimum altitude according as $\partial^2 z/\partial x^2$ and $\partial^2 z/\partial y^2$ are negative or positive, *i.e.* (Art. 67) according as the vertical sections parallel to the planes zx and zy are convex or concave upwards. If we imagine the axes of x, y to be rotated in their own plane, we can infer that *every* vertical section through P is in this case convex upwards, or concave upwards, respectively.

But if

$$\frac{\partial^2 z}{\partial x^2}\frac{\partial^2 z}{\partial y^2} < \left(\frac{\partial^2 z}{\partial x \partial y}\right)^2, \quad \ldots\ldots\ldots\ldots\ldots(12)$$

the roots of (10) are real and distinct. The contour-line has a node at P, the two branches separating the parts of the surface which lie above the tangent plane from those which lie below.

If

$$\frac{\partial^2 z}{\partial x^2}\frac{\partial^2 z}{\partial y^2} = \left(\frac{\partial^2 z}{\partial x \partial y}\right)^2, \quad \ldots\ldots\ldots\ldots\ldots(13)$$

the roots of (10) are real and coincident. The contour-line has in general a cusp at P, and the question as to whether the altitude at P is a maximum or minimum cannot be determined without further investigation.

Ex. 1. Let $\qquad z = x^3 - 3ax^2 - 4ay^2 + C. \quad \ldots\ldots\ldots\ldots\ldots(14)$

This makes $\qquad \dfrac{\partial z}{\partial x} = 3x(x-2a), \quad \dfrac{\partial z}{\partial y} = -8ay, \quad \ldots\ldots\ldots\ldots(15)$

$$\frac{\partial^2 z}{\partial x^2} = 6(x-a), \quad \frac{\partial^2 z}{\partial x \partial y} = 0, \quad \frac{\partial^2 z}{\partial y^2} = -8a. \quad \ldots\ldots\ldots\ldots(16)$$

The conditions (9) are satisfied by $x = 0$, $y = 0$, and also by $x = 2a$, $y = 0$. The former solution satisfies the inequality (11), and since $\partial^2 z/\partial y^2$ is negative, z is a *maximum*. The latter solution comes under (12); z is then neither a maximum nor a minimum. The contour-lines for this case are shewn in Fig. 69, p. 286.

Ex. 2. Let $\qquad z = (x^2 + y^2)^2 - 2a^2(x^2 + y^2) + C. \quad \ldots\ldots\ldots\ldots(17)$

We find $\qquad \dfrac{\partial z}{\partial x} = 4x(x^2 + y^2 - a^2), \quad \dfrac{\partial z}{\partial y} = 4y(x^2 + y^2 + a^2), \quad \ldots\ldots\ldots(18)$

$$\frac{\partial^2 z}{\partial x^2} = 4(3x^2 + y^2 - a^2), \quad \frac{\partial^2 z}{\partial x \partial y} = 8xy, \quad \frac{\partial^2 z}{\partial y^2} = 4(x^2 + 3y^2 + a^2). \quad (19)$$

The real solutions of (9) are, in this case, $x = 0$, $y = 0$, and $x = \pm a$, $y = 0$. The former values fulfil the relation (12), so that z is neither a maximum nor a minimum. The solutions $x = \pm a$, $y = 0$ satisfy (11), and, since they make $\partial^2 z/\partial x^2$ positive, z is then a *minimum*. The contour-lines of the surface (17) are shewn in Fig. 106, p. 321. The surface has two symmetrical hollows with the 'bar' between them. If we reverse the sign of the right-hand side of (17), we get two peaks, with the 'pass' between them.

197. Conditional Maxima and Minima.

The problem is to find the maxima and minima, or rather the stationary, values of a given function of n variables which are not all independent, but are connected by m given relations ($n > m$). The question whether these correspond to maxima or minima, as well as the discrimination, is usually decided on grounds independent of the present investigation.

Theoretically, we might eliminate m of the variables by means of the given relations, and so express the given function in terms of $n - m$ really independent variables. This procedure would however often prove cumbrous, if not impracticable. To meet this difficulty, the method of (in the first instance) 'undetermined' multipliers was devised by Lagrange. In cases where the given function, and the given relations, possess a more or less symmetrical character, this method is especially convenient.

The following cases will sufficiently elucidate the method.

1°. Suppose that we have a function

$$u = \phi(x, y, z), \quad \ldots\ldots\ldots\ldots\ldots\ldots(1)$$

where x, y, z are subject to the relation

$$f(x, y, z) = 0. \quad \ldots\ldots\ldots\ldots\ldots\ldots(2)$$

Expressing that $\delta u = 0$, we have

$$\frac{\partial \phi}{\partial x} \delta x + \frac{\partial \phi}{\partial y} \delta y + \frac{\partial \phi}{\partial z} \delta z = 0.\ldots\ldots\ldots\ldots\ldots(3)$$

The infinitesimal variations δx, δy, δz are not independent, but are connected by the relation

$$\frac{\partial f}{\partial x} \delta x + \frac{\partial f}{\partial y} \delta y + \frac{\partial f}{\partial z} \delta z = 0, \quad \ldots\ldots\ldots\ldots\ldots(4)$$

since $\delta f = 0$. From these we might eliminate δz; and since δx, δy may then be regarded as independent, we might, in the result, equate their coefficients separately to zero. A more symmetrical procedure is to form from (3) and (4) the equation

$$\left(\frac{\partial \phi}{\partial x} - \lambda \frac{\partial f}{\partial x}\right) \delta x + \left(\frac{\partial \phi}{\partial y} - \lambda \frac{\partial f}{\partial y}\right) \delta y + \left(\frac{\partial \phi}{\partial z} - \lambda \frac{\partial f}{\partial z}\right) \delta z = 0. \ldots(5)$$

So far, λ may have any value, but we now suppose it to be determined so as to make the coefficient of one of the variations, say δz, vanish. Since there is no necessary relation between δx and δy, their coefficients must also vanish. We are thus led to the equations

$$\frac{\partial \phi}{\partial x} = \lambda \frac{\partial f}{\partial x}, \quad \frac{\partial \phi}{\partial y} = \lambda \frac{\partial f}{\partial y}, \quad \frac{\partial \phi}{\partial z} = \lambda \frac{\partial f}{\partial z}. \quad \dots\dots\dots(6)$$

These, together with (2), constitute a system of four equations determining the four unknowns x, y, z, λ.

2°. Let the function be

$$u = \phi(x, y, z), \quad \dots\dots\dots\dots\dots(7)$$

where the variables are subject to the *two* relations

$$F(x, y, z) = 0, \quad f(x, y, z) = 0. \quad \dots\dots\dots(8)$$

Proceeding as before, we form the equation

$$\left(\frac{\partial \phi}{\partial x} - \lambda \frac{\partial F}{\partial x} - \mu \frac{\partial f}{\partial x}\right) \delta x + \left(\frac{\partial \phi}{\partial y} - \lambda \frac{\partial F}{\partial y} - \mu \frac{\partial f}{\partial y}\right) \delta y$$

$$+ \left(\frac{\partial \phi}{\partial z} - \lambda \frac{\partial F}{\partial z} - \mu \frac{\partial f}{\partial z}\right) \delta z = 0, \quad \dots\dots(9)$$

where λ, μ are at present undetermined multipliers. We may suppose these chosen so as to make the coefficients of δy and δz to vanish. The coefficient of δx must then also vanish. We thus obtain the equations

$$\frac{\partial \phi}{\partial x} = \lambda \frac{\partial F}{\partial x} + \mu \frac{\partial f}{\partial x}, \quad \frac{\partial \phi}{\partial y} = \lambda \frac{\partial F}{\partial y} + \mu \frac{\partial f}{\partial y}, \quad \frac{\partial \phi}{\partial z} = \lambda \frac{\partial F}{\partial z} + \mu \frac{\partial f}{\partial z}. \quad \dots(10)$$

These, together with (8), determine the five unknowns x, y, z, λ, μ.

Ex. 1. To find the stationary values of

$$u = x^2 + y^2 \dots\dots\dots\dots\dots\dots(11)$$

subject to the condition

$$Ax^2 + 2Hxy + By^2 = 1. \quad \dots\dots\dots\dots(12)$$

This is the problem of finding the principal axes of a conic whose centre is at the origin.

The method gives

$$x = \lambda(Ax + Hy), \quad y = \lambda(Hx + By). \quad \dots\dots\dots(13)$$

Multiplying these by x, y, respectively, and adding, we have

$$u = \lambda. \quad \dots\dots\dots\dots\dots\dots(14)$$

Hence $\quad (Au - 1)x + Huy = 0, \quad Hux + (Bu - 1)y = 0. \quad \dots\dots(15)$

Eliminating the ratio $x : y$, we have

$$(AB - H^2) u^2 - (A + B) u + 1 = 0. \quad\ldots\ldots\ldots\ldots(16)$$

Again, eliminating u,

$$H (x^2 - y^2) = (A - B) xy. \quad\ldots\ldots\ldots\ldots\ldots(17)$$

Ex. 2.   To find the stationary values of

$$u = x^2 + y^2 + z^2, \quad\ldots\ldots\ldots\ldots\ldots\ldots(18)$$

under the conditions

$$Ax^2 + By^2 + Cz^2 = 1, \quad lx + my + nz = 0. \quad\ldots\ldots\ldots(19)$$

The problem is that of finding the principal axes of a central section of a quadric surface.

We find

$$x = \lambda Ax + \mu l, \quad y = \lambda By + \mu m, \quad z = \lambda Cz + \mu n. \quad\ldots\ldots(20)$$

Multiplying by x, y, z, respectively, and adding, we find

$$u = \lambda. \quad\ldots\ldots\ldots\ldots\ldots\ldots\ldots(21)$$

Hence

$$x + \frac{\mu l}{Au - 1} = 0, \quad y + \frac{\mu m}{Bu - 1} = 0, \quad z + \frac{\mu n}{Cu - 1} = 0. \quad\ldots(22)$$

If we multiply these by l, m, n, respectively, and add, we find

$$\frac{l^2}{Au - 1} + \frac{m^2}{Bu - 1} + \frac{n^2}{Cu - 1} = 0, \quad\ldots\ldots\ldots\ldots(23)$$

which is, virtually, a quadratic in u.   If u be either root of this, the corresponding values of the ratios $x : y : z$ are given by (22); thus

$$x : y : z = \frac{l}{Au - 1} : \frac{m}{Bu - 1} : \frac{n}{Cu - 1}. \quad\ldots\ldots\ldots(24)$$

198.   Envelopes.

A similar treatment applies to the finding of the envelope of a curve whose equation involves n parameters connected by $n - 1$ relations.

For instance, suppose that it is required to find the envelope of

$$\phi (x, y, \alpha, \beta) = 0, \quad\ldots\ldots\ldots\ldots\ldots\ldots(1)$$

where the parameters α, β are connected by the relation

$$f(\alpha, \beta) = 0. \quad\ldots\ldots\ldots\ldots\ldots\ldots(2)$$

At the intersection of (1) with a consecutive curve we have

$$\phi (x, y, \alpha + \delta\alpha, \beta + \delta\beta) - \phi (x, y, \alpha, \beta) = 0, \quad\ldots\ldots(3)$$

or

$$\frac{\partial\phi}{\partial\alpha} \delta\alpha + \frac{\partial\phi}{\partial\beta} \delta\beta = 0, \quad\ldots\ldots\ldots\ldots\ldots(4)$$

ultimately. The variations $\delta\alpha$, $\delta\beta$ are connected by the relation

$$\frac{\partial f}{\partial \alpha} \delta\alpha + \frac{\partial f}{\partial \beta} \delta\beta = 0. \quad \dots\dots\dots\dots\dots(5)$$

Hence $\qquad \left(\frac{\partial\phi}{\partial\alpha} - \lambda\frac{\partial f}{\partial\alpha}\right)\delta\alpha + \left(\frac{\partial\phi}{\partial\beta} - \lambda\frac{\partial f}{\partial\beta}\right)\delta\beta = 0. \quad \dots\dots\dots(6)$

If we determine λ so that the coefficient of $\delta\beta$ shall vanish, that of $\delta\alpha$ must also vanish. We have then

$$\frac{\partial\phi}{\partial\alpha} - \lambda\frac{\partial f}{\partial\alpha} = 0, \quad \frac{\partial\phi}{\partial\beta} - \lambda\frac{\partial f}{\partial\beta} = 0. \quad \dots\dots\dots\dots(7)$$

The locus of ultimate intersections is found by eliminating α, β, λ, between (1), (2), and (7).

Ex. 1. To find the envelope of the line

$$\frac{x}{a} + \frac{y}{\beta} = 1, \quad \dots\dots\dots\dots\dots\dots\dots(8)$$

where a, β are subject to the condition

$$a^2 + \beta^2 = a^2. \quad \dots\dots\dots\dots\dots\dots\dots(9)$$

The method gives

$$\frac{x}{a^2} = \lambda a, \quad \frac{y}{\beta^2} = \lambda\beta. \quad \dots\dots\dots\dots\dots\dots(10)$$

Hence $\qquad \lambda(a^2 + \beta^2) = \frac{x}{a} + \frac{y}{\beta} = 1,$

or $\qquad\qquad \lambda = 1/a^2. \quad \dots\dots\dots\dots\dots\dots\dots(11)$

Hence $\qquad a^3 = a^2x, \quad \beta^3 = b^2y, \quad \dots\dots\dots\dots\dots\dots(12)$

and, substituting in (9), we find

$$x^{\frac{2}{3}} + y^{\frac{2}{3}} = a^{\frac{2}{3}}.$$

Cf. Art. 145, Ex. 2.

Ex. 2. To find the envelope of

$$ax + \beta y = 1 \quad \dots\dots\dots\dots\dots\dots\dots(13)$$

under the condition $\qquad a\beta + Aa + B\beta + C = 0. \quad \dots\dots\dots\dots\dots(14)$

We have to eliminate a, β, λ between these and

$$x = \lambda(\beta + A), \quad y = \lambda(a + B). \quad \dots\dots\dots\dots(15)$$

Eliminating λ, we have

$$ax - \beta y = Ay - Bx,$$

which, combined with (13), gives

$$ax = \tfrac{1}{2}(Ay - Bx + 1), \quad \cdot\beta y = \tfrac{1}{2}(Bx - Ay + 1). \quad \dots\dots(16)$$

Substituting in (14), we find

$$(Bx - Ay)^2 + 4Cxy + 2(Ay + Bx) + 1 = 0, \quad \dots\dots .(17)$$

which is the equation of the required envelope.

199. Applications of Partial Differentiation.

Numerous problems of partial differentiation present themselves in geometrical and physical questions. As a rule, they are best dealt with as they arise; but we give one or two simple cases which may serve to elucidate the chief points to be attended to.

1°. Let
$$u = \phi(v), \qquad \qquad \dots\dots\dots\dots\dots(1)$$

where v is a function of the independent variables x, y; and let it be required to form the successive partial derivatives of u with respect to these variables.

By Art. 32 we have
$$\frac{\partial u}{\partial x} = \phi'(v)\frac{\partial v}{\partial x}, \quad \frac{\partial u}{\partial y} = \phi'(v)\frac{\partial v}{\partial y}. \quad \dots\dots\dots(2)$$

Again,
$$\frac{\partial^2 u}{\partial x^2} = \frac{\partial}{\partial x}\phi'(v)\cdot\frac{\partial v}{\partial x} + \phi'(v)\frac{\partial^2 v}{\partial x^2} = \phi''(v)\left(\frac{\partial v}{\partial x}\right)^2 + \phi'(v)\frac{\partial^2 v}{\partial x^2}, \quad \dots(3)$$

$$\frac{\partial^2 u}{\partial x \partial y} = \frac{\partial}{\partial y}\phi'(v)\cdot\frac{\partial v}{\partial x} + \phi'(v)\frac{\partial^2 v}{\partial x \partial y} = \phi''(v)\frac{\partial v}{\partial x}\frac{\partial v}{\partial y} + \phi'(v)\frac{\partial^2 v}{\partial x \partial y},$$
$$\dots\dots\dots(4)$$

$$\frac{\partial^2 u}{\partial y^2} = \frac{\partial}{\partial y}\phi'(v)\cdot\frac{\partial v}{\partial y} + \phi'(v)\frac{\partial^2 v}{\partial y^2} = \phi''(v)\left(\frac{\partial v}{\partial y}\right)^2 + \phi'(v)\frac{\partial^2 v}{\partial y^2}, \quad \dots(5)$$

and so on.

2°. Let
$$u = \phi(x, y), \qquad \qquad \dots\dots\dots\dots\dots(6)$$

where x, y are given functions of the independent variable t; and let it be required to calculate the derivatives of u with respect to t. We have, by Art. 59, 1°,
$$\frac{du}{dt} = \frac{\partial\phi}{\partial x}\frac{dx}{dt} + \frac{\partial\phi}{\partial y}\frac{dy}{dt}. \qquad \dots\dots\dots\dots(7)$$

Differentiating again, we find
$$\frac{d^2 u}{dt^2} = \frac{\partial\phi}{\partial x}\frac{d^2 x}{dt^2} + \frac{\partial\phi}{\partial y}\frac{d^2 y}{dt^2} + \frac{d}{dt}\left(\frac{\partial\phi}{\partial x}\right)\frac{dx}{dt} + \frac{d}{dt}\left(\frac{\partial\phi}{\partial y}\right)\frac{dy}{dt}. \quad \dots(8)$$

Now, by the theorem referred to,
$$\frac{d}{dt}\left(\frac{\partial\phi}{\partial x}\right) = \frac{\partial}{\partial x}\left(\frac{\partial\phi}{\partial x}\right)\frac{dx}{dt} + \frac{\partial}{\partial y}\left(\frac{\partial\phi}{\partial x}\right)\frac{dy}{dt},$$

and
$$\frac{d}{dt}\left(\frac{\partial\phi}{\partial y}\right) = \frac{\partial}{\partial x}\left(\frac{\partial\phi}{\partial y}\right)\frac{dx}{dt} + \frac{\partial}{\partial y}\left(\frac{\partial\phi}{\partial y}\right)\frac{dy}{dt}.$$

Substituting in (8), and recalling the commutative property established in Art. 193, we have

$$\frac{d^2u}{dt^2} = \frac{\partial\phi}{\partial x}\frac{d^2x}{dt^2} + \frac{\partial\phi}{\partial y}\frac{d^2y}{dt^2} + \frac{\partial^2\phi}{\partial x^2}\left(\frac{dx}{dt}\right)^2 + 2\frac{\partial^2\phi}{\partial x\partial y}\frac{dx}{dt}\frac{dy}{dt} + \frac{\partial^2\phi}{\partial y^2}\left(\frac{dy}{dt}\right)^2.$$

$$\dots\dots\dots(9)$$

The process might be continued, but it is seldom necessary to proceed beyond this stage.

This process is sometimes required when we transform the coordinates in a dynamical problem. Thus, to change from rectangular to polar coordinates in two dimensions, we have

$$x = r\cos\theta, \quad y = r\sin\theta,$$

and the above method enables us to express d^2x/dt^2 and d^2y/dt^2 in terms of the differential coefficients of r and θ with respect to t.

Ex. Let $z = \phi(x - ct) + \chi(x + ct), \dots\dots\dots\dots\dots(10)$

where the variables x and t are independent.

Putting $x - ct = u, \quad x + ct = v,$

for shortness, we find

$$\frac{\partial z}{\partial x} = \phi'(u) + \chi'(u), \quad \frac{\partial z}{\partial t} = -c\phi'(u) + c\chi'(u), \dots\dots(11)$$

$$\frac{\partial^2 z}{\partial x^2} = \phi''(u) + \chi''(u), \quad \frac{\partial^2 z}{\partial t^2} = c^2\phi''(u) + c^2\chi''(u). \dots\dots(12)$$

Hence $$\frac{\partial^2 z}{\partial t^2} = c^2\frac{\partial^2 z}{\partial x^2}. \dots\dots\dots\dots(13)$$

200. Differentiation of Implicit Functions.

Let y be a function of x, defined 'implicitly' by the equation

$$\phi(x, y) = 0; \dots\dots\dots\dots\dots(1)$$

it is required to calculate the successive derivatives of y with respect to x.

We have, as in Art. 59,

$$\frac{\partial\phi}{\partial x} + \frac{\partial\phi}{\partial y}\frac{dy}{dx} = 0. \dots\dots\dots(2)$$

If we differentiate this with respect to x, we find

$$\frac{d}{dx}\left(\frac{\partial\phi}{\partial x}\right) + \frac{d}{dx}\left(\frac{\partial\phi}{\partial y}\right)\frac{dy}{dx} + \frac{\partial\phi}{\partial y}\frac{d^2y}{dx^2} = 0. \dots\dots(3)$$

Now, by Art. 59,

$$\frac{d}{dx}\left(\frac{\partial\phi}{\partial x}\right) = \frac{\partial}{\partial x}\left(\frac{\partial\phi}{\partial x}\right) + \frac{\partial}{\partial y}\left(\frac{\partial\phi}{\partial x}\right)\frac{dy}{dx},$$

$$\frac{d}{dx}\left(\frac{\partial\phi}{\partial y}\right) = \frac{\partial}{\partial x}\left(\frac{\partial\phi}{\partial y}\right) + \frac{\partial}{\partial y}\left(\frac{\partial\phi}{\partial y}\right)\frac{dy}{dx}.$$

Hence (3) becomes

$$\frac{\partial^2\phi}{\partial x^2} + 2\frac{\partial^2\phi}{\partial x\partial y}\frac{dy}{dx} + \frac{\partial^2\phi}{\partial y^2}\left(\frac{dy}{dx}\right)^2 + \frac{\partial\phi}{\partial y}\frac{d^2y}{dx^2} = 0. \quad\ldots\ldots\ldots(4)$$

If we substitute the value of dy/dx from (2), we find

$$\frac{d^2y}{dx^2} = -\frac{\phi_{xx}\phi_y^2 - 2\phi_{xy}\phi_x\phi_y + \phi_{yy}\phi_x^2}{\phi_y^3}. \quad\ldots\ldots\ldots(5)$$

Again, by differentiation of (4), and substitution, we could find d^3y/dx^3 and so on.

The formula (5) leads to an expression for the curvature of a curve whose equation in rectangular coordinates is given by (1), viz.

$$\frac{1}{\rho} = -\frac{\phi_{xx}\phi_y^2 - 2\phi_{xy}\phi_x\phi_y + \phi_{yy}\phi_x^2}{(\phi_x^2 + \phi_y^2)^{\frac{3}{2}}}. \quad\ldots\ldots\ldots\ldots(6)$$

The condition for a point of inflexion is obtained by equating the numerator to zero.

It may be noticed that (4) is included in Art. 199 (9) by putting $t = x$, $u = 0$.

201. Change of Variable.

1°. Let it be required to interchange the dependent and independent variables in the case of a function of a single variable.

We have (Art. 33)

$$\frac{dy}{dx} = \left(\frac{dx}{dy}\right)^{-1}. \quad\ldots\ldots\ldots\ldots\ldots\ldots(1)$$

Hence
$$\frac{d^2y}{dx^2} = \frac{d}{dx}\left(\frac{dx}{dy}\right)^{-1} = \frac{d}{dy}\left(\frac{dx}{dy}\right)^{-1}\cdot\frac{dy}{dx}$$

$$= -\left(\frac{dx}{dy}\right)^{-3}\frac{d^2x}{dy^2}; \quad\ldots\ldots\ldots\ldots\ldots\ldots(2)$$

and so on.

2°. Let
$$u = \phi(\xi, \eta), \quad\ldots\ldots\ldots\ldots\ldots(3)$$

where ξ, η are given functions of the independent variables x and y, and let it be required to calculate the second partial derivatives of u with respect to x and y.

We have

$$\frac{\partial u}{\partial x} = \frac{\partial u}{\partial \xi}\frac{\partial \xi}{\partial x} + \frac{\partial u}{\partial \eta}\frac{\partial \eta}{\partial x}, \quad \frac{\partial u}{\partial y} = \frac{\partial u}{\partial \xi}\frac{\partial \xi}{\partial y} + \frac{\partial u}{\partial \eta}\frac{\partial \eta}{\partial y}. \quad\ldots\ldots\ldots(4)$$

Hence

$$\frac{\partial^2 u}{\partial x^2} = \frac{\partial u}{\partial \xi}\frac{\partial^2 \xi}{\partial x^2} + \left(\frac{\partial^2 u}{\partial \xi^2}\frac{\partial \xi}{\partial x} + \frac{\partial^2 u}{\partial \xi \partial \eta}\frac{\partial \eta}{\partial x}\right)\frac{\partial \xi}{\partial x}$$

$$+ \frac{\partial u}{\partial \eta}\frac{\partial^2 \eta}{\partial x^2} + \left(\frac{\partial^2 u}{\partial \xi \partial \eta}\frac{\partial \xi}{\partial x} + \frac{\partial^2 u}{\partial \eta^2}\frac{\partial \eta}{\partial x}\right)\frac{\partial \eta}{\partial x}$$

$$= \frac{\partial^2 u}{\partial \xi^2}\left(\frac{\partial \xi}{\partial x}\right)^2 + 2\frac{\partial^2 u}{\partial \xi \partial \eta}\frac{\partial \xi}{\partial x}\frac{\partial \eta}{\partial x} + \frac{\partial^2 u}{\partial \eta^2}\left(\frac{\partial \eta}{\partial x}\right)^2$$

$$+ \frac{\partial u}{\partial \xi}\frac{\partial^2 \xi}{\partial x^2} + \frac{\partial u}{\partial \eta}\frac{\partial^2 \eta}{\partial x^2}. \quad\ldots\ldots\ldots\ldots(5)$$

In like manner we should find

$$\frac{\partial^2 u}{\partial x \partial y} = \frac{\partial^2 u}{\partial \xi^2}\frac{\partial \xi}{\partial x}\frac{\partial \xi}{\partial y} + \frac{\partial^2 u}{\partial \xi \partial \eta}\left(\frac{\partial \xi}{\partial x}\frac{\partial \eta}{\partial y} + \frac{\partial \xi}{\partial y}\frac{\partial \eta}{\partial x}\right) + \frac{\partial^2 u}{\partial \eta^2}\frac{\partial \eta}{\partial x}\frac{\partial \eta}{\partial y}$$

$$+ \frac{\partial u}{\partial \xi}\frac{\partial^2 \xi}{\partial x \partial y} + \frac{\partial u}{\partial \eta}\frac{\partial^2 \eta}{\partial x \partial y}, \quad\ldots\ldots\ldots\ldots(6)$$

$$\frac{\partial^2 u}{\partial y^2} = \frac{\partial^2 u}{\partial \xi^2}\left(\frac{\partial \xi}{\partial y}\right)^2 + 2\frac{\partial^2 u}{\partial \xi \partial \eta}\frac{\partial \xi}{\partial y}\frac{\partial \eta}{\partial y} + \frac{\partial^2 u}{\partial \eta^2}\left(\frac{\partial \eta}{\partial y}\right)^2$$

$$+ \frac{\partial u}{\partial \xi}\frac{\partial^2 \xi}{\partial y^2} + \frac{\partial u}{\partial \eta}\frac{\partial^2 \eta}{\partial y^2}. \quad\ldots\ldots\ldots\ldots\ldots(7)$$

Ex. 1. In the case of a particle moving under a resistance proportional to the cube of the velocity, we have

$$\frac{d^2 x}{dt^2} = -k\left(\frac{dx}{dt}\right)^3. \quad\ldots\ldots\ldots\ldots\ldots(8)$$

It follows at once from (2), with a mere difference in the notation, that

$$\frac{d^2 t}{dx^2} = k. \quad\ldots\ldots\ldots\ldots\ldots(9)$$

Hence

$$t = \tfrac{1}{2}kx^2 + Ax + B. \quad\ldots\ldots\ldots\ldots(10)$$

Ex. 2. To change from rectangular to polar coordinates in the expression

$$\frac{\partial^2 u}{\partial x^2} + \frac{\partial^2 u}{\partial y^2}. \quad\ldots\ldots\ldots\ldots\ldots(11)$$

Putting

$$x = r\cos\theta, \quad y = r\sin\theta, \quad\ldots\ldots\ldots\ldots(12)$$

we find

$$\left.\begin{aligned}
\frac{\partial u}{\partial r} &= \frac{\partial u}{\partial x}\frac{\partial x}{\partial r} + \frac{\partial u}{\partial y}\frac{\partial y}{\partial r} = \cos\theta\,\frac{\partial u}{\partial x} + \sin\theta\,\frac{\partial u}{\partial y}, \\
\frac{\partial u}{\partial \theta} &= \frac{\partial u}{\partial x}\frac{\partial x}{\partial \theta} + \frac{\partial u}{\partial y}\frac{\partial y}{\partial \theta} = r\left(-\sin\theta\,\frac{\partial u}{\partial x} + \cos\theta\,\frac{\partial u}{\partial y}\right),
\end{aligned}\right\} \quad\ldots\ldots(13)$$

whence

$$\frac{\partial u}{\partial x} = \cos \theta \frac{\partial u}{\partial r} - \sin \theta \frac{\partial u}{r \partial \theta}, \quad \frac{\partial u}{\partial y} = \sin \theta \frac{\partial u}{\partial r} + \cos \theta \frac{\partial u}{r \partial \theta}. \quad \dots (14)$$

Hence

$$\left.\begin{aligned}\frac{\partial^2 u}{\partial x^2} &= \left(\cos \theta \frac{\partial}{\partial r} - \sin \theta \frac{\partial}{r \partial \theta}\right)\left(\cos \theta \frac{\partial u}{\partial r} - \sin \theta \frac{\partial u}{r \partial \theta}\right), \\ \frac{\partial^2 u}{\partial y^2} &= \left(\sin \theta \frac{\partial}{\partial r} + \cos \theta \frac{\partial}{r \partial \theta}\right)\left(\sin \theta \frac{\partial u}{\partial r} + \cos \theta \frac{\partial u}{r \partial \theta}\right).\end{aligned}\right\} \quad \dots \dots (15)$$

It is not necessary to perform all the operations indicated, as several of the terms will cancel when we form the sum (11). The remaining terms give

$$\frac{\partial^2 u}{\partial x^2} + \frac{\partial^2 u}{\partial y^2} = \frac{\partial^2 u}{\partial r^2} + \frac{1}{r}\frac{\partial u}{\partial r} + \frac{1}{r^2}\frac{\partial^2 u}{\partial \theta^2}. \quad \dots \dots \dots \dots (16)$$

EXAMPLES. LXIV.

(Partial Differentiation; Exact Differentials.)

1. If

$$u = \frac{xy}{x+y},$$

verify the relations

$$x\frac{\partial u}{\partial x} + y\frac{\partial u}{\partial y} = u,$$

$$x^2 \frac{\partial^2 u}{\partial x^2} + 2xy\frac{\partial^2 u}{\partial x \partial y} + y^2 \frac{\partial^2 u}{\partial y^2} = 0.$$

2. If

$$z = x^2 \tan^{-1}\frac{y}{x} - y^2 \tan^{-1}\frac{x}{y},$$

prove that

$$\frac{\partial^2 z}{\partial x \partial y} = \frac{x^2 - y^2}{x^2 + y^2}.$$

3. If

$$z = F(x) + f(y),$$

prove that

$$\frac{\partial^2 z}{\partial x \partial y} = 0.$$

Conversely shew that, if $\partial^2 z / \partial x \partial y = 0$, z must have the above form.

4. Prove that the equation

$$\frac{\partial^2 \phi}{\partial r^2} + \frac{1}{r}\frac{\partial \phi}{\partial r} + \frac{1}{r^2}\frac{\partial^2 \phi}{\partial \theta^2} = 0$$

is satisfied by

$$\phi = \left(Ar^n + \frac{B}{r^n}\right)\cos n\,(\theta - a).$$

5. Prove that any differential equation of the type

$$F(x^2 + y^2)(x\,dx + y\,dy) + f\left(\frac{y}{x}\right)(x\,dy - y\,dx) = 0$$

becomes exact on division by $x^2 + y^2$.

6. Prove that $\qquad \dfrac{\sinh y\,dx - \sin x\,dy}{\cosh y - \cos x}$

is an exact differential of a function u; and find u.

7. If $\qquad \dfrac{\partial^2 \phi}{\partial x^2} + \dfrac{\partial^2 \phi}{\partial y^2} = 0,$

prove that a function ψ exists such that

$$\frac{\partial \phi}{\partial x} = -\frac{\partial \psi}{\partial y}, \qquad \frac{\partial \phi}{\partial y} = \frac{\partial \psi}{\partial x},$$

and that $\qquad \dfrac{\partial^2 \psi}{\partial x^2} + \dfrac{\partial^2 \psi}{\partial y^2} = 0.$

8. If $\qquad \dfrac{\partial^2 \phi}{\partial r^2} + \dfrac{1}{r}\dfrac{\partial \phi}{\partial r} + \dfrac{1}{r^2}\dfrac{\partial^2 \phi}{\partial \theta^2} = 0,$

prove that a function ψ exists such that

$$\frac{\partial \phi}{\partial r} = -\frac{\partial \psi}{r\partial \theta}, \qquad \frac{\partial \phi}{r\partial \theta} = \frac{\partial \psi}{\partial r},$$

and that ψ satisfies the same partial differential equation as ϕ.

9. If $\qquad \dfrac{\partial^2 \phi}{\partial x^2} + \dfrac{\partial^2 \phi}{\partial y^2} + \dfrac{1}{y}\dfrac{\partial \phi}{\partial y} = 0,$

prove that a function ψ exists such that

$$\frac{\partial \phi}{\partial x} = -\frac{1}{y}\frac{\partial \psi}{\partial y}, \qquad \frac{\partial \phi}{\partial y} = \frac{1}{y}\frac{\partial \psi}{\partial x},$$

and that $\qquad \dfrac{\partial^2 \psi}{\partial x^2} + \dfrac{\partial^2 \psi}{\partial y^2} - \dfrac{1}{y}\dfrac{\partial \psi}{\partial y} = 0.$

EXAMPLES. LXV.

(Maxima and Minima.)

1. Prove that in the surface

$$az = x^2 - y^2,$$

the ordinate (z) is stationary, but not a maximum or minimum, when $x = 0$, $y = 0$. Sketch the contour-lines of the surface.

2. Prove by means of the rule of Art. 196 that the parallelepiped of least surface for a given volume is a cube.

3. If A, B, C be the vertices of a triangle, and P a variable point, the sum

$$PA^2 + PB^2 + PC^2$$

is a minimum when P coincides with the mean centre of A, B, C

4. With the same notation, the sum

$$m_1 . PA^2 + m_2 . PB^2 + m_3 . PC^2$$

is a minimum when P coincides with the mass-centre of three particles m_1, m_2, m_3 situate at A, B, C.

5. Find for what values of x, y the ordinate of the surface

$$z = x^3 + y^3 - 3axy$$

is stationary.

[The values are a, a, and 0, 0. The latter do not make z a maximum or minimum.]

6. Prove that the ordinate of the surface

$$c^2 z = ay^2 - x^3$$

is stationary, but not a maximum or minimum, when $x = 0$, $y = 0$. Sketch the contour-lines.

7. Find for what values of x, y the function

$$x^4 + y^4 - 2 (x - y)^2$$

is stationary.

[There is a stationary value when $x = 0$, $y = 0$, and two minima when $x = \pm \sqrt{2}$, $y = \mp \sqrt{2}$.]

8. Prove that the function

$$(x^2 + y^2) e^{x^2 - y^2}$$

has a minimum value when $x = 0$, $y = 0$, and a stationary value which is neither a maximum nor a minimum when $x = 0$, $y = \pm 1$.

9. Prove that the ordinate of the surface

$$z = f (ax^2 + 2hxy + by^2)$$

is in general stationary when $x = 0$, $y = 0$, and examine whether it is a maximum or minimum. Sketch the contour-lines in the several cases.

10. Find the stationary points of the function

$$(x^2 - a^2)^2 + (x^2 - a^2) (y^2 - b^2) + (y^2 - b^2)^2,$$

and examine their nature.

Sketch the contour-lines of the function.

11. If $x_1, x_2, \ldots, x_n$ be n positive quantities subject to the relation

$$x_1 + x_2 + \ldots + x_n = \text{const.},$$

their product is greatest when they are all equal.

Hence shew that the arithmetic mean of n positive quantities exceeds their geometric mean unless they are all equal.

12. Prove that the rectangular parallelepiped of greatest volume for a given surface is a cube.

13. Prove that the greatest rectangular parallelepiped which can be inscribed in a given sphere is a cube.

EXAMPLES. LXVI.

(Change of Variable, &c.)

1. If $x = \sin\theta$, the equation

$$(1 - x^2)\frac{d^2y}{dx^2} - x\frac{dy}{dx} + a^2y = 0$$

transforms into

$$\frac{d^2y}{d\theta^2} + a^2y = 0.$$

2. If $x^2 = 4t$, the equation

$$\frac{d^2y}{dx^2} + \frac{1}{x}\frac{dy}{dx} + y = 0$$

transforms into

$$t\frac{d^2y}{dt^2} + \frac{dy}{dt} + y = 0.$$

3. If

$$ax^2 + 2hxy + by^2 + 2gx + 2fy + c = 0,$$

prove that

$$p - \frac{3q^2}{r} = \frac{(ab - h^2)y + af - gh}{(ab - h^2)x + bg - hf},$$

where

$$p = dy/dx, \quad q = d^2y/dx^2, \quad r = d^3y/dx^3.$$

4. If

$$u = f\left(\frac{y}{x}\right),$$

prove that

$$x\frac{\partial u}{\partial x} + y\frac{\partial u}{\partial y} = 0,$$

and

$$\frac{\partial^2 u}{\partial x^2} + \frac{\partial^2 u}{\partial y^2} = \frac{1}{x^4}\left\{(x^2 + y^2)f''\left(\frac{y}{x}\right) + 2xyf'\left(\frac{y}{x}\right)\right\}.$$

5. If

$$z = f(x^2 + y^2),$$

prove that

$$\frac{\partial^2 z}{\partial x^2} + \frac{\partial^2 z}{\partial y^2} = 4(x^2 + y^2)f''(x^2 + y^2) + 4f'(x^2 + y^2).$$

6. If
$$u = f(r),$$
where $r = \sqrt{(x^2 + y^2)}$, prove that
$$\frac{\partial^2 u}{\partial x^2} + \frac{\partial^2 u}{\partial y^2} = f''(r) + \frac{1}{r} f'(r).$$

7. If ∇_1^2 stand for the operator $\partial^2/\partial x^2 + \partial^2/\partial y^2$, prove that
$$\nabla_1^2 \log r = 0,$$
where
$$r = \sqrt{\{(x-a)^2 + (y-\beta)^2\}}.$$

8. If
$$u = f(x^2 + y^2 + z^2),$$
prove that
$$\frac{\partial^2 u}{\partial x^2} + \frac{\partial^2 u}{\partial y^2} + \frac{\partial^2 u}{\partial z^2} = 4(x^2 + y^2 + z^2)f''(x^2 + y^2 + z^2) + 6f'(x^2 + y^2 + z^2).$$

9. If
$$u = f(r),$$
where $r = \sqrt{(x^2 + y^2 + z^2)}$, prove that
$$\frac{\partial^2 u}{\partial x^2} + \frac{\partial^2 u}{\partial y^2} + \frac{\partial^2 u}{\partial z^2} = f''(r) + \frac{2}{r} f'(r).$$

10. If ∇^2 stand for the operator $\partial^2/\partial x^2 + \partial^2/\partial y^2 + \partial^2/\partial z^2$, prove that
$$\nabla^2 r = \frac{2}{r}, \qquad \nabla^2 \frac{1}{r} = 0,$$
where
$$r = \sqrt{\{(x-a)^2 + (y-\beta)^2 + (z-\gamma)^2\}}.$$

11. With the same meaning of ∇^2, prove that if
$$\nabla^2 u = 0, \quad \nabla^2 v = 0, \quad \nabla^2 w = 0,$$
and
$$\frac{\partial u}{\partial x} + \frac{\partial v}{\partial y} + \frac{\partial w}{\partial z} = 0,$$
then
$$\nabla^2(xu + yv + zw) = 0.$$

12. If u, v be two functions of x, y, z satisfying the equations $\nabla^2 u = 0$, $\nabla^2 v = 0$, and if v be a function of u, then v must be of the form $Au + B$.

13. If
$$x = r \cos \theta, \quad y = r \sin \theta,$$
where r, θ are functions of t, prove that
$$\frac{d^2 x}{dt^2} \cos \theta + \frac{d^2 y}{dt^2} \sin \theta = \frac{d^2 r}{dt^2} - r\left(\frac{d\theta}{dt}\right)^2,$$
$$-\frac{d^2 x}{dt^2} \sin \theta + \frac{d^2 y}{dt^2} \cos \theta = \frac{1}{r} \frac{d}{dt}\left(r^2 \frac{d\theta}{dt}\right).$$

14. If
$$u = \frac{1}{r}\,\phi\,(ct - r) + \frac{1}{r}\,\chi\,(ct + r),$$

prove that
$$\frac{\partial^2 u}{\partial t^2} = c^2 \left(\frac{\partial^2 u}{\partial r^2} + \frac{2}{r}\,\frac{\partial u}{\partial r} \right).$$

15. If
$$u = t^{-\frac{1}{2}}\,e^{-x^2/4\kappa t},$$

prove that
$$\frac{\partial u}{\partial t} = \kappa\,\frac{\partial^2 u}{\partial x^2}.$$

16. If
$$u = t^{-\frac{3}{2}}\,e^{-r^2/4\kappa t},$$

prove that
$$\frac{\partial u}{\partial t} = \kappa\left(\frac{\partial^2 u}{\partial r^2} + \frac{2}{r}\,\frac{\partial u}{\partial r} \right).$$

17.
$$u = x^n \phi\left(\frac{y}{x} \right),$$

verify that
$$x\,\frac{\partial u}{\partial x} + y\,\frac{\partial u}{\partial y} = nu,$$

$$x^2\,\frac{\partial^2 u}{\partial x^2} + 2xy\,\frac{\partial^2 u}{\partial x \partial y} + y^2\,\frac{\partial^2 u}{\partial y^2} = n\,(n-1)\,u.$$

18. If
$$y - nz = f\,(x - mz),$$

prove that
$$m\,\frac{\partial z}{\partial x} + n\,\frac{\partial z}{\partial y} = 1.$$

19. If
$$z - \gamma = (x - a) f\left(\frac{y - \beta}{x - a} \right),$$

prove that
$$(x - a)\,\frac{\partial z}{\partial x} + (y - \beta)\,\frac{\partial z}{\partial y} = z - \gamma.$$

20. If $\quad x = c \cosh \xi \cos \eta, \quad y = c \sinh \xi \sin \eta,$
prove that
$$\frac{\partial^2 u}{\partial \xi^2} + \frac{\partial^2 u}{\partial \eta^2} = \tfrac{1}{2}c^2 (\cosh 2\xi - \cos 2\eta) \left(\frac{\partial^2 u}{\partial x^2} + \frac{\partial^2 u}{\partial y^2} \right).$$

21. Prove that if at a point on the curve
$$\phi\,(x,\,y) = 0$$
we have
$$\phi_x = 0, \quad \phi_y = 0,$$
simultaneously, then two (real or imaginary) branches of the curve pass through that point, whose directions are given by the quadratic

$$\phi_{xx} + 2\phi_{xy}\,\frac{dy}{dx} + \phi_{yy}\left(\frac{dy}{dx} \right)^2 = 0.$$

Hence shew that the point is a node, a cusp, or an isolated point, according as
$$(\phi_{xy})^2 \gtreqless \phi_{xx}\phi_{yy}.$$

APPENDIX

NUMERICAL TABLES

A. Squares of Numbers from 10 to 100.

	0	1	2	3	4	5	6	7	8	9
1	100	121	144	169	196	225	256	289	324	361
2	400	441	484	529	576	625	676	729	784	841
3	900	961	1024	1089	1156	1225	1296	1369	1444	1521
4	1600	1681	1764	1849	1936	2025	2116	2209	2304	2401
5	2500	2601	2704	2809	2916	3025	3136	3249	3364	3481
6	3600	3721	3844	3969	4096	4225	4356	4489	4624	4761
7	4900	5041	5184	5329	5476	5625	5776	5929	6084	6241
8	6400	6561	6724	6889	7056	7225	7396	7569	7744	7921
9	8100	8281	8464	8649	8836	9025	9216	9409	9604	9801

B 1. Square-Roots of Numbers from 0 to 10, at Intervals of ·1.

	·0	·1	·2	·3	·4	·5	·6	·7	·8	·9
0	0	·316	·447	·548	·632	·707	·775	·837	·894	·949
1	1·000	1·049	1·095	1·140	1·183	1·225	1·265	1·304	1·342	1·378
2	1·414	1·449	1·483	1·517	1·549	1·581	1·612	1·643	1·673	1·703
3	1·732	1·761	1·789	1·817	1·844	1·871	1·897	1·924	1·949	1·975
4	2·000	2·025	2·049	2·074	2·098	2·121	2·145	2·168	2·191	2·214
5	2·236	2·258	2·280	2·302	2·324	2·345	2·366	2·387	2·408	2·429
6	2·449	2·470	2·490	2·510	2·530	2·550	2·569	2·588	2·608	2·627
7	2·646	2·665	2·683	2·702	2·720	2·739	2·757	2·775	2·793	2·811
8	2·828	2·846	2·864	2·881	2·898	2·915	2·933	2·950	2·966	2·983
9	3·000	3·017	3·033	3·050	3·066	3·082	3·098	3·114	3·130	3·146
10	3·162									

B 2.   Square-Roots of Numbers from 10 to 100, at Intervals of 1.

	0	1	2	3	4	5	6	7	8	9
1	3·162	3·317	3·464	3·606	3·742	3·873	4·000	4·123	4·243	4·359
2	4·472	4·583	4·690	4·796	4·899	5·000	5·099	5·196	5·292	5·385
3	5·477	5·568	5·657	5·745	5·831	5·916	6·000	6·083	6·164	6·245
4	6·325	6·403	6·481	6·557	6·633	6·708	6·782	6·856	6·928	7·000
5	7·071	7·141	7·211	7·280	7·348	7·416	7·483	7·550	7·616	7·681
6	7·746	7·810	7·874	7·937	8·000	8·062	8·124	8·185	8·246	8·307
7	8·367	8·426	8·485	8·544	8·602	8·660	8·718	8·775	8·832	8·888
8	8·944	9·000	9·055	9·110	9·165	9·220	9·274	9·327	9·381	9·434
9	9·487	9·539	9·592	9·644	9·695	9·747	9·798	9·849	9·899	9·950

C.   Reciprocals of Numbers from 10 to 100, at Intervals of ·1.

	·0	·1	·2	·3	·4	·5	·6	·7	·8	·9
1	1·000	·909	·833	·769	·714	·667	·625	·588	·556	·526
2	·500	·476	·455	·435	·417	·400	·385	·370	·357	·345
3	·333	·323	·313	·303	·294	·286	·278	·270	·263	·256
4	·250	·244	·238	·233	·227	·222	·217	·213	·208	·204
5	·200	·196	·192	·189	·185	·182	·179	·175	·172	·169
6	·167	·164	·161	·159	·156	·154	·152	·149	·147	·145
7	·143	·141	·139	·137	·135	·133	·132	·130	·128	·127
8	·125	·123	·122	·120	·119	·118	·116	·115	·114	·112
9	·111	·110	·109	·108	·106	·105	·104	·103	·102	·101

D.   Circular Functions at Intervals of One-Twentieth of the Quadrant.

$\theta/\frac{1}{2}\pi$	$\sin\theta$	$\operatorname{cosec}\theta$	$\tan\theta$	$\cot\theta$	$\sec\theta$	$\cos\theta$	
·0	0	∞	0	∞	1·000	1·000	1·00
·05	·078	12·745	·079	12·706	1·003	·997	·95
·10	·156	6·392	·158	6·314	1·012	·988	·90
·15	·233	4·284	·240	4·165	1·028	·972	·85
·20	·309	3·236	·325	3·078	1·051	·951	·80
·25	·383	2·613	·414	2·414	1·082	·924	·75
·30	·454	2·203	·510	1·963	1·122	·891	·70
·35	·522	1·914	·613	1·632	1·173	·853	·65
·40	·588	1·701	·727	1·376	1·236	·809	·60
·45	·649	1·540	·854	1·171	1·315	·760	·55
·50	·707	1·414	1·000	1·000	1·414	·707	·50
	$\cos\theta$	$\sec\theta$	$\cot\theta$	$\tan\theta$	$\operatorname{cosec}\theta$	$\sin\theta$	$\theta/\frac{1}{2}\pi$

E. Exponential and Hyperbolic Functions of Numbers from 0 to 2·5, at Intervals of ·1

x	e^x	e^{-x}	cosh x	sinh x	tanh x
0	1·000	1·000	1·000	0	0
·1	1·105	·905	1·005	·100	·100
·2	1·221	·819	1·020	·201	·197
·3	1·350	·741	1·045	·305	·291
·4	1·492	·670	1·081	·411	·380
·5	1·649	·607	1·128	·521	·462
·6	1·822	·549	1·185	·637	·537
·7	2·014	·497	1·255	·759	·604
·8	2·226	·449	1·337	·888	·664
·9	2·460	·407	1·433	1·027	·716
1·0	2·718	·368	1·543	1·175	·762
1·1	3·004	·333	1·669	1·336	·801
1·2	3·320	·301	1·811	1·509	·834
1·3	3·669	·273	1·971	1·698	·862
1·4	4·055	·247	2·151	1·904	·885
1·5	4·482	·223	2·352	2·129	·905
1·6	4·953	·202	2·577	2·376	·922
1·7	5·474	·183	2·828	2·646	·935
1·8	6·050	·165	3·107	2·942	·947
1·9	6·686	·150	3·418	3·268	·956
2·0	7·389	·135	3·762	3·627	·964
2·1	8·166	·122	4·144	4·022	·970
2·2	9·025	·111	4·568	4·457	·976
2·3	9·974	·100	5·037	4·937	·980
2·4	11·023	·091	5·557	5·466	·984
2·5	12·182	·082	6·132	6·050	·987

F.   Logarithms to Base e.

	·0	·1	·2	·3	·4	·5	·6	·7	·8	·9
1	0	·095	·182	·262	·336	·405	·470	·531	·588	·642
2	·693	·742	·788	·833	·875	·916	·956	·993	1·030	1·065
3	1·099	1·131	1·163	1·194	1·224	1·253	1·281	1·308	1·335	1·361
4	1·386	1·411	1·435	1·459	1·482	1·504	1·526	1·548	1·569	1·589
5	1·609	1·629	1·649	1·668	1·686	1·705	1·723	1·740	1·758	1·775
6	1·792	1·808	1·825	1·841	1·856	1·872	1·887	1·902	1·917	1·932
7	1·946	1·960	1·974	1·988	2·001	2·015	2·028	2·041	2·054	2·067
8	2·079	2·092	2·104	2·116	2·128	2·140	2·152	2·163	2·175	2·186
9	2·197	2·208	2·219	2·230	2·241	2·251	2·262	2·272	2·282	2·293

$$\log 10 = 2 \cdot 303, \quad \log 10^2 = 4 \cdot 605, \quad \log 10^3 = 6 \cdot 908.$$

INDEX